D1673847

Mixed-Valence Systems

Fundamentals, Synthesis, Electron Transfer, and Applications

Edited by Yu-Wu Zhong, Chun Y. Liu, and Jeffrey R. Reimers

Editors

Prof. Yu-Wu Zhong
Institute of Chemistry, CAS
Laboratory of Photochemistry
2 Bei Yi Jie
Zhong Guan Cun
100190 Beijing
China

Prof. Chun Y. Liu
Jinan University
Department of Chemistry
601 Huang-Pu Avenue West
510632 Guangzhou
China

Prof. Jeffrey R. Reimers
Shanghai University
International Centre for Quantum and
Molecular Structures and School of
Physics
99 Shangda Lu
200444 Shanghai
China

and

University of Technology Sydney
Department of Mathematical & Physical
Sciences
15 Broadway
2007 Sydney
Australia

Cover Image: Courtesy of Prof. Tianfei Liu, Nankai University, Tianjin, China

All books published by **WILEY-VCH** are carefully produced. Nevertheless, authors, editors, and publisher do not warrant the information contained in these books, including this book, to be free of errors. Readers are advised to keep in mind that statements, data, illustrations, procedural details or other items may inadvertently be inaccurate.

Library of Congress Card No.:
applied for

British Library Cataloguing-in-Publication Data
A catalogue record for this book is available from the British Library.

Bibliographic information published by the Deutsche Nationalbibliothek
The Deutsche Nationalbibliothek lists this publication in the Deutsche Nationalbibliografie; detailed bibliographic data are available on the Internet at <http://dnb.d-nb.de>.

© 2023 WILEY-VCH GmbH, Boschstr. 12, 69469 Weinheim, Germany

All rights reserved (including those of translation into other languages). No part of this book may be reproduced in any form – by photoprinting, microfilm, or any other means – nor transmitted or translated into a machine language without written permission from the publishers. Registered names, trademarks, etc. used in this book, even when not specifically marked as such, are not to be considered unprotected by law.

Print ISBN: 978-3-527-34980-7
ePDF ISBN: 978-3-527-83526-3
ePub ISBN: 978-3-527-83527-0
oBook ISBN: 978-3-527-83528-7

Typesetting   Straive, Chennai, India
Printing and Binding   CPI Group (UK) Ltd, Croydon, CR0 4YY

Dedicated to Noel S. Hush AO FAA FRS (1924–2019)

Noel Hush completed his BSc (1945) and MSc (1948) at the University of Sydney, moving then to the United Kingdom to a lectureship at Manchester and then readership at Bristol, completing a DSc under the guidance of C. Longuet-Higgins and M.H.L. Pyrce. His career spanned over 70 years, from his first publication in 1947 in Nature *on reduction reactions to his last paper in 2019 in* J. Chem. Phys. *on a simple theory for understanding attosecond spectroscopy.*

In 1952–1953, Hush introduced the use of harmonic diabatic potential-energy surfaces for the description of electron-transfer reactions, these being much simpler than the diabatic Morse potentials that had dominated the previous two decades, in which focus had been primarily on the understanding of dissociative chemical reactions. Then, in 1958, he introduced the notion that intramolecular vibrations within molecules, or else the inner coordination spheres of ions, were just as important as long-range solvent interactions. Controversially, he then also introduced the concept that quantum mechanics was needed to understand electron-transfer reactions, focusing on the continuous transfer of charge that takes place along the reaction coordinate through a transition state. This led to another controversial proposition, that the burgeoning Molecular Orbital Theory could be used to make quantitative predictions concerning reaction and spectroscopic outcomes. Given the widespread application of Quantum Mechanics, supported by high-level calculations of electronic structure that pervade this book, and indeed all modern literature, it is hard to imagine that these basics were not immediately embraced by the community.

His theories concerning electron transfer became accepted following his 1967 paper in Prog. Inorg. Chem. *that linked the reactivity and spectroscopy of mixed-valence compounds. This led to simple explanations for observed properties, paralleling the Robin–Day classification scheme introduced in the same year. The synthesis of the Creutz–Taube ion two years later, which demonstrated unprecedented chemical features that were only interpretable using quantum mechanics, was central to the acceptance of Hush's theories. It led to the award of the* Nobel Prize *to Taube in 1982, with the works of Hush, and Robin and Day defining much of the field of Mixed-Valence Chemistry unto this day.*

During the 1970s, Hush pioneered computational methods that allowed the responses of quantum systems to an externally applied electric field to be studied, facilitating quantitative analysis of the Stark effect. During the 1980s, his focus was on the interpretation of new experiments demonstrating intramolecular photoinduced charge separation, works that were to form a platform for understanding the quantum aspects of biological photosynthesis. This led, during the next two decades, to quantitative theories for the operation of primary charge separation in purple bacteria, as well as to the development of the field of Molecular Electronics. These and many parallel achievements have brought the field of Mixed-Valence Chemistry to where it is today, providing a backbone for the understanding of many commercial processes including dyes, sensors, medicines, light-harvesting, and electron-transport systems.

Contents

Preface *xiii*

1   **Introduction and Fundamentals of Mixed-Valence Chemistry** *1*
Chun Y. Liu and Miao Meng
1.1   Introduction *1*
1.2   Brief History *4*
1.3   Diversity of Mixed-Valence Systems – Some Examples *6*
1.4   Characterization and Evaluation of Mixed-Valence Systems *12*
1.4.1   Electron Paramagnetic Resonance Spectroscopy *12*
1.4.2   Electrochemical Methods *13*
1.4.3   Optical Analysis *14*
1.5   Important Issues in Mixed-Valence Chemistry *16*
1.5.1   System Transition in Mixed Valency from Localized to Delocalized *16*
1.5.2   Solvent Control of Electron Transfer *17*
1.6   Theoretical Background *18*
1.6.1   Potential Energy Surfaces from Classical Two-State Model *18*
1.6.2   Quantum Description of the Potential Energy Surfaces *20*
1.6.3   Reorganization Energies *24*
1.6.4   Electronic Coupling Matrix Element and the Transition Moments *26*
1.6.5   The Generalized Mulliken–Hush Theory (GMH) *27*
1.6.6   Analysis of IVCT Band Shape *28*
1.6.7   Rate Constant Expressions of Electron Transfer Reaction – The Marcus Theory *30*
1.6.8   McConnell Superexchange Mechanism and the CNS Model *32*
1.7   Conclusion and Outlook *35*
       Acknowledgments *35*
       References *36*

2		**Conceptual Understanding of Mixed-Valence Compounds and Its Extension to General Stereoisomerism** *45*
		Jeffrey R. Reimers and Laura K. McKemmish
2.1		Introduction *45*
2.2		Modeling MV and Related Chemistry *47*
2.2.1		Origins Within Chemical Bonding Theory *47*
2.2.2		Coupled Harmonic Oscillator Model *48*
2.2.3		Intermolecular and Intramolecular Contributions to the Reorganization Energy *55*
2.2.4		Effects of Electric Fields on MV Optical Band Shapes *56*
2.2.5		Non-adiabatic Effects *58*
2.2.6		MV Complexes as Potential Quantum Qubits *60*
2.2.7		Entanglement as a Measure of the Failure of the BO Approximation *64*
2.2.8		Further Reading *65*
2.3		Some Traditional Mixed-Valence Example Molecules and Iconic Model Systems *65*
2.3.1		Photochemical Charge Separation *66*
2.3.2		MV Excited States in a Bis–Metal Complex *66*
2.3.3		Hole Transport in a Molecular Conducting Material *68*
2.3.4		Ground-State Delocalization in the Creutz–Taube Ion *68*
2.3.5		Photochemical Charge Separation During Bacterial Photosynthesis *70*
2.3.6		Prussian Blue *73*
2.4		Applications to Stereoisomerism *73*
2.4.1		Breakdown of Aromaticity in the (π,π^*) $^3A_1$ Triplet Ground State of Pyridine *74*
2.4.2		Isomerism of BNB *75*
2.4.3		Isomerism of Ammonia and Related Molecules *75*
2.4.4		Proton Transfer in $[NH_3 \cdot H \cdot NH_3]^+$ *79*
2.4.5		Aromaticity in Benzene *80*
2.5		Conclusion and Outlook *81*
		References *82*
3		**Quantum Chemical Approaches to Treat Mixed-Valence Systems Realistically for Delocalized and Localized Situations** *93*
		Martin Kaupp
3.1		Introduction and Scope *93*
3.2		How Did We Start *94*
3.3		Moving to Transition Metal MV Systems, Getting into Conformational Aspects *97*
3.4		More Recent Work on Organic MV Systems and More General Use for Charge Transfer Questions *99*
3.5		More Recent Insights into Conformational Aspects for Transition Metal Complexes *100*
3.6		Other Applications to Organometallic MV Systems *103*

3.7	Limitations of the Simple Computational Protocols, Gas-Phase Benchmarks, and Improved Electronic Structure Methods	*104*
3.8	More Advanced Treatments of Environmental Effects	*109*
3.9	Conclusion and Outlook	*111*
	Acknowledgement	*112*
	References	*113*

4 Mixed Valency in Ligand-Bridged Diruthenium Complexes *121*

Sanchaita Dey, Sudip Kumar Bera, Wolfgang Kaim, and Goutam Kumar Lahiri

4.1	Introduction	*121*
4.2	$Ru^{II}Ru^{III}$ Mixed-Valent Systems	*124*
4.2.1	Pyrazine-Derived Bridges	*124*
4.2.2	Other Bridging Ligands	*130*
4.3	$Ru^{III}Ru^{IV}$ Mixed-Valent Systems	*135*
4.4	$Ru^{II}Ru^{I}$ and $Ru^{I}Ru^{0}$ Mixed-Valent Systems	*139*
4.5	Conclusion and Outlook	*141*
	Acknowledgment	*142*
	References	*142*

5 Electronic Communication in Mixed-Valence (MV) Ethynyl, Butadiynediyl, and Polyynediyl Complexes of Iron, Ruthenium, and Other Late Transition Metals *151*

Sheng Hua Liu, Ya-Ping Ou, and František Hartl

5.1	Introduction	*151*
5.2	Iron–Ethynyl Complexes	*152*
5.2.1	Dinuclear Iron–Ethynyl Complexes with Butadiynediyl Bridge	*153*
5.2.2	Dinuclear Iron–Ethynyl Complexes with Diynediyl, Polycyclic Aromatic Hydrocarbons and Heterocycles in the C_4 Bridge Core	*154*
5.2.3	Dinuclear Iron–Ethynyl Complexes with Non-conjugated C_4 Bridge Core	*156*
5.2.4	Functionalized Dinuclear Iron–Ethynyl Complexes	*157*
5.3	Ruthenium–Ethynyl Complexes	*158*
5.3.1	Dinuclear Ruthenium–Ethynyl Complexes with $Cp'(L_2)$Ru-Based Termini	*158*
5.3.2	Dinuclear Ruthenium–Ethynyl Complexes with $Ru(dppe)_2$X-Based Termini	*163*
5.3.3	Ruthenium–Ethynyl Complexes with Alternating Polyyndiyl and Capped Ru–Ru Units	*165*
5.3.4	Ruthenium–Ethynyl Complexes with Other Ruthenium–Ethynyl Termini and Core Units	*166*
5.4	Other Transition Metal–Ethynyl Complexes	*168*
5.4.1	Dinuclear Group 6 (Cr and Mo) Metal–Ethynyl Complexes	*168*

5.4.2	Dinuclear Group 7 (Mn and Re) Metal–Polyynediyl Complexes	169
5.4.3	Dinuclear Group 8 (Os) and Group 9 (Co) Metal–Polyyndiyl Complexes	170
5.5	Concluding Remarks and Outlook	171
	Acknowledgment	172
	References	172

6 Electron Transfer in Mixed-Valence Ferrocenyl-Functionalized Five- and Six-Membered Heterocycles *181*
Peter Frenzel and Heinrich Lang

6.1	Introduction	181
6.2	Ferrocenyl-Functionalized Five-Membered Heterocycles	182
6.2.1	Five-Membered Heterocyclic Compounds with Group 13 Elements	183
6.2.2	Five-Membered Heterocyclic Compounds with Group 14 Elements	183
6.2.3	Five-Membered Heterocyclic Compounds with Group 15 Elements	185
6.2.4	Five-Membered Heterocyclic Compounds with Group 16 Elements	201
6.2.5	Five-Membered Heterocyclic Compounds with Transition Metal Elements	213
6.3	Ferrocenyl-Functionalized Six-Membered Heterocycles	217
6.4	Conclusion and Outlook	218
	Acknowledgment	219
	References	220

7 Electronic Coupling and Electron Transfer in Mixed-Valence Systems with Covalently Bonded Dimetal Units *229*
Chun Y. Liu, Nathan J. Patmore, and Miao Meng

7.1	Introduction	229
7.2	Synthesis and Characterization	233
7.3	d(δ)(M$_2$)-p(π)(Ligand) Conjugation	235
7.4	Electronic and Intervalence Transitions and DFT Calculations	238
7.5	Transition in Mixed Valency Between Robin–Day Classes	240
7.6	Distance Dependence of Electronic Coupling and Electron Transfer	247
7.7	Conformational Effects of Electronic Coupling and Electron Transfer	252
7.8	Class III and Beyond	256
7.9	Cross-Conjugation and Quantum Destructive Effect	257
7.10	Electronic Coupling and Electron Transfer Across Hydrogen Bonds	258
7.11	Mixed-Valence Diruthenium Dimers	260
7.12	Conclusions and Outlook	262
	Acknowledgments	263
	References	263

8 Mixed-Valence Electron Transfer of Cyanide-Bridged Multimetallic Systems 269
Shao-Dong Su, Xin-Tao Wu, and Tian-Lu Sheng

8.1 Introduction 269
8.2 Dinuclear Cyanide-Bridged Mixed-Valence Complex 272
8.3 Trinuclear Cyanide-Bridged Mixed-Valence Complex 276
8.4 Tetranuclear and Higher Nuclear Cyanide-Bridged Mixed-Valence Complex 284
8.5 Conclusion and Outlook 290
Acknowledgment 290
References 291

9 Organic Mixed-Valence Systems: Toward Fundamental Understanding of Charge/Spin Transfer Materials 297
Akihiro Ito

9.1 A Brief Sketch of the History of Organic Mixed-Valence Systems 297
9.2 A Glossary for This Chapter 299
9.2.1 Hush Analysis 300
9.2.2 Mulliken–Hush Two-State Analysis 301
9.2.3 Mulliken–Hush Two-Mode Analysis 301
9.2.4 Generalized Mulliken–Hush Three-State Analysis 302
9.3 Relationship Between Bridging Units and Electronic Coupling 304
9.4 Where to Attach Redox Centers 310
9.5 Through-Bond or Through-Space? 311
9.6 Control of Spin States Through Mixed-Valence States 314
9.7 Future Prospects 315
Acknowledgment 316
References 316

10 Mixed-Valence Complexes in Biological and Bio-mimic Systems 323
Xiangmei Kong, Yixin Guo, Zijie Zhou, and Tianfei Liu

10.1 Introduction 323
10.2 Mixed-Valence Iron–Sulfur Clusters in Biological and Bio-mimic Systems 325
10.2.1 Basic FeS Clusters 325
10.2.2 [FeFe]-Hydrogenase 326
10.2.3 Nitrogenases 328
10.2.4 Carbon Monoxide Dehydrogenase 329
10.3 Mixed-Valence Systems in Multiheme and Other Multiiron-Contained Biological Systems and Their Mimics 331

10.4	Mixed-Valence Multicopper Cofactors in Biological and Mimicking Systems *332*
10.5	OEC and Other Mixed-Valence Multimanganese Cofactors *336*
10.6	Summary *339*
	Acknowledgement *339*
	References *340*

11 Control of Electron Coupling and Electron Transfer Through Non-covalent Interactions in Mixed-Valence Systems *349*

Zijie Zhou, Yixin Guo, Xiangmei Kong, Ying Wang, and Tianfei Liu

11.1	Introduction *349*
11.2	Electronic Coupling Through Hydrogen Bonds *350*
11.2.1	Electronic Coupling Between Transition Metal Centers Through Hydrogen Bonds *350*
11.2.2	Electronic Coupling Between Organic Fragments Through Hydrogen Bonds *353*
11.3	Modulation of Electronic Coupling via Host–Guest or Through-Space Interaction *356*
11.4	Conclusion *361*
	Acknowledgment *361*
	References *361*

12 Stimulus-Responsive Mixed-Valence and Related Donor–Acceptor Systems *365*

Jiang-Yang Shao and Yu-Wu Zhong

12.1	Introduction *365*
12.2	Photoswitchable Compounds *365*
12.3	Anion-Responsive Compounds *378*
12.4	Proton-Responsive Compounds *380*
12.5	Conclusion and Outlook *385*
	Acknowledgement *385*
	References *385*

13 Mixed Valency in Extended Materials *393*

Harrison S. Moore, Eleanor R. Kearns, Martin P. van Koeverden, and Deanna M. D'Alessandro

13.1	Introduction *393*
13.1.1	Fundamental Aspects of Mixed Valency in the Solid State *393*
13.1.2	Quantum Mechanical Considerations in Mixed Valency and IVCT *394*
13.1.3	Marcus–Hush Theory and the Quantification of CT *395*
13.1.4	Classifications of Mixed Valency *395*
13.1.5	Organic Mixed Valency *396*
13.2	Electron Transfer in Extended MV Materials *397*
13.2.1	Introduction to Extended Materials *397*
13.2.2	Organic-Based Mixed Valency in Extended Frameworks *397*

13.2.2.1	Thiazolo[5,4-d]thiazole-Based Compounds *397*
13.2.2.2	Tetrathiafulvalene (TTF)-Based Compounds *399*
13.2.2.3	Tetraoxolene-Based Compounds *400*
13.2.2.4	Naphthalenediimide (NDI)-Based Compounds *405*
13.2.2.5	Phenalenyl-Based Compounds *406*
13.2.2.6	Covalent-Organic Frameworks (COFs) *407*
13.2.3	Metal-Based Mixed Valency *408*
13.2.3.1	First-Row Transition Metals *408*
13.2.3.2	Other Metals *414*
13.2.3.3	Catalysis in Uncoupled MV Systems *414*
13.3	Conclusion *418*
	References *419*

14 Near-Infrared Electrochromism Based on Intervalence Charge Transfer *431*

Ying Han, Xiaohua Cheng, Yu-Wu Zhong, and Bin-Bin Cui

14.1	Introduction *431*
14.2	Near-Infrared Electrochromic Materials *432*
14.2.1	Inorganic NIR Electrochromic Materials *433*
14.2.2	Organic NIR Electrochromic Materials *435*
14.2.2.1	Viologen Derivatives *435*
14.2.2.2	Triphenylamine Derivatives *437*
14.2.2.3	Organic Conducting Polymers *439*
14.2.2.4	Covalence-Organic Framework (COF) *442*
14.2.3	Organic–Inorganic Hybrid NIR Electrochromic Materials *444*
14.2.3.1	Metal Complexes *444*
14.2.3.2	Conducting Polymers of Metal Complexes *447*
14.2.3.3	Monolayer and Multilayer Assembled Films *452*
14.3	Potential Applications of NIR Electrochromic Materials *453*
14.3.1	Smart Windows *453*
14.3.2	Molecular Logic Gates and Optical Storage *453*
14.3.3	Optical Communication *453*
14.3.4	Military Camouflage *454*
14.4	Summary and Outlook *454*
	Acknowledgment *455*
	References *455*

15 Manipulation of Metal-to-Metal Charge Transfer Toward Switchable Functions *463*

Wen Wen, Yin-Shan Meng, and Tao Liu

15.1	Introduction *463*
15.2	Switchable Cyanide-Bridged MMCT Systems *465*
15.3	Cyanide-Bridged MMCT Complexes Showing Switchable Functional Properties *472*
15.3.1	Modulating Molecular Nanomagnet Behavior *472*

15.3.2 Modulating Molecular Electric Dipole *474*
15.3.3 Modulating Thermal Expansion Behavior *478*
15.3.4 Modulating Photochromic Behavior *480*
15.4 Conclusion and Outlook *483*
References *484*

Index *492*

Preface

In the simplest form, mixed-valence (MV) compounds refer to redox-active molecular systems in which the same chemical element is present in different oxidation states. In more generalized forms, differing ions, or indeed differing chemical groups, may be involved, the core idea being the presence of two groups that may exchange electron(s) internally to create isomeric forms of the compound. One famous example is the Creutz–Taube ion, $\{[(Ru(NH_3)_5](\mu\text{-pz})[(Ru(NH_3)_5]\}^{5+}$ (pz = 1,4-pyrazine), which, applying classical valence theory, can be envisaged as containing one Ru^{2+} (d^6) and one Ru^{3+} (d^5) center connected via a pyrazine bridge. The principle applies not just to compounds but also to materials, the classic example being the dye Prussian blue that contains charge-localized Fe^{2+} and Fe^{3+} centers bridged by cyanide ligands.

MV compounds provide simple model system for the examination of the fundamental electron-transfer (ET) processes in donor–bridge–acceptor molecular systems. A prominent feature of MV compounds is the observation of the intervalence charge-transfer (IVCT) transition in a broad range from the visible to infrared region, depending on the nature of the system and the strength of electronic coupling. By analyzing the IVCT band, important parameters of the ET process between mixed-valent redox sites can be derived, including the reorganization energy (λ), the electronic coupling parameter (H_{ab}), and the thermal activation barrier (ΔG^*). Along with these studies, enormous information on the influence of the chemical structures and environments on ET processes is obtained. This field has become increasingly important when molecular electronics and artificial photosynthesis emerge as the research frontiers in physical sciences across the globe.

In 1979 and 1990, two monographs with the title *Mixed-Valence Compounds: Theory and Applications in Chemistry, Physics, Geology, and Biology* and *Mixed Valency Compounds: Applications in Chemistry, Physics and Biology* were published as the NATO Advanced Science Institutes Series. From then on, no monograph on the topic of MV compounds has been published. Considering that a great deal of progress has been made in recent decades and this field has received continuous interest to date, we felt it necessary and important to organize an updated book to reflect the current state of studies on MV compounds. We proposed this book in December 2020 and started to invite contributions a few months later. Thanks to the professional, enthusiastic, and timely contributions from the co-authors, we've been able to complete it on schedule.

This edited book invited contributions from the main experts who are currently actively working in the field of MV compounds. Chapters 1–3 describe the fundamentals and recent theoretical progress on the understanding and analysis of MV compounds. Chapters 4–9 present updated results, in particular the ET properties, of covalently connected MV compounds, including bridged diruthenium complexes and metal alkynyls, ferrocenyl-functionalized heterocycles, covalently bonded dimetal (M_2) complexes, cyanide-bridged multimetallic systems, and organic MV systems. Chapters 10–15 describe various nonclassical aspects of MV compounds, including the MV complexes in biological and biomimetic systems, the ET of noncovalent systems, materials with stimulus-responsive IVCT or metal-to-metal charge transfer absorptions, mixed valency in extended materials, and the applications of MV compounds in near infrared (NIR) electrochromism. These topics cover the important advances in the theory, synthesis, ET, and application of conventional and nonclassical MV systems. We believe this book will be an essential reference for a wide range of scientific researchers and graduate students interested in MV systems and electron-transfer studies.

July 2022

Yu-Wu Zhong (Beijing)
Chun Y. Liu (Guangzhou)
Jeffrey R. Reimers (Sydney)

1

Introduction and Fundamentals of Mixed-Valence Chemistry

Chun Y. Liu and Miao Meng

Jinan University, College of Chemistry and Materials Science, Department of Chemistry, 601 Huang-Pu Avenue West, Guangzhou 510632, China

1.1 Introduction

The term mixed valence (MV) is used to describe chemical systems in condensed media and solids in which the same chemical element exists in different oxidation states [1–3]. Thus, MV compounds refer to the category of unimolecular systems consisting of more than one redox center derived from the same element but formally having different oxidation levels in the ground state. In this context, molecules or solids having the same chemical constitutions but different oxidation states for the nonequivalent atoms should be viewed as distinct chemical identities or materials, but those having the same oxidation level are chemically identical. Prussian blue, the prototype of MV compound, is identical to Turnbull's blue [4]. It should be addressed that in MV compounds, the oxidation states of individual redox-active atoms that share the same elemental redox potential depend upon the electronic properties of the chemically bonded atoms or groups. For example, a high oxidation level is given to a redox center surrounded by more or stronger electron-withdrawing atoms or groups, and vice versa, a lesson learned from text book chemistry. However, mixed valency of MV compounds, which concerns charge distribution over the molecular ground state, is a very comprehensive issue pertaining to electrons and nuclei in motion that compasses a number of fundamental chemical problems, including energetic, dynamic, kinetic, and mechanistic of chemical transformations [5–8]. Moreover, MV compounds possess a unique optical property resulting from charge transfer between the spatially separated (chemically bonded or nonbonded) atoms with different valence electron shells. The interplays of electronic and nuclear dynamics within the molecule and between molecules (MV molecules and solvent molecules) are implicated through their optical behaviors, which are translated into the dynamics and energetics of the interpenetrated chemical and physical systems. With its enriched scientific contents, mixed-valent chemistry has evolved into one of the major playgrounds in modern chemistry in its own right for experimental and theoretical practitioners [6–10].

Mixed-Valence Systems: Fundamentals, Synthesis, Electron Transfer, and Applications, First Edition.
Edited by Yu-Wu Zhong, Chun Y. Liu, and Jeffrey R. Reimers.
© 2023 WILEY-VCH GmbH. Published 2023 by WILEY-VCH GmbH.

The attraction of mixed-valence systems is largely enforced by the fact that the valences of the discrete redox centers are intramolecularly self-exchangeable, thus representing the most elementary chemical reaction: intramolecular electron transfer (ET). In the middle of last century, the theoretical framework for ET was constructed and expanding rapidly, as marked by a series of profound progresses made in a relatively short period of time. Kubo and Toyozawa derived the general expression of activation energy (1955) [11]; Levich and Dogonadze presented the rate equation for ET reaction in the nonadiabatic limit (1960) [12, 13]; Marcus introduced the dielectric continuum model of solvation and the classical ET kinetic formalism (1956) [14, 15]; McConnell developed the superexchange model (1961) [16]; and Hush described the intramolecular effects using coupled harmonic surfaces (1958) [17] and calculations of the electronic coupling integral from intervalence optical parameters (1967) [5]. In the two-state description, the energy profiles of initial and final states of the system are approximated with a harmonic oscillator, which models the incorporated electron–nuclei dynamics in chemical transformation from reactant to product along the reaction coordinate. This simplified theoretical model on ET demands an experimental model that has single transferring electrons and well-defined electronic configuration. Thus, research work on MV chemistry gained a strong impetus to experimentally monitor the ET processes and to validate the semiclassical theories.

The follow-up experimental study was pioneered by Taube and Creutz with the elegantly designed, pyrazine (pz)-bridged diruthenium complex (**I**), $\{[(Ru(NH_3)_5](\mu\text{-pz})[(Ru(NH_3)_5]\}^{5+}$, known as the Creutz–Taube ion [18], in which the two bridged Ru ions have formal oxidation numbers +2 and +3.

I

In a formal sense, the $Ru^{2+}(d^6)$ and $Ru^{3+}(d^5)$ centers in **I** serve the electronic donor (D) and acceptor (A), respectively, and electron self-exchange crossing the pz bridge (B) occurs without change of the free energy ($\Delta G = 0$). In the mixed-valent D–B–A molecular system, electron migrating from D to A and nuclear motion conform energetically and dynamically to the semiclassical two-state models [19, 20]. The Creutz–Taube ion allowed the first observation of Frank–Condon transition that induces ET between two metal centers in a molecular complex, namely, intervalence charge transfer or IVCT [18, 21]. Inspired by the Creutz–Taube complex, a large number of MV compounds in form of D–B–A with various transition metal complex and organic charge-bearing units for the D and A sites have been synthesized, and studied in terms of electronic coupling (EC) and ET [6, 8, 22–24].

Electron transfer in MV systems may proceed via one of the two reaction pathways, thermal or optical [6, 22, 25–27]. By thermal ET pathway, the system overcomes the thermal energy barrier (ΔG^*) and reaches the transition state through thermal fluctuations. In the transition state, designated as $[D-B-A]^{\neq}$ and $[A-B-D]^{\neq}$ in Figure 1.1 for the forward and reverse reactions, respectively, the system has an averaged nuclear configuration for the MV molecule (the activated complex) and solvation. From the reactant to the product, the system experiences an adiabatic process. Optical ET in MV compounds is initiated by vertical transition of the reactant state (with the extra electron on the donor) to the vibrational excited states of the product (with the extra electron transferred to the acceptor) (Figure 1.1). This transition occurs between two diabatic states and is governed by the Frank–Condon principle. Radiationless relaxation of the system from the nuclear excited state to the ground states completes the ET process [6, 19, 25].

For the ET event to occur, no matter which pathway is taken, the donor and acceptor electronic states must be coupled. It is the extent of coupling that controls the ET dynamics and kinetics, which is quantified by the coupling matrix element in quantum mechanics, i.e. H_{ab}. Hush demonstrated that this crucial quantity can be derived from the IVCT spectrum of the MV compound [5, 6, 9, 19]. The Hush model connects the spectral data (transition energy, intensity, and absorption bandwidth) of the molecular system and the energetic parameters of the ET reaction (coupling integral and thermal ET barrier), and paves the way to optical determination of ET rate constant (k_{ET}). This optically determined coupling integral (H_{ab}) can be incorporated into adiabatic and nonadiabatic ET kinetic expressions in the classical and semiclassical formalisms, which have been successfully applied in strongly and weakly coupled MV systems, respectively. Advances in time-resolved spectroscopic techniques allow the photoexcited states to be monitored, thus providing a powerful means for study of the photoinduced ET process in systems involving electronic excited states, D*–B–A or D–B–A*. Optical study of MV compounds and transient spectroscopic investigations of photoinitiated ET are complemented in development, validation, and refinement of the contemporary ET theories [19, 20, 25]. The gained

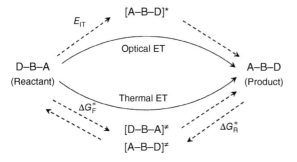

Figure 1.1 Optical (top) and thermal (bottom) ET pathways in mixed-valence D–B–A compounds. [A–B–D]* represents the vibrational excited state of the product. E_{IT} is the intervalence charge transfer transition energy. $[D-B-A]^{\neq}$ and $[A-B-D]^{\neq}$ refers to the transition complex for the reactant and product, respectively. ΔG^*_F and ΔG^*_R is activation energy of the forward and reverse ET reaction, respectively.

understanding allows control of electron (charge) transfer in molecular systems and elucidation of the long-range charge transport processes in biological system and is beneficial to development of innovative technologies such as conductive materials, molecular electronics, and catalysts for solar–chemical energy conversion.

1.2 Brief History

Historically, mixed-valence solids were found several centuries ago in various minerals, such as metal oxides, sulfates, and phosphates, in which the metal elements exist in different valence states [1, 3, 28]. These minerals usually show intense colors. The coloration of vivianite crystal with the chemical formula $Fe_3(PO_4)_2 \cdot 8H_2O$ is one of the interesting examples [5]. Vivianite is colorless when freshly exposed, as expected for the Fe^{2+} ion; after being exposed to air, it shows varying colors from light blue, light green, to dark blue or green, depending on the length of exposure due to oxidation of Fe^{2+} to Fe^{3+}. As early as in the eighteenth century, it was realized that the blue color of ceramic glaze on vases was produced from ferrous iron (Fe^{3+}) in reducing conditions. In nearly the same period of time, Prussian blue became a popular pigment for artists, which contains Fe^{3+} ions and negatively charged hexacyanoferrate ions $[Fe(CN)_6]^{4-}$, formulated as $Fe_4[Fe(CN)_6]_3$, as described in the chemistry text book at entry level. However, chemists at that time were unable to explain the coloration of this material because both ferrous and ferric ions in aqueous solution do not show strong absorptions in this particular spectral range. It was generally observed that solutions or solids containing an element in two different valence states often exhibit unusual intense coloration, which does not appear when either of the elements is present alone [3]. This recognition connected coloration of the complexes to the valences of its ingredients [1], importantly, beginning to be aware that the distribution of oxidation states within the molecule can exchange under the influence of light so as to produce the light absorption and hence the color. For these systems, "valency oscillation" and "resonant valency" were proposed to describe the physical origin of intense color presented by one element in different valence states [29], which is more or less close to today's understanding. In 1950s, long-distance electron transfer between metal ions was assumed to explain "valency oscillation." Weyl first noted that light absorption in MV systems is related to the interactions between two valence states of the same element [30].

Mixed-valence phenomena are also widely seen in enzymes and cofactors of biological systems where the active sites consist of multiple metal centers in variable oxidization states. For example, naturally occurring photosynthesis produces energy materials from low-potential molecules such as H_2O and CO_2 by absorption of visible light. The energy conversion processes involve two protein cofactor complexes, namely, photosystems (PS) II and I. In PS II, the oxygen-evolving complex (OEC), which conducts oxidation of water to molecular oxygen, is an ox-tetramanganese cluster with the Mn atoms in different oxidation levels [31]. In PS I, ferredoxin $\{(Cys)_2Fe^{II}\text{-}(\mu\text{-}S)2Fe^{III}(Cys)_2\}^+$ ([2Fe2S]) is the electron carrier that transports electrons to the enzymatic reductive reaction center, the ferredoxin([2Fe2S])-nicotinamide adenine dinucleotide (NADP/H) reductase (FNR) [32]. Accomplishments of these biochemical reactions depend on intramolecular and intermolecular electron transfer with the driving force ultimately from sunlight [33].

In 1960–1970s, three important publications by Hush (1967) [5], Robin and Day (1968) [34], and Creutz and Taube (1969) [35] marked the cornerstone in development of mixed-valence chemistry. Based on the Mulliken charge transfer theory, Hush demonstrated [36, 37] that the electronic coupling matrix element (H_{ab}) can be calculated from the IVCT parameters [5–7], transition energy E_{IT}, molar extinction coefficient ε_{IT}, and half-height bandwidth $\Delta v_{1/2}$ (Eq. 1.1).

$$H_{ab} = 2.06 \times 10^{-2} \frac{(E_{IT}\varepsilon_{IT}\Delta v_{1/2})^{1/2}}{r_{ab}} \tag{1.1}$$

where E_{IT} and $\Delta v_{1/2}$ are in wavenumber (cm^{-1}) and r_{ab} is the effective electron transfer distance in angstrom (Å). This Mulliken–Hush expression, developed in the pure classical two-state regime, can be used in broad range of double-well charge transfer systems. The Hush model also reveals the correlation between optical (radiative) and thermal (radiationless) electron transfer for symmetric MV systems [5, 7, 14, 38].

$$E_{IT} = 4\Delta G^* \tag{1.2}$$

Equation (1.2) suggests that the kinetics and energetics for the ET process can be described through optical analysis of the intervalence charge transition of MV compounds [19]. It is interesting that this fundamental energetic relationship concerning activation energy of ET reaction was revealed by Kubo and Toyozawa [11], Marcus [14], and Hush [5] from their independent works.

Robin and Day provided a scheme that classifies the MV compounds in terms of the extent of electronic coupling [34]. According to them, within the semiclassical framework, there are three regimes that MV compounds in different coupling strength belong to, that is, noncoupled (fully localized) Class I, strongly coupled or fully delocalized Class III, and the intermediate Class II that encompasses systems from weakly to moderately strongly coupled. In Robin–Day's classification [34], for MV compounds in Class I, thermal exchange of the oxidation states for the element in different sites is very slow, and the optical transition occurs by weak absorption of high-energy photons, while in Class III compounds, the valence states for the element in the multiple sites are averaged and thus crystallographically indistinguishable [1]. Class II compounds are those for which the electronic wave functions of the ground state and the excited state are significantly mixed, and the valence states are interchangeable in response to external stimulations, such as light and heat [1, 34].

Synthesis of the Creutz–Taube complex in 1969 initiated experimental studies of intramolecular EC and ET. For the Creutz–Taube ion, a broad, asymmetric absorption band was observed at 6369 cm^{-1}, which was attributed to electron transfer from Ru^{2+} to Ru^{3+} crossing the pyrazine molecule. However, it took many years to characterize this MV compound in terms of the Robin–Day's scheme, that is, whether it belongs to localized Class II with +2 for one Ru center and +3 for the other or to delocalized Class III with an averaged oxidation state of +2.5 for each of the two Ru centers. Now, it is generally accepted that the Creutz–Taube ion is best placed on the Class II–III borderline [39, 40].

A prominent feature of MV compounds is the observation of characteristic IVCT transition that occurs in a broad region from visible to infrared depending on the strength of electronic coupling between the redox centers. By analyzing the IVCT band, important parameters of the ET process between mixed-valent redox sites can be extracted, including the reorganization energy (λ), the electronic coupling parameter (H_{ab}), and the thermal activation barrier (ΔG^*). These concepts arose from the seminal works of Kubo and Toyozawa, Marcus, and Hush concerning the activation energy, Marcus and Hush regarding the intermolecular and intramolecular contributions to the reorganization energy, and Hush and Levich and Dogonadze to the electronic coupling [12, 41]. McConnell's theory is then applied to understand the dependence of the coupling, and hence the spectra, on systematic extension of the separation between the mixed-valence centers. These pioneering works established the theoretic framework of mixed-valence chemistry, which has inspired and guided research in this field for a half century [1, 2, 6, 8].

1.3 Diversity of Mixed-Valence Systems – Some Examples

Following the Creutz–Taube ion, various mixed-valence D–B–A compounds were synthesized with different d^{5-6} transition metal ions (Ru, Os, and Fe) by substituting the auxiliary ligands NH_3 with inorganic anions, e.g. Cl^-, CN^-, or organic molecules, e.g. bipyridine (bpy) and terpyridine (tpy), or by modifying the bridging ligand (BL) [6, 8, 40]. For these analogues, broad, low-energy IVCT absorptions are observed typically in the near-infrared region. Asymmetrical compounds derived from heterodinuclear metal centers [42], or from homodinuclear metal ions coordinatively saturated with different supporting ligands [43], exhibited distinct energetic profiles in the two-state framework, i.e. $\Delta G^0 \neq 0$, and attracted significant attention. However, for dinuclear d^{5-6} systems with building blocks that have distorted octahedral geometry, the intervalence spectrum must be carefully assigned because the d orbitals between the two metal centers interact through $d\pi$–$d\pi$ conjugations across BL [39, 40]. As a result, multiple electronic transitions, including three intervalence transitions (IT) and two interconfigurational (IC) transitions occurring at the acceptor, may appear, as shown in Figure 1.2, and may overlap with each other [40]. In this case, only the lowest energy IT band, IT(1) in Figure 1.2, arises from pure donor–acceptor ET that accounts for the reorganization energy (λ) [39].

When multidentate pyridyl and phenyl ligands are used, the degeneracy of d orbitals in an octahedral field is removed, which gives rise to single intervalence band for the mixed-valence complexes. Organometallic Ru (II/III) building blocks prepared with multidentate phenyl ligands feature five-membered ring structures

Figure 1.2 Multiple transitions occurring in d^{5-6} mixed-valence M–BL–M systems (M = Ru and Os).

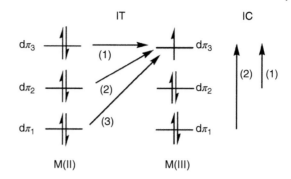

involving a Ru—C bond (**II**). The cyclometalated Ru redox centers are able to increase the molecular rigidity and stability of the assembled D–B–A complexes and to strengthen the d(Ru)–π(phenyl ligand) orbital interactions [44]. Aligning the Ru—C bonds on the donor and acceptor sites with the IVCT axis substantially enhances the electronic coupling by favoring the BL-mediated hole transfer pathway [45].

The first reported dinuclear MV complex with nonoctahedral redox sites was the biferrocenium cation $[(C_5H_5)Fe-(C_5H_4-C_5H_4)-Fe(C_5H_5)]^+$ (**III**), synthesized by Cowan and coworkers in 1973 [46]. This molecule exhibits a broad IVCT band at 1900 nm. With different BLs, a series of ferrocenium MV organometallics have been studied in terms of electronic coupling [47]. In earlier studies, few other transition metals were used to construct binuclear MV systems. For example, transition metal ions in group VIB of the periodic table were exploited as the redox centers, typically, $[M(CO)_3(PR_3)_2]_2(\mu\text{-pz})$ (M = Mo and W) (**IV**) [48, 49] and $[Mo(tp^*)(NO)Cl]_2(\mu\text{-BL})]$ (tp* = tris(3,5-dimethylpyrazolyl)hydroborate) (**V**) [50]. These dinuclear organometallic complexes feature an 18 e⁻ configuration for each metal center and present redox properties sensitive to the coordination environment due to the strong π back-bonding from the ligand to the metal center [51]. $\{[Mo(CO)_3(PR_3)_2]_2(\mu\text{-pz})\}^+$ with an electronic configuration $4d^5/4d^6$ exhibits a relatively narrow ($\Delta v_{1/2}$ 700 cm⁻¹), symmetric IVCT band (4650 cm⁻¹) in the near-IR region [49].

In the 1990s, unimolecular mixed-valence D–B–A systems were extended from metal-containing inorganic complexes to pure organic compounds. The bistriarylamine (**VI**) [52, 53] and bishydrazine (**VII**) [54, 55] derivatives involving redox-active sp^3 nitrogen atoms are the prototypes of organic MV D–B–A systems, which were studied systematically by Nelsen and Lambert, respectively. Following these works, another organic radical system, namely, D–BL–D$^{·+}$ (**VIII**), was developed in Kochi's group, with a redox-active group 2,5-dimethoxy-4-methylphenyl (D) as the donor and acceptor [56]. By employing redox-active organic groups, the concepts of mixed-valence chemistry are generalized. Compared to metal complex systems, the organic systems possess several features that favor study of EC. In these radical systems, electronic coupling and electron transfer involve single electrons that are specified with respect to orbital and electronic state, which facilitates the assignment and analysis of the IVCT bands. Study of organic systems concerns all aspects of mixed-valence chemistry, which has contributed to advance our knowledge in this field [24, 57].

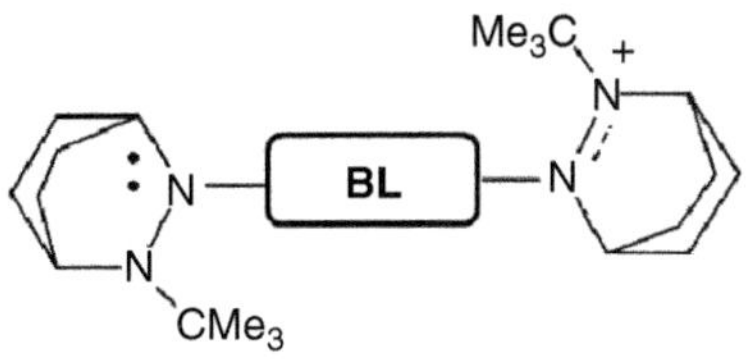

VII

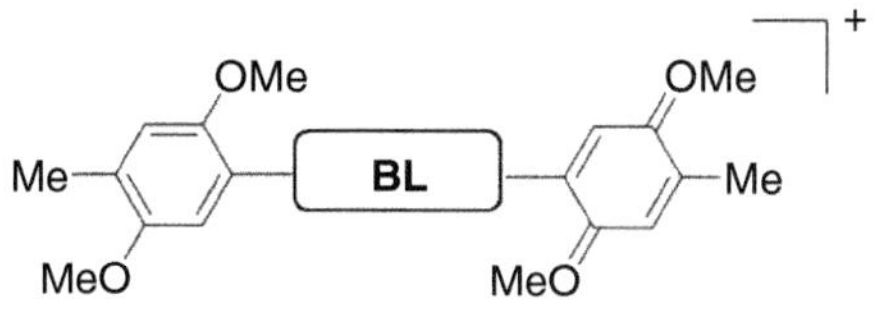

VIII

The family of bridged MV compounds was further expanded with involvement of redox-active metal clusters containing more than one metal atom, used as a whole to be a building block for assembling the D–B–A molecule. Linking two quadruply-bonded dimetal units, M_2 (M = Mo and W), with a tetradentate bridging ligand was first achieved in Chisolm's group in 1989 [58], yielding the *dimers of dimers* of form of M_2–BL–M_2, which has a formal oxidation state +4 for each of the M_2 centers, as shown in **IX**. The MV complexes $[M_2–BL–M_2]^+$ are prepared by one-electron oxidation using appropriate oxidizing reagents [58, 59]. Cotton and coworkers optimized the synthetic method with the designed dimolybdenum building block $[Mo_2(DAniF)_3]^+$ (DAniF = N,N'-di(p-anisyl)formamidinate) for converged assembly [60, 61], which led to the synthesis and structural characterization of many Mo_2 dimers with diverse bridging ligands. A quadruply-bonded M_2 unit has a well-defined, distinct electronic configuration, $\sigma^2\pi^4\delta^2$ [62]. For this M_2 MV complex system, electron delocalization within the M_2 unit is assumed. In a view of electron localization, the donor site (M_2^{4+}) has a close shell with the valence electrons paired in the δ orbital, while in the acceptor (M_2^{5+}), the δ orbital is singly occupied. Therefore, in the M_2 dimers, EC and ET involve only the δ electrons. Recently, Liu and coworkers systematically studied EC and ET in the Mo_2 MV systems under the contemporary ET theories [63, 64]. Covalently bonded diruthenium complexes were exploited by Ren and coworkers as the redox centers to assemble Ru_2 dimers through an axial linkage. For the Ru_2–BL–Ru_2 systems, polyyn-diyl chains $-(C_2)_n-$ are the favorable bridging ligands to link two $Ru_2(ap)_4$ (ap = 2-anilinopyridinate) complex molecules (**X**) with the number of alkynyl units (n) up to 6 [65].

The oxygen (O)-centered triruthenium cluster, [Ru$_3$O(acetate)$_6$-(CO)L$_2$], has been used as the electron donor or acceptor for construction of the MV D–B–A complexes by Ito and Kubiak [66]. In a [Ru$_3$O(acetate)$_6$-(CO)L$_2$] complex, the three Ru atoms are in formal oxidation states III, III, and II, which presumably are fully delocalized through the acetate bridging ligands. Replacing one of the L ligands with a pyridyl ligand modifies the redox potential of the Ru$_3$ center, which controls the properties of donor and acceptor; the other L position is replaced by a pyridyl bridging ligand, resulting in the *dimer of trimers* (**XI**). The mixed-valence system results from one-electron reduction of the Ru$_3$ dimers, {Ru$_3$(III, II, II)–BL–Ru$_3$(III, III, II)}$^-$. In this system, the electronic coupling can be tuned by alternation of the remaining L ligand, besides variation of the BL. Uniquely, the CO groups function as an IR probe, which has been successfully used to study the ET kinetics by analysis of vibrational band broadening of the carbonyl group [66, 67].

It is generally recognized that in MV D–B–A molecules, the bridging ligand plays a dominant role in control of D–A electron transfer. For decades, much of the work has focused on BL mediation of electronic coupling [23, 24, 47, 68]. In these studies, the main goals are to evaluate efficiency of various BL in coupling the electronic states, to determine its ability of transporting electrons, and to explore how the electronic event takes place. The study encompasses three crucial aspects of electron transfer reactions: energetics, kinetics, and mechanism. Approaches to these issues include variation of the BL structures, mainly through changes in length, conformation, and conjugation. Distance dependence of EC is the characteristic property for a given MV system [69], determined by the nature of the donor and acceptor. For a homologous series with varying BL length, the distance dependence of EC and ET can be evaluated by an attenuation factor β, which describes the exponential decay of EC constant (H_{ab}) or ET rate (k_{et}) against charge transfer distance (r_{ab}) [47, 69]. For a D–B–A series specified with the same D and A units, the magnitude of β reflects the charge transport ability of the BL type, and it is generally realized that a conjugated BL gives a small β value compared to saturated and cross-conjugated BLs. The influence of BL geometric conformation on EC has been intensively investigated in both inorganic [70] and organic MV systems [24, 71, 72]. On this issue, understanding obtained from combined experimental and theoretic work has revealed that BL conformation affects electronic coupling by changing the extent of orbital overlap between the donor (acceptor) and the bridge moiety, which seems quite obvious on the basis of quantum mechanics. Recent interest in cross-conjugated BLs for MV molecules is to verify the destructive quantum interference in charge transport of single-molecule conductance, aiming at development of molecular electronics [73, 74]. In this context, the *meta* (**XII**) and *para* phenylene pair is the prototype, which appears repeatedly in textbook to address phenomena like electronic resonance, π conjugation, and electron density distribution. The other example of cross-conjugated BL is σ-geminal-diethynylethene (**XIII**). **XII** and **XIII** have been used to link, for example, covalently bonded Mo_2 and Ru_2 units [75, 76] as well as organic triaryl groups [77] for constructing the MV systems. The decoupling effects of cross-conjugated BLs in the MV systems in solution are well established [75], in parallel with the results from the studies in terms of molecular conductivity.

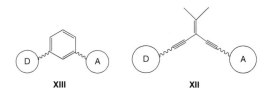

XIII   XII

Considering the relevance of ET to the biological systems, specifically DNA and proteins, hydrogen bond (HB) BLs were used to bridge donor and acceptor for MV D–B–A molecules [78]. In this regard, the HB-bridged MV systems are particularly of interest because the study can model ET induced by thermal fluctuations in the dark. The H-bond-bridged MV systems differ from those with covalent bond BLs in the molecular reflexibility and nuclear dynamics. Unfortunately, characteristic IVCT bands may not be observed in the HB-bridged systems [79, 80], being an obstacle for optical analysis. To strengthen the linkage between donor and acceptor, multiple

H-bonds are preferred [81, 82], but this seems not helpful in improving the visibility of the IVCT absorption. Study of HB D–B–A systems raise two fundamental issues, the efficiency of hydrogen bonds in transporting electrons and the coupling of electronic and nuclear degrees of freedom. Therefore, mechanistic aspect has been the focus of study in the area, specifically, how the bonded H atom(s) moves corresponding to electron transfer or vice versa. While the dominant theory is on the proton-coupled electron transfer (PCET), with respect to formation and cleavage of the bridging H-bond [83], a proton-uncoupled electron transfer (PUCT) is observed in H-bond mixed-valence system (**XIV**) recently [78]. Furthermore, in a MV D–B–A system, the bridging ligand may be functionalized so as to modulate the mixed valency in more than one state upon external stimulations such as light [84], pH, and redox reactions [85], which, therefore, are named as switching BLs. Typical examples of switching BLs are those incorporated with a dithenylethene group (**XV**) in order to make the extent of EC photo-switchable; as such, on/off status of electric conductance can be achieved by changing the radiation energy from ultraviolet to visible light [86].

1.4 Characterization and Evaluation of Mixed-Valence Systems

1.4.1 Electron Paramagnetic Resonance Spectroscopy

Mixed-valence compounds are paramagnetic with at least one unpaired electron. Thus, electron paramagnetic resonance (EPR) spectroscopy is a basic means for characterization of the electronic states and investigation of the mixed valency of the systems. For example, EPR spectroscopy is an important method used to elucidate the electronic and spin states of the Mn_4Ca cluster of the oxygen-evolving center (OEC) in photosynthesis II [87] and to characterize the mimic OEC complexes [88]. For some MV systems in which the hyperfine structures provide detailed information on the electron–nuclei coupling, EPR spectroscopy can be a powerful technique for study of the D–A electronic interaction and the ET dynamics [55, 68, 89, 90]. Particularly, the EPR timescale of 10^{-6}–10^{-8} seconds, lower than the vibrational frequencies of solvent modes in sub-picoseconds, allows the relatively slow ET processes to be probed. For weakly coupled MV compounds, whose IVCT bands are not available, EPR spectra become an important technique that complements with the optical analysis [91]. Furthermore, using the EPR methods, temperature effect of thermal ET kinetics can be readily assessed by spectral simulations [56, 91, 92].

1.4.2 Electrochemical Methods

Electrochemical methods by measuring the chemical potentials of the redox centers are widely used for characterization of MV D–B–A compounds and for semi-quantitative evaluation of the coupling strength. Separation of half-wave potentials between two successive one-electron oxidations (or reduction) occurring on each of the two redox centers, $\Delta E_{1/2}$, is recorded by cyclic voltammetry (CV) or differential pulse voltammetry (DPV) [93]. The magnitude of $\Delta E_{1/2}$ dictates the coupling extent between the D and A sites. For the members in a homologous series sharing a common donor and acceptor, the larger the $\Delta E_{1/2}$ value, the stronger the EC. A superficial explanation for this is that electronic coupling, through either electrostatic or electronic resonant effects, brings the positive charge from the site where electron is removed to the other site such that its oxidation potential increases, as a result, enlarging the $\Delta E_{1/2}$ value. In fact, there are more factors that affect the magnitude of $\Delta E_{1/2}$. From $\Delta E_{1/2}$, the free energy change (ΔG_c) and equilibrium constant K_C for comproportionation of the mixed-valence species can be determined (Eq. 1.3c–d) [8].

$$[D-B-A]^n + [D-B-A]^{(n+2)} \rightleftharpoons 2[D-B-A]^{(n+1)} \qquad (1.3a)$$

or,

$$[D-B-A]^n + [D-B-A]^{(n-2)} \rightleftharpoons 2[D-B-A]^{(n-1)} \qquad (1.3b)$$

$$\Delta G_C = -RT \ln K_C \qquad (1.3c)$$

$$K_C = \exp(\Delta E_{1/2}/25.69) \qquad (1.3d)$$

where $\Delta E_{1/2}$ is measured in millivolts (mV) at 25 °C. The magnitude of ΔG_C, which measures the thermodynamic stability of the MV species, is the sum of several energetic factors [8, 94]

$$\Delta G_c = \Delta G_e + \Delta G_r + \Delta G_i + \Delta G_s \qquad (1.4)$$

Of the four terms that contribute to ΔG_c, ΔG_e (electrostatic effect) and ΔG_r (electronic resonance effect) jointly account for the strength of electronic interaction between the D and A sites. The other two terms, statistic ($\Delta G_s = 4$) and inductive factors (ΔG_i), are generally small and similar in value for homologous systems;

therefore, ΔG_c can be divided into the resonance and nonresonance two components [25], i.e.

$$\Delta G_c = \Delta G_r + \Delta G_{nr} \quad (1.5)$$

ΔG_{nr} includes all the nonresonance contributions to the overall free energy change, i.e. ΔG_e, ΔG_i, and ΔG_s. ΔG_r is dominated by orbital interactions between the donor and acceptor. The contributions of these two terms to the comproportionation equilibrium are expressed by the following reactions,

$$[D-B-A]^n + [D-B-A]^{(n+2)} \rightleftharpoons 2\{[D-B-A]^{(n+1)}\}_0 \quad \Delta G_{nr} \quad (1.6a)$$

$$2\{[D-B-A]^{(n+1)}\}_0 \rightleftharpoons 2[D-B-A]^{(n+1)} \quad \Delta G_r \quad (1.6b)$$

In Eq. (1.6), $\{[D-B-A]^{(n+1)}\}_0$ represents the zero-interaction (charge localized) MV compound, and the second reaction shows explicitly that ΔG_r is the stabilization energy of two moles of the Class II or III MV complex by the D–A electronic interaction (delocalization) [25]. However, it should be noted that for those with different donors and acceptors, the $\Delta E_{1/2}$ values are not correlated to the EC content. Furthermore, application of electrochemical method has a low potential limit. The minimum of $\Delta E_{1/2}$ is 36.5 mV, when the compropotionation constant (K_C) reaches the statistical factor of 4 [95, 96]. For instance, this limit is approached with increasing the bridge length and lowering the symmetry compatibility between the donor (acceptor) and the BL. Evaluation of EC for a MV system may be performed at two levels in terms of accuracy, electrochemical measurements of $\Delta E_{1/2}$ for assessment of the electronic interaction and optical analysis of H_{ab} to quantitatively determine the degree of EC from the Mulliken–Hush expression (Eq. 1.1).

1.4.3 Optical Analysis

The most striking optical phenomenon for mixed-valence molecules is the absorption arising from the charge transfer from the donor to the acceptor (IVCT). Compared to the electronic transitions, the IVCT absorbance features low transition energy (E_{IT}) (usually appearing in the near-infrared or lower region) and low molar extinction coefficient (ε_{IT}) and broadness characterized by the half-height bandwidth ($\Delta v_{1/2}$); an asymmetric spectral profile is commonly found for moderately to strongly coupled MV systems. Analysis of IVCT absorption based on the Hush model (Eq. 1.1) gives rise to important chemical physical parameters in regard of D–A EC and ET. However, for Class I, the IVCT energy is high and the absorption is fairly weak, which may not be detectable spectroscopically. The basic relationship of energy conservation for the ET reaction in MV systems is given by [5, 17]

$$E_{IT} = \lambda + \Delta G^0 \quad (1.7)$$

where λ and ΔG^0 is the total reorganization energy and free energy change for the ET reaction, respectively. For symmetrical MV systems consisting of two electronically identical redox centers, $\Delta G^0 = 0$, and thus, $E_{IT} = \lambda$, which gives the thermal activation energy equaling one quarter of the transition energy (or reorganization energy), i.e. $\Delta G^* = \lambda/4$ (Eq. 1.2) [97, 98]. Based on the Mulliken formalism, Hush showed that the electronic coupling integral H_{ab} can be calculated from the IVCT spectral parameters [5]. The Mulliken–Hush expression (Eq. 1.1) is applicable for Class II compounds in a broad range of extent of EC and has been widely used in MV systems with various redox centers [6–8, 10, 20, 23, 40]. However, it should be noted that the effective ET distance, r_{ab} in Eq. 1.1, is usually shorter than the geometric distance between the redox centers due to orbital overlap between the D (A) and BL moieties [99]. Furthermore, for sufficiently strong coupling systems, the IVCT band is narrowed as the low-energy side of the absorption envelope is cutoff [9, 10, 53, 57]. Therefore, direct measurements of the r_{ab} from the molecular structure and of the $\Delta v_{1/2}$ from IVCT band would bring significant errors to estimation of H_{ab}. The effective r_{ab} is related to the dipole moment change induced by the intervalence transition, which can be probed experimentally through electroabsorption (Stark effect) [100, 101]. In case where this technique is not available, judicial determination of r_{ab} with chemical tuition might be helpful. For strongly coupled systems, simulation of a Gaussian-shaped full IVCT band is necessary to accurately measure $\Delta v_{1/2}$ for calculation of H_{ab} using the Mulliken–Hush expression (Eq. 1.1). For delocalized Class III systems, $H_{ab} = E_{IT}/2$ [17, 19]. With the classical two-state treatment, electronic coupling leading to electron delocalization lowers the adiabatic potential minima by H^2_{ab}/λ relative to those in the diabatic system [25]. In symmetrical Class II system, from the comproportionation equilibrium, we have

$$-\Delta G_r = 2H^2_{ab}/\lambda$$
$$= 2H^2_{ab}/E_{IT} \tag{1.8}$$

Similarly, for symmetrical Class III compounds, the resonance stabilization energy is the difference between the energies of two moles of fully delocalized compound (H_{ab}) and of the noninteraction mixed valence ($\lambda/4$), that is,

$$-\Delta G_r = 2(H_{ab} - \lambda/4)$$
$$= E_{IT} - \lambda/2 \tag{1.9}$$

For borderline Class III systems [25], or Class II–III and Class III systems characterized by $H_{ab} \approx \lambda/2$, $-\Delta G_r \to \lambda/2$, while for very strong coupled systems, Class III with $H_{ab} \gg \lambda/2$, $-\Delta G_r \to \lambda$. Therefore, the strongly and very strongly coupled systems can be distinguished by comparison of the magnitudes of $\Delta E_{1/2}$ and E_{IT}. For borderline Class III compounds, $\Delta E_{1/2} \approx E_{IT}/2$, and for fully delocalized systems, $\Delta E_{1/2} \approx E_{IT}$. Knowing that for strong coupling systems $\Delta G_c \approx \Delta G_r$, and thus, $-\Delta G_r = \Delta E_{1/2}$, then the reorganization energy can be estimated by $\lambda = 2(E_{IT} - \Delta E_{1/2})$. These energetical correlations that combine electrochemical and spectral data are very useful for characterization of strongly coupled MV compounds [102].

1.5 Important Issues in Mixed-Valence Chemistry

1.5.1 System Transition in Mixed Valency from Localized to Delocalized

Mixed valency is designated to determine the extent of mixing oxidation states of the spatially separated atoms of the same element or the degree of electronic coupling between the bridged redox centers. The Robin–Day's classification of MV compounds has been widely employed to characterize individual MV compounds and to map the full landscape of MV systems in terms of mixed valency. The Robin–Day's three classes of MV compounds are distinguished by the IVCT spectral features and can be schemed by physical parameters, vibrational timescale, and coupling integral and reorganization energies. Fully localized Class I compounds with $H_{ab} \leq 10\,cm^{-1}$ exhibit a high-energy IVCT band usually beyond the visible region [1]. The IVCT bands for Class II compounds ($2H_{ab} < \lambda$) may appear in visible, near-infrared region, even the infrared region, depending on the degree of EC and the nature of the system. With increasing electronic coupling, the IVCT transition energy decreases, the bandwidth becomes narrow attributed to the cutting-off phenomenon, and the band shape is more asymmetric. Within the two-state theoretic framework, the IVCT bands for the borderline Class III compounds ($2H_{ab} \approx \lambda$) are expected to be sharp [9, 10, 57]. For the very strongly coupled compounds with $2H_{ab} \gg \lambda$, the valence electrons of the redox centers are fully delocalized so that an averaged oxidation state should be assigned to each of them. The charge transfer transition (IVCT) is transformed into electronic resonance between delocalized molecular orbitals where the valence electrons reside. In this case, the Robin–Day's classification is no longer applicable [1]; the "IVCT" band, as so called, is high in energy and more symmetric, showing the transition from vibronic to electronic [103]. From the distinct electronic structure, optical behavior, and energetic correlations for this category of MV compounds, *genuine delocalized systems* or Class IV are suggested in the literature [40, 102–104].

However, the Mulliken–Hush formalism is incapable of predicting the cutoff phenomenon of IVCT absorption and earlier work did not elucidate the band asymmetry of Class II compounds, which is the most important, observable feature for moderately strongly coupled systems. More specifically, how to understand the pronounced variations of band shape and intensity as intervalence system approaches the Class II–III borderline, and how to optically characterize the transition from localized to delocalized system are the key issues in MV chemistry, which have attracted significant attention in both experimental and theoretical studies for the recent decades [9, 10, 39, 40, 55, 57]. Both metal complex and organic intervalence compounds have been exploited to resolve these problems [39, 53, 105]. Experimental approaches to this issue is to create homologous series of MV compounds, (i) having different charge-bearing units but the same BL [39, 40] (ii) sharing the same redox sites linked with different ligands [53, 106], and (iii) alternation of the auxiliary ligands and coordinating atoms [66, 105], so that the optical parameters are mapped with increasing EC. Examples in the d^{5-6} dinuclear metal complexes, such as [(bpy)$_2$ClRu(pz)RuCl(bpy)$_2$]$^{3+}$ and [(bpy)$_2$ClOs(pz)OsCl(bpy)$_2$]$^{3+}$, [(NH$_3$)$_5$Ru(pz)Ru(NH$_3$)$_5$]$^{5+}$ [(NH$_3$)$_5$Ru(4,4'-bpy)Ru(NH$_3$)$_5$]$^{5+}$, and [(NH$_3$)$_5$Os(NN)Os(NH$_3$)5]$^{5+}$, [(NH$_3$)$_5$Ru(pz)Ru(NH$_3$)$_5$]$^{5+}$, and [(bpy)$_2$ClRu(pz)RuCl(bpy)$_2$]$^{3+}$, show that with similar molecular structures, changes in metal and ligands can significantly alter the extent of electronic delocalization of the odd electron, leading to transition of the system from one MV regime to another [39]. In M$_2$–BL–M$_2$ (M = Mo and W) complex systems, it is reported that transition of mixed valency can be realized by changing the nuclearity of the dimetal units [104] or by alternation of the chelating atoms of BL [105, 107], while the molecular structures remain similar.

With intense investigations in various MV systems, the IVCT cutoff phenomenon is well understood in the two-state model framework. For strongly coupled symmetrical MV systems ($\Delta G^0 = 0$), the vibrational levels are unevenly populated in the vicinity of the equilibrium configuration of the reactant on the lower adiabatic surface because of the low activation energy (ΔG^*). This non-Boltzmann distribution of the vibrational states of the reactant ground state eliminates the spectral lines of Frank–Condon transition in low energy, consequently truncating the Gaussian-shaped band profile on the low-energy side of the IVCT band [24, 52, 57]. Therefore, for moderately strongly coupled Class II systems, an asymmetric IVCT band is observed with the bandwidth significantly narrower than the Gaussian-shaped spectrum with a half-height bandwidth $\Delta \nu_{1/2}$ (HTL) at the high-temperature limit (HTL) given by Eq. (1.10) [9, 10, 40, 57].

$$\Delta \nu_{1/2}(\text{HTL}) = \sqrt{16 \ln 2 k_B T \nu_{max}} \qquad (1.10)$$

The IVCT band gets more asymmetric as the coupling integral H_{ab} increases. Ideally, for systems on the Class II–III borderline, the IVCT band features the lowest transition energy ($E_{IT} = 2H_{ab}$) and a half-cut absorption at $2H_{ab}$ [9, 10, 105]. This unique optical feature leads to a proposal that defines a new class of mixed valency, namely, Class II–III [9, 39, 40, 53, 108].

1.5.2 Solvent Control of Electron Transfer

The three Robin–Day classes of MV compounds differ in the magnitude of H_{ab} and the ET rate (k_{et}) [109]. While Class I and III are considered to be fully electron localized and electron delocalized, respectively, electron transfer in Class II compounds can be very slow (<10^{-6} seconds) or picosecond fast (10^{-12} seconds). Solvent molecules respond to change of charge density on the donor and acceptor by reorienting their electric dipoles. Here, the key issue is the timescale of electron migration relative to that of solvent molecule motions (stretching, rotational, and translational), which governs the ET dynamics [24]. Orientational energy of solvent molecules is in the range of 30–100 cm^{-1}; the vibrational transmission frequency ν_n is taken as 5 × 10^{12} s^{-1} on average [6]. In Class II systems, when the ET scale is substantially smaller than ν_n, solvent molecules can quickly reorient their dipoles corresponding to the charge redistribution of the solute molecule, being the case where both electron and solvent are localized and ET proceeds nonadiabatically. Solvents differing in dipole polarity are characterized by static dielectric constant (ε_s). Polar solvents exert strong electrostatic interactions on the charged solute molecules, which increases the outer-sphere (solvent) reorganization energy. Solvent effects on EC and ET in Class II MV regime are predicted by the Marcus dielectric solvation continuum theory [6, 8, 14, 15, 38]. In Class III, electron exchange between the donor and acceptor occurs at a frequency close to nuclear vibrational transmission. Variation of charge density on the two charge-bearing units is "invisible" to the solvent molecules. Solvent dipoles surrounding a discrete MV molecule are randomly arranged, being the case of electron delocalized and solvent averaged [24]. A complicated situation is encountered when the system approaches Class III or on the borderline Class II–III, where ET dynamics are in the timescale of picoseconds ($10^{12}\,\text{s}^{-1}$). In Class II–III, while electron exchange proceeds rapidly, solvent molecules, with slightly slower timescale, are unable to respond promptly. In this case, the MV system remains electron localized, but the solvent effect is averaged [110]. The distinct optical behaviors and solvation properties for the borderline species give the reasons that a new mixed-valency regime, Class II–III, is defined [9, 40]. After examining the electronic dynamics of a series of Ru_3–Ru_3 compounds in various solvents and under variable temperatures, Kubiak and coworkers found that solvent dynamic properties (relaxation time τ), rather than the dielectric property of solvent, exert a particular impact on electronic dynamics of the Class II–III borderline systems [108]. The study points out that in Class II–III system, ET is controlled by time-dependent parameters (i.e. solvent relaxation times and moments of inertia) of solvents, but not by the static parameters of solvents.

1.6 Theoretical Background

1.6.1 Potential Energy Surfaces from Classical Two-State Model

The semiclassical two-state model was first used as a theoretical model to elucidate chemical transformation between particles (atoms, ions, and colloids) in earlier 1930s by Landau & Zener [111, 112]. It was adopted to illustrate the kinetics and energetics of electron transfer by Kubo and Toyozawa [11], Marcus [14, 15], Levich & Doganadze [12, 13, 41], and by Hush to derive the calculation equation of coupling constant H_{ab} from intervalence parameters in MV systems [5]. The two-state model can be interpreted in classical and quantum mechanics ways. The overall energy for a solvated molecule is contributed by electron and nuclei in motion, and the interactions between them. It changes as the electronic and nuclear configurations of the system (molecule + surrounding solvent molecules) change. Typically, in a mixed-valence system, removal of an electron from the donor and simultaneously addition of an electron to the acceptor induce variations of the nuclear configurations, and back and forth movements of the odd electron between the donor and acceptor result in energy fluctuation of the system. In the classical Marcus theory [113], the reactant and product potential energies are modeled by a harmonic oscillator. On this potential curve, the ordinate represents the potential

Figure 1.3 Diabatic free energy surfaces of the reactant (G_R) and the product (G_P) against the reaction coordinate for the ET reaction in asymmetric MV D–B–A system with $E_{IT} = \lambda + \Delta G°$ and $\Delta G^* = \lambda/4$.

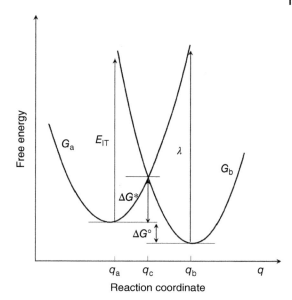

energy that becomes free energy with ignoring entropy changes (ΔS), and the abscissa measures the change of nuclear configuration or nuclear coordinate q. The potential energy of this oscillator, G_a for the reactant and G_b for the product, varies as a function of deviation ($q-q_c$) from nuclear equilibrium position (q_c) caused by electron transfer, which is then described pictorially by a parabolic curve and mathematically by a quadratic equation.

$$G = 1/2f(q_0 - q)^2 \tag{1.11}$$

where f is the oscillator strength or force constant. The diabatic potential energy surfaces for electron transfer system is constructed by the reactant (G_a) and product (G_b) potential curves crossing each other at the equal nuclear configuration q^*, as shown in Figure 1.3.

$$G_a = 1/2f(q_a - q)^2 \tag{1.12a}$$

$$G_b = 1/2f(q - q_b)^2 \tag{1.12b}$$

$$\lambda = 1/2f(q_a - q_b)^2 \tag{1.13}$$

Eq. (1.13) was used by Marcus [113, 114] and Hush [17] to define the reorganization energy for electron transfer. When the separation of the minima of the noninteracting reactant and product parabolas, ($q_a - q_b$) is replaced by a_0, and the displacement along the reaction coordinate by x, the reaction process is scaled by a dimensionless coordinate X ($=x/a_0$). X varies from 0 ($x = 0$) to 1 ($x = 1$) as the reaction proceeds from reactant to product [7]. Then, G_a and G_b are given with respect to X (Eq. (1.14a) and (1.14b)).

$$G_a = fx^2/2 = \lambda X^2 \tag{1.14a}$$

$$G_b = f(x-a_0)^2/2 + \Delta G^0 = \lambda(X-1)^2 + \Delta G^0 \tag{1.14b}$$

$$(G_b - G_a) = (\lambda + \Delta G^0) - 2\lambda X \tag{1.14c}$$

or,

$$X^2 = [\lambda + \Delta G^0 - (G_b - G_a)]^2/4\lambda^2 \tag{1.14d}$$

where ΔG^0 is the change of free energy for the electron transfer reaction. G_a and G_b cross at the transition state X^*, where $E^*_R = E^*_P$ and $X^* = (\lambda + \Delta G^0)/2\lambda$. Substituting X^* to Eq. (1.14a) and (1.14b), we obtain the free energy barrier for asymmetric electron transfer, $\Delta G^* (= G^*_a = G^*_b)$ [11, 17, 38, 97, 98],

$$\Delta G^* = (\lambda + \Delta G^0)^2/4\lambda \tag{1.15}$$

For symmetric mixed-valence systems, $\Delta G^0 = 0$, and then, $\Delta G^* = \lambda/4$. In the Marcus theory, this is the energy barrier for thermal electron transfer at the nonadiabatic limit [14, 15]. In an exothermal electron transfer reaction, $\Delta G^0 < 0$, which is the case described by Figure 1.3, while an endothermic reaction has $\Delta G^0 > 0$. The reaction driving force is conventionally designated by $-\Delta G^0$. The quadratic nature of Eq. (1.15) gives rise to three scenarios, known as the Marcus defined three regions [7, 115], as shown schematically in Figure 1.4. In the different region, the ΔG^* and thus the ET rate (k_{et}) correlate with the magnitude of ΔG^0 relative to λ in different ways. In the normal region, where $-\Delta G^0 < \lambda$, increasing the driving force lowers the energy barrier, and as a result, ET is speeded up. Further increasing the driving force to $-\Delta G^0 = \lambda$, the system enters the barrierless region ($\Delta G^* = 0$). When $-\Delta G^0 > \lambda$, from Eq. (1.15), large driving leads to increase of the energy barrier and reduction of the ET rate, which is called the inverted region.

1.6.2 Quantum Description of the Potential Energy Surfaces

Alternatively, the diabatic potential surfaces for electron transfer process can be built by quantum mechanics from the diabatic initial (ϕ_I) and final (ϕ_F) electronic states, in which the transferring electron resides on the donor and acceptor, respectively [19]. Suitable electronic wave functions that define the diabatic states can be obtained by interpreting results obtained by making electronic structure calculations within the Born–Oppenheimer framework, that is

$$\psi(r, Q) = \phi(r, Q)\chi(Q) \tag{1.16}$$

The implication of using the reduced wave functions $\phi(r, Q)$ is that the electronic wave functions for the electron in motion are instantaneously adjusted to change of the nuclear coordinate (Q) caused by nuclear dynamics, such as molecular vibrations, bond break and form. Diabatic wave functions can be constructed as sums of wave functions of that form so as to allow the transferring charge to become exposed for study. Each diabatic state has an associated potential energy varying as a function of Q due to nuclear motion,

$$E(Q) \equiv \langle \phi_I(Q) | H | \phi_F(Q) \rangle \tag{1.17}$$

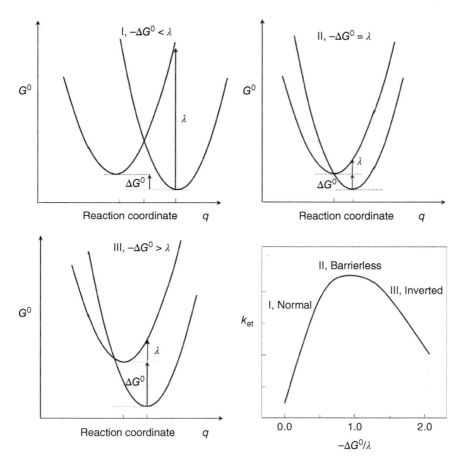

Figure 1.4 Plot of the free energy (G^0) versus the reaction coordinate q for three different values of ΔG^0 that corresponds to the three regions of Marcus, I, II, and III, respectively.

where **H** is the electronic Hamiltonian with Q dependence for the initial and final diabatic states [19].

Interaction of initial diabatic state with the final one with respect to the zeroth-order electronic Hamiltonian **H** of the system gives rise to the diabatic representation of potential energy profiles for electron transfer along the reaction coordinate Q that connects the minima of the reactant (G_a) and product (G_b) wells.

$$H_{aa} = \langle \phi_a | H | \phi_a \rangle = G_a \tag{1.18a}$$

$$H_{bb} = \langle \phi_b | H | \phi_b \rangle = G_b \tag{1.18b}$$

The eigenvalues (G_a) and (G_b) cross at the intersection Q^* that defines the diabatic activation energy ΔG^* for thermal electron transfer for Class I, in accordance with the classical description (Figure 1.3). The interaction between the reactant and product states produces the electron transfer integral H_{ab}

$$H_{ab} = \langle \phi_a | H | \phi_b \rangle \tag{1.19}$$

Linear combination of the two diabatic states ϕ_a and ϕ_b generates the adiabatic ground (lower) ψ_g and excited (upper) ψ_e states [9, 97],

$$\psi_g = c_a\phi_a + c_b\phi_b \tag{1.20a}$$

$$\psi_e = c_a\phi_a - c_b\phi_b \tag{1.20b}$$

The mixing coefficients are normalized, i.e.

$$c_a^2 + c_b^2 = 1 \tag{1.20c}$$

where c_a^2 and c_b^2 are the fraction of the transferring charge on the donor and acceptor, respectively, at any given nuclear configuration. The energies of the adiabatic states, G_g and G_e, and the interaction integral H_{ab} can be obtained by solving the two-state secular determinant:

$$\begin{vmatrix} H_{aa} - G & H_{ab} \\ H_{ab} & H_{bb} - G \end{vmatrix} = 0 \tag{1.21}$$

where $H_{aa} = G_a$ and $H_{bb} = G_b$, if the overlap integral S_{ab} is neglected. The electronic coupling between the two diabatic states leads to the formation of two new surfaces, the first-order or adiabatic states of the system [7].

$$G_g = 1/2\left\{(G_a + G_b) - [(G_b - G_a)^2 + 4H_{ab}^2]^{1/2}\right\} \tag{1.22a}$$

$$G_g = 1/2\left\{(G_a + G_b) + [(G_b - G_a)^2 + 4H_{ab}^2]^{1/2}\right\} \tag{1.22b}$$

The adiabatic potential surfaces are constructed along with the reaction coordinate X.

$$G_g = \frac{[\lambda(2X^2 - 2X + 1) + \Delta G^0]}{2} - \frac{[(\lambda(1 - 2X) + \Delta G^0)^2 + 4H_{ab}^2]^{1/2}}{2} \tag{1.23a}$$

$$G_e = \frac{[\lambda(2X^2 - 2X + 1) + \Delta G^0]}{2} + \frac{[(\lambda(1 - 2X) + \Delta G^0)^2 + 4H_{ab}^2]^{1/2}}{2} \tag{1.23b}$$

With G_g and G_e determined, the coefficients c_a and c_b can be obtained by solving the equations [116]

$$\begin{pmatrix} H_{aa} - G_g & H_{ab} \\ H_{ab} & H_{bb} - G_g \end{pmatrix} \begin{pmatrix} c_a \\ c_b \end{pmatrix} = 0$$

$$\begin{pmatrix} H_{aa} - G_e & H_{ab} \\ H_{ab} & H_{bb} - G_e \end{pmatrix} \begin{pmatrix} -c_b \\ c_a \end{pmatrix} = 0$$

Then, the product of the mixing coefficients is given by Eq. (1.24).

$$c_a c_b = H_{ab}/(G_e - G_g) = H_{ab}/E_{IT} \tag{1.24a}$$

$$c_a c_b = \frac{1}{2}\left[1 - \left(\frac{G_b - G_a}{G_e - G_g}\right)^2\right]^{1/2} \tag{1.24b}$$

Figure 1.5 plots the variations of the diabatic and adiabatic free energies along the reaction coordinate and the correlations between them.

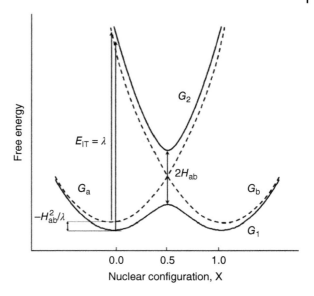

Figure 1.5 Plots of the diabatic (dashed line) and adiabatic free energy curves for ET reaction in a symmetric MV D–B–A system ($\Delta G° = 0$).

Equation (1.23) dictates that for a symmetric system, there are two energy minima ($G_g = G_e = 0$) at $X = 0$ and $X = 1$, if $H_{ab} = 0$, which is the diabatic representation. As a result of electronic coupling, H_{ab} increases, the minima for the adiabatic ground state (G_g) is lowered in energy by H^2_{ab}/λ, and shifted toward the half way ($X^* = 0.5$), getting close to each other, as shown in Figure 1.5. Then, the adiabatic minima are located at

$$\frac{1}{2}\left\{1 - \left[1 - 4(H_{ab}/\lambda)^2\right]^{1/2}\right\}, \quad \frac{1}{2}\left\{1 + \left[1 - 4(H_{ab}/\lambda)^2\right]^{1/2}\right\} \quad (1.25)$$

compared to 0 and 1, respectively, for the diabatic surfaces [9].

The difference between the adiabatic energies is given by

$$(G_e - G_g) = \left[(G_b - G_a)^2 + 4H^2_{ab}\right]^{1/2}$$
$$= \left\{[\lambda(1 - 2X + \Delta G°)]^2 + 4H^2_{D,A}\right\}^{1/2} \quad (1.26a)$$
$$(G_e + G_g) = (G_b + G_a)^2$$
$$= \lambda(2X^2 - 2X + 1) + \Delta G° \quad (1.26b)$$

Equation (1.26a) shows that the intervalence transition maximum occurs at the reactant minimum ($X = 0$) with $E_{IT} = \lambda$ if $H_{ab} = 0$ and that the adiabatic states are separated at $X = 0.5$ by $2H_{ab}$ (Figure 1.5). In a symmetric adiabatic system, the vertical transition at the equilibrium configuration of the reactants (or products) remains equal to λ regardless of the magnitude of the electronic coupling as long as the system remains valence trapped.

The reaction coordinate X is related to c_b^2, which equals the charge transferred to the acceptor from the donor by

$$c_b^2 = \frac{1}{2}\left[1 - \frac{(1 - 2X)}{\{[(1 - 2X) + \Delta G/\lambda]^2 + 4H_{ab}/\lambda^2\}^{1/2}}\right] \quad (1.27)$$

Both X and c_b^2 measure the progress of the electron transfer by nuclear coordinate and electronic coordinate, respectively. From Eq. (1.27), the two coordinates are not linearly related until $H_{ab}/\lambda \geq 1$. Also, in the diabatic limit ($H_{ab} = 0$), there is no electron density transferred until $X = 0.5$ at which the electron "suddenly" jumps over the intersection [19]. In this case, c_b^2 is not a continuous function of X: instead $c_b^2 = 0$ for all $X < 1/2$ and $c_b^2 = 1$ for $X > 1/2$. As H_{ab} increases, the charge fraction transferred to the acceptor increases gradually with variation of nuclear coordinate from $X = 0$ to $X = 1$. At the transition state ($X = 1/2$), c_b^2 always equals $1/2$. It is easy to prove that for symmetrical system with $2H_{ab} < \lambda$ (Class II), the transferred charge fraction at the equilibrium is given by

$$(c_b^2)_{eq} = \frac{1}{2}\left[1 - \left(1 - \frac{4H_{ab}^2}{\lambda^2}\right)^{1/2}\right] \qquad (1.28)$$

For a borderline Class III system ($2H_{ab} = \lambda$), $c_b^2 = 1/2$, indicating electron delocalization.

For the symmetric MV systems ($\Delta G^0 = 0$) [25], the free energy of activation for the adiabatic ET reaction (thermal exchange) is given by Eq. (1.19)

$$\Delta G^* = \lambda/4 - H_{ab} + H_{ab}^2/\lambda$$
$$= (\lambda - 2H_{ab})^2/4\lambda \qquad (1.29)$$

Equation (1.29) shows that adiabatic ΔG^* is a sum of three terms. The first term is the ΔG^* in nonadiabatic limit, the second term is from the coupling integral that contributes to lower the activation energy by half-splitting of the upper and lower surfaces, and the last term, H_{ab}^2/λ, increases the reaction barrier through stabilization of the reactant (Figure 1.5). Equation (1.29) holds for a broad range of Class II compounds for which the self-exchange reaction is described by a double-well potential, that is, from weakly coupled Class II to the borderline of Class II–III. When further increasing the donor–acceptor coupling to $2H_{ab} > \lambda$, the lower energy surface features a single well at $X = 1/2$. This is the case of electron delocalization ($\Delta G^* = 0$), Class III. For the asymmetrical adiabatic system ($\Delta G^0 \neq 0$), the free energy of activation is given by Eq. (1.30) [25],

$$\Delta G^* = \frac{\lambda}{4} + \frac{\Delta G^0}{2} + \frac{(\Delta G^0)^2}{4(\lambda - 2H_{ab})} - H_{ab} + \frac{(H_{ab})^2}{4(\lambda + \Delta G^0)} \qquad (1.30)$$

1.6.3 Reorganization Energies

Equation (1.7) is the basic and important energetic relationship for electron transfer reaction in MV chemistry. This equation, $E_{IT} = \lambda + \Delta G^0$, dictates that the absorbed photon energy (E_{IT}) is used to rearrange the nuclear configuration for ET (λ) and to change the free energy of the sysytem (ΔG^0). For an endothermic reaction, $\Delta G^0 > 0$, and for an exothermic reaction, $\Delta G^0 < 0$. The reorganization energy is to overcome the Frank–Condon barrier of intervalence transition. This portion of energy is dispatched through radiationless relaxation after the electron transfer process and does not contribute to free energy of the system. For ET in condensed medium, the total reorganization energy (λ) was divided into two portions [15, 38]. One portion is for adjustment of the intramolecular nuclear configuration, namely, inner reorganization energy, (λ_{in}), and the other is to control the intermolecular nuclear configuration, (λ_{out}) [7, 14, 17, 97, 98].

$$\lambda = \lambda_{in} + \lambda_{out} \tag{1.31}$$

λ_{in} and λ_{out} are dominated by the high-frequency vibration modes (>1000 cm^{-1}) of the molecule and the low-frequency vibrational modes (<100 cm^{-1}) of solvent molecules, respectively, therefore, being termed sometimes as λ_v and λ_s, respectively. Although λ_{in} and λ_{out} are conceptually well defined, practically, splitting of the total reorganization remains challenging. For most of mixed-valence systems, there are many vibrational modes contributing to the inner reorganization energy; it is difficult to identify the dominant models and determine their frequencies. In the 1950s, Marcus developed the dielectric continuum theory (Eq. 1.32), which originally was for out-sphere electron transfer reaction in solution [14, 15] that is applicable to determine the quantity of outer reorganization energy in the MV systems [6, 7, 25, 98].

$$\lambda_{out} = (\Delta e)^2 \left\{ \frac{1}{2a_1} + \frac{1}{2a_2} - \frac{1}{R} \right\} \left\{ \frac{1}{\varepsilon_\infty} - \frac{1}{\varepsilon_0} \right\} \tag{1.32}$$

With contribution of inner vibrations proposed originally by Hush, the Marcus–Hush theory was framed for electron transfer in the adiabatic limit [97, 98]. Without approximation, Eq. (1.32) was derived by Li using nonequilibrium statistical mechanics recently [117, 118]. Using the dielectric continuum model to treat a MV D–B–A system, both the donor and acceptor are viewed as a rigid sphere with radii a_1 and a_2, separated by R, and the molecule as a whole is bathed in solvent molecules with dielectric continuum. In Eq. (1.33), ε_∞ and ε_c are the optical and static dielectric constants of the medium, respectively, and Δe is the amount of charge transferred. For systems in which the electronic coupling is significant, electron delocalization causes the transferred charge be less than a unit. In this case, we can calculate the "real" charge transferred from the mixing coefficients (c_a and c_b) of the diabatic states. As a consequence of electron delocalization, the charge reduced at the equilibrium configuration of the donor is

$$\Delta q = \left(c_a^2 - c_b^2\right)_{eq} = \left(1 - 2c_b^2\right)_{eq} \tag{1.33}$$

From Eq. (1.28),

$$\left[\left(1 - c_b^2\right)_{eq}\right]^2 = \left(1 - \frac{4H_{ab}^2}{\lambda^2}\right) \tag{1.34}$$

Thus, the "real" charge-transferred Δe is $e\Delta q$, where e is the charge of one single electron. Accordingly, the solvent reorganization λ_{out} in Eq. (1.32) should be scaled by $(1 - 4H^2{}_{ab}/\lambda^2)$, giving

$$\lambda'_{out} = \lambda_{out} \left(1 - 4H_{ab}^2/\lambda^2\right) \tag{1.35a}$$

Similarly,

$$\lambda'_{in} = \lambda_{in} \left(1 - 4H_{ab}^2/\lambda^2\right) \tag{1.35b}$$

Consequently,

$$\lambda' = \lambda'_{in} + \lambda'_{out} = \lambda \left(1 - 4H_{ab}^2/\lambda^2\right) \tag{1.36a}$$

or,

$$\lambda = \lambda' + 4H_{ab}^2/\lambda \tag{1.36b}$$

Equations (1.35) and (1.36) show that partial electron delocalization lowers the reorganization energy to λ' but stabilizes the ground state and destabilizes the excited state, which increases the reorganization by $4H_{ab}^2/\lambda$. For Class II systems, these two effects cancel each other so that the optical transition energy (E_{IT}) remains equal to λ regardless of the degree of localization.

1.6.4 Electronic Coupling Matrix Element and the Transition Moments

In the previous section, we have defined the electronic matrix element H_{ab} that controls the thermal electron transfer between the donor and acceptor through electronic interactions between the diabetic reactant and product states (Eq. 1.19). In a similar vein, the intensity of optical electron transfer at the adiabatic equilibrium of the reactant is governed by the transition dipole moment, μ_{ge}.

$$\mu_{ge} \equiv \langle \psi_g | \mu | \psi_e \rangle \tag{1.37a}$$

where μ is the electronic dipole operator. ψ_g and ψ_g are the linear combinations of the diabatic states ϕ_a and ϕ_b with a mixing coefficient c_a and c_b (Eq. 1.20), respectively. Substitution for ψ_g and ψ_e from Eqs. 1.20a and 1.20b gives

$$\mu_{ge} = c_a c_b (\mu_b - \mu_a) \tag{1.37b}$$

From $c_a c_b = H_{ab}/E_{IT}$ (Eq. 1.24a), it follows that

$$\frac{H_{ab}}{E_{IT}} = \left| \frac{\mu_{ge}}{\mu_b - \mu_a} \right| \tag{1.38}$$

Defining the effective electron transfer distance $r_{ab} \equiv |(\mu_b - \mu_a)|/e$, the coupling integral H_{ab} is related to transition dipole moment μ_{ge} by

$$H_{ab} = \frac{E_{IT}|\mu_{ge}|}{er_{ab}} \tag{1.39}$$

Expression of Eq. (1.39) may be considered to be a generalization of the Marcus–Hush theory (GMH) [119]. Hush showed that μ_{ge} is related to oscillator strength f and energy of the transition by [3]

$$|\mu_{ge}|^2 = f/1.085 \times 10^{-5} E_{IT} \tag{1.40a}$$

Assuming a Gaussian IVCT band for the weakly coupled Class II systems,

$$f = 4.6 \times 10^{-9} \varepsilon_{IT} \Delta v_{1/2} \tag{1.40b}$$

Combining and rearranging these equations gives the square of H_{ab}

$$(H_{ab})^2 = \frac{4.24 \times 10^{-4} E_{IT} \varepsilon_{IT} \Delta v_{1/2}}{(r_{ab})^2} \tag{1.41}$$

which is exactly the same as Eq. (1.1), known as the Mulliken–Hush expression.

By definition, the effective ET distance r_{ab} is determined by the dipole displacement of the diabatic states when a unit of charge (e) is transferred from the donor to the acceptor. r_{ab} is usually much less than r_0, the distance separating the localized charge centroids of the donor and acceptor because of electron delocalization. The diabatic dipole moment difference, $(\mu_b - \mu_a)$ is related to the ground dipole moment change $\Delta\mu$ upon electron transfer and the transition dipole moment μ_{ge} by

$$(\mu_b - \mu_a) = \left[\Delta\mu^2 + 4(\mu_{ge})^2\right]^{1/2} \tag{1.42a}$$

Measurements of $\Delta\mu$ and μ_{ge} can be achieved by electroabsorption (second-order Stark) spectroscopy, which probes the extent of charge redistribution with the charge transfer transition [101, 120]. Combining Eqs. (1.42a) with (1.39) gives

$$\Delta\mu^2 = (\mu_b - \mu_a)^2 \left[1 - 4(H_{ab}/E_{IT})^2\right] \tag{1.42b}$$

According to Eq. (1.39) and Eq. (1.42b), for a Class III system ($E_{IT} = 2H_{ab}$), $\Delta\mu = 0$, and $\mu_{ge} = (\mu_b - \mu_a)/2$, while for Class II,

$$\Delta\mu = (\mu_b - \mu_a)\left[1 - 4(H_{ab}/E_{IT})^2\right]^{1/2} \tag{1.42c}$$

Equation (1.42c) shows the dependence of $\Delta\mu/(\mu_b - \mu_a)$ on H_{ab} for symmetrical systems. For a Class II system, μ_{ge} varies as a function of H_{ab}; contrarily, for a Class III system ($2H_{ab} > \lambda$), μ_{ge} is independent of H_{ab}.

1.6.5 The Generalized Mulliken–Hush Theory (GMH)

Equation (1.39) shows that the diabatic coupling integral H_{ab} and the difference of the diabatic dipole moments ($\Delta\mu_{ab} \equiv \mu_b - \mu_a$) are correlated to the adiabatic transition dipole moment (μ_{ge}) and transition energy (E_{IT}). This transformation is implemented based on the assumption that transition moments connecting the diabatic states localized on different sites are zero ($\mu_{ab} = 0$). This transformation, which allows diabatic states to be defined in terms of purely adiabatic quantities, has general significance. As shown above, in the two-state limit, Eq. (1.39) turns to Eq. (1.1) or Eq. (1.42), the usual form of Mulliken–Hush expression. Eq. (1.39) actually provides a general method for determination of the off-diagonal matrix element from either experimental or theoretical data for the corresponding adiabatic systems, therefore, denoted as the generalized Mulliken–Hush (GMH) model. The GMH method is applicable for both symmetric and asymmetric mixed valence systems involving multiple electronic states and for various electron transfer processes (thermal, optical, and photoinduced) [121]. In application of the GMH model (Eq. 1.39) for Class II systems with a Gaussian IVCT band, the transition dipole moment μ_{ge} can be calculated from spectral parameters by

$$\mu_{ge} = 2.06 \times 10^{-2}(\varepsilon_{IT}\Delta v_{1/2}/E_{IT})^{1/2} \tag{1.43}$$

For mixed-valence systems presenting an asymmetrical IVCT band, the transition dipole moment is generally calculated from the integrated band area by [24, 116]

$$|\mu_{ge}|^2 = \frac{3hc\varepsilon_0 \ln 10}{2\pi^2 N_A} \int \frac{\varepsilon(\bar{v})}{\bar{v}} d\bar{v} \tag{1.44a}$$

$$|\mu_{ge}|^2 = 4.0 \times 10^{-4} \int \frac{\varepsilon(\bar{v})}{\bar{v}} d\bar{v} \tag{1.44b}$$

according to Eq. (1.44a) and (1.44b). With the μ_{ge} value, the difference of diabatic dipole moments $\Delta\mu_{ab}$ ($=\mu_b - \mu_a$) can be calculated from Eq. (1.42a) [119], in which the dipole moment change $\Delta\mu$ induced by electron transfer is determined by electrospectroscopy (Stark effect) [100, 101], or from quantum chemical calculations [99], or alternatively by estimation of the effective electron transfer distance ($\Delta\mu \equiv er_{ab}$) from the molecular geometry.

1.6.6 Analysis of IVCT Band Shape

In the high-temperature limit of classical treatment, the molar absorptivity of a charge transfer transition at a given transition energy ($h\nu$) is determined by the transition probability with Boltzmann distribution of the vibrational states over the energy surface of the ground-state configuration. Then, the absorption intensity $\varepsilon(\nu)$ relative to the maximum absorption ε_{max} at λ can be calculated from the energy difference relative to the equilibrium of the ground-state configuration from Eq. (1.45)

$$\varepsilon(\nu) = \varepsilon_{max} \exp[-(G_g - G_{g,eq})/RT] \tag{1.45a}$$

In considering the zeroth order interaction, $G_b = G_a$ and $G_a = \lambda X^2$ (Eq. 1.14a), and Eq. (1.45a) is converted into Eq. (1.45b).

$$\varepsilon(\nu) = \varepsilon_{max} \exp[-\lambda X^2/RT] \tag{1.45b}$$

Since $(G_2 - G_1) = h\nu$ and $\lambda + \Delta G^0 = h\nu_0$, X^2 in Eq. (1.14d) is expressed by

$$X^2 = [h\nu_0 - h\nu]^2/4\lambda^2 \tag{1.46}$$

Therefore,

$$\varepsilon(\nu) = \varepsilon_{max} \exp[-(h\nu_0 - h\nu)/4\lambda RT] \tag{1.47}$$

Equation (1.47) predicts a Gaussian-shaped band profile that can be obtained by plotting $\varepsilon(\nu)$ against $h\nu$, and the band maximum ε_{max} appears at $h\nu_0 = h\nu_{max} = \lambda + \Delta G^0$. The half-bandwidth (full band at half-height) $\Delta\nu_{1/2}$ measures the energy separation between $(h\nu_0 + h\nu)$ and $(h\nu_0 - h\nu)$, or $2h\nu$, at which $\varepsilon(\nu)/\varepsilon_{max} = 1/2$. Therefore, theoretical prediction of $\Delta\nu_{1/2}$ for the Gaussian-shaped band is given by Eq. (1.10) or

$$\Delta\nu_{1/2} = \Delta\nu_{high} + \Delta\nu_{low} = 2[4\ln(2)\lambda RT]^{1/2} \tag{1.48a}$$

where $\Delta\nu_{high} = \Delta\nu_{low} = [4\ln(2)\lambda RT]^{1/2}$ [9]. At room temperature, Eq. (1.49a) is reduced to

$$\Delta\nu_{1/2}^{II} = (2310\lambda)^{1/2} \tag{1.48b}$$

Equation (1.48) is applicable for weakly coupled Class II systems. As electronic coupling increases, the cutoff phenomenon occurs, resulting in asymmetric IVCT band with the low-energy side being truncated, and the measured $\Delta\nu_{1/2}$ will be significantly narrower than the calculated value. In fact, Class II defined by $2H_{ab} < \lambda$ covers a broad range of MV compounds from weakly to moderately strongly coupled that shows significant asymmetry of the IVCT absorption. For those with $\lambda > 2H_{ab} < (\lambda - \Delta v_{1/2})$, the cutoff is small, and $\Delta v_{1/2}^{II}$ is determined by Eq. (1.48). But further increasing coupling until $\lambda > 2H_{ab} > (\lambda - \Delta v_{1/2})$, the cutoff reduces the low-energy side from Δv_{lo} to $(\lambda - 2H_{ab})$, while Δv_{hi} on the high-energy side remains unchanged; thus, the half-height bandwidth is calculated by [9]

$$\Delta v_{1/2}^{II} = \Delta v_{hi} + (\lambda - 2H_{ab})$$
$$= [4\ln(2)\lambda RT]^{1/2} + (\lambda - 2H_{ab}) \qquad (1.48c)$$

In the Class II–III limit ($2H_{ab} = \lambda$), or the borderline Class III [9], a half-cutoff at $2H_{ab}$ is expected, and then

$$\Delta V_{1/2}^{II-III} = [4\ln(2)\lambda RT]^{1/2} \qquad (1.49a)$$

After entering the Class III region, when $\lambda < 2H_{ab} < (\lambda + \Delta v_{1/2}/4)$, the half-height bandwidth is calculated by Eq. (1.49). In the case of Class III, Eq. 1.49b indicates a sharp IVCT band because $2H_{ab} > \lambda$, which makes the $\Delta v_{1/2}^{III} < \Delta v_{hi}$ ($=[4\ln(2)\lambda RT]^{1/2}$). For very strongly coupled Class III systems, the bandwidth is given by

$$\Delta v_{1/2}^{III} = -(2H_{ab} - \lambda) + \left[(2H_{ab} - \lambda)^2 + 4\ln(2)\lambda RT\right]^{1/2} \qquad (1.49b)$$

Figure 1.6 shows the IVCT band shapes for MV compounds in different regimes of mixed valency [9, 10, 57]. Note that intensity cutoff caused by strong coupling is usually rounded off so that the band profile may not show clearly where it is being cut off as predicted. In the extreme of strong coupling, i.e. $2H_{ab}/\lambda \gg 1$, the "IVCT" band arises from electronic transition between the bonding and antibonding molecular orbitals. Thus, the dipole transition is more electronic but less vibronic in character; therefore, the observed absorption is narrow but more symmetric. This might be

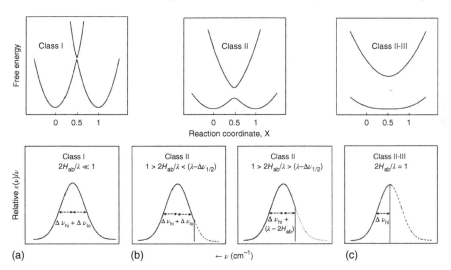

Figure 1.6 Adiabatic potential energy surfaces (top) and IVCT band shape predictions (bottom) for symmetric MV D-B-A compounds in Robin-Day's classification (a, Class I; b, Class II; c, Class III) from the two-state treatment.

thought of being the extreme of delocalization, for which treatment of the two-state model is no longer appropriate and Robin–Day's classification is inapplicable.

1.6.7 Rate Constant Expressions of Electron Transfer Reaction – The Marcus Theory

In the classical formalism, the reaction kinetics for thermal electron transfer (electron self-exchange) in D–B–A mixed-valence compounds are usually described by the transition-state theoretic model (TST). The transition state of electron transfer occurs at the crossing between the reactant and product diabatic states, or at the avoided region in the adiabatic potential surfaces, thus, at $X = 0.5$. In this situation, the ET reaction is governed by first-order kinetics. Therefore, the rate constant may be represented as [7]

$$k_{ET} = (\nu_n \kappa_{el}) \kappa_n \tag{1.50}$$

where ν_n is the effective nuclear frequency, κ_{el} is the electronic transmission factor, and κ_n is the nuclear factor. In Eq. (1.50), κ_n represents the effective fraction of the reactant species in the transition state, while the fraction of the systems that successfully pass through the crossing region and on to the products per unit time is given by the product of ν_n and κ_{el}. For classical nuclear motion,

$$\kappa_n = \exp(-\Delta G^*/k_B T) \tag{1.51}$$

where k_B is the Boltzmann constant and ΔG^* the activation energy or the energy barrier for electron transfer. Then, the ET rate can be expressed by [122]

$$k_{ET} = \nu_n \kappa_{el} \exp(-\Delta G^*/k_B T) \tag{1.52}$$

in accordance with the familiar Arrhenius equation with the prefactor equaling $(\nu_n \kappa_{el})$. For thermal ET in solution, ν_n is controlled by the low-frequency vibrational modes (stretching, rotation, and translation) of the solvent molecules with a time scale of 1 – 10 ps, i.e. $10^{-12} - 10^{-13}$ seconds. Usually, an average nuclear frequency $\nu_n = 5 \times 10^{12} \, s^{-1}$ is adopted for determination of k_{ET} [6].

In the zero or weakly coupling situation, the system (Class I) can be treated in the nonadiabatic limit. Electron exchange between D and A occurs by crossing the diabatic intersection with energy barrier determined by Eq. (1.15) in the Marcus theory; for symmetric system ($\Delta G^0 = 0$), $\Delta G^* = \lambda/4$. However, the electronic transmission coefficient is very low, that is, $\kappa_{el} \ll 1$, because there is a small probability that the fraction of the reactant species in the transition state crosses the intersection, becoming the product. Adiabatic potential surfaces evolve with increasing the electronic coupling of the system. As a result, the thermal exchange barrier is lowered by the splitting between the upper and lower curves, i.e. $2H_{ab}$. In the adiabatic limit, ΔG^* is calculated from Eq. (1.29) for symmetric systems ($\Delta G^0 = 0$) or Eq. (1.30) for asymmetrical systems ($\Delta G^0 \neq 0$), both of which include contribution of H_{ab}. In this scenario, $\kappa_{el} \approx 1$ because nearly all the reactant species that reach the transition state are able to smoothly pass on to the product. Then, Eq. (1.53) is applied with the prefactor $(v_n \kappa_{el}) = v_n$ by assuming $\kappa_{el} = 1$. This adiabatic treatment is applicable for Class II MV systems in a broad range of coupling strength, which are characterized by solvent-controlled electron transfer kinetics.

In condensed phase, electron transfer kinetics is governed by the interplay of the atomic (nuclear) and electronic dynamics of the system (including medium). Comparison between the electron hopping frequency (v_{el}) and nuclear vibrational frequency (v_n) determines ET in the adiabatic and nonadiabatic limits, that is

$$\text{adiabatic}: v_{el} \gg v_n; \text{nonadiabatic}: v_{el} \ll v_n$$

while v_n is averaged by the solvent modes and v_{el} is a reflection of the joint influence of physical parameters H_{ab}, λ, and T, as indicated by [7, 19, 41, 64, 122]

$$v_{el} = \frac{2H_{ab}^2}{h} \sqrt{\frac{\pi^3}{\lambda k_B T}} \tag{1.53}$$

Semiclassical Landau–Zener models deal with the weakly coupled intermediate systems in the near-adiabatic regime, in which electron transfer crossing the intersection through nonadiabatic transition [19]. To quantitatively distinguish the nonadiabatic and adiabatic limits, the Landau–Zener models [111, 112] define two parameters: adiabatic parameter γ (Eq. 1.54a) and transition probability P_0 (Eq. 1.54b) with the exponent term being the nonadiabatic transition contribution, which gives the electronic transmission coefficient κ_{el} from Eq. (1.54c).

$$\gamma = \frac{H_{ab}^2}{2hv_n} \sqrt{\frac{\pi}{\lambda k_B T}} \tag{1.54a}$$

$$P_0 = 1 - \exp(-2\pi\gamma) \tag{1.54b}$$

$$\kappa_{el} = 2P_0/(1 + P_0) \tag{1.54c}$$

When $\gamma \gg 1$, the adiabatic limit is realized, and for thermal ET, $\kappa_{el} \approx 1$, while the nonadiabatic limit prevails with $\gamma \ll 1$. By definition of γ, it is clear that nonadiabatic transition is governed by the electronic and nuclear factors, represented by H_{ab} and v_n, respectively.

In the nonadiabatic limit, as characterized by $v_{el} \ll v_n$, the electron hopping frequency (v_{el}) dominates the ET process, and Eq. (1.52) is replaced by Eq. (1.55a)

$$k_{ET} = v_{el} \exp(-\Delta G^*/k_B T) \tag{1.55a}$$

The ET rate constant can be calculated by the Levich–Dogonadze–Marcus expression (Eq. 1.55b) [13, 38, 41]

$$k_{et} = \frac{2H_{ab}^2}{h} \sqrt{\frac{\pi^3}{\lambda k_B T}} \exp\left(-\frac{\lambda}{4k_B T}\right) \tag{1.55b}$$

Recent study [64] on MV complex system shows that both nonadiabatic (Eq. 1.55b) and adiabatic (Eq. 1.53) expressions work equally well for the intermediate systems.

1.6.8 McConnell Superexchange Mechanism and the CNS Model

Superexchange model has been widely accepted for interpretation of bridge-mediated electronic coupling and electron transfer through orbital interactions. It was first proposed by McConnell based on the simple two-state, one-electron approximation [16]. In McConnell's approach, the donor (ϕ_d) and acceptor (ϕ_a) states are separated by a bridge possessing n equivalent localized states ϕ_n. By neglecting direct donor–acceptor interactions, two new states ψ_S (symmetrical) and ψ_A (antisymmetrical) involving only the localized donor and acceptor states are formed

$$\psi_S = 1/\sqrt{2}(\phi_d + \phi_a) \tag{1.56a}$$

$$\psi_A = 1/\sqrt{2}(\phi_d - \phi_a) \tag{1.56b}$$

Here, the D–A coupling (H_{ab}) is small and neglectable, and other direct Hamiltonian matrix elements are zero by construction. However, the two states ϕ_d and ϕ_a interact with the nearest bridge states, and two nearest neighboring bridge states are allowed to interact with each other [16, 19]. Consequently, the degeneracy of ψ_S and ψ_A is removed, yielding an energy gap ΔE between them.

$$\Delta E = E_S + E_A = 2\sum_{n=1}^{n} \frac{\langle \phi_d | H | \psi_n \rangle \langle \psi_n | H | \phi_a \rangle}{E_1 - E_n} \tag{1.57a}$$

where ψ_n and E_n are the eigenfunctions and eigenvalue of the nth unit in bridge block, which are identical for all units of the bridge. Mathematic treatments of Eq. (1.57a) with appropriate assumptions give rise to the McConnell superexchange expression (Eq. 1.57b) [19].

$$\Delta E = -(2T^2/E_{CT})(-t/E_{CT})^{n-1} \tag{1.57b}$$

where T is the matrix element accounting for the coupling between the donor and the first bridge unit ($n = 1$) or between the last bridge unit (n) and the acceptor, t is the matrix element that concerns the coupling between two neighboring bridge units, and E_{CT} is the energy difference between the donor (or acceptor) and bridge states [16].

It should be noted that the McConnell superexchange expression was derived originally in the nonadiabatic limit where the donor–acceptor coupling is extremely weak, by working on the example α,ω-diphenylalkanes [16]. Since there are no low-lying bridge orbitals available, involvement of virtual high-lying orbitals, for example, 3d orbitals of the sp^3 C chain, is assumed. However, the superexchange concept and expression (Eq. 1.57b) work equally well for transition metal and pure organic mixed-valence compounds with conjugated bridges. In these systems, the low-lying empty π^* and high-lying filled π orbitals are exploitable, which provide the superexchange pathways for effective electron and hole transfer, respectively. Extended application of the McConnell equations is justified by the theoretic work from Reimers and Hush in 1994 [123], which demonstrated that the superexchange mechanism is appropriate for σ- or π-bonding bridged donor–acceptor systems

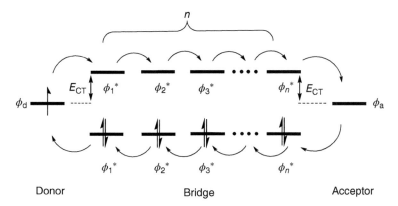

Figure 1.7 Schematic illustration of superexchange mechanism for bridge-mediated through-bond electron transfer. Electron transfer proceeds from the donor to the acceptor (from left to right) by the high-lying unoccupied bridge orbitals and hole transfer takes place through low-lying filled bridge orbitals.

in both resonant and near-resonant situations. The superexchange mechanism for long-distance through-bond electron transfer is schematically described in Figure 1.7. As a result of increasing the electronic coupling, the charge transfer system turns to be adiabatic. This is generally achieved by introducing conjugated bridge and reducing the number bridge units (n). In the adiabatic limit, Eq. (1.57a) gives $\Delta E = 2H_{ab}$, in accordance with $2H_{ab}/\lambda = 1$ from the semiclassical two-state model for the borderline Class II–III and Class III limit.

Superexchange and sequential electron (hole) hopping are the two pathways for electron transfer in the adiabatic and nonadiabatic limits, respectively. The former interprets the efficient mediation of bridge across a short distant by providing energy- and symmetry-compatible orbitals, while the latter has received broad success in interpretation of long-distance electron transfer. However, detailed investigations demonstrated that these two ET mechanisms may not be alternative but are combined or mixed to different extents [91, 124]. According to Eq. (1.57) and Figure 1.7, both pathways are conceptually the same in terms of orbital interactions. These two pathways are frequently characterized by an exponential decay parameter β that describes the different distance dependence of the coupling constant (H_{ab}), which gives exponential correlation of electron transfer rate constant k_{ET} with distance. From Eq. (1.57), letting $\beta r_{ab} = (n-1)$, we have Eq. (1.58)

$$H_{ab} = H_{ab}^0 \exp\left(-\frac{\beta}{2} r_{ab}\right) \tag{1.58}$$

where $H^0{}_{ab}$ is the electronic coupling at direct contact distance between the donor and acceptor [121]. The β value is expected to vary with the donor (acceptor)-bridge energy gap as well as with the coupling strength between the neighboring units of the bridge. When the subunit couplings are small compared to the energy gap (E_{CT}), from Eq. (1.57b), β is approximated by

$$\beta = \frac{2}{r_0} \ln\left(\frac{E_{CT}}{t}\right) \tag{1.59}$$

where r_0 is the length of one subunit [125]. For a given system, a larger value of β is found in the superexchange region, but a smaller β (weak distance dependence) is found for the sequential hopping mechanism. Ideally, for the series of MV compounds with a homologous bridge, an appreciable mechanistic transition can be probed by change of the β value with stepwise increasing the bridge length or the number of bridging units.

The McConnell superexchange formalism of ET has led Creutz, Newton, and Sutin to develop an alternative method to calculate the metal–metal coupling element $H_{MM'}$ for mixed-valence complexes M–BL–M′, known as the CNS model [10, 25, 116]. Following the same principle applied in development of the superexchange expression (Eq. 1.57b) [16, 19], CNS proposed calculation of $H_{MM'}$ from the metal–ligand (H_{ML} and $H_{M'L}$) and ligand–metal (H_{LM} and $H_{LM'}$) coupling elements by Eq. (1.60a).

$$H_{MM'} = \frac{H_{ML}H_{M'L}}{2\Delta E_{ML}} + \frac{H_{LM}H_{LM'}}{2\Delta E_{LM}} \tag{1.60a}$$

where the metal–ligand coupling elements are calculated from the Mulliken–Hush expression (Eq. 1.1) and the effective metal–ligand energy gap (ΔE_{ML}) and the effective ligand–metal energy gap (ΔE_{LM}) are calculated from Eq. (1.60b) and Eq. (1.60c), respectively.

$$\frac{1}{\Delta E_{ML}} = \frac{1}{2}\left(\frac{1}{\Delta E_{MLCT}}\right) + \left(\frac{1}{\Delta E_{MLCT} - \Delta E_{MM'CT}}\right) \tag{1.60b}$$

$$\frac{1}{\Delta E_{LM}} = \frac{1}{2}\left(\frac{1}{\Delta E_{LMCT}}\right) + \left(\frac{1}{\Delta E_{LMCT} - \Delta E_{MM'CT}}\right) \tag{1.60c}$$

In Eqs. (1.60b) and (1.60c), ΔE_{MLCT} and ΔE_{LMCT} are the measured MLCT and LMCT energies, respectively, from the electronic spectra of the singly reduced and oxidized mixed-valence compound, respectively, and $\Delta E_{MM'CT}$ refers to the IVCT transition energy (E_{IT}). The physical implication of the CNS model (Eq. 1.60) is that electron transfer and hole transfer pathways may be superpositioned, mutually contributing to the thermal electron exchange reaction. However, in application, one of the two pathways may be dominant over the other, depending on electronic structure of the bridge. Then, $H_{MM'}$ may be determined by one term of the two in Eq. (1.60a), for which mixing of metal center with ligand states is pronounced. Provided with electronically well-defined MV systems that permit accurate optical analysis, the H_{ab} values determined from the CNS model are in good agreement with the data from the Mulliken–Hush expression [68]. For weakly coupled [{Ru^{II}(NH$_3$)$_5$}(μ-4,4′-bpy){Ru^{III}(NH$_3$)$_5$}]$^{5+}$, H_{ab} = 900 cm^{-1} is comparable with $H_{MM'}$ = 800 cm^{-1} [25]. In the Mo$_2$–BL–Mo$_2$ mixed-valence systems, similar $H_{MM'}$ and H_{ab} values were found [63, 126].

1.7 Conclusion and Outlook

Mixed-valence chemistry was considered to be a ramification of inorganic chemistry half century ago when it was first introduced. Now, it expands into a huge body of molecular systems with the redox centers involving inorganic, organometallic, and pure organic units or groups linked by diverse chemical moieties. Research endeavor of decades has validated and resolved, to a large extent, the fundamental issues of mixed valency, typically, the physical origin of coloration of MV compounds, classification of MV classes, transition between the classes and ET energetics and kinetics, based on the semiclassical two-state and vibronic theories. However, these issues in MV chemistry concern electronic and nuclear degrees of freedom and the interplay between them, spreading far beyond the redox and spectroscopic properties presented by these molecules, but nearly every aspect of molecular science. Particularly, MV compounds, with the molecules in form of D–B–A as the prototype of study, are the excellent experimental models for which conventional spectral methods can be used advantageously to probe the electronic states and to monitor the vibronic dynamics. In this context, the wealth of knowledge of MV compounds has deepened our understanding of chemistry in general, and in particular, contributed to validation and refinement of the electron transfer theories.

Because of the broad interest and the specific concerns on electron transfer, MV chemistry will continue to serve an interplaying platform for both experimental and theoretical chemists who devote to gain new knowledge in this field. Given the controllable optical, electronic, and magnetic properties and great diversity in molecular assembly, MV compounds bear a great treasure of application potentials in the area of molecular optoelectronic and/or optomagnetic materials and devices. This should keep driving the study toward rational design and synthesis/fabrication of the functional molecular components and better microscopic understanding of the chemical principles. Elucidation of the functionalities of enzyme cofactors containing multiple metal centers in biological systems is the additional impetus of MV chemistry, which is helpful to development of biomimetic catalysts for solar energy conversion.

Acknowledgments

We thank the National Science Foundation of China (nos. 21971088, 20871093, 90922010, 21371074, and 21301070), the Natural Science Foundation of Guangdong Province (2018A030313894), and the Jinan University and Tongji University for financial support. We are grateful to Prof. Jeffrey Reimers (University of Technology Sydney), Prof. Yu-Wu Zhong (Institute of Chemistry, Chinese Academy of Sciences, Beijing), and Prof. Nathan J. Patmore (University of Huddersfield, UK) for their comments and suggestions on the manuscript.

References

1 Brown, D.B. (1980). *Mixed-Valence Compounds Theory and Applications in Chemistry, Physics, Geology, and Biology*. Dordrecht, Holland: D. Reidel Publishing Company.
2 Prassides, K. (1991). *Mixed Valency Systems: Applications in Chemistry*. Physics and Biology: Kluwer Academic Publishers.
3 Allen, G.C. and Hush, N.S. (1967). Intervalence-transfer absorption. Part 1. Qualitative evidence for intervalence-transfer absorption in inorganic systems in solution and in the solid state. *Prog. Inorg. Chem.* 8: 357–385.
4 Hansen, L.D., Litchman, W.M., and Daub, G.H. (1969). Turnbull's blue and Prussian blue: $KFe(III)[Fe(II)(CN)_6]$. *J. Chem. Educ.* 46: 46.
5 Hush, N.S. (1967). Intervalence-transfer absorption. II. theoretical considerations and spectroscopic data. *Prog. Inorg. Chem.* 8: 391–444.
6 Creutz, C. (1983). Mixed valence complexes of d^5-d^6 metal centers. *Prog. Inorg. Chem.* 30: 1–72.
7 Sutin, N. (1983). Theory of electron transfer reactions: insights and hindsights. *Prog. Inorg. Chem.* 30: 441–498.
8 Crutchley, R.J. (1994). Intervalence charge transfer and electron exchange studies of dinuclear ruthenium complexes. *Adv. Inorg. Chem.* 41: 273–325.
9 Brunschwig, B.S., Creutz, C., and Sutin, N. (2002). Optical transitions of symmetrical mixed-valence systems in the Class II–III transition regime. *Chem. Soc. Rev.* 31: 168–184.
10 D'Alessandro, D.M. and Keene, F.R. (2006). Current trends and future challenges in the experimental, theoretical and computational analysis of intervalence charge transfer (IVCT) transitions. *Chem. Soc. Rev.* 35: 424–440.
11 Kubo, R. and Toyozawa, Y. (1955). Application of the method of generating function to radiative and non-radiative transitions of a trapped electron in a crystal. *Prog. Theor. Phys.* 13: 160–182.
12 Levich, V.G. and Dogonadze, R.R. (1960). An adiabatic theory of electron processes in solutions. *Dokl. Akad. Nauk SSSR* 133: 158–161.
13 Levich, V.G. and Dogonadze, R.R. (1959). Theory of rediationless electron transitions between ions in solution. *Dokl Akad Nauk SSSR Ser Fiz Khim.* 124: 23–126.
14 Marcus, R.A. (1956). On the theory of oxidation-reduction reactions involving electron transfer. I. *J Chem. Phys.* 24: 966–978.
15 Marcus, R.A. (1957). On the theory of oxidation-reduction reactions involving electron transfer. II. Applications to data on the rates of isotopic exchange reactions. *J Chem. Phys.* 24: 867–871.
16 McConnell, H.M. (1961). Intramolecular charge transfer in aromatic free radicals. *J. Chem. Phys.* 35: 508–515.
17 Hush, N.S. (1958). Adiabatic rate processes at electrodes. I. Energy-Charge Relationships. *J. Chem. Phys.* 28: 962–972.

18 Creutz, C. and Taube, H. (1969). A direct approach to measuring the Franck-Condon barrier to electron transfer between metal ions. *J. Am. Chem. Soc.* 91: 3988–3989.

19 Newton, M.D. (1991). Quantum chemical probes of electron-transfer kinetics: the nature of donor-acceptor interactions. *Chem. Rev.* 91: 767–792.

20 Barbara, P.F., Meyer, T.J., and Ratner, M.A. (1996). Contemporary issues in electron transfer research. *J. Phys. Chem.* 100: 13148–13168.

21 Creutz, C. and Taube, H. (1973). Binuclear complexes of ruthenium ammines. *J. Am. Chem. Soc.* 95: 1086–1094.

22 Meyer, T.J. (1978). Optical and thermal electron transfer in metal complexes. *Acc. Chem. Res.* 11: 94–100.

23 Kaim, W. and Lahiri, G.K. (2007). Unconventional mixed-valent complexes of ruthenium and osmium. *Angew. Chem. Int. Ed.* 46: 1778–1796.

24 Heckmann, A. and Lambert, C. (2012). Organic mixed-valence compounds: a playground for electrons and holes. *Angew. Chem. Int. Ed.* 51: 326–392.

25 Brunschwig, B.S. and Sutin, N. (1999). Energy surfaces, reorganization energies, and coupling elements in electron transfer. *Coord. Chem. Rev.* 187: 233–254.

26 Newton, M.D. (2003). Thermal and optical electron transfer involving transition metal complexes: insights from theory and computation. *Coordin. Chem. Rev.* 238: 167–185.

27 Elliott, C.M., Derr, D.L., Matyushov, D.V., and Newton, M.D. (1998). Direct experimental comparison of the theories of thermal and optical electron-transfer: studies of a mixed-valence dinuclear iron polypyridyl complex. *J. Am. Chem. Soc.* 120: 11714–11726.

28 MacCarthy, G.R. (1926). Iron-stained sands and clays. *J. Geol.* 34: 352–360.

29 Hofmann, K.A. and Höschsele, K. (1915). Das Magnesiumchlorid als Mineralisator. II.: Das Urancerblau und das Wesen der konstitutiven Färbung. *Das Magnesiarot und das Magnesiagrün. K.* 48: 20–28.

30 Weyl, W.A. (1951). Light absorption as a result of an interaction of two states of valency of the same element. *J. Phys. Chem.* 55: 507–512.

31 Umena, Y., Kawakami, K., Shen, J.R., and Kamiya, N. (2011). Crystal structure of oxygen-evolving photosystem II at a resolution of 1.9 Å. *Nature* 473: 55–60.

32 Carrillo, N. and Ceccarelli, E.A. (2003). Open questions in ferredoxin-NADP+ reductase catalytic mechanism. *Eur. J. Biochem.* 270: 1900–1915.

33 Listorti, A., Durrant, J., and Barber, J. (2009). Solar to fuel. *Nature Mater.* 8: 929–930.

34 Robin, M.B. and Day, P. (1968). Mixed valence chemistry-a survey and classification. *Adv. Inorg. Chem. Radiochem.* 9: 247–422.

35 Creutz, C. and Taube, H. (1969). Direct approach to measuring the Franck-Condon barrier to electron transfer between metal ions. *J. Am. Chem. Soc.* 91: 3988–3989.

36 Mulliken, R.S. (1950). Structures of complexes formed by halogen molecules with aromatic and with oxygenated solvents. *J. Am. Chem.Soc.* 72: 600–608.

37 Mulliken, R.S. (1952). Molecular compounds and their spectra. III. The interaction of electron donors and acceptors. *J. Phys. Chem.* 56: 801–822.

38 Marcus, R.A. and Sutin, N. (1985). Electron transfers in chemistry and biology. *Biochim. Biophys. Acta* 811: 265–322.

39 Concepcion, J.J., Dattelbaum, D.M., Meyer, T.J., and Rocha, R.C. (2008). Probing the localized-to-delocalized transition. *Phil. Trans. R. Soc.* A366: 163–175.

40 Demadis, K.D., Hartshorn, C.M., and Meyer, T.J. (2001). The localized-to-delocalized transition in mixed-valence chemistry. *Chem. Rev.* 101: 2655–2685.

41 Levich, V.G. (1966). Present state of the theory of oxidation-reduction in solution (bulk and electrode reactions). *Adv. Electrochem. Electrochem. Eng.* 4: 249–371.

42 Dowling, N., Henry, P.M., Lewis, N.A., and Taube, H. (1981). Heteronuclear mixed-valence ions containing ruthenium and ferrocene centers. *Inorg. Chem.* 20: 2345–2348.

43 Hage, R., Haasnoot, J.G., Nieuwenhuis, H.A. et al. (1990). Synthesis, X-ray structure, and spectroscopic and electrochemical properties of novel heteronuclear ruthenium-osmium complexes with an asymmetric triazolate bridge. *J. Am. Chem. Soc.* 112: 9245–9251.

44 Zhong, Y.W., Gong, Z.L., Shao, J.Y., and Yao, J. (2016). Electronic coupling in cyclometalated ruthenium complexes. *Coord. Chem. Rev.* 312: 22–40.

45 Patoux, C., Launay, J.-P., Beley, M. et al. (1998). Long-range electronic coupling in bis(cyclometalated) ruthenium complexes. *J. Am. Chem. Soc.* 120: 3717–3725.

46 Cowan, D.O., LeVanda, C., Park, J., and Kaufman, F. (1973). Organic solid state. VIII. Mixed-valence ferrocene chemistry. *Acc. Chem. Res.* 6: 1–7.

47 Ribou, A.C., Launay, J.P., Sachtleben, M.L. et al. (1996). Intervalence electron transfer in mixed valence diferrocenylpolyenes. Decay law of the metal–metal coupling with distance. *Inorg. Chem.* 35: 3735–3740.

48 Bruns, W. and Kaim, W. (1990). Bindungscharakteristik der H,-koordinierenden Fragmente $W(CO)_3(PR_3)_2$.Aussergewiihnliche optische und elektrochemische Eigenschaften zweikerniger Pyrazinkomplexe. *J. Organomet. Chem.* 390: C45–C49.

49 Bruns, W., Kaim, W., Waldhör, E., and Krejčik, M. (1993). Spectroelectrochemical characterization of a pyrazine-bridged mixed-valent ($4d^5/4d^6$) organometallic analogue of the Creutz–Taube ion. *Chem. Commun.* 24: 1868–1869.

50 NavasáBadiola, J.A. and Michael, D. (1993). Metal–metal interactions across symmetrical bipyridyl bridging ligands in binuclear seventeen-electron molybdenum complexes. *J. Chem. Soc. Dalton.* 5: 681–686.

51 Ward, M.D. (1995). Metal-metal interactions in binuclear complexes exhibiting mixed valency; molecular wires and switches. *Chem. Soc. Rev.* 24: 121–134.

52 Bonvoisin, J., Launay, J.P., Vanderauweraer, M., and Deschryver, F.C. (1994). Organic mixed-valence systems-intervalence transition in partly oxidized aromatic polyamines-electrochemical and optical studies. *J. Phys. Chem. C* 98: 5052–5057.

53 Lambert, C. and Nöll, G. (1999). The class II/III transition in triarylamine redox systems. *J. Am. Chem. Soc.* 121: 8434–8442.

54 Nelsen, S.F., Ismagilov, R.F., and Powell, D.R. (1996). Charge localization in a dihydrazine analogue of tetramethyl-p-phenylenediamine radical cation. *J. Am. Chem. Soc.* 118: 6313–6314.

55 Nelsen, S.F., Ismagilov, R.F., and Trieber, D.A. (1997). Adiabatic electron transfer: comparison of modified theory with experiment. *Science* 278: 846–849.

56 Lindeman, S.V., Rosokha, S.V., Sun, D., and Kochi, J.K. (2002). X-ray structure analysis and the intervalent electron transfer in organic mixed-valence crystals with bridged aromatic cation radicals. *J. Am. Chem. Soc.* 124: 843–855.

57 Nelsen, S.F. (2000). "Almost delocalized" intervalence compounds. *Chem. Eur. J.* 6: 581–588.

58 Cayton, R.H. and Chisholm, M.H. (1989). Electronic coupling between covalently linked metal-metal quadruple bonds of molybdenum and tungsten. *J. Am. Chem. Soc.* 111: 8921–8923.

59 Cayton, R.H., Chisholm, M.H., Huffman, J.C., and Lobkovsky, E.B. (1991). Metal-metal multiple bonds in ordered assemblies. 1. Tetranuclear molybdenum and tungsten carboxylates involving covalently linked metal-metal quadruple bonds. Molecular models for subunits of one-dimensional stiff-chain polymers. *J. Am. Chem. Soc.* 113: 8709–8724.

60 Cotton, F.A., Donahue, J.P., Murillo, C.A., and Pérez, L.M. (2003). Polyunsaturated dicarboxylate tethers connecting dimolybdenum redox and chromophoric centers: absorption spectra and electronic structures. *J. Am. Chem. Soc.* 125: 5486–5492.

61 Cotton, F.A., Liu, C.Y., Murillo, C.A. et al. (2004). Strong electronic coupling between dimolybdenum units linked by the N, N′-dimethyloxamidate anion in a molecule having a heteronaphthalene-like structure. *J. Am. Chem. Soc.* 126: 14822–14831.

62 Cotton, F.A., Murillo, C.A., and Walton, R.A. (2005). Introduction and survey. In: *Multiple Bonds Between Metal Atoms* (ed. F.A. Cotton, C.A. Murillo and R.A. Walton), 1–21. Boston, MA: Springer.

63 Liu, C.Y., Xiao, X., Meng, M. et al. (2013). Spectroscopic study of δ electron transfer between two covalently bonded dimolybdenum units via a conjugated bridge: adequate complex models to test the existing theories for electronic coupling. *J. Phys. Chem. C.* 117: 19859–19865.

64 Zhu, G.Y., Qin, Y., Meng, M. et al. (2021). Crossover between the adiabatic and nonadiabatic electron transfer limits in the Landau-Zener model. *Nat. Commun.* 12 (456): 1–10.

65 Xu, G.L., Zou, G., Ni, Y.H. et al. (2003). Polyyn-diyls capped by diruthenium termini: a new family of carbon-rich organometallic compounds and distance-dependent electronic coupling therein. *J. Am. Chem. Soc.* 125: 10057–10065.

66 Ito, T., Hamaguchi, T., Nagino, H. et al. (1999). Electron transfer on the infrared vibrational time scale in the mixed valence state of 1,4-pyrazine- and 4,4'-bipyridine-bridged ruthenium cluster complexes. *J. Am. Chem. Soc.* 121: 4625–4632.

67 Ito, T., Hamaguchi, T., Nagino, H. et al. (1997). Effects of rapid intramolecular electron transfer on vibrational spectra. *Science* 277: 660–663.

68 Rosokha, S.V., Sun, D.-L., and Kochi, J.K. (2002). Conformation, distance, and connectivity effects on intramolecular electron transfer between phenylene-bridged aromatic redox centers. *J. Phys. Chem. A.* 106: 2283–2292.

69 Launay, J.-P. (2001). Long-distance intervalence electron transfer. *Chem. Soc. Rev.* 30: 386–397.

70 Chen, H.W., Mallick, S., Zou, S.F. et al. (2018). Mapping bridge conformational effects on electronic coupling in Mo_2–Mo_2 mixed-valence systems. *Inorg. Chem.* 57: 7455–7467.

71 Bursten, B.E., Chisholm, M.H., Clark, R.J. et al. (2002). Perfluoroterephthalate bridged complexes with M– M quadruple bonds:$(t\text{-}BuCO_2)_3M_2(\mu\text{-}O_2CC_6F_4CO_2)M_2(O_2CtBu)_3$, where M= Mo or W. Studies of solid-state, molecular, and electronic structure and correlations with electronic and Raman spectral data. *J. Am. Chem. Soc.* 124: 12244–12254.

72 Nelsen, S.F., Konradsson, A.E., and Teki, Y. (2006). Charge-localized naphthalene-bridged bis-hydrazine radical cations. *J. Am. Chem. Soc.* 128: 2902–2910.

73 Kaliginedi, V., Moreno-García, P., Valkenier, H. et al. (2012). Correlations between molecular structure and single-junction conductance: a case study with oligo (phenylene-ethynylene)-type wires. *J. Am. Chem. Soc.* 134: 5262–5275.

74 Venkataraman, L., Klare, J.E., Nuckolls, C. et al. (2006). Dependence of single-molecule junction conductance on molecular conformation. *Nature* 442: 904–907.

75 Gao, H., Mallick, S., Cao, L. et al. (2019). Electronic coupling and electron transfer between two Mo_2 units through meta-and para-phenylene bridges. *Chem. Eur. J.* 25: 3930–3938.

76 Forrest, W.P., Choudhuri, M.M., Kilyanek, S.M. et al. (2015). Synthesis and electronic structure of $Ru_2(Xap)_4$(Y-*gem*-DEE) type compounds: effect of cross-conjugation. *Inorg. Chem.* 54: 7645–7652.

77 Göransson, E., Emanuelsson, R., Jorner, K. et al. (2013). Charge transfer through cross-hyperconjugated versus cross-π-conjugated bridges: an intervalence charge transfer study. *Chem. Sci.* 4: 3522–3532.

78 Cheng, T., Shen, D.X., Meng, M. et al. (2019). Efficient electron transfer across hydrogen bond interfaces by proton-coupled and-uncoupled pathways. *Nat. Commun.* 10 (1053): 1–10.

79 Tadokoro, M., Inoue, T., Tamaki, S. et al. (2007). Mixed-valence states stabilized by proton transfer in a hydrogen-bonded biimidazolate rhenium dimer. *Angew. Chem. Int. Ed.* 119: 6042–6046.

80 Wilkinson, L.A., Vincent, K.B., Meijer, A.J., and Patmore, N.J. (2016). Mechanistic insight into proton-coupled mixed valency. *Chem. Commun.* 52: 100–103.

81 Sun, H., Steeb, J., and Kaifer, A.E. (2006). Efficient electronic communication between two identical ferrocene centers in a hydrogen-bonded dimer. *J. Am. Chem. Soc.* 128: 2820–2821.

82 Ghaddar, T.H., Castner, E.W., and Isied, S.S. (2000). Molecular recognition and electron transfer acrossa hydrogen bonding interface. *J. Am. Chem. Soc.* 122: 1233–1234.

83 Hammes-Schiffer, S. and Stuchebrukhov, A.A. (2010). Theory of coupled electron and proton transfer reactions. *Chem. Rev.* 110: 6939–6960.

84 Liu, Y., Lagrost, C., Costuas, K. et al. (2008). A multifunctional organometallic switch with carbon-rich ruthenium and diarylethene units. *Chem. Commun.* 46: 6117–6119.

85 Chisholm, M.H., Feil, F., Hadad, C.M., and Patmore, N.J. (2005). Electronically coupled MM quadruply-bonded complexes (M= Mo or W) employing functionalized terephthalate bridges: toward molecular rheostats and switches. *J. Am. Chem. Soc.* 127: 18150–18158.

86 Wenger, O.S. (2012). Photoswitchable mixed valence. *Chem. Soc. Rev.* 2012 (41): 3772–3779.

87 Peloquin, J.M., David Britt, R., Peloquin, J.M., and Britt, R.D. (2001). EPR/ENDOR characterization of the physical and electronic structure of the OEC Mn cluster Biochim. *Biophys. Acta* 1503: 96–111.

88 Zhang, C., Chen, C., Dong, H. et al. (2015). A synthetic Mn4Ca-cluster mimicking the oxygen-evolving center of photosynthesis. *Science* 348: 690–693.

89 Nelsen, S.F., Ismagilov, R.F., and Powel, D.R. (1997). Charge-localized p-phenylenedihydrazine radical cations: ESR and optical studies of intramolecular electron transfer rates. *J. Am. Chem. Soc.* 119: 10213–10222.

90 Chisholm, M.H., Pate, B.D., Wilson, P.J., and Zaleski, J.M. (2002). On the electron delocalization in the radical cations formed by oxidation of MM quadruple bonds linked by oxalate and perfluoroterephthalate bridges. *Chem. Commun.* 10: 1084–1085.

91 Lloveras, V., Vidal-Gancedo, J., Figueira-Duarte, T.M. et al. (2011). Tunneling versus hopping in mixed-valence oligo-p-phenylenevinylene polychlorinated bis(triphenylmethyl) radical anions. *J. Am. Chem. Soc.* 133: 5818–5833.

92 Lancaster, K., Odom, S.A., Jones, S.C. et al. (2009). Intramolecular electron-transfer rates in mixed-valence triarylamines: measurement by variable-temperature ESR spectroscopy and comparison with optical data. *J. Am. Chem. Soc.* 131: 1717–1723.

93 Richardson, D.E. and Taube, H. (1981). Determination of $E_2°$-$E_1°$ in multistep charge transfer by stationary-electrode pulse and cyclic voltammetry: application to binuclear ruthenium ammines. *Inorg. Chem.* 20: 1278–1285.

94 Evans, C.E., Naklicki, M.L., Rezvani, A.R. et al. (1998). An investigation of superexchange in dinuclear mixed-valence ruthenium complexes. *J. Am. Chem. Soc.* 120: 13096–13103.

95 Richardson, D.E. and Taube, H. (1984). Mixed-valence molecules: electronic delocalization and stabilization. *Coord. Chem. Rev.* 60: 107–129.

96 Low, P.J. and Brown, N.J. (2010). Electronic interactions between and through covalently bonded polymetallic complexes. *J. Clust. Sci.* 21: 235–278.

97 Hush, N.S. (1961). Adiabatic theory of outer sphere electron-transfer reactions in solution. *Trans. Faraday Soc.* 57: 557–580.

98 Marcus, R.A. (1964). Chemical and electrochemical electron transfer theory. *Annu. Rev. Phys. Chem.* 15: 155–196.

99 Nelsen, S.F. and Newton, M.D. (2000). Estimation of electron transfer distances from AM1 calculations. *J. Phys. Chem. A* 104: 10023–10031.

100 Reimers, J.R. and Hush, N.S. (1991). Electronic properties of transition-metal complexes determined from electroabsorption (Stark) spectroscopy. 2. Mononuclear complexes of ruthenium(II). *J. Phys. Chem.* 95: 9773–9781.

101 Brunschwig, B.S., Creutz, C., and Sutin, N. (1998). Electroabsorption spectroscopy of charge transfer states of transition metal complexes. *Coord. Chem. Rev.* 177: 61–79.

102 Cheng, T., Tan, Y.N., Zhang, Y. et al. (2015). Distinguishing the strength of electronic coupling for Mo_2-containing mixed-valence compounds within the class III regime. *Chem. Eur. J.* 21: 2353–2357.

103 Tan, Y.N., Cheng, T., Meng, M. et al. (2017). Optical behaviors and electronic properties of Mo2–Mo2 mixed-valence complexes within or beyond the Class III regime: testing the limits of the two-state model. *J. Phys. Chem. C* 121: 27860–27873.

104 Lear, B.J. and Chisholm, M.H. (2009). Oxalate bridged MM (MM = Mo_2, MoW, and W_2) quadruply bonded complexes as test beds for current mixed valence theory: looking beyond the intervalence charge transfer transition. *Inorg. Chem.* 48: 10954–10971.

105 Wu, Y.Y., Meng, M., Wang, G.Y. et al. (2017). Optically probing the localized to delocalized transition in Mo_2–Mo_2 mixed-valence systems. *Chem. Comm.* 53: 3030–3033.

106 Lambert, C., Amthor, S., and Schelter, J. (2004). From valence trapped to valence delocalized by bridge state modification in bis (triarylamine) radical cations: evaluation of coupling matrix elements in a three-level system. *J. Phys. Chem. A* 108: 6474–6486.

107 Xiao, X., Liu, C.Y., He, Q. et al. (2013). Control of the charge distribution and modulation of the Class II–III transition in weakly coupled Mo_2–Mo_2 systems. *Inorg. Chem.* 52: 12624–12633.

108 Lear, B.J., Glover, S.D., Salsman, J.C. et al. (2007). Solvent dynamical control of ultrafast ground state electron transfer: implications for Class II–III mixed valency. *J. Am. Chem. Soc.* 129: 12772–12779.

109 Chen, P.Y. (1998). Medium effects on charge transfer in metal complexes. *Chem. Rev.* 98: 1439–1478.

110 Demadis, K.D., Neyhart, G.A., Kober, E.M. et al. (1999). Intervalence transfer at the localized-to-delocalized, mixed-valence transition in osmium polypyridyl complexes. *Inorg. Chem.* 38: 5948–5959.

111 Landau, L.D. (1932). Zur theorie der energieübertragung bei stössen. *Phys. Z. Sowjetunion* 1: 88–98.

112 Zener, C. (1932). Non-adiabatic crossing of energy levels. *Proc. R. Soc. Lond. A* 137: 696–702.

113 Marcus, R.A. (1960). Exchange reaction and electron transfer reactions including isotopic exchange. Theory of oxidation-reduction reactions involving electron transfer. *Disc. Faraday. Soc.* 29: 21–31.

114 Marcus, R.A. (1965). On the theory of electron transfer reactions. VI. Unified treatment for homogeneous and electrode Reactions. *J. Chem. Phys.* 43: 679–701.

115 Marcus, R.A. (1993). Electron transfer reactions in chemistry: theory and experiment. *Angew. Chem. Int. Ed. Engl.* 32: 1111–1121.

116 Creutz, C., Newton, M. D., Sutin, N. (1994). Metal-ligand and metal-metal coupling elements. *J. Photochem. Photobiol. A: Chem.*, 82, 47-59.

117 Wu, H.-Y., Ren, H.-S., Zhu, Q., and Li, X.-Y. (2012). A modified two-sphere model for solvent reorganization energy in electron transfer. *Phys. Chem. Chem. Phys.* 14: 5538–5544.

118 Li, X.-Y. (2015). An overview of continuum models for nonequilibrium solvation: popular theories and new challenge. *Int. J. Quantum Chem.* 115: 700–721.

119 Cave, R.J. and Newton, M.D. (1996). Generalization of the Mulliken-Hush treatment for the calculation of electron transfer matrix elements. *Chem. Phys. Lett.* 249: 15–19.

120 Silverman, L.N., Kanchanawong, P., Treynor, T.P., and Boxer, S.G. (2008). Stark spectroscopy of mixed-valence systems. *Phil. Trans. R. Soc. A* 366: 33–45.

121 Amthor, S. and Lambert, C. (2006). [2.2] Paracyclophane-bridged mixed-valence compounds: application of a generalized Mulliken-Hush three-level model. *J. Phys. Chem. A* 110: 1177–1189.

122 Newton, M.D. and Sutin, N. (1984). Electron transfer reactions in condensed phases. *Ann. Rev. Phys. Chem.* 35: 437–480.

123 Reimers, J.R. and Hush, N.S. (1994). Electron transfer and energy transfer through bridged systems III. Tight-binding linkages with zero or non-zero asymptotic band Gap. *J. Photochem. Photobiol. A: Chem.* 82: 31–46.

124 Lambert, C., Nöll, G., and Schelter, J. (2002). Bridge-mediated hopping or superexchange electron-transfer processes in bis (triarylamine) systems. *Nat. Mat.* 1: 69–73.

125 Eng, M.P. and Albinsson, B. (2006). Non-exponential distance dependence of bridge-mediated electronic coupling. *Angew. Chem. Int. Ed.* 45: 5626–5629.

126 Xiao, X., Meng, M., Lei, H., and Liu, C.Y. (2014). Electronic coupling and electron transfer between two dimolybdenum units spaced by a biphenylene group. *J. Phys. Chem. C* 118: 8308–8315.

2

Conceptual Understanding of Mixed-Valence Compounds and Its Extension to General Stereoisomerism

Jeffrey R. Reimers[1,2] *and Laura K. McKemmish*[3]

[1] *Shanghai University, International Centre for Quantum and Molecular Structures and School of Physics, 99 Shangda Road, Shanghai 200444, China*
[2] *University of Technology Sydney, School of Mathematical and Physical Sciences, Broadway, Ultimo NSW, 2007 Australia*
[3] *University of New South Wales Sydney, School of Chemistry, Anzac Parade, Kensington, NSW, 2052 Australia*

2.1 Introduction

Pioneering works [1] that drove the field of mixed-valence (MV) chemistry forward included the classification scheme developed by Robin and Day [2] and the chemical interpretations provided by Hush [3, 4]. Iconic MV complexes (Figure 2.1) include Prussian blue **10**, a coordination polymer involving Fe(II)–CN–Fe(III) linkages [2, 3, 5], and the Creutz–Taube ion [Ru(NH$_3$)$_5$ – pyrazine – Ru(NH$_3$)$_5$]$^{5+}$ dissolved in solution [6], **5**. A central feature is that MV systems involve two (or more) metal ions, each of which can exist in multiple valence states. In the case of Prussian blue, excited states can be formed as Fe(III)–CN–Fe(II) that have very different properties to the ground state, whereas the Creutz–Taube ion could exist as two stereoisomers Ru^{2+}(NH$_3$)$_5$ – pyrazine – Ru^{3+}(NH$_3$)$_5$ and Ru^{3+}(NH$_3$)$_5$ – pyrazine – Ru^{2+}(NH$_3$)$_5$ or an intermediary aromatic-like "delocalized" isomer Ru$^{2.5+}$(NH$_3$)$_5$ – pyrazine – Ru$^{2.5+}$(NH$_3$)$_5$. The central question of interest for any MV system concerns how the charge distributes among the available centers and then its effect on isomerism, isomerization, and excited-state structure and dynamics. Key system properties may be detected by many means, including X-ray spectroscopy, electrochemistry, spin resonance spectroscopy, and optical spectroscopic techniques such as NMR, IR, Raman, and UV/visible techniques. Also relevant is the Stark effect upon any of these properties. Systems of interest include not only coordination polymers and solvated ions but also general molecular solids, periodic arrays of centers with multiple valencies, polymer and protein systems, etc. Typically, MV chemical entities involve metal ions, but any chemical entity that can sustain multiple charged states can also be considered within the MV framework. Therefore, MV species comprises a significant subset of chemistry, typified by low-energy isomerization processes that interact strongly with the external environment, allowing control of one

Mixed-Valence Systems: Fundamentals, Synthesis, Electron Transfer, and Applications, First Edition.
Edited by Yu-Wu Zhong, Chun Y. Liu, and Jeffrey R. Reimers.
© 2023 WILEY-VCH GmbH. Published 2023 by WILEY-VCH GmbH.

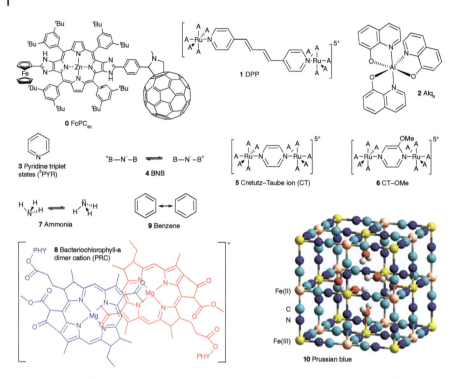

Figure 2.1 Some sample molecular systems with electronic states that can be described using two coupled diabatic potential energy surfaces. OMe is methoxy, PHY is phytyl; ^{t}Bu is tertiary butyl; PRC is the photosynthetic reaction center of *Rhodobacter sphaeroides*; A is ammonia; FcPC$_{60}$ is zinc, [[5,10,16,21-tetrakis[3,5-bis(1,1-dimethylethyl)phenyl]-13-[4-(1′,5′-dihydro-1′-methyl-2′H-[5,6]fullereno-C$_{60}$-I_h-[1,9-c]pyrrol-2′-yl)phenyl]-1,12-dihydro-23H,25H-diimidazo[4,5-b:4′,5′-l]porphin-2-yl-κN^{23},κN^{24},κN^{25},κN^{26}]ferrocenato(2-)]-, (*SP*-4-1); Alq3 is *mer*-tris(8-hydroxyquinolinato)aluminum(III); DPP is ruthenium(5+), decaammine[μ-[4,4′-[(1*E*,3*E*)-1,3-butadiene-1,4-diyl]bis[pyridine-κ*N*]]]di- (9Cl); only the Fe$_7$(CN)$_{18}$(H$_2$O)$_6$ framework of Prussian blue is indicated, there being water molecules, and possibly also dissociated salts such as KOH, located in its pores.

type of property by the manipulation of another. Hence, MV species and associated theory have wide applications throughout biochemistry and chemical technology.

This chapter reviews the chemical interpretations developed by Hush [3, 4] for MV compounds and applies them to isomerization processes involving covalent-bonding rearrangements. MV reactions involve primarily the transfer of a single electron (or electron hole), presenting a simplicity that has been amenable to extensive analytical development. This research has led to the understanding of classical reactions that proceed smoothly over a transition state [7], quantum reactions that proceed non-adiabatically [8], and the quantum entanglement that develops during chemical reactions [9]. A central feature in each case is that connections are developed between isomerization reactions and the spectroscopic properties of the isomers.

To effectively discuss these chemical interpretations, Section 2.2 reviews basic concepts and historical development pertinent to MV and related chemistry. Section 2.3 then reviews some iconic reference problems and applications, including molecules such as the Creutz–Taube ion **5** and related molecules **1** and **6**, optically significant materials such as Prussian blue **10**, organic conductors **2**, and natural **8** or artificial **0** light-harvesting systems. Finally, Section 2.4 interprets chemical properties of stereoisomers not considered as being formally "mixed valence" that nevertheless can be interpreted using the same conceptual basis.

The diversity of compounds considered in Section 2.4 is worth describing in depth here to illustrate the broad applications of the MV basic principles and analytical results. First considered is the triplet excited-state manifold of pyridine **3**, showing its analogy to an asymmetric MV compound. Next considered is BNB **4**, which can be thought of a covalently bound molecule or else represented in terms of charged valence-bond structures such as $\cdot B^+ - \ddot{N}^- - B:$ or $\cdot B = N^+ = B:^-$ that make the analogy to electron transfer apparent, except that many electrons other than the primary one are affected by the electron transfer process. The two boron atoms may have equivalent or inequivalent structures, a situation analogous to charge localization and charge delocalization in MV compounds. Next, stereoisomers of ammonia **7**, and indeed the whole XH_3 series, are considered that are related by the inversion reaction. This chemistry involves the rearrangement of non-bonding electrons, but the same basic question arises as to how the observed structure varies between the localized pyramidal possibilities and the symmetric planar structure favored by resonance interactions. Analysis unifies the properties of various excited states with those of the ground state, identifies 86.7° as the natural HXH bond angle in the absence of resonance for the XH_3 series and 101.5° for $[NH_3]^+$, and identifies the cause of the large change in properties found between NH_3 and PH_3 owing to an interchange in the ordering of the XH antibonding orbitals and the lowest Rydberg orbitals. Finally, aromaticity in benzene **9** is considered. This involves four π-electrons distributed among the doubly degenerate benzene HOMO and LUMO orbitals and thus involves complex electron rearrangements, yet the same question arises – is the structure delocalized (aromatic), or do the bonds localize to stabilize cyclohexatriene Kekulé structures? The analysis unifies many observed spectroscopic properties of benzene with the properties expected for hexatriene.

2.2 Modeling MV and Related Chemistry

2.2.1 Origins Within Chemical Bonding Theory

The advent of quantum mechanics in the mid-1920s quickly led to the development of basic concepts of chemical bonding. These concepts were later to be included as critical principles needed to interpret the properties of MV compounds. In 1927, Born and Oppenheimer (BO) introduced a method for separating the effects of nuclear and electronic motions [10], Hund provided the basic description for chemical reactions in terms of transformations between local minima on potential energy surfaces [11], and Heitler and London provided a description of covalent chemical bonding in terms of linear combinations of atomic orbitals [12]. There was a need to find simple ways to describe otherwise very complex quantum chemical phenomena, and in 1929, von Neumann and Wigner introduced the idea that simple chemical structures could be used to describe concepts such as reactants and products, with these structures interacting through quantum resonance to account for chemical reactivity and other properties [13]. The simple structures we call "diabatic" ones and in terms of MV theory depict standard localized valence structures in which metal atoms and other functional units adopt purely integral charges. Application of the BO approximation to a chemical model comprising coupled diabatic states leads to the adiabatic description of chemical structure, reactivity, and spectroscopy.

In modern times, adiabatic descriptions of chemical and spectroscopic phenomena can be obtained from *ab initio*, density functional theory (DFT), and other approaches that directly solve the BO equations, without the need to envisage simpler diabatic surfaces. The concept of transition-state theory (TST) is an important one that flows from adiabatic descriptions of chemical reactivity and was developed in 1935 by Evans and Polanyi and independently by Eyring [14–16]. Nevertheless, diabatic surfaces continue to provide the conceptual framework needed to interpret modern electronic structure computations, enabling intuitive predictions of the effects of chemical and other perturbations on a system.

It is notable that if von Neumann and Wigner's resonance energy [13] becomes too small, then the BO approximation can break down. When this happens, the diabatic states themselves provide more realistic treatments of the reactants and products than do the adiabatic states. In this case, chemical reactions proceed by crossing between diabatic surfaces. These critical ideas were first captured by London [17] in 1932 who provided a unified description of both adiabatic processes, i.e. those that occur on a single potential energy surface, and non-adiabatic processes, i.e. those that cannot be described in terms of dynamics on just a single potential energy surface. MV compounds provide just one example of these basic chemical features.

2.2.2 Coupled Harmonic Oscillator Model

During the 1930s and 1940s, theory development focused mainly on reactions involving bond breakage and formation. The use of Morse oscillators [18] to describe dissociative phenomena was common, with a major success being the coupled Morse oscillator diabatic London–Eyring–Polanyi–Sato (LEPS) potential energy surface for triatomic molecular reactivity [19]. Nevertheless, many chemical processes, including those occurring within MV compounds, occur without bond breakage and reformation. To describe proton and hydrogen transfer reactions, Horiuti and Polanyi [20] introduced the use of harmonic diabatic surfaces in 1935, and for other atom transfer reactions and electron transfer reactions, Hush introduced a similar approach in 1952–1953 [21, 22].

In a MV complex, two metal ions (or analogous redox functionalities) occur in which one would normally assign different valence states to each ion, even if the ions are symmetrically related to each other. A classic example of this is the Creutz–Taube ion **5** [6] $[Ru(NH_3)_5 - pyrazine - Ru(NH_3)_5]^{5+}$. Normal chemical valence rules would assign valences of Ru(II) and Ru(III) to the two metal atoms, yet the atoms appear to be symmetrically related. If indeed such valence assignments are appropriate, then the atomic structures on the two sides must differ, and hence, the symmetry must be reduced to something lower than what it could be. The two different ways in which charges can be so assigned within the molecule lead to two charge-localized diabatic states, one, called "A," depicting the properties of $Ru^{2+}(NH_3)_5$ – pyrazine – $Ru^{3+}(NH_3)_5$, and the other, called "B," depicting the properties of $Ru^{3+}(NH_3)_5$ – pyrazine – $Ru^{2+}(NH_3)_5$.

Many geometrical properties differ between the two ends of these structures, but mostly, one can envisage a single representative nuclear coordinate Q that embodies all of these effects. This is an *antisymmetric* variable, i.e. the equilibrium geometry of $Ru^{2+}(NH_3)_5$ – pyrazine – $Ru^{3+}(NH_3)_5$ is taken to be at $Q = -Q_m$, while the equilibrium geometry of $Ru^{3+}(NH_3)_5$ – pyrazine – $Ru^{2+}(NH_3)_5$ is taken to be at $Q = Q_m$. Specification of the harmonic oscillators also requires knowledge of the force constant k, and if the reaction is asymmetric, e.g. if one Ru atom is substituted by Fe or Os, then the free energy difference E_0 between the two diabatic states is also required. Quantum mechanics allows these two diabatic structures to interact with each other, and this is manifested through a resonance energy H_{ab}. Given the effective mass μ of the oscillators, it is then convenient to introduce the vibration frequency $\omega = (k/\mu)^{1/2}$ and take the coordinates Q to be dimensionless quantities expressed in terms of the zero-point vibration length of the oscillators, $(\mu\omega/\hbar)^{1/2}$. These parameters then conveniently defined the coupled harmonic oscillator Hamiltonian model:

$$\mathbf{H}^{loc} = \begin{bmatrix} V_A + T & H_{ab} \\ H_{ab} & V_B + T \end{bmatrix} \quad (2.1)$$

where V_A and V_B are harmonic diabatic surfaces representing the charge-localized structures

$$V_A = \frac{\hbar\omega}{2}(Q + Q_m)^2 \quad \text{and} \quad V_B = \frac{\hbar\omega}{2}(Q - Q_m)^2 + E_0 \quad (2.2)$$

and

$$T = -\frac{\hbar\omega}{2}\frac{\partial^2}{\partial Q^2} \quad (2.3)$$

is the nuclear kinetic energy operator.

Some illustrative examples of diabatic surfaces (in red and blue) representing charge-localized states are shown in Figure 2.2. These depict five *coupling regimes*, as listed in Table 2.1, that are subsequently quantitatively defined following Hush [3, 4, 7, 8, 23]. Shown in the figure are the adiabatic potential energy surfaces (PES) ε_- depicting the ground state (GS) and ε_+ depicting the excited state (ES)

$$\varepsilon_\pm = \frac{E_0}{2} + \frac{\lambda}{4} + \frac{\hbar\omega}{2}Q^2 \pm \left[\left(\frac{E_0}{2} - \hbar\omega Q_m Q\right)^2 + H_{ab}^2\right]^{1/2} \quad (2.4)$$

(in magenta and green) that result through application of the BO approximation to solve Eq. (2.1). In this equation appears the reorganization energy

$$\lambda = 2\hbar\omega Q_m^2 \quad (2.5)$$

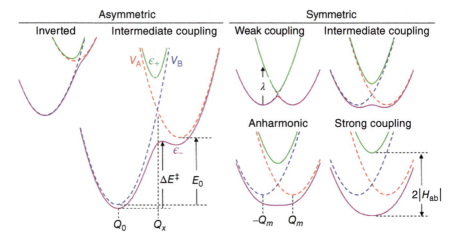

Figure 2.2 Sketches of potential energy surfaces (energy vs. nuclear coordinate Q) pertaining to the coupled harmonic oscillator model. Shown in red and blue are harmonic diabatic surfaces representing charge-localized states V_A and V_B, respectively, while in purple and green are shown the corresponding adiabatic potential energy surfaces $\varepsilon_\pm$, obtained by application of the BO approximation. The asymmetric surfaces have $2|H_{ab}|/\lambda = 0.3$ and either $E_0/\lambda = 0.3$ (intermediate coupling) or else $E_0/\lambda = 1.3$ (inverted), while the symmetric surfaces have $2|H_{ab}|/\lambda = 0.1$ (weak coupling), 0.5 (intermediate coupling), 1 (anharmonic), and 1.5 (strong coupling).

Table 2.1 Regimes appropriate to electron transfer reactions appropriate to MV complexes and other systems.

Coupling regime	Robin–Day classification	Requirement	Typical for symmetric MV	Properties
Weak	I	$\|Q_c\| < \frac{1}{\sqrt{8}}$ (equiv to $H_{ai}^2 < \hbar\omega\lambda/16$)	$2\|H_{ab}\|/\lambda \ll 1$	Non-adiabatic so that the BO approximately and TST fail
Intermediate	I or II	GS PES contains a local maximum	$2\|H_{ab}\|/\lambda < 1$	Standard chemical regime
Anharmonic	II-III	If double welled, then $\hbar\omega/2 > \Delta E^\ddagger$ If single welled, then $E_H/E_T < 0.75$	$2\|H_{ab}\|/\lambda \approx 1$	Vibrational quantum effects are important
Strong	III	$\left(\frac{E_0}{\lambda}\right)^{\frac{2}{3}} + \left(\frac{2H_{ab}}{\lambda}\right)^{\frac{2}{3}} \geq 1$ and $2\|H_{ab}\| > \|E_0\|$	$2\|H_{ab}\|/\lambda > 1$	Not describable as a classical MV structure
Inverted		$\left(\frac{E_0}{\lambda}\right)^{\frac{2}{3}} + \left(\frac{2H_{ab}}{\lambda}\right)^{\frac{2}{3}} \geq 1$ and $2\|H_{ab}\| < \|E_0\|$	n.a.	Non-adiabatic, reaction rate decreases as exothermicity increases

used to represent the energy cost needed to transfer a system from one diabatic state at its equilibrium geometry to the other, without change in nuclear structure. The reorganization energy is a convenient parameter for the representation of the harmonic oscillator model as it often can be determined directly from experimental data. It also depicts the heat released during some process of interest, a quantity that is critical to the understanding of properties such as device efficiency, quantum dephasing, quantum entanglement, etc. Convenient parameters for the specification of the coupled harmonic oscillator model then become [8]

$$\frac{2|H_{ab}|}{\lambda}, \quad \frac{E_0}{\left(4H_{ab}^2 + \lambda^2\right)^{1/2}} \text{ or } \frac{E_0}{\hbar\omega}, \text{ and } \frac{\hbar\omega}{\left(4H_{ab}^2 + \lambda^2\right)^{1/2}} \quad (2.6)$$

that describe the strength of the resonance coupling, the degree of asymmetry (in two possible non-equivalent ways), and the vibrational energy scale compared to the electronic energy scale, respectively.

Four of the five regimes presented in Table 2.1 and Figure 2.2 apply to symmetric problems ($E_0 = 0$), and the relationships between them are sketched in Figure 2.3 using the variables $\frac{2|H_{ab}|}{\lambda}$ and $\frac{E_0}{\left(4H_{ab}^2 + \lambda^2\right)^{1/2}}$ to specify the available parameter space. A qualitative summary of all five regimes follows, followed by quantitative analysis.

In the intermediate-coupling regime, the GS PES shows a double-minimum structure depicting two stereoisomers separated by a transition state, see Figure 2.2. This is the standard regime in which most chemical processes operate. The BO approximation applies, with methods such as TST and classical molecular dynamics (MD) on the GS PES being descriptive of reactivity, structure, and function. In the qualitative Robin–Day classification [2], the intermediate-coupling regime is referred to as "Class II."

In the weak-coupling regime, Figure 2.2 shows that the transition state appears very abruptly, instead of the smooth evolution seen for the intermediate-coupling examples. This abruptness leads to failure of the BO approximation. As TST is based on this approximation, it breaks down in this region, often considerably overestimating the rates, and MD also fails to describe dynamical processes. Quantum methods describing both electronic structure and nuclear motion must be used for the interpretation of spectra and reactivity in the weak-coupling regime. In the qualitative Robin–Day classification [2], the weak-coupling regime is referred to as "Class I."

In the strong-coupling and inverted regimes, the GS PES no longer displays a transition state, depicting instead single-welled structures. Nevertheless, these two regimes represent quite different chemical features.

In the strong-coupling regime (Figure 2.2), the GS PES minimum depicts the average properties of the reactant and product diabatic states. In covalent molecule chemistry, this effect is well known, leading to aromaticity, e.g. benzene **9** has a structure that is the average of its two diabatic Kekulé representations [24, 25]. In MV chemistry, the effect is similar, producing the envisaged $Ru^{2.5+}(NH_3)_5$ – pyrazine – $Ru^{2.5+}(NH_3)_5$ structure of the CT ion **5**. In it, the valence electrons are said to *delocalize* over both metal centers, just as if a Ru–Ru covalent bond was formed, despite the presence of the pyrazine ligand in what would be the middle of the bond. These features indicate that novel electronic structures are

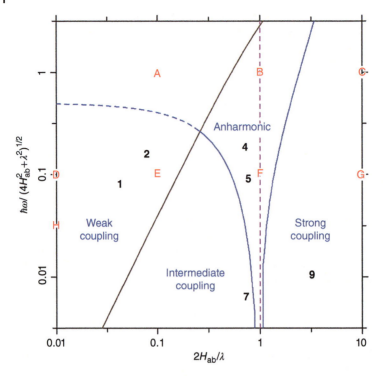

Figure 2.3 Regimes of the parameter space for symmetric reactions (Table 2.1 and Figure 2.2), with bounding lines: magenta – Eq. (2.11), discriminating single vs. double wells; blue – defining the surrounding changeover region in which anharmonic effects can be critical, as defined by Eq. (2.12) (in particular, $2\Delta E^{\ddagger}/\hbar\omega = 1$ whenever $\frac{2\hbar\omega}{(4H_{ab}^2+\lambda^2)^{1/2}} = \left(1 - \frac{2|H_{ab}|}{\lambda}\right)^2 \left(1 + \frac{4H_{ab}^2}{\lambda^2}\right)^{-1/2}$); and brown – Eqs. (2.28)–(2.30), describing the requirement for a full non-adiabatic quantum treatment. The locations on this space for appropriate example systems **0–9** and models **A–H** are indicated.

generated in the strong-coupling regime, but the BO GS PES remains characteristic, with standard chemical theories and classical MD remaining appropriate to describe spectroscopy and function. In the qualitative Robin–Day classification [2], the strong-coupling regime is referred to as "Class III."

In the inverted region (Figure 2.2), the reaction asymmetry $|E_0|$ becomes very large, preventing the expected transition state from forming. Classical MD and TST again fail in this regime, and only coupled electronic and nuclear quantum methods can be applied to model reactivity. This regime is not addressed by the Robin–Day classification scheme.

The fifth regime indicated is called the "anharmonic" regime (Figure 2.2), which, as Figure 2.3 shows for symmetric reactions, sits at the boundaries of the strong-coupling and intermediate-coupling regimes, extending also into the weak-coupling regime. Quantum mechanics mostly does not support abrupt changes in properties within some parameter space, and this is indeed what is found for the coupled harmonic oscillator model. The anharmonic regime separates regimes in which MD and TST work from those in which full quantum treatment of the coupled electronic and nuclear motions is required. In it, only quantum treatment of the nuclear motion on a single BO PES is required. It applies to situations in which there is a shallow double well that does not support zero-point motion or else quantum tunneling effects or ultrafast classical reactions are paramount, as well as to the region in which there is a single well but the GS PES is anharmonic. This regime has overlap with the "Class II–III" Robin–Day-type description of symmetric mixed-valence [26, 27], but that class is perceived as being classical in nature, associated with fast reactions over low barriers, whereas the intent of the "anharmonic" regime defined here encompass broader aspects of anharmonicity and quantum vibrational effects that provide for smooth passage between the intermediate-coupling, strong-coupling, and inverted regimes.

Quantitative analysis pertaining to many properties of the GS and ES PES has been established. For symmetric compounds, in the weak-coupling and intermediate-coupling regimes, the adiabatic vertical transition energy (the center of the absorption band in MV and other compounds) is given by [4]

$$\hbar\omega_{max} = \Delta E_v = \lambda \tag{2.7}$$

and in the strong-coupling regime by

$$\hbar\omega_{max} = \Delta E_v = 2|H_{ab}| \tag{2.8}$$

For symmetric potentials, the adiabatic reaction free energy is given by [7]

$$\Delta E^{\ddagger} = \frac{(\lambda - 2|H_{ab}|)^2}{4\lambda} \tag{2.9}$$

For asymmetric reactions in the limit of $2|H_{ab}|/\lambda \ll 1$, the activation energy was determined independently by Kubo and Toyozawa [28] and by Hush [29] as

$$\Delta E^{\ddagger} = \frac{(\lambda + E_0)^2}{4\lambda} \tag{2.10}$$

Note that these equations reduce to the expression $\Delta E^{\ddagger} = \lambda/4$ developed by Marcus [30] for symmetric reactions in the weak-coupling limit.

TST can only be applied to understand chemical reactivity if the GS PES is double welled, containing minima corresponding to both reactants and products. Alternatively, the GS PES displays a single well whenever [31]

$$\left(\frac{E_0}{\lambda}\right)^{\frac{2}{3}} + \left(\frac{2H_{ab}}{\lambda}\right)^{\frac{2}{3}} \geq 1 \tag{2.11}$$

Figure 2.2 shows that this can be induced either by large coupling ($2|H_{ab}| \geq \lambda$), leading to the strong-coupling regime [4, 32], or else by a large energy offset ($|E_0| \geq \lambda$), leading to the inverted regime. In the inverted regime, when $E_0 < -\lambda$, Eq. (2.10) indicates that reactions become slower, owing to increasing activation energy, as they become more exothermic, an effect extensively pursued by Marcus [33]. Also, TST will only be useful when the double well supports zero-point vibration.

When the sum in Eq. (2.11) only slightly exceeds 1, the GS PES is single welled but presents a small curvature, allowing anharmonic terms to dominate the low-energy region of the PES to control properties. Similarly, when the sum is just less than 1, the GS PES is double welled, but the well depth is too shallow to support zero-point vibration. Hence, a regime exists in which anharmonic vibrational effects can control observed properties. This regime is bounded by

$$\frac{2\Delta E^{\ddagger}}{\hbar \omega} < 1 \quad \text{if double welled, and}$$

$$\frac{E_H}{E_T} < 0.75 \quad \text{if single welled,} \tag{2.12}$$

where E_T is the minimum energy increase above GS minimum at displacements of $Q = \pm 1$ and E_H is the corresponding energy predicted by a harmonic model of the GS PES.

Analytical expressions are also available for many other system properties, including equilibrium geometries, vibrational transition energies, line intensities, electronic band widths, and the curvature at the transition state; all these can be very useful in interpreting experimental data pertaining to MV compounds [7, 8]. Also, electric field effects can be introduced as modifications to E_0, allowing vibrational and electronic polarizabilities, hyperpolarizabilities, and Stark effects to be modeled [34, 35]. A particularly relevant expression for MV compounds is that for the intensity of the electronic transition in the weak- and intermediate-coupling cases, with the associated transition moment given by [3]

$$M = n^{1/2} \frac{H_{ab}}{\Delta E_v} eR \tag{2.13}$$

where R is the distance between the redox centers and $n = 1$ or 2 is the number of electrons that can make the transition. For one electron and arbitrary coupling, this equation generalizes to [23, 34]

$$M = eR \left(1 - \frac{E_0}{\Delta E_v}\right)^{1/2} \tag{2.14}$$

with the change in dipole moment upon excitation, $\Delta \mu$, being given through

$$(\Delta \mu)^2 = e^2 R^2 - 4M^2 \tag{2.15}$$

The dipole moment change of a transition can be directly measured by Stark spectroscopy (see later), while the transition moment can be obtained from the area of the absorption band [3] as

$$|M| = 0.0206 \sqrt{\int_{\text{band}} \frac{\varepsilon(\nu)}{\nu} d\nu} \approx 0.0206 \left(\frac{\nu_{\text{FWHM}} \varepsilon_{\text{max}}}{\nu_{\text{max}}}\right)^{1/2} \tag{2.16}$$

where M is in eÅ and ε is the molar extinction coefficient in M^{-1} cm^{-1}, with ν_{FWHM} being the band full width at half maximum and ν_{max} and ε_{max} being the frequency and extinction at the band maximum, respectively.

2.2.3 Intermolecular and Intramolecular Contributions to the Reorganization Energy

In the previous subsection, the critical antisymmetric coordinate Q was described in terms of the geometrical differences around the Ru atoms at each end of the Creutz–Taube ion. In 1958, Hush recognized that these differences can be classified into two distinct types, *intramolecular* contributions associated with Ru—N bond length changes, as well as associated bond angle and torsional angle changes, and *intermolecular* contributions associated with the rearrangements of the surrounding liquid, solid, or protein environment [29]. The intramolecular changes may be associated with high-frequency motions of energy much greater than the available thermal energy, while the solvent motions will typically be of low frequency and hence be in thermal equilibrium at room temperature. Often in chemical theories, very different treatments of these types of motion are warranted, as can be seen, for example, in the 1970's semi-classical electron transfer theories developed by Duyne and Fischer [36], Efrima and Bixon [37], Jortner [38, 39], and Warshel [40]. In simple models, differing thermal effects are usually ignored, with the effects of the intramolecular and intermolecular contributions combined together.

How this is achieved in practice is through utilization of the reorganization energies λ_i and λ_s associated with the intramolecular and intermolecular (solvent) degrees of freedom, respectively. For a MV complex, distinction between intramolecular and intermolecular effects is usually clear. For a metal ion in solution, it is customary to define the "molecule" as the metal ion plus its inner coordination sphere of solvent molecules, and the "solvent" as the remaining solvent molecules, leading to the notion of "inner-sphere" and "outer-sphere" reorganization and reactivity [41]. The reorganization energy is the energy released upon relaxation following the transfer of a system from one diabatic state to the other at the geometry of the initial diabatic state, and simple models [29] set the total reorganization energy as the sum of the two contributions, $\lambda = \lambda_i + \lambda_s$. In principle, both the intramolecular and intermolecular reorganizations can be adjusted using subtle chemical and solvent control, respectively. If such subtleties act to change the electron transfer regime, then they could have profound effect on system properties.

An important result pertaining to solvent control, interpretable in terms of the coupled harmonic oscillator model, is the dumbbell model of non-equilibrium solvation developed by Marcus in 1956 [30]

$$\lambda_s = \frac{(\Delta q)^2}{2} \left(\frac{1}{\varepsilon_{op}} - \frac{1}{\varepsilon_s} \right) \left(\frac{1}{R_A} + \frac{1}{R_B} - \frac{2}{R} \right) \tag{2.17}$$

where Δq is the charge transferred, R_A and R_B are effective radii of A and B, R is the distance between them, and ε_s and ε_{op} are the static and optical dielectric constants of the solvent, respectively. The core idea of this and related [42, 43] dielectric solvation models is embedded into many modern self-consistent reaction field (SCRF) approaches to quantitative electronic structure modeling. In recent years,

Marcus' equation has been improved using non-equilibrium statistical mechanics by Li et al. [44–46] to give

$$\lambda_s = \frac{(\Delta q)^2}{2} \left(\frac{1}{\varepsilon_{op}} - \frac{1}{\varepsilon_s} \right) \frac{\varepsilon_s - \varepsilon_{op}}{\varepsilon_{op}(\varepsilon_s - 1)} \left(\frac{1}{R_A} + \frac{1}{R_B} - \frac{2}{R} \right) \quad (2.18)$$

2.2.4 Effects of Electric Fields on MV Optical Band Shapes

Internal electric fields are inherent at chromophores inside proteins, as well as in most materials. In addition, external electric fields can be applied to proteins, materials, and liquids, with the works of Boxer et al. providing access to unprecedentedly high applied field strengths [47–49]. Electric fields modify the energy differences between diabatic states, introducing the change

$$\Delta E_0 = -\Delta \boldsymbol{\mu} \cdot \mathbf{F} \quad (2.19)$$

where $\mathbf{F}$ is the applied electric field vector and $\Delta \boldsymbol{\mu}$ is the dipole moment difference between the diabatic states. In modern experiments, the applied electric field is often oscillated so as to prevent short circuiting, resulting in a signal that is proportional to the square of the applied field strength. Also studied samples are often isotropically distributed rather than systematically aligned, demanding averaging over all directions of the applied field. The response of the molar extinction coefficient $\varepsilon(F)$ as a function of the field strength F can then be interpreted using the equations of Liptay [50, 51] as

$$\frac{1}{F^2 R(\chi)} \left(\frac{\varepsilon(F)}{\omega} - \frac{\varepsilon(0)}{\omega} \right) = \left(\frac{A^2}{M} + \frac{2B}{M} \right) \frac{\varepsilon(0)}{\omega} + \left(\frac{2A\Delta\mu}{M} + \frac{\Delta\alpha}{2} \right) \frac{d\left(\frac{\varepsilon(0)}{\omega}\right)}{\hbar d\omega}$$

$$+ \left(\frac{\Delta\mu}{2} \right)^2 \frac{d^2\left(\frac{\varepsilon(0)}{\omega}\right)}{\hbar^2 d\omega^2} \quad (2.20)$$

in terms of the following parameters: $\Delta\mu$ – the change in dipole moment magnitude on excitation, $\Delta\alpha$ – the change in dipole polarizability magnitude on excitation, A – the transition moment polarizability, and B – the transition moment hyperpolarizability, as well as an optical geometrical factor $R(\chi)$.

This analysis could be relevant to the understanding of MV compounds in many ways, but here, it is considered an example in which a compound containing two metal centers has a symmetric structure in the ground state, with optical excitation providing for the possibility of MV excited-state species in which the excitation preferentially localizes on one of the two metal centers. In this case, the two adiabatic states $\varepsilon_\pm$ introduced in Eq. (2.4) depict two different excited states that excitation could occur to.

Figure 2.4 shows simulated spectra obtained by solving the coupled harmonic oscillator problem numerically without use of the BO approximation [23, 52–54]. Predicted absorption spectra, and associated Stark spectra, are shown as a function of the change in energy $h\nu = \hbar\Delta\omega$ away from the ground state to non-interacting diabatic state, as a function of the interaction energy H_{ab}, for a reorganization energy

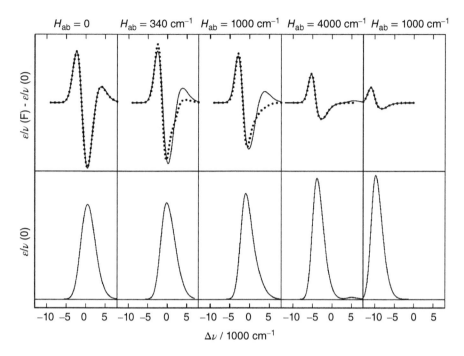

Figure 2.4 Field-free absorption spectrum and the finite field difference spectrum (ε is the molar extinction coefficient at frequency $v = 2\pi\omega$ and F is the applied field strength), evaluated from the model for bis-metal complexes for different values of the coupling H_{ab} using $\lambda = 10\,000$ cm^{-1}. The lower panels are the calculated absorption spectra, while the top panels are the calculated Stark difference spectra (solid lines) and the best possible fit to a function of the classical Liptay form, Eq. (2.20), (dots). A non-classical behavior is observed for intermediate-coupling strengths. Source: Reimers and Hush [35] with permission of Springer Nature.

common in MV compounds of $\lambda = 10\,000$ cm^{-1}. In the weak-coupling ($H_{ab} = 0$) and intermediate-coupling ($H_{ab} \ll 340$ cm^{-1}) regimes, the calculated spectra take forms typical of Franck–Condon progressions, and the Stark spectra appear as the second derivative of the absorption spectra. Through the Liptay equation above, this allows immediate extraction of $\Delta\mu$ from the observed data and hence provides critical information concerning the nature of the MV complex through, for example, Eqs. (2.14) and (2.15). In the strong-coupling regime ($H_{ab} = 10\,000$ cm^{-1}), the predicted absorption spectra again take on the appearance of a Franck–Condon progression, although a much narrower one than that found for weak-coupling regime owing to suppression of the antisymmetric vibration [23], but now, the Stark spectrum takes on a first derivative appearance so that fitting with the Liptay equation reveals primarily the change in polarizability $\Delta\alpha$ between the ground state and the allowed component of the excited state. In this case, the electron involved in the optical transition becomes delocalized over both metal centers, and the transition induces no major charge rearrangement. Approaching the anharmonic regime ($H_{ab} = 4000$ cm^{-1}), the calculated spectrum is distorted away from that of a Franck–Condon progression, but the Liptay form realistically fits the calculated Stark profile, revealing contributions from both $\Delta\mu$ and $\Delta\alpha$. Nevertheless, in the intermediate region of 340 to 1000 cm^{-1}, not only is the predicted spectrum distorted but also the Stark spectrum cannot be interpreted adequately using the Liptay form. The Stark spectrum therefore highly reveals the detailed properties of a MV compound.

2.2.5 Non-adiabatic Effects

The BO approximation allows electronic wavefunctions (and associated energies $\varepsilon_\pm$) to be determined parametrically at each nuclear geometry Q to give an approximate solution to the coupled nuclear electronic Hamiltonian, Eq. (2.1). In terms of the electronic basis set presented by these wavefunctions, the original Hamiltonian may be rewritten without approximation as [8]

$$\mathbf{H}^{\text{adiab}} = \begin{bmatrix} \varepsilon_- + \Delta H^{\text{DC}} + T & \Delta P^{\text{DC}} \frac{\partial}{\partial Q} + \Delta H^{\text{SD}} \\ \Delta P^{\text{DC}} \frac{\partial}{\partial Q} + \Delta H^{\text{SD}} & \varepsilon_+ + \Delta H^{\text{DC}} + T \end{bmatrix} \quad (2.21)$$

manifesting three corrections to the BO approximation: ΔH^{DC}, known as the *diagonal correction*, which equally modifies both BO PES in a mass-dependent way; $\Delta P^{\text{DC}} \frac{\partial}{\partial Q}$, known as the *first derivative* or *momentum* correction, which induces dependence of the electronic wavefunctions on nuclear momentum; and ΔH^{SD}, known as the *second derivative* correction, which induces dependence of the electronic wavefunctions on the nuclear kinetic energy. Application only of the diagonal correction retains the original separation introduced between the nuclear and electronic variables, with the resultant modified PES $\varepsilon_\pm + \Delta H^{\text{DC}}$ being known as Born–Huang PES [55–58]. Alternatively, the momentum and kinetic energy corrections couple the electronic states together, and hence, the resulting wavefunctions are non-adiabatic. The coupling terms are most simply expressed [8] in terms of the geometry Q_x at which the localized diabatic harmonic oscillators cross (see Figure 2.2)

$$Q_x = \frac{E_0}{\lambda} Q_m \quad (2.22)$$

and the cusp radius

$$Q_c = \frac{2|H_{ab}|}{\lambda} Q_m \quad (2.23)$$

The latter quantity is thus called as the BO approximation results in a cusp when $H_{ab} = 0$ that introduces a derivative discontinuity into the BO PES. As a result, quantum dynamics viewed from within the BO description becomes chaotic as the cusp is approached [59–62], making the solution of Eq. (2.21) by numerical means technically difficult [8]. Note that Eq. (2.11) can be rewritten in these variables as $|Q_x|^{2/3} + |Q_c|^{2/3} \geq Q_m^{2/3}$, expressing the strong-coupling regime simply as occurring whenever $|Q_c| \geq Q_m$ and the inverted limit as occurring whenever $|Q_x| \geq Q_m$. In terms of these new variables, the corrections to the BO approximation become [8]

$$\Delta P^{\text{FD}} = \frac{-\hbar\omega}{2} \frac{Q_c}{Q_c^2 + (Q - Q_x)^2} \quad (2.24)$$

$$\Delta H^{DC} = \frac{1}{2\hbar\omega}(\Delta P^{FD})^2, \text{ and} \tag{2.25}$$

$$\Delta H^{SD} = \frac{2}{\hbar\omega}\frac{Q-Q_x}{Q_c}(\Delta P^{FD})^2 \tag{2.26}$$

The first derivative correction is strongly spiked around the intersection geometry, and the other two corrections scale as its square. In computational treatments of non-adiabatic coupling, it is often assumed that the non-adiabatic corrections are small and hence dominated by the first-order correction ΔP^{FD}, but such a hierarchical treatment is not valid when considering dynamics around a cusp, and it has been found that accurate solutions of Eq. (2.21) require the treatment of all three correction terms [8].

In Figure 2.5, the impact of the three corrections on the BO description of the chemistry is indicated for cusp radii of $Q_c = 1/2, 1/8^{1/2}$, and $1/4$. This includes presentation of the relative shapes of the first derivative and second derivative correction profiles, plus the Born–Huang PES made by adding the diagonal correction to the BO PES. The effect of the corrections change from minor to significant to dominant over just this small range in Q_c. The significance of the central value in this range is that

$$\Delta H^{DC}\big|_{max} = \hbar\omega \text{ when } |Q_c| = \frac{1}{\sqrt{8}} \tag{2.27}$$

The figure shows that, when the diagonal correction exceeds the vibrational energy spacing, non-adiabatic effects become important [8], an effect apparently seen from the figure. The weak-coupling regime in which non-adiabatic effects become critical is therefore defined by

$$|Q_c| < \frac{1}{\sqrt{8}} \tag{2.28}$$

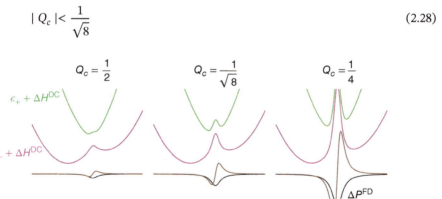

Figure 2.5 Non-adiabatic corrections appropriate to the asymmetric reaction shown in Figure 2.2 with $2H_{ab}/\lambda = 0.3$ and $E_0/\lambda = 0.3$ for various values of the cusp radius Q_c. Shown in magenta and green are the Born–Huang PES made by adding the diagonal correction to the BO PESs, as well as the first derivative and second derivative corrections in blue and brown, respectively. Note that at constant coupling H_{ab} and reorganization energy λ, the vibration frequency ω varies with Q_c as $\hbar\omega = \frac{2H_{ab}^2}{\lambda Q_c}$, as does the equilibrium geometry $Q_m = Q_c\left(\frac{\lambda}{2|H_{ab}|}\right)$, and also only relative position is indicated.

This equation encompasses the arguments made originally by Born and Oppenheimer [10] that non-adiabatic effects are dependent on the oscillator mass as, from Eq. (2.23), Q_c depends on Q_m, which is the geometry change in zero-point units and thus is mass dependent, but through Eq. (2.23), it also shows a strong dependence on the strength of the coupling $2|J|/\lambda$, as has been recognized in some form since the first treatments of non-adiabatic effects on chemical reactions [17, 63–66]. It is common to consider properties predicted by the coupled diabatic model in terms of the reduced variables such as the relative coupling strength $2|J|/\lambda$ and the relative vibrational energy $\hbar\omega/(4H_{ab}^2 + \lambda^2)^{1/2}$. In terms of these variables, Eq. (2.28) can be rewritten as

$$\frac{\hbar\omega}{(4H_{ab}^2 + \lambda^2)^{1/2}} > \frac{4\left(\frac{2|H_{ab}|}{\lambda}\right)}{\left[1 + \left(\frac{\lambda}{2H_{ab}}\right)^2\right]^{1/2}} \qquad (2.29)$$

An alternative way to write these results is that non-adiabatic weak-coupling limit applies whenever

$$H_{ab}^2 < \frac{\hbar\omega\lambda}{16} \qquad (2.30)$$

When the diagonal correction is significant, trajectories on the GS BO PES that cross the transition state to yield a chemical reaction get blocked by the diagonal correction, effectively trapping the system in its reactant state. The first derivative and second derivative couplings induce tunneling underneath this barrier and thus facilitate the reaction to proceed, but the net effect is that the rate is reduced below that expected by TST. In the weak-coupling limit, assuming a classical distribution of vibrational oscillators at temperature T, the reaction rate across a barrier, utilizing Eq. (2.10), was determined by Levich and Dogonadze [63–66] to be

$$k = \frac{2\pi H_{ab}^2}{\hbar(4\pi\lambda k_\beta T)^{1/2}} \exp\frac{-(\Delta G_0 + \lambda)^2}{4\lambda k_\beta T} \qquad (2.31)$$

This rate can be expressed [8] as the rate predicted by TST, scaled by a transmission coefficient κ given by

$$\kappa = \left(\frac{\pi^3 \lambda}{k_\beta T}\right)^{\frac{1}{2}} Q_c^2 \qquad (2.32)$$

indicating again the critical importance of the cusp radius to the understanding of non-adiabatic effects.

2.2.6 MV Complexes as Potential Quantum Qubits

Quantum information systems are most commonly built around *quantum qubits* in which a wavefunction of two variables cannot be written as a simple product of wavefunctions of each individual variable. In this case, the quantum wavefunction contains more information than would be present in an analogous classical description of the problem, known as *quantum entanglement* [67]. There are many ways in which two variables can be coupled in useful ways, and indeed, all chemical reactions embody quantum entanglement between the electronic and nuclear degrees of freedom. The question of interest concerns whether or not this entanglement can be harnessed to make functionally relevant quantum qubits, with the ammonia inversion reaction being initially proposed as a possible candidate [68]. The coupled harmonic oscillator model used herein to model chemistry is analogous to the spin-boson model applied in many areas of physics that is known to manifest entanglement [69]. In principle, any amount of entanglement could present a qubit useful in some situation [70], but in practice, qubits in which the entanglement persists in the wavefunction at finite temperature are of the greatest interest. To be useful, quantum qubits must also be writable and readable externally, with the read operation resulting in the collapse of a quantum wavefunction to produce a classical measurement. This measurement needs to be sensitive to one of the two variables that are entangled together in the wavefunction, and for chemical qubits, spatially distinct detectors that recognize molecular geometry could detect which diabatic state (reactant or product) that the system is in.

Figure 2.6 presents a general description of entanglement across the parameter space of the coupled harmonic oscillator model. For the examples indicated by the red arrows, it also shows the BO GS and ES PES, as well as the vibrational density associated with the lowest energy vibronic wavefunction. Spatially based detectors can only register persistent entanglement if the vibrational density is bimodal (i.e. manifests local maxima in each of the A and B diabatic wells separated by a local minimum) [73–75], as found for the three examples on the left-hand side of the figure. Hence, reactions in either the weak-coupling or intermediate-coupling regimes could lead to large entanglement, with reactions in the unimodal strong-coupling regime (see the right-hand side of the figure) expected to develop very little entanglement.

In this figure, the von Neumann definition of entanglement S [67, 73] is utilized. This is based on the wavefunction density projected onto the basis states representing either of the two entangled variables; in this case, either the electronic wavefunctions or the nuclear wavefunctions [9]. Both approaches yield the same value for the entanglement, despite being technically very different to apply. Taking the conceptually simpler approach, the electronic density of the ground-state vibronic wavefunction, obtained from full quantum solution to Eq. (2.1), can be expressed as

$$\rho = \begin{bmatrix} \rho_{AA} & \rho_{AB} \\ \rho_{AB} & \rho_{BB} \end{bmatrix} \qquad (2.33)$$

and has eigenvalues of

$$\rho_\pm = \frac{\rho_{AA} + \rho_{BB}}{2} \pm \frac{1}{2}\left[(\rho_{AA} - \rho_{BB})^2 + 4\rho_{AB}^2\right]^{1/2} \qquad (2.34)$$

where $\rho_{AA} + \rho_{BB} = \rho_- + \rho_+ = 1$. The entanglement of the zero-point level is then given by

$$S_0 = -\rho_- \log_2 \rho_- - \rho_+ \log_2 \rho_+ \qquad (2.35)$$

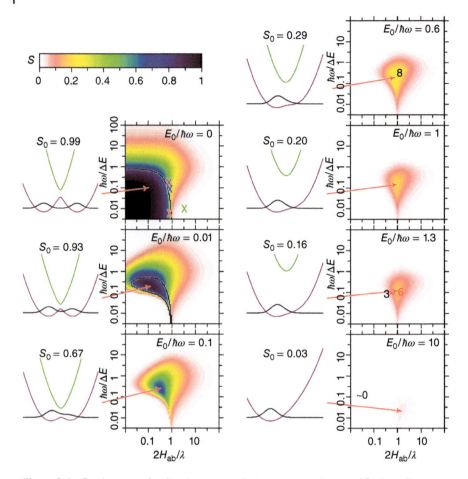

Figure 2.6 Persistent vs. fragile, degeneracy-induced, entanglement. 2D plots: Electron nuclear entanglement quantified by von Neumann entropy, S, is shown as a contour plot (see key) for the vibronic ground-state wavefunctions as a function of $2|H_{ab}|/\lambda$ and $\frac{\hbar\omega}{(4H_{ab}^2+\lambda^2)^{1/2}} = \Delta E$ at various values of $E_0/\hbar\omega$. The black lines denote regions in which the ground-state vibrational probability density is bimodal or unimodal (Eq. (2.11)), while the numbers indicate the parameters appropriate to the molecules listed in Table 2.2 and shown in Figure 2.1. PES: For the parameters indicated by the red arrows, shown are the ground-state BO PES ε_- (purple) and related excited-state BO PES ε_+ (green) are sketched, along with the vibrational density of the vibronic ground-state wavefunction (black). Source: McKemmish et al. [9] with permission of the American Institute of Physics.

Figure 2.6 presents seven sets of results, at increasing values of $E_0/\hbar\omega$. For the set pertaining to $E_0/\hbar\omega = 0$, the BO GS and ES PES are shown for the case with $2|H_{ab}|/\lambda = \hbar\omega/(4H_{ab}^2+\lambda^2)^{1/2} = 0.1$. In this example, the ground-state wavefunction consists of two nearly Gaussian, non-overlapping, components centered near the diabatic surface minima, leading to an electronic density matrix of

$$\boldsymbol{\rho} \approx \begin{bmatrix} 1/2 & 0 \\ 0 & 1/2 \end{bmatrix} \tag{2.36}$$

and hence, the entanglement is close to the maximum value of $S_0 = 1$. Alternatively, for $E_0/\hbar\omega = 0$ in the strong-coupling situation with $2|H_{ab}|/\lambda = 1.5$ and $\hbar\omega/(4H_{ab}^2 + \lambda^2)^{1/2} = 0.01$ highlighted in Figure 2.2, the single-welled GS PES results in a vibrational density that is approximately describable by a single Gaussian component centered at $Q = 0$, making the density matrix

$$\rho \approx \begin{bmatrix} 1/2 & 1/2 \\ 1/2 & 1/2 \end{bmatrix} \quad (2.37)$$

and hence, the entanglement vanishes, $S_0 \approx 0$. At a crude level, the entanglement for symmetric reactions can be categorized as

$$S = 1 \text{ if } \frac{2|H_{ab}|}{\lambda} < 1 \text{ and } \frac{\hbar\omega}{(4H_{ab}^2 + \lambda^2)^{1/3}} < 1$$

$$= 0 \text{ otherwise} \quad (2.38)$$

indicating that maximal entanglement arises only within subsets of the weak-coupling and intermediate-coupling regimes. As the bordering anharmonic regime is entered, the entanglement reduces quickly but can achieve medium-level values of c. $S = 0.5$.

The other images in Figure 2.6 depict the surfaces, vibrational densities, and entanglements as the energy asymmetry $E_0/\hbar\omega$ increases. At the small value of $E_0/\hbar\omega = 0.01$, large changes in the entanglement profile result, with the entanglement only being large in a broad region in the vicinity of $2|H_{ab}|/\lambda = 0.1$ and $\hbar\omega/(4H_{ab}^2 + \lambda^2)^{1/2} = 0.3$. The reason for this is that when the vibration frequency is small compared to the barrier height or else when the coupling is weak, small energy changes force the vibrational density to localize on one of the two diabatic states only (A for $E_0 > 0$), producing an electronic density of

$$\rho \approx \begin{bmatrix} 1 & 0 \\ 0 & 0 \end{bmatrix} \quad (2.39)$$

so that once again $S_0 \approx 0$. Entanglement in this region found for the purely symmetric situation is called *fragile* entanglement as it does not persist when energy-level degeneracy is lifted. Figure 2.6 shows that as $E_0/\hbar\omega$ increases, significant contraction occurs to the chemical space that supports persistent entanglement, with entanglement localizing to the region near $2|H_{ab}|/\lambda = 1$ and $\hbar\omega/(4H_{ab}^2 + \lambda^2)^{1/2} = 0.1$. Operational qubits need to be located within some external environment, and environmental fluctuations induce asymmetry E_0. Chemical systems featuring $2|H_{ab}|/\lambda \approx 1$ and $\hbar\omega/(4H_{ab}^2 + \lambda^2)^{1/2} \approx 0.1$ will therefore develop entanglement that will persist despite environmental fluctuations; otherwise, the entanglement will be fragile and environment sensitive. The design of molecules for chemical quantum qubits should therefore focus on the development of systems near this region of the parameter space. Also of interest are the border zones of the anharmonic region as, even though the maximum entanglement is significantly reduced from the desired value of 1, it persists throughout environmental fluctuations.

2.2.7 Entanglement as a Measure of the Failure of the BO Approximation

In the previous section, entanglement was calculated using a diabatic-state basis set, solving Eq. (2.1) directly, evoking spatially localized detectors that can discriminate between the diabatic states A and B. Such entanglement is, in principle, experimentally measurable. Now, Eq. (2.1) is first transformed (without approximation) into Eq. (2.21), wherein the electronic wavefunctions are represented using a BO basis set. The entanglement in this basis set can be obtained again using Eqs. (2.33)–(2.34), but the quantity thus defined has no physical meaning as the BO basis set is purely a convenient mathematical construct. What it provides is insight into the degree to which the BO approximation breaks down. Normally, the degree of breakdown is accessed with regard to how errant the value predicted by the BO approximation for the observable of interest is, which has manifold variations, whereas the entanglement is a universal measure. It tells how quantum a chemical system is.

Figure 2.7 shows the entanglement of the BO electronic states determined by numerical solution of Eq. (2.21) obtained for the lowest energy vibronic wavefunction and for the wavefunction nearest in energy to the transition state. For the lowest energy vibronic wavefunction, the entanglement becomes appreciable only when the vibrational energy scale exceeds the electronic energy scale. As this is a very rare circumstance, the BO approximation usually performs well in describing

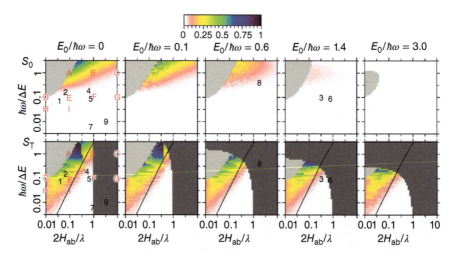

Figure 2.7 Entanglement of the BO states in the exact wavefunction, obtained solving Eq. (2.21), for the ground vibronic state, S_0, and for the vibronic wavefunction of closest energy to the transition state, S_T, as a function of $2|H_{ab}|/\lambda$ and $\dfrac{\hbar\omega}{(4H_{ab}^2+\lambda^2)^{1/2}} = \Delta E$ at various values of $E_0/\hbar\omega$. Light gray – regions for which a stable solution were not obtained. Dark gray – regions for which no transition-state exists (see Eq. (2.11)). **1–9** indicate the molecules listed in Table 2.2 and Figure 2.1, with **A–H** being related to iconic reference systems. The marked line differentiates the weak-coupling scenario, in which the BO approximation fails, from the other regions (see Eqs. (2.28)–(2.30)). Source: McKemmish et al. [31] with permission from the Royal Society of Chemistry.

the properties molecules at zero temperature. When a system is heated sufficiently to allow reactions over the transition-state barrier, the situation changes considerably, however. The figure shows that reactive trajectories will develop significant entanglement whenever the weak-coupling condition, (equivalently represented as Eqs. (2.28)–(2.30)), is satisfied, showing a direct link between entanglement and the breakdown of TST.

2.2.8 Further Reading

The aspects of MV theory and the basic coupled harmonic oscillator model have been widely applied throughout chemistry, and many reviews are available. Some reviews from over the decades include the review of chemical reactivity by van Voorhis et al. [76]; the review pertaining to the equal coupling regime by Demadis et al. [26]; the books by Kuznetsov [77] and by Kuznetsov and Ulstrup [78]; the chemical reviews of Nelsen et al. [79], Richardson [80], Hush [81], and Brunschwig et al. [27, 82]; the MV computational methods review by Kaupp and Parthey [83]; and the Stark spectroscopy review by Bublitz and Boxer [84]. Related work includes the review on pseudo Jahn–Teller theory [32] by Bersuker [85]. Other references include the classic books of Brown [86] and Prassides [87] and the modern book of Duarte and Kamerlin [88]. In addition, some creative modern applications are discussed in this current book.

2.3 Some Traditional Mixed-Valence Example Molecules and Iconic Model Systems

Figure 2.1 shows 10 molecular examples that may be treated using coupled diabatic harmonic oscillators **0–9**. These, as well as eight iconic model systems **A–H**, are used to exemplify the properties of MV compounds and related chemical systems. For each, adiabatic PES, vibrational energy levels and the GS vibrational wavefunctions are displayed in Figure 2.8. Their results are also indicated in Figure 2.3 (symmetric coupling regimes), Figure 2.6 (ground-state entanglement), and Figure 2.7 (BO entanglement). Three parameters are needed to define each system, but, as the previously developed equations indicate, there is no single representation of these parameters that simultaneously depict all properties of interest. Figures 2.3, 2.6–2.8 are presented as a function of the variables $E_0/\hbar\omega$, $2|H_{ab}|/\lambda$, and $\hbar\omega/(4H_{ab}^2 + \lambda^2)^{1/2}$; for **0–9** and **A–H**, these parameters are listed in Table 2.2, along with their various combinations pertinent to the regime-defining equations, Eqs. (2.9)–(2.12), (2.23), (2.28–2.30). These parameters allow each example to be categorized into the regimes described in Figures 2.2 and 2.3 and Table 2.1, with the results given in Table 2.2.

Model systems **A–C** pertain to the situation in which the vibrational energy spacing is as large as the electronic energy features, a somewhat unusual situation. In these cases, the ground-state vibrational wavefunction delocalizes over both valence states, and so, many properties will mimic a simple single chemical species. Nevertheless, the excited state is very close and strongly coupled for **A**, so that many properties will in fact not obey relationships expected for classical chemicals. Next,

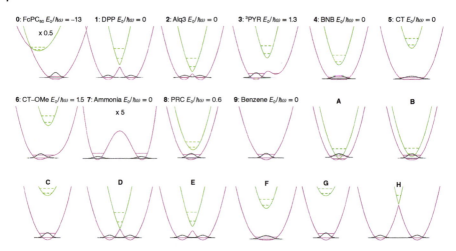

Figure 2.8 Properties of sample compounds **0–9** and reference systems **A–H** (see Figure 2.1 and Tables 2.1 and 2.2). Purple – lower adiabatic BO PES ε_-; green – upper adiabatic BO PES ε_+; purple and green horizontal lines – lowest two vibrational energy levels; and black – vibrational density of the lowest energy vibronic wavefunction.

model systems **D–G** span the range from weak to strong coupling for the common scenario found in many technologically interesting molecules and materials in which the vibrational energy scale is one tenth of the electronic energy scale. Finally, model **H** is typical of many biochemical and technological processes in which charge is transferred slowly (i.e. ms–ns timescale) between deep double wells.

2.3.1 Photochemical Charge Separation

Molecule **0**, FcPC$_{60}$, is a ferrocene–porphyrin–fullerene triad used to model photochemical primary charge separation [89–91]. For this system, MV is generated after photoexcitation, with the ferrocene and C$_{60}$ groups displaying two possible charge states. The ratio of the electronic coupling to the reorganization energy, $2|H_{ab}|/\lambda$, is found to be 0.029 (Table 2.2), which is relatively large given the donor to acceptor separation owing to the conductivity of the bridging porphyrin. Nevertheless, it is very small compared to the energy asymmetry E_0 and also results in the low value of $Q_c = 0.06$. Hence, Figure 2.7 indicates that the rate of primary charge separation is controlled by non-adiabatic effects, as expressed through Eqs. (2.28)–(2.30). As its energy asymmetry is large with $|E_0|/|2H_{ab}| = 67$, this system is classified in Table 2.2 as being in the inverted region. Its observed electron transfer rate is much slower than would be expected based on TST and can only be interpreted in terms of the quantum kinetics equation, Eq. (2.31).

2.3.2 MV Excited States in a Bis–Metal Complex

A typical MV compound is **1**, a dipyridyl polyene DPP with a charge of +5e distributed among two ruthenium atoms, thereby presenting two traditional valence

Table 2.2 Properties of the example compounds **0–9** (see Figure 2.1) and the model systems **A–H** considered [8].

Molecule or model	$\frac{2|H_{ab}|}{\lambda}$	$\frac{\hbar\omega}{\Delta E}$ [a]	$\frac{16H_{ab}^2}{\hbar\omega\lambda}$	$|Q_c|$	$\frac{E_0}{\hbar\omega}$	$\left|\frac{E_0}{2H_{ab}}\right|$	$\frac{2\Delta E^{\ddagger}}{\hbar\omega}$	$\frac{E_H}{E_T}$	$(E_0/\lambda)^{\frac{2}{3}} + (2H_{ab}/\lambda)^{\frac{2}{3}}$	Coupling regime	Entanglement
0 FcPC$_{60}$	0.029	0.15	0.02	0.06	−13	67	0	0.99	1.7	Inverted	Very low
1 DPP	0.043	0.08	0.09	0.11	0	0	5.7	0.99	0.12	Weak	High, fragile
2 Alq$_3$	0.08	0.16	0.16	0.14	0	0	2.6	1.00	0.19	Weak	High
3 ^{3}PYR	0.3	0.095	3.6	0.67	1.3	0.42	3.8	0.99	0.70	Intermediate	Very low
4 BNB[b]	0.74	0.18	10	1.11	0	0	0.15	0.72	0.82	Anharmonic	Medium
5 CT	0.80	0.089	22	1.68	0	0	0.18	0.70	0.86	Anharmonic	Medium
6 CT-OMe	0.80	0.089	22	1.68	1.5	0.21	0	0.90	1.17	Strong	Low
7 NH$_3$	0.80	0.006	330	6.45	0	0	2.6	0.89	0.86	Intermediate	Med., fragile
8 PRC	1.8	0.41	15	1.39	0.6	0.28	0	0.85	2.1	Strong	Low
9 benzene	3.3	0.010	1300	12.6	0	0	0	1.00	2.2	Strong	Very low
A	0.1	1	0.04	0.07	0	0	0.40	1.00	0.22	Weak	Medium
B	1	1	2.8	0.59	0	0	0	0.00	1.00	Anharmonic	Medium
C	10	1	40	2.23	0	0	4.0	1.00	4.6	Strong	Very low
D	0.01	0.1	0.004	0.02	0	0	4.9	0.99	0.05	Weak	High, fragile
E	0.1	0.1	0.4	0.22	0	0	4.0	1.00	0.22	Weak	High, fragile
F	1	0.1	28	1.88	0	0	0	0.00	1.00	Anharmonic	Medium
G	10	0.1	400	7.05	0	0	40	1.00	4.6	Strong	Very low
H	0.01	0.032	0.01	0.04	0	0	16	0.99	0.05	Weak	High, fragile

a) $\Delta E = \left(4H_{ab}^2 + \lambda^2\right)^{\frac{1}{2}}$.
b) Based on the calculations of Stanton [71]; other calculations depict $2|H_{ab}|/\lambda > 1$ [72].
Source: Adapted from Reimers [8].

structures for the description of its ground state [92, 93]. The electronic coupling can be adjusted by varying the length of the bridging polyene, with, at this length, the coupling being only weak, with $2|H_{ab}|/\lambda = 0.043$ and $Q_c = 0.11$ (Table 2.2). Hence, this symmetric system is in the weak-coupling regime (Figure 2.3 and Table 2.2), with a significant double well (Figure 2.8). Hence, any reaction following the preparation of DPP in one of its localized valence states leading to the other valence state must proceed non-adiabatically via Eq. (2.31). This process will embody significant entanglement between the BO states (Figure 2.7). Figure 2.6 shows that this reaction manifests high entanglement between localized diabatic states, but that this entanglement is fragile not expected to persist throughout environmental fluctuations. The interpretation of transition moment data using Eq. (2.13) is central to the experimental property determination discussed herein [93].

2.3.3 Hole Transport in a Molecular Conducting Material

MV compounds can also be arranged into functional materials, and **2**, mer-tris(8-hydroxyquinolinato)aluminum(III) (Alq$_3$), is useful in hole transport applications [94]. For individual hops of charge from one molecule to another, the molecules can adopt one of two charge states and hence mimic MV compounds. For such processes, $2|H_{ab}|/\lambda = 0.08$ and $Q_c = 0.14$ (Table 2.2) remain low, and so, this symmetric system is in the weak-coupling regime (Figure 2.3 and Table 2.2), and the hole transport proceeds non-adiabatically according to Eq. (2.31). Of interest, this system is expected to display significant entanglement in its ground-state vibronic wavefunction that is expected to persist throughout environmental fluctuations (Figure 2.6). Indeed, this is the only one of the examples considered that has both high and persistent entanglement, making it of interest to chemical quantum qubit design. Experimental data analysis for this system involves the interpretation of observed charge mobilities given the extended 3D transport network present in the molecular crystal [94].

2.3.4 Ground-State Delocalization in the Creutz–Taube Ion

One of the most iconic applications of the adiabatic electron transfer theory to MV compounds is the example of the CT ion [6] [Ru(NH$_3$)$_5$ – pyrazine – Ru(NH$_3$)$_5$]$^{5+}$ discussed earlier, **5** (CT). This compound was important in that early interpretations placed it as being in the strong-coupling regime, with the net charge shared equally by both metal atoms, hence depicting a chemical with an unprecedented structure. These observations vindicated the adiabatic electron transfer theories that were developed in the 1950s that allowed charge to be shared in MV compounds. Hush had derived Eq. (2.10) parametrically as a function of the shared electron density, directly implementing second-quantization and density-functional concepts [29], the fundamental underlying concepts of which had been strongly criticized by Marcus [95–97], while Levich and Dogonadze derived the prefactor in Eq. (2.31) using analogous wavefunction-based theory. Evidence supporting its delocalized nature of CT includes the symmetry found in its X-ray structure [98, 99] and infrared spectrum [100–103] with no tunneling vibrations apparent [104], results from Mössbauer and electron paramagnetic spectroscopy (EPR) [102], as well as the sharpness of its electronic absorption band [105], the effect of chemical variation on it [106], and its Stark spectrum [107]. Nevertheless, some Raman spectra can only be interpreted in terms of localized charges [108, 109].

These and other seemingly conflicting experimental results led Demadis, Hartshorn, and Meyer [26] to propose a new Robin–Day-like classification "Class II–III" to account for molecules with unusual properties, the characteristic feature being "In Class II–III, the solvent is averaged and the exchanging electron is localized." A classical way in which this could happen is if the reaction rate for the intramolecular interconversion of the two localized isomers over their interconnecting transition state is much faster than the timescale for solvent reorganization. How this could arise physically, for the case of the Creutz–Taube ion, is not clear, however, as the minimum possible reaction lifetime according to TST is $k_B/h = 0.16$ ps, whereas strongly coupled solvent dynamics in water occurs throughout the 0.05–1.5 ps range and the intramolecular vibrational periods of interest are of order 0.08 ps. Our interpretation is that CT presents a double-welled potential that does not support zero-point vibration [110, 111], thus providing an example of a molecule in the anharmonic regime (Table 2.2). Such a system cannot be considered in terms of rapid chemical reactions between localized structures but instead must be considered as a non-classical structure in which the electron of interest is delocalized across the double-welled potential energy surface. In either case, solvent fluctuations control the observed properties, with the distribution of the total solvent reorganization energy among modes of different frequencies being critical [110]. Hence, observed properties may show variations between those expected for intermediate coupling and for strong coupling, as dictated by their nature. Table 2.2 lists $2|H_{ab}|/\lambda = 0.80$, $\hbar\omega/(4H_{ab}^2 + \lambda^2)^{1/2} = 0.089$, giving $2\Delta E^{\ddagger}/\hbar\omega = 0.18$, indicating a shallow double well (Figure 2.8), but, as $Q_c = 1.68$, the properties of this double-well are not expected to be dominated by non-adiabatic effects (Figure 2.7). Of interest, its ground state generates a significant amount of entanglement that is not fragile (Figure 2.6).

In the anharmonic regime, vibrational quantum effects can dominate properties, with such effects often depending significantly on mode frequency. As many contributions arise to the reorganization energy for CT, and as these have varying frequency scales, the use of a single-mode model like that depicted by Eq. (2.2) may be inappropriate. Hence, modeling of the properties of the CT ion requires full numerical simulations involving both intramolecular and intermolecular degrees of freedom [110]. Of note, **6** involves a peripheral chemical substitution to CT, making CT-OMe, with parameters differing by only the introduction of an energy asymmetry $E_0/\hbar\omega=1.5$ (Table 2.2). This seemingly small change has a profound impact on properties [112], as is anticipated by the coupled harmonic oscillator model, transferring it into the strong-coupling regime (Table 2.2).

2.3.5 Photochemical Charge Separation During Bacterial Photosynthesis

The last molecular system from Table 2.2 and Figure 2.1 that can readily be interpreted as being akin to a MV compound is the bacteriochlorophyll dimer radical cation from the bacterial photosynthetic reaction center (PRC) of *Rhodobacter sphaeroides* [113, 114]. Following irradiation and exciton transport to this dimer, an electron is ejected to form the radical cation shown in Figure 2.1. Formally, the induced charge may localize on either of the two bacteriochlorophylls, making an effective MV system. Analysis of its electrochemical potential, spin distribution, and absorption spectrum leads to the conclusion that it presents strong coupling and a single-welled GS PES (Figure 2.8), with $2|H_{ab}|/\lambda = 1.8$ and $E_0/\hbar\omega = 0.6$, leading to $Q_c = 1.39$ and $|E_0/2H_{ab}| = 0.28$ (Table 2.2). A key tool leading to this analysis was site-directed mutagenesis, in which neighboring residues are mutated so as to modify the internal electric field felt by the dimer and hence E_0 via Eq. (2.19). The total charge densities ρ_L and ρ_M located on each bacteriochlorophyll molecules (named after the L and M protein chains) are related by [115]

$$\frac{\rho_L}{\rho_M} - \left(\frac{\rho_M}{\rho_L}\right)^{\frac{1}{2}} = \frac{E_0}{|H_{ab}|} \tag{2.40}$$

and, given the GS energy stabilization induced by resonance, the electrochemical potential E_m of the dimer can be correlated with the spin density as

$$FE_m = \text{constant} - |H_{ab}|\left(\frac{\rho_M}{\rho_L}\right)^{1/2} \tag{2.41}$$

where F is the Faraday. Hence, by plotting electrochemical potential vs. the spin density ratio obtained for various mutant reaction centers, the resonance coupling between the special pair can be determined, as shown in Figure 2.9.

Bacteriochlorophyll molecules contain large macrocyclic rings throughout which the charge distributes, leading to very low reorganization energies for charge transport. This property makes them ideal for use in light-harvesting systems as the reorganization energy λ is usually lost as heat during charge transport processes, wasting the absorbed energy. As Figure 2.1 shows, the rings overlap poorly within the dimer, meaning that the resonance energy H_{ab} is also relatively low, resulting in $2|H_{ab}|/\lambda = 1.8$. Hence, the electronic transition energy ΔE_v is also very low, and the transition is in fact observed centered amidst the infrared absorption lines at 0.33 eV (2700 cm^{-1}). This results in strong intensification of the infrared absorption lines associated with vibrations coupled to the reaction coordinate Q. Indeed, three very intense infrared lines are observed in the spectrum of the special pair radical cation that has no counterparts observed in the spectra of bacteriochlorophyll, its cation, or its dimer, being essentially the ring-breathing modes of pseudo g symmetry that are not normally expected to manifest in infrared spectra. These lines are known as *phase phonon* lines [117], and their observation provides direct evidence supporting the unusual quantum properties predicted by the coupled harmonic oscillator model [113]. Hence, despite being in the strong-coupling regime, effects pertaining to the

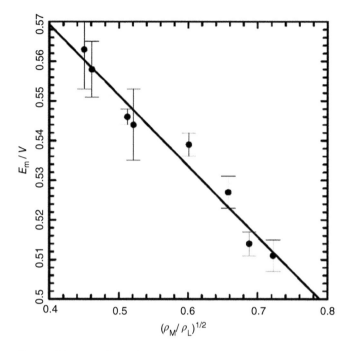

Figure 2.9 The effect of site-directed mutagenesis on the observed [116] electrochemical midpoint potentials E_m and the ratio ρ_M/ρ_L of the spin densities on the two bacteriochlorophyll molecules that comprise the special pair involved in photochemical charge separation in the PRC of *Rhodobacter sphaeroides*. Source: Reproduced from Reimers et al. [115] with permission of the American Chemical Society.

breakdown of the BO approximation can be observed, although they would most commonly be interpreted as an extreme example of the Herzberg–Teller effect [118].

The special pair radical cation also supports low-energy excitations from the SHOMO orbital of each bacteriochlorophyll into the half-occupied HOMO orbital, and these transitions overlap with the MV optical excitation [119]. Simulation of the intervalence spectrum therefore requires inclusion of these transitions as well. This was done by expanding the traditional two diabatic-state model to include four coupled diabatic states. To account for all the associated vibrational changes, a 70-mode model was developed, with most of the parameters in it determined using DFT calculations [120]. Then, seven free parameters were fitted to a wide range of spectroscopic, electrochemical, and spin density data for the wild-type reaction center [113]. This model was then able to predict property variations for 29 mutant reaction centers [114], forming a comprehensive understanding of the operation of the reaction center and how its properties have been optimized by evolution. This included interpretation of the observed Stark spectrum of the wild type [114], as shown in Figure 2.10. Subsequently, predictions were made for the Stark spectra of three mutant reaction centers, and these predictions, which included profound spectral changes, were later found to agree very well with observations, as shown in the figure [121].

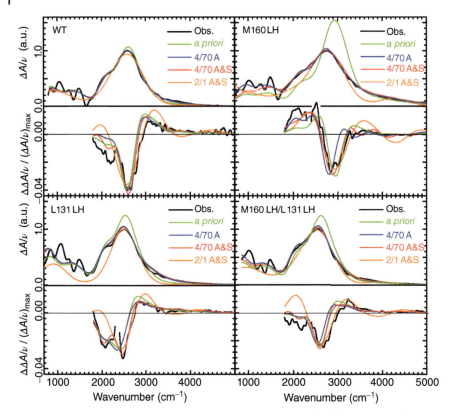

Figure 2.10 For the wild-type (WT) and M160LH, L131LH, and M16LH/L131LH mutants of the *Rhodobacter sphaeroides*, PRC shows the observed light minus dark (cation minus neutral) difference spectra ΔA and Stark spectra $\Delta\Delta A$ (black) of the special pair radical cation [121]. These are compared with predicted spectra based on a previous a priori analysis [122] of the WT (green) based on a 4-state 70-mode coupled diabatic model. In addition, three fits to the observed spectra are shown in orange, red, and blue based on various spectral models. The spectra are displayed weighted by frequency $v = 2\pi\omega$. Source: Reproduced from Kanchanawong et al. [121] with permission of the American Chemical Society.

Mostly, biology operates according to classical equations of motion describing tunneling-free nuclear motion on mass-independent BO potential energy surfaces. Standard MD simulations can, in principle, reproduce most phenomena and indeed conform to this description. Common exceptions involve quantum effects such as zero-point energy and vibrational tunneling, effects that still require only a single BO surface in order to manifest. The operation of non-adiabatic coupling introduces non-trivial quantum effects, with no single PES being prescriptive. That the deduced value of Q_c for the PRC is 1.39 (Table 2.2) suggests that some adiabatic scheme based on the vibrational tunneling within the BO approximation could account for most observations, as indeed is vindicated by the correlation shown in Figure 2.9. However, as the generic example in Figure 2.4 demonstrates, no such scheme could account for the details of the absorption spectrum of the PRC or explain its Stark spectroscopy. Primary charge separation in bacterial photosynthesis, a process

2.3.6 Prussian Blue

Prussian blue is an iconic MV material associated with the critical discoveries that established MV compounds as an important research field [2, 3, 5]. Synthesized by Diesback in 1706, it was the word's first synthetic dye, with 12 000 tonnes synthesized annually [123]. Currently, it has many technological and medicinal uses, most of which rely on its optical properties, as well as being iconic in purely scientific research [124, 125]. It is categorized as a coordination polymer in which Fe(II) and Fe(III) ions are bound in a lattice by coordinating cyanide bridges, but if cyanide is re-categorized as an organic ligand rather than an inorganic one, then Prussian blue would become the iconic metal–organic framework solid. It is the unexpectedness of its intense absorption in the red centered at c. 1.7 eV, a transition not present in analogous pure Fe(II) or Fe(III) materials, that is attributed to the intervalence transition in the MV material.

Figure 2.1 and Table 2.2 do not feature Prussian blue; however, as even 60 years after the critical discoveries were made as to the MV nature of the transition, modern quantitative analysis has not been performed. The initial characterizations were based on the then known X-ray structure of Prussian blue [126], which identified the composition $Fe_2(CN)_6 \cdot K \cdot nH_2O$, with octahedral coordination of CN around alternating Fe(II) and Fe(III) ions, with water and potassium ions located in the framework's cavities. While the revealed Fe(II)–CN–Fe(III) structural arrangement that is central to the initial analyses is a robust feature, modern interpretations [125, 127–130] consider the material as embodying the basic form $Fe_3^{2+}Fe_4^{3+}(CN^-)_{18}(H_2O)_6 \cdot mKOH \cdot nH_2O$, and it is this framework that is shown in Figure 2.1. In this framework, one in four Fe(II) ions are missing from the lattice without long-range order, with water molecules entering the inner coordination sphere of some of the Fe(III) ions. Hence, many inner-coordination environments are present for the Fe(II)–CN–Fe(III) units, each of which would be naively expected to generate a different excitation energy. In addition, Prussian blue is a coordination polymer, so that the iron orbitals that contribute to an intervalence transition across one cyanide ligand also contribute to transitions across different ligands, raising the possibility of strong exciton coupling within the material. Finally, its composition and spectral properties are sensitive to environment and preparation conditions [125]. The detailed identity of the chromophore responsible for the intervalence transition therefore remains unknown. Nevertheless, the band intensity analysis performed by Hush [3] is expected to remain robust, placing Prussian blue as a MV compound in the intermediate-coupling regime.

2.4 Applications to Stereoisomerism

Stereoisomers may form in many ways and often can be interconverted by unimolecular reaction mechanisms that simply change one geometrical structure to another. Isomerization of MV compounds indeed provides an example of this process.

Generally, isomers can take on localized structures akin to the localized diabatic states sketched in Figure 2.2 that are interconvertible by concerted unimolecular reaction mechanisms like that envisaged for MV compounds. A classic example of this is pyramidal inversion of ammonia **7**. Also, aromaticity in benzene **9** is a direct analog of delocalization in MV compounds. Hence, the infrastructure built to understand MV compounds applies, in principle, to stereoisomerism in general. Recognition of this comes through the "twin state" concepts developed by Shaik et al. for aromatic molecules [131–135] that apply the rhetoric developed to describe MV compounds to qualitatively understand aromatic molecules.

Isomerization of MV compounds usually involves just the rearrangement of a single electron or hole, whereas many electrons can be involved in general. Hydrogen bonding rearrangements and ammonia inversion involve the rearrangement of an electron lone pair, while aromaticity involves at least the rearrangement of the four electrons in the degenerate or near-degenerate HOMO and SHOMO orbitals among the degenerate or near-degenerate LUMO and SLUMO orbitals [7]. For general reactions, the diabatic basis states used must be able to accommodate rearrangements of all of the involved electrons, meaning that in general, more than two diabatic states will be required in the description. It is therefore necessary to either adapt or generalize Eq. (2.1) to forms appropriate for each stereoisomerization of interest.

2.4.1 Breakdown of Aromaticity in the (π,π^*) $^3A_1$ Triplet Ground State of Pyridine

Pyridine **3** is an aromatic molecule conceived by replacement in benzene of CH with N (Figure 2.1). As discussed later for benzene, aromaticity arises through strong coupling between localized states, known as Kekulé states, involving alternating C—C (or C—N) single and double bonds. The properties of such compounds are controlled by the functionalities of the four electrons in the near-degenerate HOMO and SHOMO orbitals and their excitations into the near-degenerate LUMO and SLUMO orbitals. The lowering of symmetry from benzene to pyridine splits these paired orbitals in a similar way, with the higher energy orbitals having a_2 symmetry and the lower energy ones having b_1 symmetry [136]. As a result, the $a_1 \rightarrow a_1$ and $b_2 \rightarrow b_2$ triplet single excitations have similar energy but become decoupled from all other single and higher excitations, resulting in an effective two-state model. The situation is therefore similar to the asymmetric MV examples sketched in Figure 2.2. For asymmetric reactions, the symmetry does not change during displacement along the reaction coordinate Q, and hence, this motion is totally symmetric, a_1 symmetry in this case; in effect, the triplet excitation induces the Kekulé distortion. Fitted to calculated and observed data, the deduced parameters for the $^3A_1$ state of pyridine **3** are $2|H_{ab}|/\lambda = 0.30$, $\hbar\omega/(4H_{ab}^2 + \lambda^2)^{1/2} = 0.095$, and $E_0/\hbar\omega = 1.3$ (Table 2.2), leading to $Q_c = 0.67$ (the coupling is not weak), $(E_0/\lambda)^{\frac{2}{3}} + (2H_{ab}/\lambda)^{\frac{2}{3}} = 0.70$ (the coupling is neither strong nor inverted), and $2\Delta E^{\ddagger}/\hbar\omega$ with $E_H/E_T = 0.99$ (outside of the anharmonic regime); hence, this system is in the intermediate-coupling regime. Nevertheless, it is near the borders to both the weak-coupling and inverted regimes so that non-adiabatic effects may manifest in its spectroscopic and kinetic properties. Figure 2.6 shows that its vibrational ground state develops very little quantum entanglement.

2.4.2 Isomerism of BNB

The BNB molecule 4 is a free radical in which the radical may localize on either of the two boron atoms, and its electronic structure can be written in terms of resonance such as ·B⁺ − Ṅ⁻ − B : and ·B = N⁺ = B:⁻. If only one type of resonance structure is relevant, then the two resulting diabatic configurations take the same form as those in a MV complex in which boron is the "metal." As the electron transfer affects other bonds between the B and N atoms as well, a manifold of diabatic states is required to properly describe the bonding. Nevertheless, a crude mapping onto a two-state model has been shown to be effective in describing critically observed properties [71], and this approach is taken herein. Analysis [9] of high-level calculations [71] that predict charge localization yields $2|H_{ab}|/\lambda = 0.74$ and $\hbar\omega/(4H_{ab}^2 + \lambda^2)^{1/2} = 0.18$ so that $2\Delta E^{\ddagger}/\hbar\omega = 0.15$ (Table 2.2). Hence, these calculations predict that BNB displays a shallow well in the anharmonic regime, resulting in its unusual properties [71]. Entanglement in its ground-state BO vibronic wavefunction (Figure 2.6) is of a medium level but is persistent. The BO approximation shows little entanglement (Figure 2.7), indicating that quantum effects are confined to be within the vibrational analysis. The actual nature of BNB remains under discussion, however, with recent calculations predicting a single well (i.e. $2|H_{ab}|/\lambda > 1$), also claiming that such results are consistent with its observed unusual spectroscopic properties [72].

2.4.3 Isomerism of Ammonia and Related Molecules

There is no direct analogy of the localized diabatic states presented in Figure 2.2 to those for pertaining to the ammonia inversion reaction, but diabatic states are not unique and so can be transformed into many equivalent representations. A representation that is especially useful is the delocalized diabatic description in which the electronic basis states for A and B in Eq. (2.1) are rotated uniformly by 45° [23]. This produces a "spectroscopic" description of the problem in terms of equivalent harmonic oscillators representing the GS and ES of the system, providing in fact a pertinent zeroth-order model for the BO adiabatic surfaces shown in Figure 2.2 in the strong-coupling limit. In MV compounds, these surfaces differ by the transfer of a single electron. In situations with multiple electrons involved in the isomerization process, the delocalized diabatic states can be generalized into a ladder, with each step in the ladder representing the transfer of one electron. Once the complete Hamiltonian is constructed in this basis, it can be back-transformed to yield the required number (in this case, three) of localized diabatic surfaces akin to A and B [7]. The extended localized diabatic Hamiltonian is then identified as being [7]

$$\mathbf{H}^{loc} = \begin{bmatrix} \frac{\hbar\omega}{2}(Q + \sqrt{2}Q_m)^2 + T & -\sqrt{2}H_{ab} & 0 \\ -\sqrt{2}H_{ab} & \frac{\hbar\omega}{2}Q^2 + T & -\sqrt{2}H_{ab} \\ 0 & \sqrt{2}H_{ab} & \frac{\hbar\omega}{2}(Q + \sqrt{2}Q_m)^2 + T \end{bmatrix} \quad (2.42)$$

Figure 2.11 shows the surfaces that result when this approach is applied to describe the isomerization of NH_3, showing the above three localized diabatic states L, C, and R; the equivalent three delocalized diabatic states G, S, and D (named for the ground state, singly excited state, and doubly excited state); and their associated adiabatic PES g, s, and d obtained by applying the BO approximation to Eq. (2.42). The L and R diabatic surfaces correspond to the A and B diabatic surfaces for mixed-valence compounds, but their displacement is rescaled by a factor of $2^{1/2}$, while the C (central) state is new and involves no displacement. The electronic couplings are both phased and scaled with respect to those in MV compounds. The surfaces in the figure were obtained [7] by fitting the observed BO GS potential as deduced by Swalen and Ibers [137].

This diabatic description of ammonia inversion allows the process to be catalogued, as molecule 7 in Table 2.2 and Figure 2.1, alongside MV and analogous compounds, with $2|H_{ab}|/\lambda = 0.80$ and $\hbar\omega/(4H_{ab}^2 + \lambda^2)^{1/2} = 0.006$, making $Q_c = 6.45$, $2\Delta E^{\ddagger}/\hbar\omega = 2.6$, and $E_H/E_T = 0.89$. This locates ammonia inversion as a process with intermediate coupling (Figure 2.8) that is very well described using the BO approximation (Figure 2.7). Nevertheless, it is close to the anharmonic region, and so, some quantum vibrational properties can be expected, leading to the ammonia maser. Stereoisomerization of ammonia was the first reaction proposed for use as a chemical qubit [68], but in Figure 2.6, it is seen to develop only a medium level of entanglement and also sit very close to the region in which persistent entanglement disappears. Hence, ammonia itself may prove unsuitable, but there may be related molecules that could display better performance.

The inversion reaction in the analogous XH_3 molecules PH_3, AsH_3, SbH_3, and BiH_3 can also be investigated, with good fits obtained to the ground-state, singly excited state, and doubly excited state PES, evaluated using high-level computational methods, over a wide range of the torsional angles [138]. This required expansion of the Hamiltonian model Eq. (2.42) to include anharmonicity and small variations of the basic parameters between states. In the strong-coupling regime, the molecule would be planar with HXH bond angles of 120°. Figure 2.12 shows the calculated angles, which start at 106° for NH_3 and drop sharply to 94° for PH_3 and then slowly to 91° for BiH_3. The discontinuous change between the first- and second-row elements is typical of chemical properties throughout the periodic table. From the diabatic model fit to the wide-range computed data, values for $2|H_{ab}|$ and λ are also shown in the figure, along with their ratio. The ratio shows also a sharp decline, from 0.88 for NH_3 to 0.47 for PH_3, thence decreasing slowly to 0.22 for BiH_3, explaining the calculated change in the equilibrium bond angle of the GS PES. Most significantly, Figure 2.12 shows that it is the variation in the resonance energy rather than the reorganization energy that produces the unexpected properties of the first-row element [138].

Examining the computational results [138], the sharp decline in the resonance energy between NH_3 and PH_3 can be understood by looking at the relative energies of the σ^* NH orbital and the lowest lying Rydberg orbital, illustrated in Figure 2.12

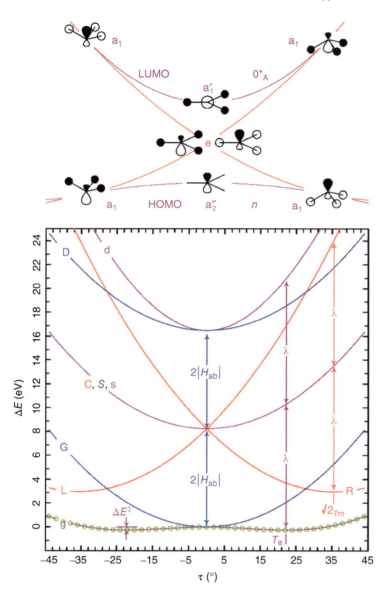

Figure 2.11 The lower figure shows the GS adiabatic PES determined by Swalen and Ibers [137] as a function of the improper torsional angle τ of NH_3 (green circles), fitted by a three-state three-parameter diabatic potential model ($|H_{ab}| = 4.13$ eV, $\lambda = 10.55$ eV, diabatic torsional angle = 25.1°) to an RMS error 1.2 meV: purple – resulting adiabatic ground-state "g," singly valence-excited state "s," and doubly valence-excited state "d"; red – localized representation of the "L," "C," and "R" diabatic potentials; blue – delocalized representation of the diabatic potentials "G," "S," and "D," The upper figure shows the adiabatic Walsh diagram for the HOMO and valence LUMO orbitals that generate these transitions (purple) and the associated localized diabatic orbitals (red). Source: Reimers et al. [7] with permission from the Royal Society of Chemistry.

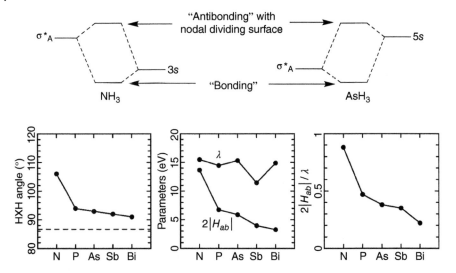

Figure 2.12 Interpretation [138] of the change in HXH bond angle in the XH$_3$ series (left) in terms of the ratio (right) of twice the resonance energy H_{ab} to the reorganization energy λ (center). The dashed line shows the bond angle of 86.7° expected for zero resonance. The top illustration sketches the relative ordering of the sigma antibonding orbital and the lowest lying Rydberg orbital, highlighting their inversion in order found for first-row elements. Source: Adapted from Reimers et al. [138] with permission from the Royal Society of Chemistry.

for NH$_3$ and AsH$_3$. The resonance interaction determines the relative energies of the bonding and antibonding orbitals, and so, its apparent value is determined by the global properties of the electronic structure, including any interactions that these orbitals may have with others. For first-row elements such as nitrogen, Rydberg orbitals are usually lower in energy than σ* orbitals, whereas for later rows, the situation reverses. As the space occupied by the lowest Rydberg orbitals overlaps with that occupied by the antibonding orbitals, there is a strong interaction, resulting in orbital pairs of mixed characteristics whose nature then affects the occupied orbitals through the orthogonality condition. Of the interacting pair, the lower energy one will adopt "bonding" characteristic in which the electrons delocalize over the overlapping spatial region, whereas the higher energy one will develop "antibonding" characteristic with a nodal surface dividing the interaction region. In the double excited state whose properties manifest the resonance energy H_{ab}, the key electrons are compressed together by the nodal surface in the case of NH$_3$, whereas they spread apart for PH$_3$–BiH$_3$ as there is no node. This sharp, qualitative change in orbital characteristic is well known and was described by Mulliken as *Rydbergization* [139, 140]. It acts as a binary switch that controls the resonance energy and hence the HXH bond angle, explaining the observed discontinuous property changes [138]. The influential work of Kutzelnigg [141] that discusses the difference between first- and higher row elements recognizes that the valence orbitals are much more extended in higher row elements but reason that this is caused by an effect of the occupied core p orbitals, whereas in effect, it is the differing properties of the Rydberg orbitals that control the size effect.

The coupled diabatic model for ammonia inversion has, as one of its properties, the HXH angle expected when the resonance energy is zero. If the trend seen in Figure 2.12 continues, then compounds of group XV elements heavier than bismuth should manifest a continued decrease in $2|H_{ab}|/\lambda$, and hence, the adiabatic angles should approach this limit, although the developing criticality of relativistic effects could change this scenario. The theory predicts a diabatic angle of [138]

$$\frac{1}{2}\left[a\cos\left(\sqrt{2}a\tan\frac{1}{2}\right) - 1\right] = 86.7° \qquad (2.43)$$

For the analogous cations $[XH_3]^+$, only one electron is involved in the isomerization process so that the situations become directly analogous to MV chemistry. Hence, Eq. (2.42) becomes replaced by Eq. (2.1), with most noticeably the loss of the scaling factors of $2^{1/2}$ seen in Eq. (2.42), reducing the energy scale. Another significant change is that the diabatic bond angle depicting the situation for zero resonance increases to

$$a\cos\frac{-1}{5} = 101.5° \qquad (2.44)$$

and the calculated BO potential energy surfaces display these modified basic properties [138]. Simple models, based on the premise that the reduction in hybridization that occurs down a column in the periodic table controls molecular geometries [141], anticipate an asymptotic bond angle near 90° but do not readily yield the corresponding angle for the cations.

2.4.4 Proton Transfer in $[NH_3 \cdot H \cdot NH_3]^+$

Proton transfer reactions have since 1935 been treated using coupled diabatic models [20, 142, 143], with important related problems including reactions with proton transfer coupled to electron transfer [144–146]. A simple proton transfer reaction is that between $[NH_4]^+$ and NH_3 [7]. Herein, this reaction is considered in the gas phase, but it serves as a paradigm for very many biochemical processes [147]. Proton transfer intrinsically involves the four nitrogen lone pair electrons, but calculations indicate that the rearrangement of two electrons within two orbitals dominates the reaction, making this system treatable using the same diabatic approach that was developed to consider ammonia inversion. The Hamiltonian model of Eq. (2.42) interprets the properties of three BO PES, with calculations revealing similar properties to those anticipated. Nevertheless, quantitative analysis demands variation of the two sets of electronic couplings revealed in the Hamiltonian, and so, the model is considered to be qualitatively descriptive but not quantitatively accurate, and hence, this system is not featured among the example systems. In summary, high-level calculations suggest that in the gas phase, this dimer has $2|H_{ab}|/\lambda = 1.05$ and thus features a delocalized proton in the anharmonic regime, explaining unusual observed spectroscopic features [7], but also makes the computation of these properties very sensitive to computational method.

2.4.5 Aromaticity in Benzene

For understanding aromaticity arising from delocalization of the electronic ground state of benzene over the two possible Kekulé diabatic surfaces, a minimum of five states is required [7]. In the delocalized diabatic representation, these correspond to the ground state, singly excited state, a component of the doubly excited state, the triply excited state, and the quadruply excited state. The situation concerning the lowest triplet state of pyridine considered earlier differs in that the asymmetry induced by nitrogen substitution breaks most of the resonances, allowing for a direct two-state diabatic analysis. The full localized diabatic Hamiltonian for benzene is [7]

$$\mathbf{H}^{\text{loc}} = \begin{bmatrix} V(-\sqrt{3}, -3) + T & \left(1 + \frac{2}{\sqrt{3}}\right) H_{ab} & 0 & \left(1 - \frac{2}{\sqrt{3}}\right) H_{ab} & 0 \\ \left(1 + \frac{2}{\sqrt{3}}\right) H_{ab} & V(-1, -1) + T & \frac{-4}{\sqrt{3}} H_{ab} & 0 & \left(-1 + \frac{2}{\sqrt{3}}\right) H_{ab} \\ 0 & \frac{-4}{\sqrt{3}} H_{ab} & V(0,0) + T & \frac{4}{\sqrt{3}} H_{ab} & 0 \\ \left(1 - \frac{2}{\sqrt{3}}\right) H_{ab} & 0 & \frac{4}{\sqrt{3}} H_{ab} & V(1, -1) + T & \left(-1 - \frac{2}{\sqrt{3}}\right) H_{ab} \\ 0 & \left(-1 + \frac{2}{\sqrt{3}}\right) H_{ab} & 0 & \left(-1 - \frac{2}{\sqrt{3}}\right) H_{ab} & V(\sqrt{3}, -3) + T \end{bmatrix} \tag{2.45}$$

where

$$V(x, y) = \frac{\hbar\omega}{2}(Q - xQ_m)^2 + \frac{y\lambda}{4} \tag{2.46}$$

Figure 2.13 shows these surfaces (the situation when $J = 0$), depicting parabolas centered at five different locations, plus solutions for the intermediate-coupling,

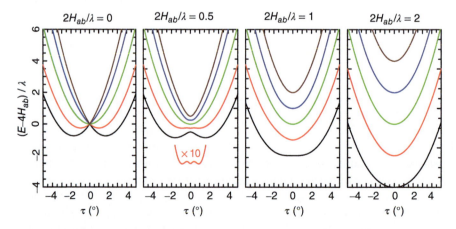

Figure 2.13 Sample BO PES (black – g, red – s, green – d, blue – t, and brown – q) from the five-state model for the Kekulé distortion of benzene, as a function of the center to carbon angular distortion ϕ obtained from Eq. (2.45). Source: Reimers et al. [7] with permission of the Royal Society of Chemistry.

equal-coupling, and strong-coupling regimes corresponding to $2|H_{ab}|/\lambda = 0.1, 1$, and 2, respectively. The geometrical variable ϕ utilized in this figure is the change in the angle of a carbon atom with respect to the ring center as a Kekulé structure is approached [7]. The different BO PESs show different inherent curvatures, and indeed, the properties of all of these excited states are linked. This explains the observations that the ground state and first excited state of benzene have connected properties, leading to the "twin state" concept within aromatic molecules [131–135], but the state that can be paired to the GS using a direct analogy to Eq. (2.4) is in fact the quadruply excited state.

Fitted to the known properties of benzene **9**, Table 2.2 lists parameters of $2|H_{ab}|/\lambda = 3.3$ and $\hbar\omega/(4H_{ab}^2 + \lambda^2)^{1/2} = 0.010$ so that $E_H/E_T = 1.00$, placing this system in the strong-coupling regime; Figure 2.8 indicates a delocalized GS, Figure 2.6 indicates minimal quantum entanglement, and Figure 2.7 indicates that the BO approximation is expected to work very well. The value of the angle ϕ at the diabatic minimum of the lowest energy localized diabatic surface is, from Eq. (2.45), equal to $3^{1/2}$ times the angle ϕ_m that would be appropriate if benzene was not cyclic. The value from the fit to observed benzene electronic spectroscopic properties hence yields $\phi_m = 1.29°$, close to the value of 1.1° used in Kekulé structure geometrical models [148]. To reproduce standard bond lengths of isolated carbon single and double bonds, an angle of $\phi_m = 2.6°$ is required, which is substantially different from the values for conjugated systems, as typified here by benzene.

Traditional analyses of the properties of benzene and other "aromatic" or "antiaromatic" molecules [24, 25] focus on the concept of "resonance" within the π system as a controlling effect [149]. Modern approaches recognize the critical importance of the σ system as well [150–153], often focusing on the interpretation of formation energies [154]. The approach reviewed herein acknowledges the important role of the π system in that the diabatic states used explicitly depict π excitation. Yet, these states fully embody the effects of the σ system on the reorganization energy and the resonance energy. It therefore provides the simplest possible description of the critical chemical features of interest.

2.5 Conclusion and Outlook

The field of MV complexes, and associated fields involving electron transfer reactions, sits at the core of many modern technological and biochemical applications. Much remains to be discovered, however, even for iconic systems such as Prussian blue that were instrumental in making early critical discoveries. New applications are constantly being developed and are in need of simple qualitative explanations, basic semi-quantitative ones, and exhaustive quantitative modeling.

Herein, the coupled harmonic oscillator model from adiabatic electron transfer theory is expounded, listing some commonly applied analytical results that the model delivers, with many other results also being available. Observed properties discussed include equilibrium atomic structures, ground-state and excited-state vibrational properties, electronic spectroscopy, electrochemical potentials, atomic charge distributions, Stark spectroscopy, the effect of protein mutagenesis, and chemical kinetics. The basic model provides a qualitative framework within which observed results can be discussed, while the equations provide the means for enabling semi-quantitative analysis. Novel in this review is the classification of systems at the boundaries of traditional analyses, identifying a fifth "anharmonic" regime added to the traditional "weak-coupling," "intermediate-coupling," "strong-coupling," and "inverted" regimes.

The extensions presented move far away from electron transfer reactions and MV complexes to provide a conceptual basis for many unimolecular stereoisomerization reactions. The variety of chemical systems described indicates that the conceptual basis developed to explain common phenomena of MV complexes, which challenge basic chemical reasoning, is widely applicable. Indeed, aromaticity in benzene is a similar phenomenon to the non-classical structures found for MV compounds, and herein, these fields become united. Previously, parallels between the semi-quantitative analyses developed for MV compounds and observed features of aromatic compounds was known, but the material reviewed herein provides the ability to make semi-quantitative analyses of aromaticity that manifest the perceived parallels. That hydrogen bonding properties are analogous to MV properties has been long known, and again, we see now how these relationships can be made semi-quantitative. To date, the greatest success found by considering general unimolecular isomerization reactions in MV terms comes from the application to the inversion reaction in the XH_3 series, with identification of the processes responsible for resonance leading to a solution to the century-old riddle as to why the properties of the first row of the periodic table differ so much from the properties of the later rows.

Despite 65 years of development, it would seem that many applications of the basic theory are yet to be developed. New developments will be needed to interpret ever new experiments, as well as to interpret well-known basic chemical properties in new ways.

References

1 Day, P., Hush, N.S., and Clark, R.J.H. (2008). Mixed valence: origins and developments. *Phil. Trans. Roy. Soc. A* 366: 5–14.
2 Robin, M.B. and Day, P. (1967). Mixed valence chemistry-a survey and classification. *Adv. Inorg. Chem. Radiochem.* 10: 247–422.
3 Hush, N.S. (1967). Intervalence-transfer absorption. II. Theoretical considerations and spectroscopic data. *Prog. Inorg. Chem.* 8: 391–444.
4 Hush, N.S. (1975). Inequivalent XPS [x-ray photoelectron spectroscopy] binding energies in symmetrical delocalized mixed-valence complexes. *Chem. Phys.* 10: 361.
5 Robin, M.B. (1962). Color and electronic configuration of Prussian blue. *Inorg. Chem.* 1: 337.
6 Creutz, C. and Taube, H. (1969). Direct approach to measuring the Franck-Condon barrier to electron transfer between metal ions. *J. Am. Chem. Soc.* 91: 3988–3989.

7 Reimers, J.R., McKemmish, L., McKenzie, R.H., and Hush, N.S. (2015). A unified diabatic description for electron transfer reactions, isomerization reactions, proton transfer reactions, and aromaticity. *Phys. Chem. Chem. Phys.* 17: 24598–25617.

8 Reimers, J.R., McKemmish, L., McKenzie, R.H., and Hush, N.S. (2015). Non-adiabatic effects in thermochemistry, spectroscopy and kinetics: the general importance of all three Born-Oppenheimer breakdown corrections. *Phys. Chem. Chem. Phys.* 17: 24641–24665.

9 McKemmish, L.K., McKenzie, R.H., Hush, N.S., and Reimers, J.R. (2011). Quantum entanglement between electronic and vibrational degrees of freedom in molecules. *J. Chem. Phys.* 135: 244110.

10 Born, M. and Oppenheimer, R. (1927). Zur Quantentheorie der Molekeln. On the quantum theory of molecules. *Ann. Phys.* 389: 457–484.

11 Hund, F. (1927). Zur Deutung der Molekelspektren. I. *Zeitschrift für Physik* 40: 742–764.

12 Heitler, W. and London, F. (1927). *Z. Angew. Phys.* 44: 455.

13 von Neumann, J. and Wigner, E. (1929). On the behaviour of eigenvalues in adiabatic processes. *Phys. Z.* 30: 467. translated in 2000 Quantum Chemistry 2008 2025-2031.

14 Evans, M.G. and Polanyi, M. (1935). Some applications of the transition state method to the calculation of reaction velocities, especially in solution. *Trans. Faraday Soc.* 31: 875–894.

15 Eyring, H. (1935). The activated complex in chemical reactions. *J. Chem. Phys.* 3: 107–115.

16 Eyring, H. (1938). The theory of absolute reaction rates. *Trans. Faraday Soc.* 34: 0041–0048.

17 London, F. (1932). On the theory of non-adiabatic chemical reactions. *Z. Angew. Phys.* 74: 143.

18 Morse, P.M. (1929). Diatomic molecules according to the wave mechanics II. Vibrational levels. *Physiol. Rev.* 34: 57–64.

19 Sato, S. (1955). On a new method of drawing the potential energy surface. *J. Chem. Phys.* 23: 592–593.

20 Horiuti, J. and Polanyi, M. (1935). Outlines of a theory of proton transfer. *Acta Physicochim. U.R.S.S.* 2: 505–532.

21 Evans, M.G. and Hush, N.S. (1952). Ionogenic reactions involving bond breaking at electrodes. *J. Chim. Phys. Phys.- Chim. Biol.* 49: C159–C171.

22 Hush, N.S. (1953). Quantum-mechanical discussion of the gas phase formation of quinonedimethide monomers. *J. Polymer Sci.* 11: 289–298.

23 Reimers, J.R. and Hush, N.S. (1996). The effects of couplings to symmetric and antisymmetric modes and minor asymmetries on the spectral properties of mixed-valence and related charge-transfer systems. *Chem. Phys.* 208: 177.

24 Couper, A. (1858). Sur une nouvelle théorie chimidue. *C.R. Hebd. Seances Acad. Sci.* 46: 1157–1160.

25 Kekulé, A. (1865). Sur la constitution des substances aromatiques. *Bull. Soc. Chim. Fr.* 3: 98–110.

26 Demadis, K.D., Hartshorn, C.M., and Meyer, T.J. (2001). The localized to delocalized transition in mixed-valence chemistry. *Chem. Rev.* 101: 2655.

27 Brunschwig, B.S., Creutz, C., and Sutin, N. (2002). Optical transitions of symmetrical mixed-valence systems in the Class II–III transition regime. *Chem. Soc. Rev.* 31: 168–184.

28 Kubo, R. and Toyozawa, Y. (1955). Application of the method of generating function to radiative and non-radiative transitions of a trapped electron in a crystal. *Prog. Theor. Phys.* 13: 160.

29 Hush, N.S. (1958). Adiabatic rate processes at electrodes. *J. Chem. Phys.* 28: 962–972.

30 Marcus, R.A. (1956). On the theory of oxidation-reduction reactions involving electron transfer. I. *J. Chem. Phys.* 24: 966–978.

31 McKemmish, L., McKenzie, R.H., Hush, N.S., and Reimers, J.R. (2015). Electron-vibration entanglement in the Born-Oppenheimer description of chemical reactions and spectroscopy. *Phys. Chem. Chem. Phys.* 17: 24666–24682.

32 Öpik, U. and Pryce, M.H.L. (1957). Studies of the Jahn-Teller Effect. I. A survey of the static problem. *Proc. R. Soc. Lond. Ser. A, Math. Phys. Sci.* 238: 425–447.

33 Marcus, R.A. (1960). Theory of oxidation-reduction reactions involving electron transfer. IV. A statistical-mechanical basis for treating contributions from solvent, ligands, and inert salt. *Discuss. Faraday Soc.* 29: 21–31.

34 Reimers, J.R. and Hush, N.S. (1991). Electronic properties of transition-metal complexes determined from electro-absorption spectroscopy II. mono-nuclear complexes of ruthenium(II). *J. Phys. Chem.* 95: 9773.

35 Reimers, J.R. and Hush, N.S. (1991). Electric field perturbation of electronic (vibronic) absorption envelopes: application to characterization of mixed-valence states. In: *Mixed Valence Systems: Applications in Chemistry, Physics, and Biology* (ed. K. Prassides), 29–50. Dordrecht: Kluwer Acad. Publishers.

36 Van Duyne, R.P. and Fischer, S.F. (1974). A nonadiabatic description of electron transfer reactions involving large free energy changes. *Chem. Phys.* 5: 183–197.

37 Efrima, S. and Bixon, M. (1976). Vibrational effects in outer-sphere electron-transfer reactions in polar media. *Chem. Phys.* 13: 447–460.

38 Jortner, J. (1976). Temperature dependent activation energy for electron transfer between biological molecules. *J. Chem. Phys.* 64: 4860–4867.

39 Kestner, N.R., Logan, J., and Jortner, J. (1974). Thermal electron transfer reactions in polar solvents. *J. Phys. Chem.* 78: 2148–2166.

40 Warshel, A. (1982). Dynamics of reactions in polar solvents. Semiclassical trajectory studies of electron-transfer and proton-transfer reactions. *J. Phys. Chem.* 86: 2218–2224.

41 Hush, N.S. (1961). Adiabatic theory of outer-sphere electron-transfer processes in solution. *Trans. Faraday Soc.* 57: 577–580.

42 Bayliss, N.S. and McRae, E.G. (1954). Solvent effects in organic spectra: dipole forces and the Franck–Condon principle. *J. Phys. Chem.* 58: 1002–1006.

43 McRae, E.G. (1957). Theory of solvent effects on molecular electronic spectra. Frequency shifts. *J. Phys. Chem.* 61: 562–572.

44 Wu, H.-Y., Ren, H.-S., Zhu, Q., and Li, X.-Y. (2012). A modified two-sphere model for solvent reorganization energy in electron transfer. *Phys. Chem. Chem. Phys.* 14: 5538–5544.

45 Bi, T.-J., Ming, M.-J., Ren, H.-S. et al. (2014). Numerical solution of solvent reorganization energy and its application in electron transfer reaction. *Theor. Chem. Acc.* 133: 1557.

46 Li, X.-Y. (2015). An overview of continuum models for nonequilibrium solvation: Popular theories and new challenge. *Int. J. Quantum Chem.* 115: 700–721.

47 Lockhart, D.J. and Boxer, S.G. (1987). Magnitude and direction of the change in dipole moment associated with excitation of the primary electron donor in *Rhodopseudomonas sphaeroides* reaction center. *Biochem* 26: 664.

48 Lockhart, D.J., Goldstein, R.F., and Boxer, S.G. (1988). Structure-based analysis of the initial electron transfer step in bacterial photosynthesis: electric field induced fluorescence anisotropy. *J. Chem. Phys.* 89: 1408.

49 Oh, D.H., Sano, M., and Boxer, S.G. (1991). Electroabsorption (Stark effect) spectroscopy of mono- and biruthenium charge-transfer complexes: measurements of changes in dipole moments and other electrooptic properties. *J. Am. Chem. Soc.* 113: 6880–6890.

50 Liptay, W. (1969). Electrochromism and solvatochromism. *Angew. Chem. Int. Ed.* 8: 177.

51 Liptay, W. (1974). *Excited States* (ed. E.C. Lim), 129. New York: Academic.

52 Piepho, S.B., Krausz, E.R., and Schatz, P.N. (1978). Vibronic coupling model for calculation of mixed valence absorption profiles. *J. Am. Chem. Soc.* 100: 2996–3005.

53 Reimers, J.R. and Hush, N.S. (2004). Hamiltonian operators including both symmetric and antisymmetric vibrational modes for vibronic-coupling and intervalence charge-transfer applications. *Chem. Phys.* 299: 79.

54 Chappell, P.J., Fischer, G., Reimers, J.R., and Ross, I.G. (1981). Electronic spectrum of 1,5-naphthyridine: theoretical calculation of vibronic coupling. *J. Mol. Spectrosc.* 87: 316–330.

55 Born, M. and Huang, K. (1954). *Dynamical Theory of Crystal Lattices*. Oxford: Clarendon.

56 Azumi, T. and Matsuzaki, K. (1977). What does the term vibronic-coupling mean. *Photochem. Photobiol.* 25: 315–326.

57 Mielke, S.L., Schwenke, D.W., Schatz, G.C. et al. (2009). Functional representation for the Born-Oppenheimer diagonal correction and Born-Huang adiabatic potential energy surfaces for isotopomers of H_3. *J. Phys. Chem. A* 113: 4479–4488.

58 Yarkony, D.R. (1996). Diabolical conical intersections. *Rev. Mod. Phys.* 68: 985–1013.

59 Saunders, P.T. (1980). *An Introduction to Catastrophe Theory*. Cambridge: Cambridge University Press.

60 Xu, F. (1990). Application of catastrophe theory to the $\Delta^{\neq}$ to - ΔG relationship in electron transfer reactions. *Z. Phys. Chem.* 166: 79–91.

61 Krokidis, X., Silvi, B., Dezarnaud-Dandine, C., and Sevin, A. (1998). Topological study, using a coupled ELF and catastrophe theory technique, of electron transfer in the Li+Cl$_2$ system. *New J. Chem.* 22: 1341–1350.

62 Wales, D.J. (2001). A microscopic basis for the global appearance of energy landscapes. *Science* 293: 2067–2070.

63 Levich, V.G. and Dogonadze, R.R. (1959). Theory of rediationless electron transitions between ions in solution. *Dokl. Akad. Nauk. SSSR Ser. Fiz. Khim.* 124: 123–126.

64 Levich, V.G. and Dogonadze, R.R. (1960). Adiabatic theory for electron-transfer processes in solution. *Dokl. Akad. Nauk. SSSR* 133: 158–161.

65 Levich, V.G. and Dogonadze, R.R. (1961). Adiabatic theory of electron-transfer processes in solution. *Collect. Czech. Chem. Commun.* 26: 193–214.

66 Levich, V.G. (1967). Theory of macroscopic kinetics of hetero-geneous and homogeneous-heterogeneous processes. *Annu. Rev. Phys. Chem.* 18: 153–176.

67 Nielsen, M.A. and Chuang, I.L. (2000). *Quantum Computation and Quantum Information*. New York: Cambridge University Press.

68 Ferguson, A.J., Cain, P.A., Williams, D.A., and Briggs, G.A.D. (2002). Ammonia-based quantum computer. *Phys. Rev. A* 65: 034303.

69 Costi, T.A. and McKenzie, R.H. (2003). Entanglement between a qubit and the environment in the spin-boson model. *Phys. Rev. A* 68: 034301.

70 Bremner, M.J., Dawson, C.M., Dodd, J.L. et al. (2002). Practical scheme for quantum computation with any two-qubit entangling gate. *Phys. Rev. Lett.* 89: 247902.

71 Stanton, J.F. (2010). An unusually large nonadiabatic error in the BNB molecule. *J. Chem. Phys.* 133: 174309.

72 Kalemos, A. (2016). Revisiting the symmetry breaking in the X̃$^2\Sigma$u+ state of BNB. *J. Chem. Phys.* 144: 234315.

73 Hines, A.P., Dawson, C.M., McKenzie, R.H., and Milburn, G.J. (2004). Entanglement and bifurcations in Jahn-Teller models. *Phys. Rev. A* 70: 022303.

74 Englman, R. (2007). Curvature maxima in two-state systems: a semi-classical study. *Phys. Lett. A* 367: 345–350.

75 Hwang, M.-J. and Choi, M.-S. (2010). Variational study of a two-level system coupled to a harmonic oscillator in an ultrastrong-coupling regime. *Phys. Rev. A* 82: 025802.

76 Van Voorhis, T., Kowalczyk, T., Kaduk, B. et al. (2010). The diabatic picture of electron transfer, reaction barriers, and molecular dynamics. *Annu. Rev. Phys. Chem.* 61: 149–170.

77 Kuznetsov, A.M. (1995). *Charge Transfer in Physics, Chemistry, and Biology*. Reading: Gordon and Breach.

78 Kuznetsov, A. and Ulstrup, J. (1998). *Electron Transfer in Chemistry and Biology: An introduction to the theory*. Chichester: Wiley.

79 Nelsen, S.F., Ismagilov, R.F., and Trieber, D.A. (1997). Adiabatic electron transfer: comparison of modified theory with experiment. *Science* 278: 846–849.

80 Richardson, D.E. (1999). Electron-transfer reaction rate theory. In: *Inorganic Electronic Structure and Spectroscopy, V2: Applications and Case Studies* (ed. E.I. Solomon and A.B.P. Lever). Hoboken, NJ: Wiley.

81 Hush, N.S. (1999). Electron transfer in retrospect and prospect 1: adiabatic electrode processes. *J. Electroanal. Chem.* 460: 5.

82 Brunschwig, B.S. and Sutin, N. (2001). Reflections on the Two-state Electron-transfer Model. In: *Electron Transfer in Chemistry*, vol. 2 (ed. V. Balzani), 583–616. Hoboken NJ: Wiley-VCH.

83 Parthey, M. and Kaupp, M. (2014). Quantum-chemical insights into mixed-valence systems: within and beyond the Robin–Day scheme. *Chem. Soc. Rev.* 43: 5067–5088.

84 Bublitz, G.U. and Boxer, S.G. (1997). Stark spectroscopy: applications in chemistry, biology, and materials science. *Annu. Rev. Phys. Chem.* 48: 213–242.

85 Bersuker, I.B. (2013). Pseudo-Jahn-Teller effect – a two-state paradigm in formation, deformation, and transformation of molecular systems and solids. *Chem. Rev.* 113: 1351–1390.

86 Brown, D.B. (ed.) *Mixed-Valence Compounds*. Dordrecht: Springer.

87 Prassides, K. (ed.) (1991). *Mixed Valence Systems: Applications in Chemistry, Physics, and Biology*. Dordrecht: Kluwer Academic Publishers.

88 Duarte, F. and Kamerlin, S.C.L. (ed.) (2017). *Theory and Applications of the Empirical Valence Bond Approach: From Physical Chemistry to Chemical Biology*. Chichester: Wiley.

89 Lee, S.-H., Larsen, A.G., Ohkubo, K. et al. (2012). Long-lived long-distance photochemically induced spin-polarized charge separation in β,β'-pyrrolic fused ferrocene-porphyrin-fullerene systems. *Chem. Sci.* 3: 257–269.

90 Curiel, D., Ohkubo, K., Reimers, J.R. et al. (2007). Photoinduced electron transfer in a b,b'-pyrrolic fused ferrocene-(zinc porphyrin)-fullerene. *Phys. Chem. Chem. Phys.* 9: 5260–5266.

91 Lee, S.-H., Blake, I.M., Larsen, A.G. et al. (2016). Synthetically tuneable biomimetic artificial photosynthetic reaction centres that closely resemble the natural system in purple bacteria. *Chem. Sci.* 7: 6534–6550.

92 Woitellier, S., Launay, J.P., and Spangler, C.W. (1989). Invervalence transfer in pentaammineruthenium complexes of α,ω-dipyridyl polyenes. *Inorg. Chem.* 28: 758.

93 Reimers, J.R. and Hush, N.S. (1990). Electron and energy transfer through bridged systems. IV. The electronic structure of the bis-pentaammineruthenium complexes of α,ω-dipyridyl trans-polyenes in D_2O. *Inorg. Chem.* 29: 3686.

94 Brinkmann, M., Gadret, G., Muccini, M. et al. (2000). Correlation between molecular packing and optical properties in different crystalline polymorphs and amorphous thin films of mer-Tris(8-hydroxyquinoline)aluminum(III). *J. Am. Chem. Soc.* 122: 5147–5157.

95 Marcus, R.A. (1964). Chemical and electrochemical electron-transfer theory. *Annu. Rev. Phys. Chem.* 15: 155–196.

96 Marcus, R.A. (1968). Electrode reactions of organic compounds. General introduction. *Discuss. Faraday Soc.* 45: 7–13.

97 Marcus, R.A. (1968). Discussion of "Electrode reactons of organic compounds: General Introduction" by R.A. Marcus. *Discuss. Faraday Soc.* 45: 53–54.

98 Beattie, J.K., Hush, N.S., Taylor, P.R. et al. (1977). Crystal structure of .mu.-pyrazine-bis(penta-ammineruthenium) penta(bromide chloride)-water (1/4). *J. Chem. Soc., Dalton Trans.* 11: 1121–1124.

99 Hush, N.S., Edgar, A., and Beattie, J.K. (1980). Single-crystal EPR study of a mixed-valence ruthenium dimer ion: the ground state of the Creutz-Taube complex. *Chem. Phys. Lett.* 69: 128.

100 Beattie, J.K., Hush, N.S., and Taylor, P.R. (1976). Electron delocalization in the mixed-valence μ-pyrazine-decaamminediruthenium(5+) ion. *Inorg. Chem.* 15: 992.

101 Best, S.P., Clark, R.J.H., McQueen, R.C.S., and Joss, S. (1989). The near and mid-infrared spectrum of the Creutz-Taube ion in aqueous solution: an application of FTIR spectroelectrochemical techniques. *J. Am. Chem. Soc.* 111: 548.

102 Fürholz, U., Burgi, H.B., Wagner, F.E. et al. (1984). The Creutz-Taube ion revisited. *J. Am. Chem. Soc.* 106: 121.

103 Creutz, C. and Taube, H. (1973). *J. Am. Chem. Soc.* 95: 1086.

104 Krausz, E., Burton, C., and Broomhead, J. (1981). Search for intervalence tunnelling in the mixed-valence compound μ-(pyrazine)-bis(pentaammineruthenium) pentabromide. *Inorg. Chem.* 20: 434.

105 Hush, N.S. (1980). Electron delocalization, structure and dynamics in mixed-valence systems. *NATO Adv. Study Inst. Ser., Ser. C* 58: 151–188.

106 Hush, N.S. (1982). Parameters of electron-transfer kinetics. *ACS Symp. Ser.* 198: 301–332.

107 Galasso, V. (1991). Ab initio calculations on the one- and two-photon electronic transitions of cyclopentadiene, spirononatetraene, 1,4-cyclohexadiene, Dewar benzene, norbornadiene, and barrelene. *Chem. Phys.* 153: 13.

108 Powers, M.J., Salmon, D.J., Callahan, R.W., and Meyer, T.J. (1976). *J. Am. Chem. Soc.* 98: 6731.

109 Lu, H., Petrov, V., and Hupp, J.T. (1995). *Chem. Phys. Lett.* 235: 521.

110 Reimers, J.R., Wallace, B.B., and Hush, N.S. (2008). Towards a comprehensive model for the electronic and vibrational structure of the Creutz-Taube ion. *Phil. Trans. Roy. Soc. A* 366: 15–31.

111 Reimers, J.R., Cai, Z.-L., and Hush, H.S. (2005). A priori evaluation of the solvent contribution to the reorganization energy accompanying intramolecular electron transfer: predicting the nature of the Creutz-Taube ion. *Chem. Phys.* 319: 39–51.

112 Hush, N.S. (1982). Parameters of electron-transfer kinetics. In: *Mechanistic Aspects of Inorganic Reactions,* ACS Symp. Ser., vol. 198 (ed. D.B. Rorabacher and J.F. Endicott), 301–332. Washington DC: American Chemical Society.

113 Reimers, J.R. and Hush, N.S. (2003). Modelling the bacterial photosynthetic reaction centre: 7. Full simulation of the hole transfer absorption spectrum of the special-pair radical cation. *J. Chem. Phys.* 119: 3262–3277.

114 Reimers, J.R. and Hush, N.S. (2004). A unified description of the electrochemical, charge distribution, and spectroscopic properties of the special-pair radical cation in bacterial photosynthesis. *J. Am. Chem. Soc.* 126: 4132–4144.

115 Reimers, J.R., Hughes, J.M., and Hush, N.S. (2000). Modelling the bacterial photosynthetic reaction centre 3. Interpretation of the effects of site-directed mutagenesis on the special-pair mid-point potential. *Biochemistry* 39: 16185–16189.

116 Artz, K., Williams, J.C., Allen, J.P. et al. (1997). Relationship between the oxidation potential and electron spin density of the primary electron donor in reaction centres from *Rb. sphaeroides*. *Proc. Natl. Acad. Sci. U.S.A.* 94: 13582.

117 Rice, M.J. (1976). Organic linear conductors as systems for the study of electron-phonon interactions in the organic solid state. *Phys. Rev. Lett.* 37: 36–39.

118 Herzberg, G. and Teller, E. (1933). Schwingungsstruktur der Elektronenübergänge bei mehratomigen Molekülen. *Z. Phys. Chem.* 21B: 410–446.

119 Reimers, J.R., Shapley, W.A., and Hush, N.S. (2003). Modelling the bacterial photosynthetic reaction centre: 5. Assignment of the electronic transition observed at 2200 cm^{-1} in the special-pair radical-cation as a SHOMO to HOMO transition. *J. Chem. Phys.* 119: 3240–3248.

120 Reimers, J.R., Shapley, W.A., Rendell, A.P., and Hush, N.S. (2003). Modelling the bacterial photosynthetic reaction centre: 6. Use of density-functional theory to determine the nature of the vibronic coupling between the four lowest-energy electronic states of the special-pair radical cation. *J. Chem. Phys.* 119: 3249–3261.

121 Kanchanawong, O., Dahlbom, M.G., Treynor, T.P. et al. (2006). Charge delocalization in the special pair radical cation of mutant reaction centers of *Rhodobacter Sphaeroides* from stark spectra and non-adiabatic spectral simulations. *J. Phys. Chem. B* 110: 18688–18702.

122 Reimers, J.R. and Hush, N.S. (2003). Modeling the bacterial photosynthetic reaction cneter. VII. Full Simulation of the intervlaence hole-transfer absorption specturm of the special-pair radical cation. *J. Chem. Phys.* 119: 3262–3277.

123 Kraft, A. (2008). On the discovery and history of Prussian Blue. *Bull. Hist. Chem.* 33: 61–67.

124 D'Alessandro, D.M. and Keene, F.R. (2006). Intervalence charge transfer (IVCT) in trinuclear and tetranuclear complexes of iron, ruthenium, and osmium. *Chem. Rev.* 106: 2270–2298.

125 Samain, L., Grandjean, F., Long, G.J. et al. (2013). Relationship between the synthesis of Prussian Blue pigments, their color, physical properties, and their behavior in paint layers. *J. Phys. Chem. C* 117: 9693–9712.

126 Keggin, J.F. and Miles, F.D. (1936). Structures and formulae of the Prussian Blues and related compounds. *Nature* 137: 577–578.

127 Bueno, P.R., Ferreira, F.F., Giménez-Romero, D. et al. (2008). Synchrotron structural characterization of electrochemically synthesized hexacyanoferrates containing K$^+$: a revisited analysis of electrochemical redox. *J. Phys. Chem. C* 112: 13264–13271.

128 Grandjean, F., Samain, L., and Long, G.J. (2016). Characterization and utilization of Prussian blue and its pigments. *Dalton Trans.* 45: 18018–18044.

129 Simonov, A., De Baerdemaeker, T., Boström, H.L.B. et al. (2020). Hidden diversity of vacancy networks in Prussian blue analogues. *Nature* 578: 256–260.

130 Ludi, A. and Güdel, H.U. (1973). Structural chemistry of polynuclear transition metal cyanides. In: *Inorganic Chemistry. Structure and Bonding*, vol. 14 (ed. J.D. Dunitz, P. Hemmerich, J.A. Ibers, et al.), 1–21. Berlin: Springer.

131 Shaik, S., Shurki, A., Danovich, D., and Hiberty, P.C. (1996). Origins of the exsalted b2u frequency in the first excited state of benzene. *J. Am. Chem. Soc.* 118: 666–671.

132 Shaik, S., Zilberg, S., and Haas, Y. (1996). A Kekulé-crossing model for the "anomalous" behavior of the b_{2u} modes of aromatic hydrocarbons in the lowest excited $^1B_{2u}$ state. *Acc. Chem. Res.* 29: 211–218.

133 Zilberg, S., Haas, Y., Danovich, D., and Shaik, S. (1998). The twin-excited state as a probe for the transition state in concerted unimolecular reactions: the semibullvalene rearrangement. *Angew. Chem., Int. Ed.* 37: 1394–1397.

134 Shaik, S., Shurki, A., Danovich, D., and Hiberty, P.C. (2001). A different story of π-delocalization-the distortivity of π-electrons and its chemical manifestations. *Chem. Rev.* 101: 1501–1539.

135 Zilberg, S. and Haas, Y. (1998). Two-state model of antiaromaticity: the low lying singlet states. *J. Phys. Chem. A* 102: 10843–10850.

136 Cai, Z.-L. and Reimers, J.R. (2000). The low-lying excited states of pyridine. *J. Phys. Chem. A* 104: 8389.

137 Swalen, J. and Ibers, J. (1962). Potential function for the inversion of ammonia. *J. Chem. Phys.* 36: 1914.

138 Reimers, J.R., McKemmish, L., McKenzie, R.H., and Hush, N.S. (2015). Bond angle variations in XH_3 [X=N,P,As,Sb,Bi]: the critical role of Rydberg orbitals exposed using a diabatic state model. *Phys. Chem. Chem. Phys.* 17: 24618–24640.

139 Mulliken, R.S. (1976). Rydberg states and rydbergization. *Acc. Chem. Res.* 9: 7.

140 Mulliken, R.S. (1977). Rydberg and valence-shell states and their interaction. *Chem. Phys. Lett.* 46: 197–200.

141 Kutzelnigg, W. (1984). Chemical bonding in higher main group elements. *Angew. Chem. Int. Ed. Engl.* 23: 272–295.

142 Babamov, V.K. and Marcus, R.A. (1981). Dynamics of hydrogen-atom and proton-transfer reactions-symmetric case. *J. Chem. Phys.* 74: 1790–1798.

143 Babamov, V.K., Lopez, V., and Marcus, R.A. (1983). Dynamics of hydrogen atom and proton transfer reactions. Nearly degenerate asymmetric case. *J. Chem. Phys.* 78: 5621–5628.

144 Hammes-Schiffer, S. (2010). Introduction: proton-coupled electron transfer. *Chem. Rev.* 110: 6937–6938.

145 Hammes-Schiffer, S. and Stuchebrukhov, A.A. (2010). Theory of coupled electron and proton transfer reactions. *Chem. Rev.* 110: 6939–6960.

146 Parada, G.A., Goldsmith, Z.K., Kolmar, S. et al. (2019). Concerted proton-electron transfer reactions in the Marcus inverted region. *Science* 364: 471–475.

147 Zoete, V. and Meuwly, M. (2004). On the influence of semirigid environments on proton transfer along molecular chains. *J. Chem. Phys.* 120: 7085–7094.
148 Berry, R. (1961). Zero-point vibrations in benzene. *J. Chem. Phys.* 35: 2253.
149 Longuet-Higgins, H.C. and Salem, L. (1959). The alternation of bond lengths in long conjugated chain molecules. *Proc. R. Soc. Lond. Ser. A, Math. Phys. Sci.* 251: 172–185.
150 Shaik, S.S. and Hiberty, P.C. (1985). When does electronic delocalization become a driving force of molecular shape and stability? 1. The aromatic sextet. *J. Am. Chem. Soc.* 107: 3089–3095.
151 Wang, Y., Wu, J.I.C., Li, Q. et al. (2010). Aromaticity and relative stabilities of azines. *Org. Lett.* 12: 4824–4827.
152 Wu, J.I.C., Mo, Y., Evangelista, F.A., and von Ragué Schleyer, P. (2012). Is cyclobutadiene really highly destabilized by antiaromaticity? *Chem. Commun.* 48: 8437–8439.
153 Wu, J.I.C., Wang, C., McKee, W.C. et al. (2014). On the large σ-hyperconjugation in alkanes and alkenes. *J. Mol. Model.* 20: 2228.
154 Cyrański, M.K., van Schleyer, P.R., Krygowski, T.M. et al. (2003). Facts and artifacts about aromatic stability estimation. *Tetrahedron* 59: 1657–1665.

3

Quantum Chemical Approaches to Treat Mixed-Valence Systems Realistically for Delocalized and Localized Situations

Martin Kaupp

Technische Universität Berlin, Institut für Chemie, Theoretische Chemie/Quantenchemie, Sekr. C7, Strasse des 17. Juni 135, 10623 Berlin, Germany

3.1 Introduction and Scope

Both the understanding of specific mixed-valence (MV) systems and the transfer of knowledge to a general understanding or design and optimization of electron transfer in practical applications require adequate microscopic insight. This should be based on accurate quantum chemical treatments of the often rather challenging distribution of charge and spin densities in MV systems. In particular, it is by no means trivial to describe MV systems that are close to the border between class II and class III systems in the Robin–Day classification. This requires a well-balanced simultaneous description of exchange as well as static and dynamic electron correlation effects by the underlying electronic structure method as well as typically a realistic modeling of environmental effects, e.g. in solution where much of the spectroscopic work on MV systems is carried out [1]. Notably, we will go beyond the limited frameworks of the Robin–Day classes and a Mulliken–Hush treatment and take into account not only the specific conformational distributions that shape the spectroscopic observations but also the electron transfer characteristics overall. Then, a suitable computational approach that reproduces realistically the localization/delocalization characteristics of a given system, as well as its adiabatic ground- and excited-state potential energy surfaces and spectroscopic properties, becomes even more crucial if we want to establish the direct link between theory and experiment.

This chapter will lay out the current state of the art in dealing with these challenges. It will, however, not cover in any detail dynamical aspects of the MV systems or of their environment. While a substantial amount of research has been carried out on these aspects as well, most of that work used either model Hamiltonians, for example, within Mulliken–Hush or extended Mulliken–Hush frameworks, or rather approximate quantum chemical treatments have been used that do not provide the required balance between various effects near the class II/class III borderline. These works allow us to understand the technical aspects that have to be dealt

Mixed-Valence Systems: Fundamentals, Synthesis, Electron Transfer, and Applications, First Edition.
Edited by Yu-Wu Zhong, Chun Y. Liu, and Jeffrey R. Reimers.
© 2023 WILEY-VCH GmbH. Published 2023 by WILEY-VCH GmbH.

with in dynamical studies of MV systems, but they will typically fail for the critical borderline cases, unless the abovementioned requirements of electronic structure method and environmental modeling have been met. Our focus will be on the latter, and we will also specifically focus on studies where adequate methodologies to obtain a realistic description of charge localization/delocalization near the class II/III borderline have been applied.

The status of quantum chemical approaches to MV systems up to 2014, including the application to specific systems, has been comprehensively reviewed in Ref. [1]. The interested reader is pointed to that work for more details. Here, I will on the one hand report on progress made since 2014. On the other hand, I will take the liberty to report on the time line of work before 2014 from the perspective of my personal involvement in the field.

3.2 How Did We Start

My own interest in the field was initiated within a graduate research school on interactions between conjugated π-systems at Universität Würzburg (DFG GRK1221) around 2007, when my friend and colleague Christoph Lambert made me aware of the challenge of describing the transition from class II to class III in organic MV systems such as the important bis-triarylamine radical cations studied in his group for years. Given the system sizes involved, the best one could do at the time was to use semiempirical MO theories augmented by limited configuration interaction (AM1-CISD) to compute the excitation spectra [2]. While reasonable trends could be obtained, the results were far from allowing a quantitative assessment of the Robin–Day classification of a given system, at least not near the borderline.

The challenges involved became clear early on, and they pertain both to the electronic structure method and to the treatment of environmental effects: Hartree–Fock theory misses both dynamic and static electron correlation completely. It localizes charge and spin densities and will provide symmetry-broken structures even for many class III systems, unless their electronic coupling is extremely large. While multi-configuration self-consistent field (MCSCF) approaches were applied early on, these cover only static correlation contributions. However, it turns out that in many cases, dynamical correlation is more important [1]. In addition, already a typical CASSCF (complete active space SCF) calculation exhibits a heavy computational load and is limited to relatively small systems. Adding dynamical correlation, e.g. by perturbation theory (CASPT2 and NEVPT2), is even much more computationally demanding. The perturbation theoretical treatment may still be insufficient for a quantitative description. That is, even more demanding MR-CI calculations may be required (see Section 3.7). In addition, the basis set convergence of such methods is slow. On the other hand, static correlation may actually be moderate in many situations, even for the symmetrical transition state for thermal electron transfer in a class II system (see further below). Then the most accurate single-reference correlation methods such as coupled cluster theory may be sufficient for a more quantitative description. These are nevertheless computationally demanding for larger systems and have been applied only more recently and only to relatively small model systems. I will come back to coupled cluster methods further below when describing recent gas-phase benchmarking.

If single-reference methods are adequate for many cases, DFT methods would appear a natural route to pursue. This had already been tried for MV systems by many groups. It turned out, however, that standard semilocal functionals (the local spin density approximation, LSDA, the generalized gradient approximation, GGA, and the meta-GGA functionals) suffer from so-called self-interaction errors [3]. That is, the Coulomb self-interaction of an electron with its own charge cloud, which in Hartree–Fock theory is exactly cancelled by the exchange term, does not vanish completely. This causes electrons to avoid each other more than is physically needed, leading to an artificially too delocalized bonding situation. Another term for these artifacts of approximate DFT functionals is therefore "delocalization errors."[4] Indeed, semilocal DFT functionals give incorrectly delocalized structures even for many class II systems.

As mentioned above, Hartree–Fock exchange cancels exactly the Coulomb self-interaction error, but Hartree–Fock theory itself lacks any correlation contributions. One might thus assume that so-called hybrid functionals, which replace a certain fraction of semilocal exchange by Hartree–Fock-type exact exchange as part of the DFT exchange–correlation functional, might be a viable compromise route toward better methods for the description of MV systems near the class II/III borderline. However, standard (global) hybrids with relatively low (constant) exact exchange (EXX) admixture such as B3LYP (20%) or PBE0 (25%) still exhibit too large delocalization errors and thus tend to still artificially delocalize many class II systems while being relatively well suited to describe typical class III cases [1]. A systematic pragmatic tuning of global hybrids for the abovementioned bis-triarylamine radical cations in conjunction with a suitable solvent model (see below) gave about 35% EXX admixture as a good compromise [5, 6], as evidenced by the good performance of the BLYP35 functional that has since been used extensively in the field by us and by many others, or some modifications discussed also below. Newer developments using more sophisticated functionals have emerged recently and will also be discussed.

The other obstacle to obtain accurate predictions of the degree of localization/delocalization in actual MV systems of chemical interest is the fact that most experimental studies are done in the condensed phase, and thus environmental effects have to be accounted for. Indeed, most spectroscopic studies are carried out in polar solvents. The reason is that most often, MV systems are radical anions or cations, which do not dissolve readily in nonpolar solvents. A polar solvent environment, however, can have a substantial effect on the structure and electronic character of an MV system. A localized structure with concentration of charge on one of the redox centers is favored energetically by a polar environment, as this leads to a favorable orientation of the solvent dipoles around the charged center. Therefore, a computation that neglects this environment inevitably tends to give a too delocalized description. Again, this effect will of course be most detrimental in the description of class II systems reasonably close to the borderline. This factor is apparent not only from the comparison between computational and experimental data but can be seen from experimental studies in different solvents. For systems sufficiently close to the borderline, even a slight increase in solvent polarity can move the system from class III to class II. That is, the character of an MV system is determined *by the ion and its solvent environment*, and the environment should be considered as an intrinsic part of the system [1]. This has to be kept in mind when considering earlier computational studies that neglected the environment while being compared to experimental studies in solution.

When we entered the field around 2008 [5, 6], we noted that most computational studies used continuum solvent models for single-point calculations of ground-state properties and excitation spectra, at DFT, semiempirical, Hartree–Fock, or post-Hartree–Fock levels. However, the underlying molecular structures were typically optimized without consideration of environmental effects and thus were often symmetrical for systems that should have been symmetry-broken class II cases (however, see, for example, Ref. [7]). As a consequence, the computed properties and excitation spectra often reflected the delocalized nature of the input structures. Since then, it has become common practice to include environmental effects in the structure optimizations as well as in subsequent single-point computations. The simplest and computationally most expedient way to include bulk solvent effects is to use polarizable continuum solvent (PCM) models [8], available in different degrees of sophistication from COSMO to IEFPCM or SMD but essentially all based on the same principles of a polarizable dielectric continuum and in our experience performing similarly well. Noting again that we completely neglect here the actual dynamics of the solvent (see introduction), MV systems in polar aprotic solvents appear to be described reasonably well by such models. Of course, such conclusions depend on having used a reliable electronic structure method in the computational studies. Given that the effects of electronic structure method and solvent model are difficult to separate for studies of MV systems in solution, the possibility to compare to experimental or high-level computational data in the gas phase is important, and we will address this point further below.

In our first studies of dinitroaryl radical anions [9], we had the opportunity to compare with experimental ESR and UV/NIR/vis data in both aprotic and protic solvents [10]. It became quickly apparent that PCM-type models did not reproduce the transition from aprotic to protic solvent environments. Consider, for example, the solvents acetonitrile and methanol: these have closely similar dielectric constants ($\varepsilon = 36.6$ and $\varepsilon = 32.6$, respectively). A PCM-type solvent model is defined only by these values, and thus, for all practical purposes, the results will be identical. However, it is clear that hydrogen bonding to the oxygen atoms of the nitro groups in methanol will help to additionally move the system toward a localized description when compared to acetonitrile. For example, ESR and UV/NIR/vis data clearly show that 1,4-dinitrobenzene radical anion is class III in acetonitrile but class II in methanol [10]. Otherwise, consider 1,3-dinitrobenzene: it is class II in both solvents, but the energy barrier for thermal electron transfer is estimated from ESR data to be about twice larger in methanol [10]. None of these changes can be reproduced by a PCM treatment, and one clearly needs a more sophisticated approach. In Ref. [9], we used the D-COSMO-RS model with good success, as it allowed the differences between acetonitrile and methanol to be reproduced for a number of such systems. D-COSMO-RS [9, 11] is a self-consistent version of Klamt's "conductor-like screening model for real solvents," [12] which is based on integral equation theory and includes a semiempirical parameterization for hydrogen bonding. In essence, the model uses a segmentation of the COSMO (PCM-like) surface charge densities on a PCM-type cavity as input for a statistical fluid dynamics model. The self-consistent version was key to the successful description of these systems, as it allowed the structure optimization in the presence of the solvent model. The model allows a description of solvent mixtures, which we successfully tested for quinone-based MV systems for the case of a mixture between ethyl acetate and t-butanol [13]. Of course, other solvent treatments of increasing sophistication can be envisioned, and we will discuss them further below in the context of more recent work by different groups (Section 3.8).

3.3 Moving to Transition Metal MV Systems, Getting into Conformational Aspects

In view of the good results for organic MV systems, it was tempting to apply the simple pragmatic computational protocol laid out in Section 3.2, i.e. DFT with a global hybrid functional having about 35% EXX admixture (like BLYP35) as electronic structure method combined with some kind of PCM model, or potentially D-COSMO-RS in case of protic solvents, also to MV systems based on transition metal centers. For this, we teamed up with the group of Paul J. Low, at that time (around 2011/2012) still at Durham University, UK, later at the University of Western Australia. Paul is an expert in spectroelectrochemical studies on MV systems in solution as well as in studies of electrical conductance through molecules. The dinuclear organometallic Ru and Fe complexes coupled by carbon bridges, studied in the Low group, seemed to provide a particularly fertile ground to test and potentially extend the methodology, as well as to help better understand the spectroscopic observations. The initial study [14] focused on diruthenium complexes with butadiynediyl bridges, having the formula $[L_nRuC_4RuL_n]^+$ in their MV state. While the spectra placed these complexes clearly in class III, a high-energy shoulder in their main NIR band, typically associated with the intervalence charge transfer (IVCT) excitation, did not quite seem to fit into this general picture. Computational studies revealed different conformers of the redox centers around the all-carbon bridge, and a mutually perpendicular arrangement was shown by TDDFT calculations (BLYP35 + PCM protocol) to give rise to excitations at higher frequency with significant intensity [14]. These had vanishing intensity for the lower lying *cis* and *trans* conformers. This interpretation could be confirmed in Durham by the synthesis and spectroelectrochemical study of a derivative of such complexes that was conformationally restricted around the *cis* conformer. Indeed, the NIR band of this complex lacked the high-energy shoulder found for the conformationally flexible systems [14]. On the one hand, this study showed that the computational protocol could also be applied to transition metal MV complexes (including the famous class II/III Creutz–Taube ion in the same study). On the other hand, it initiated our interest in looking more closely at the influence of conformational aspects.

A completely new can of worms is opened when replacing the C_4 butadiynediyl bridge in the Ru_2 complexes by a 1,4-diethynylbenzene bridge. Here, previous spectroelectrochemical results [15] were hard to reconcile with the assumed symmetric class III situation. For example, symmetric acetylene stretching bands and an aryl breathing band should essentially be absent in the IR spectra of a centrosymmetric complex but were observed experimentally. Here, a detailed initial computational study (BLYP35 + PCM) [16] that accounted simultaneously for the conformations of the metal redox centers and of the benzene bridge in a two-dimensional conformational scan (see Figure 3.1) demonstrated that conformational motion may actually lead to the simultaneous thermal population of both charge-localized class II and charge-delocalized class III conformers. A more perpendicular orientation of the benzene ring in the bridge gives rise to reduced electronic coupling. Together with the influence of the polar solvent environment, this can lead to a class II situation. More strongly coupled conformers, however, can have class III character. The Robin–Day character may thus be a conformational average over populated class II and III situations. The occurrence of class II conformers explains both the high-energy shoulder observed in the NIR bands and the additional vibrational bands in the IR spectra. Indeed, the entire concept of Robin–Day classes becomes fuzzy for such systems, and the usual Mulliken–Hush theoretical treatment is clearly insufficient for a comprehensive understanding of electron transfer (ET), also given the rather variable types of excitations that contribute to the band shapes. These findings opened the door to more recent attempts to modify the relative population of localized and delocalized conformers by changing the steric demands of both bridging and terminal ligands. These will be discussed below. Of course, one could argue that the one-dimensional Robin–Day or Mulliken–Hush framework still holds for a given conformer. However, then the direct discussion of the ensemble average, based on quantum chemical computations of the adiabatic potential energy surface, seems to be more straightforward in such cases.

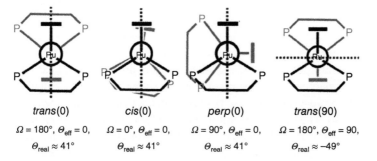

Figure 3.1 Definition of dihedral angles defining the two-dimensional conformational scan of 1,4-diethynyl-benzene-bridged dinuclear transition metal complexes. Source: Parthey et al. [16]/Reproduced with permission from John Wiley & Sons.

3.4 More Recent Work on Organic MV Systems and More General Use for Charge Transfer Questions

Our pragmatic computational protocol based on the BLYP35 global hybrid and a suitable solvent model has since been used by many researchers, sometimes modified from 35% to 30% or particularly to 40% exact exchange admixture. In this section, we mention some of the most interesting applications for purely organic MV systems, without aiming to be comprehensive. Starting with the unmodified BLYP35-based protocol, which had been assessed most carefully for triarylamine (TAA)-based MV systems (see above), and given the importance of such species, it is natural that the protocol has been applied frequently to TAA MV systems. We only mention here a few applications. One has been to the question of the role of through-bond vs. through-space electron transfer depending on the bridge [17], including paracyclophane-based bridges as models for conduction via π-stacking in important materials [18]. Others involved the tuning of oligoacene linkers [19], the reduction of reorganization energies by pre-planarized TAA moieties [20], the role of a platinum center as part of the bridge [21], charge delocalization in cross-conjugated bridges [22] and for dendritic bridging ligands [23], a comparison of Kekulé and non-Kekulé-bridged systems [24], analyses of the NIR bands in indolo-carbazole [25], and perylene-bridged [26] systems, as well as unexpected effects of solvent dynamics [27]. Angular ladder-type meta-phenylenes are another application example of the protocol [28]. Other applications of the protocol to organic MV systems include the comparison of bithiophenes and biselenophenes [29], a pyrene-based system [30], as well as tetraaza- and octaaza-paracyclophanes [31].

Aiming toward a general scheme to treat charge transfer properties on a broader basis, Herrmann and coworkers found the BLYP35-based protocol to work very well for electron transfer along rather different types of longer chains and different end groups [32]. Importantly, rather accurate predictions of the particular chain length, where charge localization takes place, could be obtained. In the context of electron transfer, this marks the chain length where the transition from tunneling to hopping transport is expected to occur [32]. Given that these materials often exhibit a stacking of the chains, Herrmann and coworkers employed continuum solvent models with the dielectric constant reflecting the nature of the bridging units. That group also used this same approach to compute actual transport properties along molecular wires [33] and deduced the transferability of a given bridging unit in the context of different electron transfer processes in different systems [34].

While a modified scheme with 30% EXX admixture was applied only to organometallic transition-metal-based systems (see below), the groups of Rathore and collaborators published extensive applications of a scheme with 40% admixture (rather than calling it BLYP40, they preferred the name B1LYP-40), again augmented by continuum solvent models and sometimes by dispersion corrections. Here, we can mention only some of the most important applications. These include studies of the role of the end-capping groups [35], the importance of the energy gap between the bridge and the end groups [36], the asymptotic dependence of the reorganization energy and the development of a multi-state Marcus model for longer chains [37], the effect of chromophore size [38], the competition between tunneling and hopping (see also above) [39], hole stabilization across stacking interfaces [40], hole delocalization in cyclic para-phenylene molecular wires [41], and general aspects of the design of charge transfer materials [42]. Related work focused on the simultaneous occurrence of different structural isomers of organic radical cations [43], Koopmans' paradigm for ionization in such systems [44], substituent control in methoxy-substituted benzo-chrysenes [45], or further work on charge delocalization in polychromophoric assemblies [46]. Overall, such computational schemes have enriched the understanding of electron transfer processes. All of the mentioned applications were done in close concert with high-level spectroscopic and other experimental approaches.

Of course, not all applications have used the BLYP35 or B1LYP-40 functionals. Other suitable functionals have been evaluated in related work on organic MV systems. This has involved in particular range-separated hybrid functionals, such as ωB97X-D, CAM-B3LYP, or LC-ωPBE [30, 47]. While the former two functionals perform generally reasonably well (consistent with their performance for a recent gas-phase benchmark, see Section 3.7 below), the latter tends to overlocalize appreciably. We note in passing that CAM-B3LYP has also been suggested as a suitable functional to treat ground and excited states for defects in solids [48], a field closely related to electron transfer and mixed valency. We will come back to what are the most promising functionals overall further below, including such "long-range-corrected" functionals and their advantages for treating charge transfer excitations [49] and also so-called local hybrid functionals with position-dependent exact exchange admixture.

3.5 More Recent Insights into Conformational Aspects for Transition Metal Complexes

In Section 3.3, we detailed how we got interested in conformational aspects of electron transfer in the context of transition-metal-capped molecular wires. The work-up to the 2014 review [1] was covered. Extensions of that work concentrated initially on further homo- and heteronuclear Ru and Fe complexes with C_4 bridges with variable auxiliary ligand sets [50, 51], unsymmetrical C_3N-bridged dinuclear complexes [52], as well as longer C_6 bridges [53]. The auxiliary ligand sphere of butadiynediyl-bridged Ru_2 complexes was altered, comparing again experimental spectroelectrochemical studies and associated BLYP35-D3-based computations, by replacing the PPh_3 ligands by dppe, or additionally the Cp ligands by Cp* [50]. Addition of D3 dispersion corrections to the earlier computational protocol was found to be important, as for larger ligands, dispersion interactions between the terminal ligands became more notable. Indeed, the larger steric bulk was found computationally to increase conformational barriers and to give more weight to conformations with roughly perpendicular arrangements of the two metal fragments. This led to more pronounced high-energy shoulders in the NIR band

and to broader peaks and the development of shoulders in the C≡C stretching bands in the IR spectra, consistent with the computed larger role of more weakly coupled conformations [50]. T-shaped stacking interactions between the phosphine ligands and the Cp or Cp* ligands of the opposite fragment also played a role in these conformational preferences. A subsequent extension to compare a series of Ru- and Fe-based homodinuclear MV complexes with butadiynediyl bridges, dppe and Cp or Cp* auxiliary ligands, confirmed both the suitability of the simplified protocol based on only three conformers and the need to use local hybrid (LH) fuctionals for the class II Fe_2 systems in computing the excitation spectra [51]. Most importantly, this work established the Fe_2 complexes both experimentally and computationally as being largely on the class II side of the border on the IR timescale (see Figure 3.2), in contrast to earlier work that had interpreted the limited available spectra on the basis of class III character [51]. That is, replacement of the 4d Ru centers by the more compact 3d Fe centers reduces the electronic coupling to the extent of changing the character of the MV complexes substantially. Notably, it is difficult to extract these differences between the largely class III Ru_2 species and the largely class II Fe_2 species from the NIR bands. These all feature similar shapes, with high-energy shoulders contributed by the more perpendicular conformers. It is the IR spectra that provide the clearest evidence for or against partial localization of charge: as we move toward more clear-cut class II situations, the intensity of symmetric CC stretching bands increases substantially relative to the asymmetric ones. This was found both computationally and experimentally. Indeed, these observations led to an estimate of the electron transfer rate in the Fe_2 complexes to be in the range 10^{-9}-10^{-13} s. Furthermore, the computations suggested IVCT bands in the IR region for both Ru_2 and Fe_2 MV complexes, which could indeed be identified experimentally [51].

Given the substantial computational effort required to sample the entire conformational landscape for more complex systems, in a subsequent study on a series of dicationic homo- and heterobinuclear Ru and Fe MV complexes with the asymmetric cyanoacetylide bridging ligand (CCCN), a simplified computational protocol was therefore established [52]. It was shown that to reproduce the high-energy shoulder in the NIR band and the additional features in the vibrational spectra due to more perpendicular conformations, it is largely sufficient to include only one conformation in the perpendicular range in addition to one *trans*- and one *cis*-type conformer. Interestingly, this study provided examples of conformationally driven redox isomerism for a few of the homodinuclear species. That is, while for most homonuclear complexes, the extra metal-centered electron hole is generated at the more electron-rich nitrogen end of the bridging ligand, for a few complexes, the conformation determines the site of the hole. These are additional examples that show the need for a more detailed analysis that goes beyond the usual Mulliken–Hush scheme for one given conformer, taking such conformational effects into account. In this work, it was first found that the usual BLYP35-based protocol fails to reproduce the NIR bands for cases with localized holes at Fe^{III} sites. This led to the first use of LHs to reduce problems with spin contamination and give improved TDDFT excitation spectra [52].

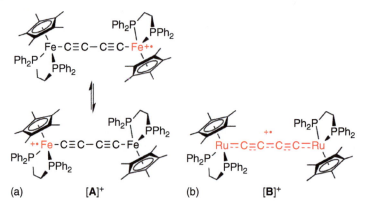

Figure 3.2 Approximate representation of the contrast between a localized class II situation for butadiyne-diyl-bridged MV Fe_2 complexes and the class III situation for the corresponding Ru_2 species. Source: Gückel et al. [53]/American Chemical Society.

Most recently, our studies were extended to related complexes featuring hexa-1,3,5-triyn-1,6-diyl bridges [53]. Again, the Fe_2 MV complexes were confirmed to be largely on the class II side, in contrast to their largely class III Ru_2 homologues. This also contrasted with earlier work on such complexes, which had the Fe_2 systems assigned as class III. Again, the IR bands provided the clearest evidence for these differences, in particular via different numbers of observable CC stretching modes related to the C_6 bridge. Of course, in all such complexes, part of the spin is on the bridging ligand, albeit with differences for different metal centers, different ligands, and different conformations. Note that in this work, the more recent LH20t functional [54] was used in the TDDFT calculations to provide accurate excitation spectra [53]. The importance of different conformations was also demonstrated, again both experimentally and computationally, for the ligand-centered mixed valency of a series of mono-oxidized ruthenium *trans*-diacetylide complexes [55].

Independent experimental work on the class II diiron complex [{Cp*(dppe)Fe−C≡C−}$_2$-μ-(1,2-C_6H_4)] highlighted the steric hindrance of the o-substituted phenylene bridge [56]. Not only does it apparently lead to an enhanced stability of the complex but it also seems to restrict the conformational space, leading to relatively narrow IVCT bands. The complex bands associated with the C≡C stretching modes of a class II Ru_2 complex bridged by a diethynyl-connected TAA unit [57] were attributed to different conformations that give rise to variable electronic coupling, in line with the above discussion. In contrast, a related complex with a larger bridge was assumed to exhibit only a single conformation. A modified BLYP30/PCM protocol was applied here to assist the spectroelectrochemical studies [57]. Multiple conformations were also invoked to explain the multiple C≡C stretching bands for a series of Ru_2 MV complexes with different bridging ligands such as biphenyl, fluorene, phenanthrene, and pyrene [58]. Again, the spectroelectrochemical studies were complemented by BLYP35/PCM-type computations. While the perpendicular conformers could not be located computationally, their existence was assumed based on the inability of the spectroscopic results for the *cis*- and *trans*-like conformers to explain the extra IR bands and the high-energy shoulders of the IVCT bands. Comparable results were obtained for MV Ru_2 complexes with different benzodithiophene bridges, again in combination between spectroelectrochemistry and BLYP35/PCM computations [59]. We mention related work on MV Ru_3 complexes with terthiophene bridges [60]. Here, the BLYP35/PCM calculations did not involve conformational searches because of system size, but the interpretation of the IR data was nevertheless based on conformational motion. We note in passing that this work also suggests some limitations of the BLYP35/PCM protocol regarding the simulation of NIR bands. We also mention a recent review that refers to conformational aspects [61].

These recent works show that, from the perspective of the auxiliary ligands, both electronic and steric aspects matter. That is, in addition to the question how much electron density is provided to the metal centers by these ligand sets, it is also crucial to know how they shape the conformational distributions. That is, bulkier ligand sets may favor more perpendicular arrangements of the metal fragments, which in turn will exhibit smaller electronic coupling, potentially even to the extent that the Robin–Day character of the various conformers may differ (see above). This will then determine the spectroscopic characteristics of a given complex. Obviously, we may expect these considerations to also extend to changes in the bridging ligand, as can already be inferred from comparing the situation for the linear C_4, C_3N, or C_6 bridges we discussed above to the diethynylbenzene bridge. In recent work [62], our team with Paul Low's group has examined the influence of substituents on the benzene ring in the latter type of bridge for a larger series of Ru_2 MV complexes. We found again a pronounced interplay between the electronic characteristics (electron donation or withdrawal by the substituent[s]) and the steric requirements, which dominate the conformational distributions and thereby the contributions to the spectroscopic characteristics provided by more delocalized and more localized conformers. Such systematic computational and experimental studies provide detailed understanding on how modifications to these molecular systems help change electron transfer character and the spectra.

3.6 Other Applications to Organometallic MV Systems

Further applications of the BLYP35/PCM or related protocols to organometallic transition metal MV systems shall be mentioned here without going into detail. Examples include a stimuli-responsive switch [63], diethynyl oligophenylamine-bridged diruthenium and diiron complexes [64], or Ru_2 MV complexes with bridging diethynyl-polyaromatic ligands [63] (all using BLYP35/PCM approaches to complement experimental studies). Other applications to related systems using the BLYP30 functional with reduced exact exchange admixture should be mentioned [57], as well as work using the ωB97X-D or the CAM-B3LYP range-separated hybrid functionals [47].

Rather different organometallic systems involving sulfur-ligand-centered radical character in a series of Ni, Pd, and Pt thiolato complexes with distinct chelate bonding modes have also been studied using the BLYP35/PCM model [65] (and additionally [66] using the related PBE0-1/3 functional [67] having 33.3% exact exchange admixture). Overall, reasonable trends of the Robin–Day characters in dependence of the metal centers and ligand framework have been found, allowing detailed analyses of the factors controlling the electronic structure and spectroscopic finger prints in such systems. The protocol overestimates somewhat the NIR excitation energies for class III cases, as expected.

3.7 Limitations of the Simple Computational Protocols, Gas-Phase Benchmarks, and Improved Electronic Structure Methods

As we extended our studies on the conformational aspects, complexes with their charge localized on Fe^{III} sites came into play, and these demonstrated limitations of the BLYP35-based pragmatic protocol (see above). That is, the low-energy electronic excitation bands in the NIR region were not properly reproduced anymore by TDDFT-BLYP35-PCM calculations [50, 52, 53]. We started to suspect the problem to originate with substantial spin contamination of the ground states of these class II systems when using a global hybrid (GH) with large EXX admixture, such as BLYP35. The spin contamination is seen from S^2 expectation values significantly above the nominal value of 0.75 for a doublet state in the unrestricted Kohn–Sham (UKS) calculations on these systems, while no such problems arose for the analogous delocalized Ru_2 MV complexes or for organic MV systems we had studied previously. Indeed, when using TDDFT-B3LYP calculations (with only 20% EXX admixture) at the BLYP35-optimized structures, we found some improvements in the agreement with the experimental spectra. Of course, ground-state optimization with B3LYP would give a delocalized class III situation and is thus not a suitable remedy. That is, relatively high global EXX admixture is needed to describe properly the class II character, but the ensuing spin contamination prevents the correct reproduction of the NIR spectra. It is fair to say that to some extent, this does already signal some multi-reference character of the systems in question.

To escape this dilemma, we resorted to using LH functionals, which our group has been developing since around 2006 (see Ref. [68] for a recent review). In these types of functionals, EXX admixture is done in a position-dependent way, governed by a so-called local mixing function (LMF). Like the abovementioned long-range corrected hybrids that vary exact exchange admixture along the interelectronic distance or the simpler global hybrids with constant exact exchange admixture, local hybrids are on the fourth or "hyper-GGA" rung of the usual ladder hierarchy of density functionals. Their more flexible form offers substantial advantages in different fields but also brings in extra requirements for the efficient implementation into computer codes, as well as some more fundamental issues around the ambiguity of exchange energy densities (the so-called gauge problem of LHs) [68].

At the time of our initial attempts for the larger Fe complexes of interest mentioned above (around 2017/2018), structure optimizations and vibrational frequency calculations with LHs were still rather computationally demanding, and so we limited our evaluations to single-point ground-state and TDDFT excited-state calculations at the BLYP35-optimized ground-state structures. This did indeed produce spectra in substantially improved agreement with experiment, and we could show that spin contamination with the applied first-generation LH (Lh12ct-SsirPW91 [69]) was substantially reduced compared to BLYP35. Of course, a full LH-based treatment of all properties of such larger MV complexes is desirable, and work along these lines has been aided by an implementation of these functionals further optimized for computational efficiency [70]. In another work, the recent next-generation LH20t functional [54] has been applied to MV iron complexes [53], as it incorporates a calibration function to deal with the so-called gauge problem and thus provides substantially improved performance for main-group thermochemistry and transition metal systems alike, and it improves particularly on weak noncovalent interactions (when combined with dispersion corrections) that become important for conformational studies on MV systems with bulky groups. LH20t does indeed provide excellent NIR bands for the abovementioned class II iron systems. It has also been used in the just published, abovementioned extensive computational and spectroscopic study of 1,4-diethynylbenzene-bridged diruthenium complexes [62]. Full use of the LH20t local hybrid for all aspects of the analyses of larger organometallic MV systems (structure optimizations, vibrational frequency analyses, and TDDFT computations of excitation spectra) has just been applied to understand the spectra and electron transfer characteristics of some class-II 1,3-diethynylbenzene-bridged diruthenium complexes [71]. In the long run, we expect that corrections for strong correlations, applied to local hybrids in the spirit of Ref. [72], may further improve such spectra and remove problems with spin contamination. Moreover, the real-space exact exchange admixture of local hybrids can be combined with long-range corrections in the interelectronic distance space (see above), leading to so-called range-separated local hybrid functionals [73]. This should help, in particular, in TDDFT treatments of longer range charge transfer excitations, as discussed above. Recent work along these lines in our group appears promising, including for MV applications.

As we try to improve our electronic structure methods for MV systems further, a frequently asked critical question is to what extent practical applications of such computational methods benefit from error compensation between the electronic structure method and the applied solvent model. Benchmarking against gas-phase data or against high-level computational data would be highly desirable. Until recently, we thought this to be elusive. The reason is that (i) reliable high-level computational benchmark data can only be generated for small systems, where we thought the electronic coupling between the redox centers to be so large as to generally lead to fully delocalized class III systems only. This would preclude studies near the class II/III borderline. In addition, (ii) typically, MV systems are studied in the condensed phase only, and spectroscopic data that inform on the localized/delocalized character of a gas-phase MV species may be hard to obtain.

Fortunately, exceptions to these assumptions exist. We came across a first example by accident: within a Berlin collaborative research center on oxide-based systems, we learned about gas-phase IRMPD studies of the radical anion $[Al_2O_4]^-$ and the fact that standard DFT methods such as B3LYP give far too few signals in the gas-phase vibrational spectra, suggesting a too symmetrical structure (D_{2h} vs. C_{2v}) [74]. In contrast, the BHLYP functional with 50% EXX admixture gave much better agreement with the spectra. This motivated us to look more closely at this system, and we found that it is indeed a small gas-phase class II MV system, with terminal oxygen atoms exchanging an electron between an oxyl-radical and an oxo anion form [75]. The computed barrier for thermal electron transfer is relatively large, about 70 kJ/mol at very high composite computational levels that correspond essentially to a CCSDT(Q) calculation at the complete basis set limit. Interestingly, the system actually exhibits a shallow additional minimum on top of this barrier. The latter corresponds to a bridge-localized state, where the electron hole is largely shared by the two bridging oxygen atoms. The small size of the system allowed an exhaustive computational evaluation of approximate methods against the generated high-level coupled cluster data (which provides an estimated energy accuracy of about 1 kJ/mol). DFT functionals with larger EXX admixtures provide a qualitatively correct energy profile, while too large EXX admixtures as in BHLYP led to an overestimate of the barrier and of the energy of the shallow bridge-localized D_{2h} minimum relative to the low-lying C_{2v} minima. Lower EXX admixtures gave only an artificially delocalized D_{2h} form where the hole is shared by the two terminal oxygen atoms. Intermediate admixtures provided the barrier and the extra D_{2h} minimum but also symmetrized the low-lying structures to D_{2h}. Apart from BLYP35, a number of GHs (e.g. PBE0-1/3 with 33.3% EXX admixture or the highly parameterized MN15 with 44%) and some range-separated hybrids (CAM-B3LYP and ωB97X-D) also provided a description of this system in reasonable agreement with the benchmark data [74]. The latter aspect is important in view of the usefulness of long-range corrections for treating longer range charge transfer excitations [49].

This provided the starting point for a more extensive benchmarking against high-level computations and gas-phase vibrational spectra for MV oxo systems. We generated the MVO-10 test set encompassing 10 oxo systems that are either class II or class III and have either metal- or oxo-centered redox character [76]. To exemplify the challenge of this test set for approximate electronic structure methods, we can restrict our discussion just to the combination of the abovementioned $[Al_2O_4]^-$ and the complex $[V_4O_{10}]^-$. The latter is currently too large for high-level benchmark computations, but gas-phase IRMPD spectra show it to be a metal-centered D_{2d}-symmetrical class III case [77], where the spin density is symmetrically delocalized over the d-orbitals of all four vanadium atoms. It is estimated that the local measurement temperature corresponds to less than 50 K, so a thermal ET barrier for a class II system would have to be below about 5 kJ/mol if it exists at all. In the initial screening of a wide variety of DFT functionals, none of these gave a simultaneously correct description of both complexes [75]: standard functionals without or with low EXX admixtures gave the correct delocalized structure for $[V_4O_{10}]^-$ but also artificially delocalized $[Al_2O_4]^-$. Functionals that gave correct results for the latter system (see above) erroneously led to a distorted C_s-symmetrical structure for the former class III system, with D_{2d} barriers between about 7 and 10 kJ/mol for some methods up to many tens of kJ/mol for others. The lowest barriers among the functionals studied in that work that at the same time gave the correct behavior for the aluminum system were provided by MN15 (6.7 kJ/mol), by a simple custom LH (8.3 kJ/mol), and by the range-separated hybrid ωB97X-D (13.3 kJ/mol). The BLYP35 functional gave a larger barrier (26.4 kJ/mol), while behaving well for the aluminum system. The LH12ct-SsirPW92 local hybrid mentioned above gets the vanadium complex correctly delocalized and finds a reasonable barrier and local bridge-localized maximum for the aluminum complex but delocalizes the low-lying minima for the latter [75]. We thus found the combination of already just these two systems to be a formidable challenge for electronic structure methods, and we applied this test also to our recently constructed LH20t functional, that we found to perform excellently for many other questions. Gratifyingly, LH20t is indeed the first functional among a wide selection of methods from all five rungs of the usual ladder that provides the correct energy profiles for both $[Al_2O_4]^-$ and $[V_4O_{10}]^{-54}$ (see Figure 3.3). As it also gives very good TDDFT excitation spectra for the abovementioned class II iron complexes, it can clearly be recommended as an overall outstanding method to treat MV systems near the divide between class II and class III. Note that the challenge of finding a suitable functional here is that a subtle balance has to be struck between reduced delocalization errors and an accurate description of left–right correlation in bonds.

Note that a recent TDDFT application of the LH12ct-SsirPW92 local hybrid to the vibronic phosphorescence spectra of tris-bipyridyl complexes demonstrated that such a functional provides a triplet excited state with a spin density properly

Figure 3.3 Spin density distributions of stationary points for $[Al_2O_4]^-$ and $[V_4O_{10}]^-$ at the LH20t/def2-TZVP level and energy profiles with relevant energy differences (in kcal/mol) compared to highly accurate coupled cluster reference data. Source: Haasler et al. [54]/Reproduced with permission from American Chemical Society.

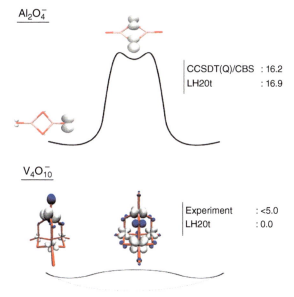

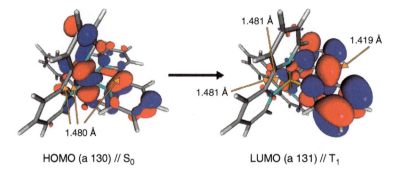

Figure 3.4 Localized nature of the lowest energy triplet excitation computed with a local hybrid for [Ru(bpy)$_3$]$^{+2}$. Source: Grotjahn and Kaupp [78]/Reproduced with permission from American Chemical Society.

localized to one of the ligands (see Figure 3.4), while standard functionals artificially delocalize it over all three ligands [78]. This shows the potential to treat MV character near the class II/III borderline also for excited states within a TDDFT framework. We note in passing that the assumption that the charge transfer character of IVCT excitations would generally preclude the use of TDDFT approaches [7] is not warranted. Long-range CT requires suitable long-range corrections, while many IVCT excitations of shorter range will also be adequately treated with some other types of functionals (see above). A recent benchmark for solid oxides should also be mentioned [79].

Another gas-phase study, in this case, on an organic MV system, the radical cation of *N,N*'-dimethylpiperidine (DMP$^+$), has been combined with high-level computational methods. Rydberg spectroscopy of neutral DMP has been interpreted in terms of the coexistence of a lower lying delocalized class III form of the cation (DMP-D$^+$, C$_{2h}$ symmetry) and a higher lying (by about 30 kJ/mol) localized class II form (DMP-L$^+$, C$_s$ symmetry), separated by a small barrier [80]. The spectra are taken to indicate that DMP-L$^+$ is formed initially but equilibrates with DMP-D$^+$ within a few ps. Here, even functionals that we would expect to clearly favor class II situations (e.g. M06-2X or M06-HF) did not provide the DMP-L$^+$ minimum, and only the BHLYP functional actually produced the barrier [81]. A controversy has developed over the question if this is a failure of the other DFT methods. Cheng et al. obtained a relatively large barrier between the two states using a PZ-SIC (Perdew–Zunger self-interaction correction [82]) DFT functional [80]. However, the reliability of this approach has been questioned. In a comment on that work, Ali et al. showed that CCSD(T)/aug-cc-pVDZ single-point calculations did not give a barrier and thus did not confirm the existence of the DMP-L$^+$ minimum (while CCSD with the same basis produced a small barrier) [83]. Cheng et al. replied that CCSD(T) may fail because of multi-reference character of the system [84] and later produced MRCI+Q/aug-cc-pVDZ data (i.e. multi-reference configuration interaction with Davidson's + Q correction and a CAS(11, 12) reference) based on single-point energies on a BHLYP/aug-cc-pVDZ energy surface, which gave a barrier [85]. They stated that this shows clearly that the metastable DMP-L$^+$

state is real. Remaining uncertainties pertain to the relatively small aug-cc-pVDZ basis sets in these computations, which certainly do not allow to recover all of the dynamical correlation, and to possible imbalances regarding the choice of active space. In addition, there remain some doubts that this system is indeed a case that requires multi-reference treatments. Of course, there is the experimental evidence of the Rydberg spectra for DMP that suggest a second metastable state [79, 80], but to this author, the matter appears far from settled.

The last example shows how involved even discussions of relatively small gas-phase MV species can get. In addition, while multi-reference character near crossing points, and by inference at electron transfer barriers, is a clear possibility one should always look for, our experience suggests the following: (i) By no means, all transition states involved have multi-reference character. In fact, in most cases we studied, the static correlation tended to be at best moderate, typically treatable at single-reference levels such as CCSD(T) and thus also accessible by appropriate DFT approaches. This may also reflect the fact that not all transition states for thermal electron transfer correspond to conical intersections [86]. In fact, often the ground- and excited-state surfaces are well separated, corresponding often to strongly avoided crossings. This then reflects clearly adiabatic processes. (ii) Importantly, the effects of dynamical correlation tend to be at least as important as static correlation or more so. This is a challenge for multi-reference ab initio methods, which excel at treating static correlation while inclusion of dynamical correlation tends to become extremely demanding in such frameworks. In particular, multi-reference CI or multi-reference perturbation theory tends to exhibit very slow basis set convergence. Therefore, calculations with smaller basis sets cannot be expected to recover dynamical correlation fully. This leads typically to an overestimate of charge localization.

In this context, it is worth mentioning theoretical treatments of another relatively simple organic MV radical cation, the 5,5'(4H,4H')-spirobi[cyclopenta[c]pyrrole]2, 2',6,6'-tetrahydro cation, often termed "spiro cation." While no experimental data are available for this system, its relatively small size and high symmetry have motivated its use as a model to evaluate different computational approaches, in particular, different multi-reference schemes. Interestingly, state-specific CASPT2 or NEVPT2 were found to provide an incorrect description of the potential energy surface near the coupling region [87]. While this was initially cured by an averaging of CASSCF orbital energies over two electronic states, or by going to the NEVPT3 level, most recently, it was shown that state-averaged multi-state approaches do not exhibit the problem [88]. Moreover, state interaction pair density functional theory (SI-PDFT) has been suggested as an approach to include dynamical correlation at moderate cost and with fast basis set convergence [87].

3.8 More Advanced Treatments of Environmental Effects

Most of the computational studies discussed above used PCM-type solvent models, which typically appear to work well for most aprotic solvents. We have already mentioned that the more refined D-COSMO-RS model has been able to describe the changes from polar aprotic to protic solvents for some organic MV systems because of its parameterized hydrogen-bonding contributions. Of course, this type of model also has its limits, e.g. in cases with other specific solute–solvent interactions, and it provides only limited insights into how the solvent distribution or, for example, entropic aspects affect the electronic structure of the solute. The most advanced treatment of the solvent environment is the explicit quantum mechanical inclusion of the solvent molecules within the framework of ab initio molecular dynamics (AIMD). Because of their intrinsically very high computational effort, such methods have so far not been pursued much for realistic MV systems. We can mention here unpublished AIMD simulations (S. Gückel, M. Kaupp, in preparation) using the BLYP35 functional on the radical anions of 1,4-dinitrobenzene and 1,3-dinitrobenzene in acetonitrile vs. methanol, meant to provide benchmark data for more approximate solvent models, such as those used in Ref. [10]. Simulations over several ps reproduced the class II character and the associated bond length differences for 1,3-dinitrobenzene in both solvents and for 1,4-dinitrobenzene in methanol, as well as the class III character of the latter radical anion in acetonitrile, in agreement with ESR data [10]. So far, no accurate estimates of the reaction barriers could be obtained at this level. Earlier SA-CASSCF-based short AIMD runs for 1,3-dinitrobenzene under vacuum [89] suggested a small ET barrier, but this is likely because of the lack of dynamical correlation. Mori found in any case that the timescale for ET found in this way is much shorter than that obtained experimentally in solution. Subsequent classical and QM/MM MD simulations in acetonitrile and methanol (with the structure of the solute pre-optimized quantum chemically and kept frozen) provided estimates of the solvent reorganization energies [88]. MD-based estimates have also been obtained for the Creutz–Taube ion, where the coupling between solute and solvent dynamics is crucial for the ET rates because of its prototypical class II/III character [7].

If one does not aim for the dynamical aspects but for a more realistic time-averaged description of solvation, less costly and more practical alternatives based on integral equation models are feasible. While the abovementioned D-COSMO-RS treatment may be categorized in this way, the more traditional statistical thermodynamics treatments of liquids offer further physical details, including an average solvent distribution. Such methods may be viewed as different manifestations of a density functional theory of classical particles. The most widely used approaches in this field are based on the reference interaction site model (RISM) theory [90]. Both 1D-RISM and 3D-RISM models and combinations thereof are available. They have been coupled to electronic structure treatments of the solute by coupling the solvent interaction terms into, for example, the self-consistent field iterations of DFT or CASSCF implementations (RISM-SCF) [91], providing a self-consistent reaction field in a similar way as done for continuum models. The solvent potential corresponds essentially to that of a classical force field including Lennard-Jones as well as electrostatic and dispersion terms. That is, RISM-SCF approaches may be viewed as providing an approximation to the long-time average of a large-scale QM/MM-MD simulation. A few applications of RISM models to questions of charge localization/delocalization in MV systems in solution are available. Yoshida et al.

used RISM-SCF theory to model the abovementioned 1,3-dinitrobenzene radical anion in acetonitrile and methanol using CASSCF wave functions [92]. Because of the lack of dynamical correlation, this computational level is expected to overestimate charge localization. However, the larger polarization and increased free energy barrier in the protic solvent was clearly confirmed, as was the reduction of the barrier with increasing temperature. Another application, in this case of DFT-based 3D-RISM-SCF, was to the localization/delocalization of the ligand-based mixed valency in two one-electron oxidized salen complexes of Ni(II) and Mn(III) [93]. The M06 functional was used in this work, which suggests a slight bias toward delocalization by the underlying DFT approach. Subsequent single-point computations of excitation energies used two-state multi-reference quasi-degenerate perturbation theory, including the previously computed 3D-RISM-SCF potential. The computations agreed with the experimentally inferred class II character of the Mn(III) complex and the class III character of the Ni(II) complex with a symmetrical salen ligand (in dichloromethane). A subsequent extension of this study to a larger set of group 7 and group 10 homologues confirmed these findings but found the Re(III) complexes to fall out of line by featuring metal-centered oxidation and no MV character [94]. Both the structure of the salen ligand and solvent polarity (e.g. CH_2Cl_2 vs. H_2O or methanol) were found to be important in determining electronic and molecular structure.

3.9 Conclusion and Outlook

If we want to understand electron transfer processes in atomistic detail, we need to have reliable quantum chemical methods that provide us with the potential energy surfaces of the underlying processes. MV systems are the basic models for electron transfer and thus a crucial starting point to develop, refine, and apply the methodology. One complication is that most experimental studies of MV systems involve polar solvent environments, and so, our simulations need to take those into account in a suitable and computationally manageable way. In fact, we should conceptually view the entirety of the MV ion and its (solvent or other) environment as the system of interest, the character of which we want to discern. The various ways to approximate the solvent environment in such studies have thus been an important topic we have discussed in this chapter. They range from computationally demanding ab initio molecular dynamics methods to the simplest dielectric continuum models, and the choice of a given model has to be made with the desired accuracy, the nature of the system in question, and the required computational effort in mind.

However, any such study needs to use an electronic structure method for the central MV part of the entire system that captures the essential balance between localization and delocalization of charge, which has been a challenge for a long time. This author thinks that suitable methodologies applicable to larger molecules have become available more than a decade ago but are still being constantly improved and refined. Such protocols are typically based on density functionals that provide a reasonable balance between low delocalization errors and an adequate simulation of both static and dynamical correlation. Such methods then have to be combined with the abovementioned suitable treatments of environmental effects. I have been careful to emphasize that static correlation effects, while important near the intersection points of ground- and excited-state surfaces of weakly coupled MV systems, play a less crucial role than often assumed in many cases with stronger electronic couplings, where the different adiabatic surfaces tend to be well separated energetically. On the other hand, some static correlation can of course become important in practical studies and then has to be accounted for. For example, spin contamination of the ground state in certain localized transition metal MV systems, when using exchange–correlation functionals with larger exact exchange admixtures, is a sign of static correlation, and I have mentioned ways to deal with this using local rather than global hybrid functionals. Ongoing methodological developments that allow an even better inclusion of static correlation into DFT calculations make me confident that such cases can often be described well even for large systems.

I have started this chapter with a historical overview of how I got involved in the field and how a pragmatic computational protocol based on a suitably tailored global hybrid functional (BLYP35) and standard polarizable continuum models allowed us to learn a lot for both organic and organometallic systems together with our experimental collaborators Christoph Lambert and Paul J. Low, culminating in an improved understanding of the influence of conformational averaging on the spectroscopic characteristics or organometallic molecular wires that is still being extended and applied by us and by other researchers in the field.

The larger part of this chapter then deals with more recent developments made since our comprehensive 2014 review. This includes a large variety of applications by many groups of the BLYP35/PCM or closely related protocols to organic and organometallic species, all the way to work on questions of charge transport along molecular wires of increasing length and different types. A deeper understanding of steric and electronic ligand effects on the conformational distributions and thus on localization/delocalization has also been achieved, and many other fruitful interactions between computational and experimental studies have emerged.

The computational methodologies are also still being refined regarding the electronic structure method (new density functional approaches as well as single- and multi-reference ab initio methods) as well as the modeling of the (solvent, solid, or surface) environment. As more complex systems come into focus, the computational efficiency will become an even more important aspect. It will be interesting to see if faster methods such as density functional tight-binding approaches may be adapted to provide a reasonable description of localization/delocalization in MV systems. In addition, I am confident that the new methodologies and insights produced for MV systems will then be transferred to other studies of important electron transfer processes.

Acknowledgement

The author is grateful to Deutsche Forschungsgemeinschaft for funding his own contributions to the field of this article.

References

1 Kaupp, M. and Parthey, M. (2014). Quantum-chemical insights into mixed-valence systems: within and beyond the Robin/Day scheme. *Chem. Soc. Rev.* 43: 5067–5088.

2 (a) Lambert, C., Amthor, S., and Schelter, J. (2004). From valence trapped to valence delocalized by bridge state modification in bis(triarylamine) radical cations: evaluation of coupling matrix elements in a three-level system. *J. Phys. Chem. A* 108: 6474–6486. (b) Nelsen, S.F., Reinhardt, L.A., Tran, H.Q. et al. (2004). Electron transfer within 2,7-dinitronaphthalene radical anion. *J. Am. Chem. Soc.* 126: 15431–15438.

3 Perdew, J.P., Ruzsinszky, A., Tao, J. et al. (2005). Prescription for the design and selection of density functional approximations: more constraint satisfaction with fewer fits. *J. Chem. Phys.* 123: 062201.

4 Cohen, A.J., Mori-Sánchez, P., and Yang, W. (2008). Insights into current limitations of density functional theory. *Science* 321: 792–794.

5 Renz, M., Theilacker, K., Lambert, C., and Kaupp, M. (2009). A reliable quantum-chemical protocol for the characterization of organic mixed-valence compounds. *J. Am. Chem. Soc.* 131: 16292–16302.

6 Kaupp, M., Renz, M., Parthey, M. et al. (2011). Computational and spectroscopic studies of organic mixed-valence compounds: where is the charge? *Phys. Chem. Chem. Phys.* 13: 16973–16986.

7 (a) Reimers, J.R., Cai, Z.-L., and Hush, N.S. (2005). A priori evaluation of the solvent contribution to the reorganization energy accompanying intramolecular electron transfer: predicting the nature of the Creutz–Taube ion. *Chem. Phys.* 319: 39–51. (b) Reimers, J.R., Wallace, B.B., and Hush, N.S. (2005). Towards a comprehensive model for the electronic and vibrational structure of the Creutz–Taube ion. *Phil. Trans. R. Soc. A* 366: 15–31.

8 Tomasi, J., Mennuci, B., and Cammi, R. (2005). Quantum mechanical continuum solvation models. *Chem. Rev.* 105: 2999–3093.

9 Renz, M., Kess, M., Diedenhofen, M. et al. (2012). Reliable quantum chemical prediction of the localized/delocalized character of organic mixed-valence radical anions. From continuum solvent models to Direct-COSMO-RS. *J. Chem. Theor. Comput.* 8: 4189–4203.

10 (a) Grampp, G., Shohoji, M.C.B.L., Herold, B.J., and Steenken, S. (1990). Solvent induced intramolecular electron exchange kinetics in 1,3-isodisubstituted aromatic radical anions II. Activation parameters. *Ber. Bunsenges.* 94: 1507–1511. (b) Telo, J.P., Grampp, G., and Shohojia, M.C.B.L. (1999). Solvent effects on intramolecular electron exchange in the 1,4-dinitrobenzene radical anion. *Phys. Chem. Chem. Phys.* 1: 99–104.

11 Sinnecker, S., Rajendran, A., Klamt, A. et al. (2006). Calculation of solvent shifts on electronic g-tensors with the conductor-like screening model (COSMO) and its self-consistent generalization to real solvents (Direct COSMO-RS). *J. Phys. Chem. A* 110: 2235–2245.

12 Eckert, F. and Klamt, A. (2002). Fast solvent screening via quantum chemistry: COSMO-RS approach. *AIChE J.* 48: 369–385.

13 Renz, M. and Kaupp, M. (2012). Predicting the localized/delocalized character of mixed-valence diquinone radical anions. Towards the right answer for the right reason. *J. Phys. Chem. A* 43: 10629–10637.

14 Parthey, M., Gluyas, J.B.G., Schauer, P.A. et al. (2013). Refining the interpretation of NIR band shapes in a polyynediyl molecular wire. *Chem. Eur. J.* 30: 9780–9784.

15 Fox, M.A., Le Guennic, B., Roberts, R.L. et al. (2011). Simultaneous bridge-localized and mixed-valence character in diruthenium radical cations featuring diethynylaromatic bridging ligands. *J. Am. Chem. Soc.* 133: 18433–18446.

16 Parthey, M., Gluyas, J.B.G., Fox, M.A. et al. (2014). Mixed-valence ruthenium complexes rotating through a conformational Robin-Day continuum. *Chem. Eur. J.* 20: 6895–6908.

17 Uebe, M., Kazama, T., Kurata, R. et al. (2017). Recognizing through-bond and through-space self-exchange charge/spin transfer pathways in bis(triarylamine) radical cations with similar geometrical arrangements. *Angew. Chem. Int. Ed.* 56: 15712–15717.

18 Kaupp, M., Gückel, S., Renz, M. et al. (2016). Electron transfer pathways in mixed-valence paracyclophane-bridged bis-triarylamine radical cations. *J. Comput. Chem.* 37: 93–102.

19 Zhang, J., Chen, Z., Yang, L. et al. (2016). Elaborately tuning intramolecular electron transfer through varying oligoacene linkers in the bis(diarylamino) systems. *Sci. Rep.* 6 (36): 310.

20 Krug, M., Fröhlich, N., Fehn, D. et al. (2021). Pre-planarized triphenylamine-based linear mixed-valence charge-transfer systems. *Angew. Chem. Int. Ed.* 60: 6771–6777.

21 Parthey, M., Vincent, K.B., Renz, M. et al. (2014). A combined computational and spectroelectrochemical study of platinum bridged bis–triarylamine systems. *Inorg. Chem.* 53: 1544–1554.

22 Gluyas, J.B.G., Manici, V., Gückel, S. et al. (2015). Cross-conjugated systems based on an (E)-hexa-3-en-1,5-diyne-3,4-diyl skeleton: spectroscopic and spectroelectrochemical investigations. *J. Organomet. Chem.* 80: 11501–11512.

23 Ou, Y.-P., Wang, A., Zhang, F., and Hu, F. (2020). Dendritic groups substituted Kekulé-Benzene-bridged bis(triarylamine) mixed-valence systems: syntheses, characterization and electronic coupling properties. *ChemistrySelect* 5: 4111–4117.

24 Uebe, M. and Ito, A. (2019). Intramolecular charge transfer in Kekulé- and non-Kekulé-bridged bis(triarylamine) radical cations: missing key compounds in organic mixed-valence systems. *Chem. Asian J.* 14: 1692–1696.

25 Zhang, J., Chen, Z., Wang, X.-Y. et al. (2017). Redox-modulated near-infrared electrochromism, electroluminochromism, and aggregation-induced fluorescence change in an indolo[3,2-b]carbazole-bridged diamine system. *Sens. Actuators, B* 246: 570–577.

26 Matsumoto, A., Suzuki, M., Hayashi, H. et al. (2016). Aromaticity relocation in perylene derivatives upon two-electron oxidation to form anthracene and phenanthrene. *Chem. Eur. J.* 22: 14462–14466.

27 Schäfer, J., Holzapfel, M., Mladenova, B. et al. (2017). Hole transfer processes in meta- and para-conjugated mixed valence compounds: unforeseen effects of bridge substituents and solvent dynamics. *J. Am. Chem. Soc.* 139: 6200–6209.

28 Boddeda, A., Hossain, M.M., Mirzaei, M.S. et al. (2020). Angular ladder-type meta-phenylenes: synthesis and electronic structural analysis. *Chem. Front.* 7: 3215–3222.

29 Jahnke, A.C., Proppe, J., Spulber, M. et al. (2014). Charge delocalization in an organic mixed valent bithiophene is greater than in a structurally analogous biselenophene. *J. Phys. Chem. A* 118: 11293–11303.

30 Merz, J., Fink, J., Friedrich, A. et al. (2017). Pyrene molecular orbital shuffle–controlling excited state and redox properties by changing the nature of the frontier orbitals. *Chem. Eur. J.* 23: 13164–13180.

31 Sakamaki, D., Ito, A., Tsutsui, Y., and Seki, S. (2017). Tetraaza[14]- and Octaaza[18]paracyclophane: synthesis and characterization of their neutral and cationic states. *J. Organomet. Chem.* 82: 13348–13358.

32 Herrmann, C. and Kröncke, S. (2019). Designing long-range charge delocalization from first-principles. *J. Chem. Theory Comput.* 15: 165–177.

33 Herrmann, C. and Kröncke, S. (2020). Toward a first-principles evaluation of transport mechanisms in molecular wires. *J. Chem. Theory Comput.* 16: 6267–6279.

34 Herrmann, C. (2019). Electronic communication as a transferable property of molecular bridges? *J. Phys. Chem. A* 123: 10205–10223.

35 Talipov, M.R., Boddeda, A., Timerghazin, Q.K., and Rathore, R. (2014). Key role of end-capping groups in optoelectronic properties of poly-p-phenylene cation radicals. *J. Phys. Chem. C* 118: 21400–21408.

36 Wang, D., Talipov, M.R., Ivanov, M.V., and Rathore, R. (2016). Energy gap between the poly-p-phenylene bridge and donor groups controls the hole delocalization in donor–bridge–donor wires. *J. Am. Chem. Soc.* 138: 16337–16344.

37 Talipov, M.R., Ivanov, M.V., and Rathore, R. (2016). Inclusion of asymptotic dependence of reorganization energy in the modified Marcus-based multistate model accurately predicts hole distribution in poly-p-phenylene wires. *J. Phys. Chem. C* 120: 6402–6408.

38 Hossain, M.M., Ivanov, M.V., Wang, D. et al. (2018). Spreading electron density thin: increasing the chromophore size in polyaromatic wires decreases interchromophoric electronic coupling. *J. Phys. Chem. C* 122: 17668–17675.

39 Wang, D. and Rathore, R. (2018). From static to dynamic: electron density of HOMO at biaryl linkage controls the mechanism of hole delocalization. *J. Am. Chem. Soc.* 140: 4765–4769.

40 Reilly, N., Ivanov, M., Uhler, B. et al. (2016). First experimental evidence for the diverse requirements of excimer vs hole stabilization in π-stacked assemblies. *J. Phys. Chem. Lett.* 7: 3042–3045.

41 Talipov, M.R., Jasti, R., and Rathore, R. (2015). A circle has no end: role of cyclic topology and accompanying structural reorganization on the hole distribution

42 Ivanov, M.V., Reid, S.A., and Rathore, R. (2018). Game of frontier orbitals: a view on the rational design of novel charge-transfer materials. *J. Phys. Chem. Lett.* 9: 3978–3986.

43 Talipov, M.R. and Steiner, E. (2019). Coexistence of structurally similar but electronically distinct isomers of delocalized cation radicals as a basis for the development of functional materials. *Phys. Chem. Chem. Phys.* 21: 10738–10743.

44 Talipov, M.R., Boddeda, A., Lindeman, S.V., and Rathore, R. (2014). Does Koopmans' paradigm for 1-electron oxidation always hold? Breakdown of IP/Eox relationship for p-hydroquinone ethers and the role of methoxy group rotation. *J. Phys. Chem. Lett.* 6: 3373–3378.

45 Ivanov, M.V., Talipov, M.R., Navale, T.S., and Rathore, R. (2018). Ask not how many, but where they are: substituents control energetic ordering of frontier orbitals/electronic structures in isomeric methoxy-substituted dibenzochrysenes. *J. Phys. Chem. C* 122: 2539–2545.

46 Ivanova, L.V., Wang, D., Lindeman, S.V. et al. (2018). Probing charge delocalization in solid state polychromophoric cation radicals using X-ray crystallography and DFT calculations. *J. Phys. Chem. C* 122: 9339–9345.

47 (a) Ding, Y., Jiang, Y., Zhang, W. et al. (2017). Influence of heterocyclic spacer and end substitution on hole transporting properties based on triphenylamine derivatives: theoretical investigation. *J. Phys. Chem. C* 121: 16731–16738; (b) Roy, S.S., Sil, A., Giri, D. et al. (2018). Diruthenium(II)-capped oligothienylethynyl bridged highly soluble organometallic wires exhibiting long-range electronic coupling. *Dalton Trans.* 47: 14304–14317; (c) Rotthowe, N., Zwicker, J., and Winter, R.F. (2019). Influence of quinoidal distortion on the electronic properties of oxidized divinylarylene-bridged diruthenium complexes. *Organometallics* 38: 2782–2799; (d) Jung, H.W., Yoon, S.E., Carroll, P.J. et al. (2020). Distance dependence of electronic coupling in rigid, cofacially compressed, π-stacked organic mixed-valence systems. *J. Phys. Chem. B* 124: 1033–1048. (e) Shen, J.-J., Shao, J.-Y., Gong, Z.-L. et al. (2015). Cyclometalated osmium–amine electronic communication through the p-oligophenylene wire *Inorg. Chem.* 54: 10776–10784; (f) Yao, C.-J., Nie, H.-J., Yang, W.-W. et al. (2015). Combined experimental and computational study of pyren-2,7-diyl-bridged diruthenium complexes with various terminal ligands *Inorg. Chem.* 54: 4688–4698.

48 Reimers, J.R., Sajid, A., Kobayashi, R., and Ford, M.J. (2018). Understanding and calibrating density-functional-theory calculations describing the energy and spectroscopy of defect sites in hexagonal boron nitride. *J. Chem. Theory Comput.* 14: 1602–1613.

49 Li, M., Kobayashi, R., Amos, R.D. et al. (2022). Density functionals with asymptotic-potential corrections are required for the simulation of spectroscopic properties of materials. *Chem. Sci.* 13: 1492–1503.

50 Gluyas, J.B.G., Gückel, S., Kaupp, M., and Low, P.J. (2016). Rational control of conformational distributions and mixed-valence characteristics in diruthenium complexes. *Chem. Eur. J.* 22: 16138–16146.

51 Gückel, S., Gluyas, J.B.G., El-Tarhuni, S. et al. (2018). Iron versus ruthenium: reconciling the spectral differences, and clarifying the electronic structure and mixed-valence characteristics of [{M(dppe)Cp′}$_2$(-CCCC)]$^+$ complexes (M = Fe, Ru). *Organometallics* 37: 1432–1445.

52 Gluyas, J.B.G., Gückel, S., Eaves, S.G. et al. (2019). A spectroscopic and computationally minimal approach to the analysis of charge-transfer processes in conformationally fluxional mixed-valence and heterobimetallic complexes. *Chem. Eur. J.* 25: 8837–8853.

53 Gückel, S., Safari, P., Ghazvini, M.H. et al. (2020). Iron versus ruthenium: evidence for the distinct differences in the electronic structures of hexa-1,3,5-triyn-1,6-diyl-bridged complexes [{Cp*(dppe)M}{μ-(C≡C)$_3$}{M(dppe)Cp*}]$^+$ (M = Fe, Ru). *Organometallics* 40: 346–357.

54 Haasler, M., Maier, T.M., Grotjahn, R. et al. (2020). A local hybrid functional with wide applicability and good balance between (de)localization and left-right correlation. *J. Chem. Theory Comput.* 16: 5645–5657.

55 Marqués-González, S., Parthey, M., Yufit, D.S. et al. (2014). Combined spectroscopic and quantum chemical study of [trans-Ru(C≡CC$_6$H$_4$R^1-4)$_2$(dppe)^2]$^{n+}$ and [trans-Ru(C≡CC$_6$H$_4$R^1-4)(C≡CC$_6$H$_4$R^2-4)(dppe)$_2$]$^{n+}$ (n = 0, 1) complexes: interpretations beyond the lowest energy conformer paradigm. *Organometallics* 33: 4947–4963.

56 Makhoul, R., Sahnoune, H., Dorcet, V. et al. (2015). 1,2-Diethynylbenzene-bridged [Cp*(dppe)Fe]$^{n+}$ units: effect of steric hindrance on the chemical and physical properties. *Organometallics* 34: 3314–3326.

57 Tang, J.H., Shao, J.-Y., He, J.-Q. et al. (2016). Transition from a metal-localized mixed-valence compound to a fully delocalized and bridge-biased electrophore in a ruthenium–amine–ruthenium tricenter system. *Chem. Eur. J.* 22: 10341–10345.

58 Zhang, J., Zhang, M.-X., Sun, C.-F. et al. (2015). Diruthenium complexes with bridgingdiethynyl polyaromatic ligands: synthesis, spectroelectrochemistry, and theoretical calculations. *Organometallics* 34: 3967–3978.

59 Ou, Y.-P., Zhang, J., Zhang, F. et al. (2016). Notable differences between oxidized diruthenium complexes bridged by four isomeric diethynyl benzodithiophene ligands. *Dalton Trans.* 45: 6503–6516.

60 Zhang, J., Sun, C.-F., Zhang, M.-X. et al. (2016). Asymmetric oxidation of vinyl- and ethynyl terthiophene ligands in triruthenium complexes. *Dalton Trans.* 45: 768–782.

61 Launay, J.-P. (2020). Mixed-valent compounds and their properties – recent developments. *Eur. J. Inorg. Chem.* 2020: 329–341.

62 Safari, P., Gückel, S., Gluyas, J.B.G. et al. (2022). The use of bridging ligand substituents to bias the population of localized and delocalized mixed-valence conformers in solution. *Chem. Eur. J.* 28: e202200926.

63 Oyama, Y., Kawano, R., Tanaka, Y., and Akita, M. (2019). Dinuclear ruthenium acetylide complexes with diethynylated anthrahydroquinone and anthraquinone frameworks: a multi-stimuliresponsive organometallic switch. *Dalton Trans.* 48: 7432–7441.

64 Zhang, J., Guo, S.-Z., Dong, Y.-B. et al. (2017). Multistep oxidation of diethynyl oligophenylamine-bridged diruthenium and diiron complexes. *Inorg. Chem.* 56: 1001–1015.

65 Mews, N.M., Hörner, G., Schubert, H., and Berkefeld, A. (2018). Tuning of thiyl/thiolate complex near-infrared chromophores of platinum through geometrical constraints. *Inorg. Chem.* 57: 9670–9682.

66 Mews, N.M., Reimann, M., Hörner, G. et al. Describing the electronic properties of radical-ligand coordination compounds by four principal parameters. *Dalton Trans.* 49: 9735–9742.

67 Guido, C.A., Brémond, E., Adamo, C., and Cortona, P. (2013). One third: a new recipe for the PBE0 paradigm. *J. Chem. Phys.* 138: 021104.

68 Maier, T.M., Arbuznikov, A.V., and Kaupp, M. (2019). Local hybrid functionals: theory, implementation, and performance of an emerging new tool in quantum chemistry and beyond. *WIREs Comp. Mol. Sci.* 9: e1378.

69 Arbuznikov, A.V. and Kaupp, M. (2012). Importance of the correlation contribution for local hybrid functionals: range separation and self-interaction corrections. *J. Chem. Phys.* 136: 014111.

70 Holzer, C. (2020). An improved seminumerical coulomb and exchange algorithm for properties and excited states in modern density functional theory. *J. Chem. Phys.* 153 (184): 115.

71 Harrison, D.P., Grotjahn, R., Mazzucato, D.M. et al. (2022). Quantum interference effects in mixed-valence complexes: tuning electronic coupling through substituent effects. *Angew. Chem., Int. Ed. Engl.* 81: e202211000.

72 Wodyński, A., Arbuznikov, A.V., and Kaupp, M. (2021). Local hybrid functionals augmented by a strong-correlation model. *J. Chem. Phys.* 155 (144): 101.

73 Haunschild, R. and Scuseria, G.E. (2010). Range-separated local hybrids. *J. Chem. Phys.* 132 (224): 106.

74 Song, X., Fagiani, M., Gewinner, S. et al. (2016). Gas phase structures and charge localization in small aluminum oxide anions: infrared photodissociation spectroscopy and electronic structure calculations. *J. Chem. Phys.* 144 (244): 305.

75 Kaupp, M., Karton, A., and Bischoff, F. (2016). $[Al_2O_4]^-$, a benchmark gas-phase class II mixed-valence radical anion for the evaluation of quantum-chemical methods. *J. Chem. Theor. Comput.* 12: 3796–3806.

76 Klawohn, S., Kaupp, M., and Karton, A. (2018). MVO-10. A gas-phase oxide benchmark for localization/delocalization in mixed-valence systems. *J. Chem. Theory Comput.* 14: 3512–3523.

77 Asmis, K.R., Santambrogio, G., Brümmer, M., and Sauer, J. (2005). Polyhedral vanadium oxide cages: infrared spectra of cluster anions and size-induced d-electron localization. *Angew. Chem. Int. Ed. Engl.* 44: 3122–3125.

78 Grotjahn, R. and Kaupp, M. (2021). A reliable TDDFT protocol based on a local hybrid functional for the prediction of vibronic phosphorescence spectra applied to Tris(2,2'-bipyridine)-metal complexes. *J. Phys. Chem. A* 125: 7099–7110.

79 Rugg, G., Genest, A., and Rösch, N. (2018). DFT variants for mixed-metal oxides. Benchmarks using multi-center cluster models. *J. Phys. Chem. A* 122: 7042–7050.

80 Deb, S., Cheng, X., and Weber, P.M. (2013). Structural dynamics and charge transfer in electronically excited N,N'-dimethylpiperazine. *J. Phys. Chem. Lett.* 4: 2780–2784.

81 Cheng, X., Zhang, Y., Jónsson, E. et al. (2016). Charge localization in a diamine cation provides a test of energy functionals and self-interaction correction. *Nat. Commun.* 7: 11013.

82 Perdew, P. and Zunger, A. Self-interaction correction to density-functional approximations for many-electron systems. *Phys. Rev. B* 23: 5048.

83 Ali, Z.A., Aquino, F.W., and Wong, B.M. (2018). The diamine cation is not a chemical example where density functional theory fails. *Nat. Commun.* 9: 4733.

84 Cheng, X., Jónsson, E., Jónsson, H., and Weber, P.M. (2018). Reply to: "The diamine cation is not a chemical example where density functional theory fails." *Nat. Commun.* 9: 5348.

85 Gałyńska, M., Ásgeirsson, V., Jónsson, H., and Bjornsson, R. (2021). Localized and delocalized states of a diamine cation: resolution of a controversy. *J. Phys. Chem. Lett.* 12: 1250–1255.

86 Blancafort, L., Jolibois, F., Olivucci, M., and Robb, M.A. (2001). Potential energy surface crossings and the mechanistic spectrum for intramolecular electron transfer in organic radical cations. *J. Am. Chem. Soc.* 123: 722–732.

87 (a) Farazdel, A., Dupuis, M., Clementi, E., and Aviram, A. (1990). Electric field induced intramolecular electron transfer in spiro a-electron systems and their suitability as molecular electronic devices. A theoretical study. *J. Am. Chem. Soc.* 112: 4206–4214; (b) Helal, W., Evangelisti, S., Leininger, T., and Maynau, D. (2009). Ab-initio multireference study of an organic mixed-valence spiro molecular system. *J. Comput. Chem.* 30: 83–92; (c) Glaesemann, K.R., Govind, N., Krishnamoorthy, S., and Kowalski, K. (2010). EOMCC, MRPT, and TDDFT studies of charge transfer processes in mixed-valence compounds: application to the spiro molecule. *J. Phys. Chem. A* 114: 8764–8771.

88 Dong, S.S., Huang, K.B., Gagliardi, L., and Truhlar, D.G. (2019). State-interaction pair-density functional theory can accurately describe a spiro mixed valence compound. *J. Phys. Chem. A* 123: 2100–2106.

89 Mori, Y. (2014). Computational study on intramolecular electron transfer in 1,3-dintrobenzene radical anion. *J. Phys. Org. Chem.* 27: 803–810.

90 (a) Chandler, D. and Andersen, H.C. (1972). Optimized cluster expansions for classical fluids. II. Theory of molecular liquids. *J. Chem. Phys.* 57: 1930–1937; (b) Hirata, F. and Rossky, P.J. (1981). An extended RISM equation for molecular polar fluids. *Chem. Phys. Lett.* 83: 329–334; (c) Hirata, F., Rossky, P.J., and Pettitt, B.M. (1983). The interionic potential of mean force in a molecular polar solvent from an extended RISM equation. *J. Chem. Phys.* 78: 4133–4144.

91 Sato, H.A. (2013). Modern solvation theory: quantum chemistry and statistical chemistry. *Phys. Chem. Chem. Phys.* 15: 7450–7465.

92 Yoshida, N., Ishida, T., and Hirata, F. (2008). Theoretical study of temperature and solvent dependence of the free-energy surface of the intramolecular electron-transfer based on the RISM-SCF theory: application to the

1,3-dinitrobenzene radical anion in acetonitrile and methanol. *J. Phys. Chem. B* 112: 433–440.

93 Aono, S., Nakagaki, M., Kurahashi, T. et al. (2014). Theoretical study of one-electron oxidized Mn(III)– and Ni(II)–salen complexes: localized vs delocalized ground and excited states in solution. *J. Chem. Theory Comput.* 10: 1062–1073.

94 Aono, S., Nakagaki, M., Fujii, H., and Sakaki, S. (2017). Theoretical study of one-electron-oxidized salen complexes of group 7 (Mn(III), Tc(III), and re(III)) and group 10 metals (Ni(II), Pd(II), and Pt(II)) with the 3D-RISM-GMC-QDPT method: localized vs. delocalized ground and excited states in solution. *Phys. Chem. Chem. Phys.* 19: 16831–16849.

4

Mixed Valency in Ligand-Bridged Diruthenium Complexes

Sanchaita Dey[1], Sudip Kumar Bera[1], Wolfgang Kaim[2], and Goutam Kumar Lahiri[1]

[1]*Indian Institute of Technology Bombay, Department of Chemistry, Powai, Mumbai 400076, India*
[2]*Universität Stuttgart, Institut für Anorganische Chemie, Pfaffenwaldring 55, D-70550 Stuttgart, Germany*

4.1 Introduction

The stabilization of different formal oxidation states of metal ions in polynuclear complexes generates mixed-valent systems, characterized by structural, electrochemical, spectral, magnetic, as well as bridge or metal–metal bond-mediated intramolecular electron transfer from the electronically rich (reduced) metal center to the electron-deficient (oxidized) metal ion [1–8]. The electron transfer process in the mixed-valent state is essentially guided by a series of interlinking factors including (i) the nature of the metal ions in the complex frameworks, (ii) the overall molecular conformation and charge, (iii) the distance between the interacting metal ions, (iv) their redox activities, (v) the nature of the spacer, (vi) the solvent polarity, and (vii) the temperature. Mixed-valent setups in biological systems such as $Mn^{III}Mn^{IV}$ clusters in the oxygen-evolving complex (OEC), Cu^I/Cu^{II} centers in cytochrome-c-oxidase, and $Fe^{II}Fe^{III}$ ions in ferredoxin are well recognized [9–16]. Similarly, spinels (Fe_3O_4, Co_3O_4, and Mn_3O_4) and Prussian blue ($Fe^{III}_4[Fe^{II}(CN)_6]_3 \cdot 14H_2O$) are widely available mixed-valent materials [17, 18].

In this regard, the Creutz–Taube ion (CT ion), i.e. the pyrazine-bridged diruthenium(II/III) complex ion [$(NH_3)_5Ru(\mu\text{-pyrazine})Ru(NH_3)_5$]$^{5+}$, represents a prototypical example of a newly synthesized stable, simple mixed-valent system [19–22]. Since the discovery of pyrazine-mediated strong intramolecular electron coupling of the mixed-valent state in the CT ion, there have been huge efforts in designing analogous diruthenium setups using modified pyrazines and other suitable moieties as the bridging units [1–8, 19–22]. The initial objectives were directed at developing fundamental issues relating to the delicate inner sphere electron transfer aspects and to expand the scope in the specific context of biological and material examples [9–18]. The inherent concept of intramolecular electron transfer in mixed-valent compounds has also been extended in designing molecular electronic systems, molecular wires, and quantum automata [23–25].

Mixed-Valence Systems: Fundamentals, Synthesis, Electron Transfer, and Applications, First Edition.
Edited by Yu-Wu Zhong, Chun Y. Liu, and Jeffrey R. Reimers.
© 2023 WILEY-VCH GmbH. Published 2023 by WILEY-VCH GmbH.

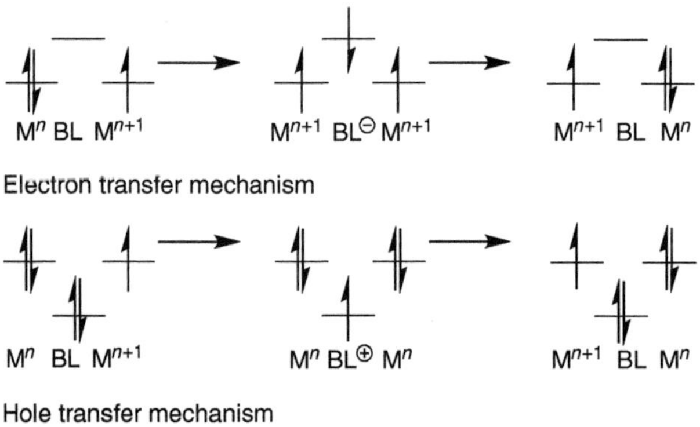

M^n BL M^{n+1} → M^{n+1} BL$^\ominus$ M^{n+1} → M^{n+1} BL M^n

Electron transfer mechanism

M^n BL M^{n+1} → M^n BL$^\oplus$ M^n → M^{n+1} BL M^n

Hole transfer mechanism

Figure 4.1 Electron transfer pathway alternatives (orbital occupation for three- and five-electron cases).

In a ligand-bridged mixed-valent diruthenium complex [(AL)M^n(μ-BL)M^{n+1}(AL)] (AL = ancillary ligand and BL = bridging ligand) with the identical ligand environment around each metal ion, electron transfer can proceed through mediation by the bridge. The overlapping of energetically compatible d-orbitals of transition metal centers and the π-orbitals of the bridge facilitates intramolecular electron transfer from the reduced metal ion (M^n) to the oxidized one (M^{n+1}) through a superexchange pathway. It can follow two directions based on the electronic nature of the bridge: (i) an electron transfer mechanism in the case of a π acceptor bridge and (ii) a hole transfer mechanism for a π electron-donating bridge (Figure 4.1).

According to Robin and Day [26], a negligible to unrecognizable interaction between metal centers leads to a class I mixed-valent situation, whereas moderate to strong interactions yield valence-localized class II ([(AL)M^n(μ-BL)M^{n+1}(AL)]) or even valence delocalized class III ([(AL)M$^{n+0.5}$(μ-BL)M$^{n+0.5}$(AL)]) mixed-valent systems. The question of valence delocalization in a mixed-valent complex can be probed by collective consideration of the electrochemical and spectroscopic features in specific relation to the isovalent congeners.

In a two-step redox reaction of the dinuclear complex (Eq. 4.1), the comproportionation constant K_c can be estimated from the potential difference ($\Delta E = E_2 - E_1$) between the

[ALM^n(μ-BL)M^{n}AL] $\xrightarrow{E_1}$ [ALM^n(μ-BL)M^{n+1}AL] $\xrightarrow{E_2}$ [ALM^{n+1}(μ-BL)M^{n+1}AL]
   Reduced form   Intermediate form   Oxidized form

K_c = {[Intermediate form]^2/[Reduced form] [Oxidized form]}

at 298 K, $Kc = 10^{\Delta E/59 \text{ mV}}$, where $\Delta E = E_2 - E_1$ (4.1)

successive reversible redox processes [2, 27, 28]. The statistical value of K_c is 4, representing a non-interacting class I system, while larger K_c values up to 10^5 and greater than that correspond to class II and class III systems, respectively. However, classification of a mixed-valent system, i.e. localized versus delocalized simply based on the electrochemically estimated K_c value, may lead to a conflicting scenario in certain instances. Intervalence charge transfer transitions (IVCT) in the low-energy near-infrared (NIR) region of the mixed-valent state need to be analyzed in conjunction with K_c.

For a valence-localized weakly coupled mixed-valent state (class II), the metal center in the lower oxidation state (electron rich) promotes the transfer of electron density to the metal in the higher oxidation state (electron deficient). This can lead to a bridge-mediated unidirectional intervalence charge transfer (IVCT) transition (Figure 4.1) in the lower energy, visible-to-NIR to even IR region [29–34]. A correlation of the IVCT transition with the extent of electronic coupling between the metal centers in the localized mixed-valent state has been made early following the Hush formula (Eq. 4.2) [35]

$$\Delta \nu_{1/2} = [2.31 \times 10^3 (E_{IT})]^{1/2} \tag{4.2}$$

where $\Delta \nu_{1/2}$ and E_{IT} correspond to the calculated bandwidth at half-height of the IVCT band and energy of the IVCT peak in cm^{-1}, respectively. In the case of a symmetric IVCT band for the class II setup, it can be correlated with the electronic coupling constant (H_{ab}/cm^{-1}) by Eq. (4.3),

$$H_{ab} = \left[2.05 \times 10^{-2} (\varepsilon_{max} \bar{\nu}_{max} \Delta \nu_{1/2})^{1/2}\right]/R \tag{4.3}$$

where ε_{max} is the molar extinction coefficient, $\bar{\nu}_{max}$ is the absorption maximum in wavenumber, and $\Delta \nu_{1/2}$ is the bandwidth at half-height in wavenumber. R represents the metal–metal distance in Å for a valence-localized system. In the case of a fully delocalized class III system, H_{ab} can be defined by $\bar{\nu}_{max}/2$.

The varying IVCT band profile including band shape, bandwidth, and intensity as a function of solvent; electronic/vibrational motion; and temperature makes its correlation with the intermetallic electronic coupling more intricate. For class III mixed-valent systems, the motions from the solvent and vibration are averaged to give rise to delocalization of electron density across the metal–bridge–metal domain, leading to a solvent-independent narrow IVCT band. On the other hand, a class II mixed-valent setup typically exhibits solvent-dependent broad IVCT bands as the motions due to the exchange of electrons and solvent is localized (large geometry changes). However, borderline hybrid class II–III systems develop when the electron exchange rate in the localized mixed-valence state is sufficiently fast to cause solvent averaging, resulting in a solvent-independent narrow IVCT band as in the class III system [1–8].

Faster vibrational spectroscopic techniques (infrared or resonance Raman) are also found to be very effective in differentiating localized versus delocalized mixed-valent situations for the selective set of molecules [29–34]. Moreover, EPR g parameters ($<g> = \{1/3(g_1^2+g_2^2+g_3^2)\}^{1/2}$ and $\Delta g = g_1 - g_3$) and DFT-calculated Mulliken spin densities of paramagnetic mixed-valent states ascertain the location of the unpaired spin either on the metal or on the ligand or as delocalized at the metal–ligand interface, particularly for redox-active ligand-derived molecular systems [7].

Although the majority of studies in this direction have been centered on the $Ru^{II}(d^6)Ru^{III}(d^5)$-derived mixed-valence systems involving diverse molecular frameworks [1–8], a relatively smaller number of well-defined ligand-bridged $Ru^{III}(d^5)Ru^{IV}(d^4)$ complexes have also been explored in recent years [3, 4, 36–40].

The divergent issues and challenges pertaining to the $Ru^{II}Ru^{III}$ and $Ru^{III}Ru^{IV}$ mixed-valent complexes are highlighted in the subsequent sections using selective sets of representative examples.

4.2 $Ru^{II}Ru^{III}$ Mixed-Valent Systems

4.2.1 Pyrazine-Derived Bridges

The discovery of the pyrazine (pz)-bridged mixed-valent $Ru^{II}Ru^{III}$ state in the Creutz–Taube ion [19–22] [[(NH$_3$)$_5$Ru(μ-pz)Ru(NH$_3$)$_5$]$^{5+}$: K_c:$10^{6.6}$; EPR: $g_1 = 2.799$, $g_2 = 2.489$, $g_3 = 1.346$, $<g> = 2.30$, $\Delta g = 1.453$; IVCT: 1570 nm ($\varepsilon = 6330\,M^{-1}\,cm^{-1}$); H_{ab}: $3185\,cm^{-1}$] with its bridge-mediated strong intermetallic electronic coupling led to the exploration of designing several pyrazine-derived mixed-valent diruthenium complexes in order to advance the fundamental understanding relating to the delicate intramolecular electron transfer aspects in the mixed-valent state. The potential of pyrazines as π conjugated redox-active "linear" bridges was recognized early [22] and has been exploited more recently in the studies of molecular magnetism. On introducing chelate coordination, the following pyrazine-based bridging chelate ligands (H$_2$BL$_1$, H$_4$BL$_2$, H$_2$BL$_3$, and tppz, Figure 4.2) were considered in recent years to highlight their impact on the intramolecular electronic coupling processes of a mixed-valent $Ru^{II}Ru^{III}$ setup.

Mixed donor–acceptor-based dianionic and bis-bidentate 2,5-pyrazine-dicarboxylate (BL$_1^{2-}$)-bridged symmetrical diruthenium complexes were investigated in combination with ancillary ligands, having various electronic features such as electron-rich acac (acetylacetonate, **1**), π-accepting bpy (2,2′-bipyridine, **2^{2+}**), and monodentate σ-donating NH$_3$ (**3^{2+}**) (Figure 4.3) [41–43]. Successive one-electron reductions of the structurally characterized isovalent diruthenium(III) complex [(acac)$_2$Ru^{III}(μ-BL$_1^{2-}$)Ru^{III}(acac)$_2$] (**1**) involving electron-rich anionic acac$^-$ ancillary ligands led to the electrochemical generation of a mixed-valent $Ru^{III}Ru^{II}$ (**1$^-$**) intermediate followed by an isovalent $Ru^{II}Ru^{II}$ (**1^{2-}**) counterpart [41]. The mixed-valent state in **1$^-$** exhibited a weak IVCT band at 1040 nm ($\varepsilon = 380\,M^{-1}\,cm^{-1}$, CH$_3$CN) corresponding to a localized class II situation with limited electronic

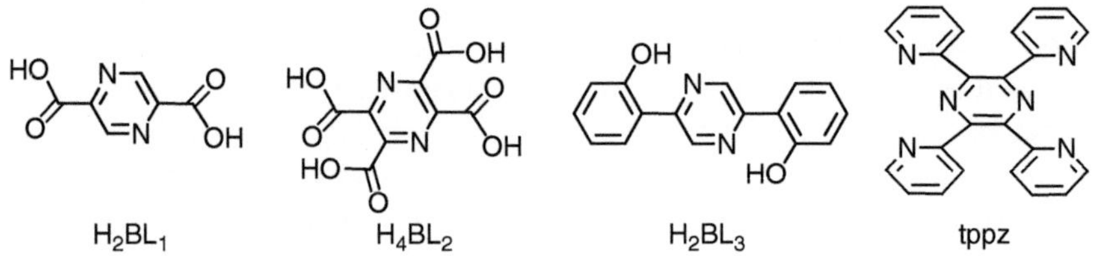

Figure 4.2 Pyrazine-based bridging chelate ligands.

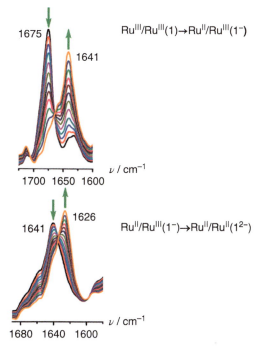

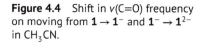

Figure 4.3 Representation of BL$_1^{2-}$- and H$_2$BL$_2^{2-}$-bridged diruthenium complexes.

Figure 4.4 Shift in ν(C=O) frequency on moving from $1 \rightarrow 1^-$ and $1^- \rightarrow 1^{2-}$ in CH$_3$CN.

coupling. However, a high K_c value of 10^7 involving the stepwise reduction processes and unsplit ν(CO) bands of the carboxylate group in the IR spectra of the mixed-valent state (1^-) as in the isovalent congeners (1, 1^{2-}) (Figure 4.4) implied equal valence configuration of the metal centers corresponding to a delocalized Ru$^{2.5}$Ru$^{2.5}$ formulation, at least at the vibrational time scale ($\approx 10^{-12}$ s) or a borderline class II–III hybrid situation. Further, significantly lower EPR g anisotropic values ($g_1 = 2.400, g_2 = 2.229, g_3 = 2.076$, and $\Delta g(g_1 - g_3) = 0.32$) of Ru^{III}Ru^{II} derived 1^-, as compared to that in the Creutz–Taube ion ($g_1 = 2.799, g_2 = 2.489, g_3 = 1.346$, and $\Delta g = 1.453$) revealed an appreciable contribution from the radical bridge to the singly occupied MO (SOMO), as expected for a partial hole transfer valence exchange mechanism for the anionic BL^{2-} bridge. The partial negative spin accumulation onto the bridge (Mulliken spin densities: Ru1, Ru2, BL, and acac of 0.655, 0.357, −0.170, and 0.157, respectively) is also in agreement with that.

On the contrary, the impact of π-acidic bpy or σ-donating monodentate NH_3 [42] ancillary ligands in the BL_1^{2-}-bridged symmetric diruthenium complexes 2^{2+} ($[(bpy)_2Ru^{II}(\mu-BL_1^{2-})Ru^{II}(bpy)_2]^{2+}$) or 3^{2+} ($[(NH_3)_4Ru^{II}(\mu-BL_1^{2-})Ru^{II}(NH_3)_4]^{2+}$), respectively (Figure 4.3), is reflected by the stabilization of the metal ions in the +2 oxidation state. The moderate K_c values of 10^3 and 10^5 along with the absence of detectable IVCT bands or a poorly resolved IVCT band at <1560 nm (ε = <450 M^{-1} cm^{-1}, D_2O) for the $Ru^{II}Ru^{III}$ mixed-valent states in 2^{3+} and 3^{3+}, respectively, suggest a rather weakly coupled localized class II mixed-valent state.

The analogous dianionic but bis-tridentate 3,6-dicarboxypyrazine-2,5-pyrazine-dicarboxylate ($H_2BL_2^{2-}$) bridge-derived diruthenium(II) complex **4** ($[(PPh_3)_2ClRu^{II}(\mu-H_2BL_2^{2-})Ru^{III}(Cl)(PPh_3)_2]^{2+}$) incorporating PPh_3/Cl ancillary ligands was structurally characterized (Figure 4.3) [43]. Electrochemically generated EPR silent one-electron oxidized 4^+ failed to display an IVCT band in the low-energy near-infrared region as otherwise expected for a ligand-bridged $Ru^{II}Ru^{III}$ mixed-valent complex. The rather complicated electronic form of 4^+ was interpreted based on the combined experimental including high-pressure electrochemical studies and theoretical investigation as the probable valence tautomeric form of charge-localized $[Ru^{III}(\mu-H_2BL^{2-})Ru^{II}]/[Ru^{III}(\mu-H_2BL^{•3-})Ru^{III}]$ or charge-delocalized $[Ru^{II.5}(\mu-H_2BL^{2-})Ru^{II.5}]$. On the other hand, deprotonation of the bridge led to the asymmetrical charge localized formulation of $[Ru^{III}(\mu-H_2BL^{3-})Ru^{II}]^0$ ($4^+ - H^+$).

The modified dianionic 2,5-bis(2-oxidophenyl)pyrazine (BL_3^{2-}) bridging ligand comprising electron-rich phenolato and electron-poor pyrazine groups was utilized in framing diruthenium complexes with bpy (5^{2+}, $[(bpy)_2Ru^{II}(\mu-BL_3^{2-})Ru^{II}(bpy)_2]^{2+}$), pap (pap = 2-phenylazopyridine, 6^{2+}, $[(pap)_2Ru^{II}(\mu-BL_3^{2-})Ru^{II}(pap)_2]^{2+}$), and acac (**7**, $[(acac)_2Ru^{III}(\mu-BL_3^{2-})Ru^{III}(acac)_2]$) (Figure 4.5) [44, 45] ancillary ligands. The complexes were further employed for investigating the impact of the ancillary ligands on the electron transfer processes of the analogous complex frameworks as well as the electronic form of the mixed-valent state. While π-acidic bpy or pap facilitated the stabilization of ruthenium(II) state in the isolated 5^{2+} or 6^{2+}, the electron-rich acac directed the isolation of the ruthenium(III)-derived **7** as in the case of **1** in Figure 4.3. The bpy-derived 5^{2+} displayed two successive oxidation

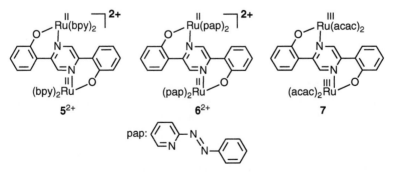

Figure 4.5 Representation of BL_3^{2-}-bridged diruthenium complexes.

processes corresponding to $Ru^{II}Ru^{II}$ to $Ru^{II}Ru^{III}$ to $Ru^{III}Ru^{III}$ with a K_c value of $10^{2.5}$ for the intermediate mixed-valent $Ru^{II}Ru^{III}$ state in 5^{3+}, which corresponded to a weakly coupled class II system as per the Robin and Day classification, and accordingly, it exhibited an exceedingly weak IVCT band in the NIR region at 2600 nm [44]. Switching from moderately π-acidic bpy ancillary ligand in 5^{2+} to a strongly π-acidic pap in 6^{2+} enhanced the Ru^{II}/Ru^{III} oxidation potential significantly from 0.571/0.720 V ($\Delta E = 0.149$ V) to 1.05/1.45 V ($\Delta E = 0.40$ V) versus SCE in CH_3CN, leading to a greater K_c value of $10^{6.7}$ for the one-electron oxidized intermediate 6^{3+} as compared to that in 5^{3+} ($K_c = 10^{2.5}$), despite the much stronger acceptor characteristic of pap, implying a radical feature of 6^{3+} instead of a conventional mixed-valent $Ru^{II}Ru^{III}$ state. The radical state of 6^{3+} was substantiated further by its single EPR signal at $g = 2.002$ in frozen CH_3CN at 4 K, revealing the electronic form of 6^{3+} as a radical-bridged isovalent system $[Ru^{II}(BL_3{}^{\bullet-})Ru^{II}]$ instead of an otherwise expected mixed-valent situation of $[Ru^{II}(BL_3{}^{2-})Ru^{III}]$ as in the case of 5^{3+} [45].

The one-electron reduction of $Ru^{III}Ru^{III}$-derived **7** (Figure 4.5) resulted in the intermediate $Ru^{III}Ru^{II}$ state in $\mathbf{7^-}$ with a K_c value of $10^{5.8}$. The appreciably higher K_c value of $\mathbf{7^-}$ than that of the corresponding bpy-derived 5^{3+} ($K_c = 10^{2.5}$) is attributed to a strongly coupled mixed-valent $Ru^{II}Ru^{III}$ situation in the former. Although $\mathbf{7^-}$ exhibits a typical Ru^{III}-based EPR profile with $g_1 = 2.316$, $g_2 = 2.150$, $g_3 = 1.760$, $\Delta g = 0.556$, and $<g> = 2.088$ in frozen CH_3CN at 4 K, it failed to display the expected IVCT band at the NIR region up to 2500 nm [45].

The mixed-valent aspects of the electrochemically generated $Ru^{II}Ru^{III}$ state using the pyrazine-derived inherently non-planar bis-tridentate tppz (2,3,5,6-tetrakis (2-pyridyl)pyrazine)-bridged diruthenium complexes of the general formula of $[(AL)ClRu^{II}(tppz)Ru^{II}Cl(AL)]^n$ ($\mathbf{8^n}$) were also explored in combination with ALs having varying electronic features (Figure 4.6) [46–58]. The detailed experimental (structure, electrochemistry, UV–vis–NIR, EPR, and IR spectroelectrochemistry) and theoretical investigations (MO compositions, DFT-calculated Mulliken spin density distribution at the paramagnetic state(s), and TD-DFT) of the isolated and electrogenerated states revealed all the three mixed-valent situations: class II, class III, and borderline class II–III as a function of variable ancillary ligands.

The collective consideration of the K_c value (10^4) and the IVCT band profile in the light of the Hush formula ($\Delta v_{1/2}/cm^{-1}$: 3552(cal) ≈ 3150(exp)) may be attributed to a typical class II mixed-valent $Ru^{II}Ru^{III}$ situation for the tppz-bridged complex $\mathbf{8a^{3+}}$, encompassing the strongly π-acidic bis(tolylamino)acenaphthene (AL_1) ancillary ligand (Figure 4.6, Table 4.1). The competitive π-accepting feature of the bridge (tppz) and the ancillary ligand (AL_1) reduced the electron density on the metal ions, which in effect exerted a diminished intramolecular electron transfer process at the mixed-valent state, leading to a class II situation [46].

On the other hand, electron-rich anionic ancillary ligands AL_2 (acac, Figures 4.6 and 4.7) and AL_3 (picolinate, Figure 4.6) as well as more donating ancillary ligands AL_4 (dipyridylamine, Figure 4.6) in $\mathbf{8b}^{n+}$, $\mathbf{8c}^{n+}$, and $\mathbf{8d}^{n+}$ [47–49] resulted in a strongly coupled valence delocalized class III mixed-valent $Ru^{II}Ru^{III}$ system as supported by the larger K_c value of 10^6–10^{10} and a smaller $\Delta v_{1/2}$ value of the IVCT band

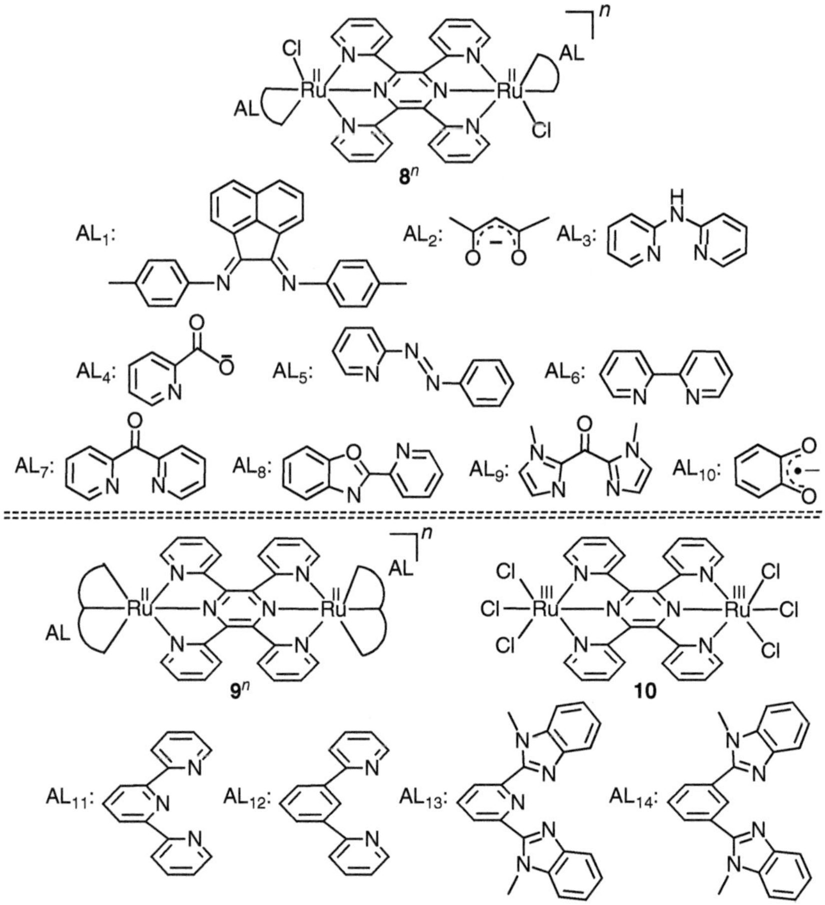

Figure 4.6 Representation of tppz-bridged diruthenium complexes.

with respect to that calculated based on the Hush formula for the localized class II system (Table 4.1 and Figures 4.6 and 4.7).

On the contrary, π-acidic ancillary ligands (AL_5 = 2-phenylazopyridine, AL_6 = 2,2′-bipyridine, AL_7 = 2,2′-dipyridylketone, AL_8 = 2-(2-pyridyl)benzoxazole, and AL_9 = 2,2′-bis(1-methylimidazolyl)ketone, Figure 4.6) led to mixed-valent $Ru^{II}Ru^{III}$ states in $8e^{3+}$-$8i^{3+}$, which exhibited moderate K_c values of 10^3–10^5 but smaller $\Delta\nu_{1/2}$ values of the IVCT band relative to that of the calculated $\Delta\nu_{1/2}$ based on the Hush formula (Table 4.1), corresponding to a hybrid class II–III system [50–53].

The bidirectional redox-active [54] benzoquinone ancillary ligand AL_{10} in Figure 4.6 yielded a stable complex **8j** ([$(Q_{sq})ClRu^{II}(\mu\text{-tppz})Ru^{II}Cl(Q_{sq})$], Figure 4.8), where the one-electron reduced benzosemiquinone form of AL_{10} stabilized the triplet ($S = 1$) ground state. Unlike the ancillary ligands AL_1–AL_9-derived Ru-tppz-Ru core in **8a**–**8i** (Figure 4.6), the experimental and theoretical events of **8j** collectively revealed the dominating contribution of the quinone-based frontier

Table 4.1 Varying K_c and IVCT as a function of AL in 8^n.

AL	K_c	ν_{IVCT}/nm[a] (ε/M^{-1}cm^{-1})	$\Delta\nu_{1/2}$/cm^{-1} (cal/exp)	$<g>/\Delta g$[b)c)]	Class	References
AL$_1$	2×10^4	1831(1900)	3552/3150	b	II	[46]
AL$_2$	10^{10}	1680(2900)	3708/1590	2.304/0.96	III	[47]
AL$_3$	2.7×10^6	1700(2250)	3680/1390	2.553/1.693	III	[48]
AL$_4$	4.0×10^6	1670(2600)	3720/1620	b	III	[49]
AL$_5$	4.0×10^3	1890(3800)	3500/1650	c	II–III	[50]
AL$_6$	8.2×10^4	1647 (c)	3745/1500	c	II–III	[51]
AL$_7$	2.5×10^4	1800(1500)	3584/700	2.193/0.597	II–III	[52]
AL$_8$	3.8×10^4	1635(1400)	3758/1790	b	II–III	[53]
AL$_9$	1.2×10^5	1637(1200)	3754/1560	b	II–III	[52]
AL$_{10}$	5.2×10^2 1.7×10^4	—	—	2.120/0.262	—	[54]
AL$_{11}$	2.6×10^5	1527(7650)	3890/970	c	II–III	[55]
AL$_{12}$	5.4×10^4	1661(5500)	3729/1033	c	II–III	[56]
AL$_{13}$	3.7×10^3	2924(960)	2810/3830	2.180/0.547	II–III	[57]
AL$_{14}$	1.2×10^5	1520(1360)	3898/1313	c	II–III	[57]
Cl$_3$	1.1×10^{12}	1800(3500) 3030(1200) 4550(250)	3584/1150 2761/1050 2254/800	c	II–III	[58]

a) CH$_3$CN (CD$_3$CN for AL$_{11}$, AL$_{12}$, and Cl$_3$).
b) Not detected at 4 K.
c) Not reported.

orbitals in the successive redox steps, except in the case of one-electron reduced $8j^-$, which could be better defined as a mixed-valent Ru^{II}Ru^{III} state in resonance with the corresponding radical form (Figure 4.8). The mixed-valent form of $8j^-$ was conceived via the redox-induced electron transfer (RIET) process [59] (Figure 4.8), i.e. one-electron reduction of each of the Q$_{sq}^{\bullet-}$ units in $8j$ to Q$_{cat}^{2-}$ complemented by one-electron oxidation of one of the metal ions (Ru^{II} to Ru^{III}) with a net 1e$^-$ reduction process.

Similarly, the insertion of N,N,N (AL$_{11}$ = 2,2':6',2''-terpyridine and AL$_{13}$ = bis(N-methylbenzimidazolyl)pyridine) and N,C,N (AL$_{12}$ = 2,6-bis(2'-pyridyl)phenyl and AL$_{14}$ = bis(N-methylbenzimidazolyl)benzene) donating tridentate ancillary ligands (Figure 4.6) in the Ru(μ-tppz)Ru diruthenium(II) core in 9^n led to borderline class II–III Ru^{II}Ru^{III} mixed-valent situations [55–57], as could be manifested by their moderate K_c values and IVCT profile, i.e. $\Delta\nu_{1/2}$(cal) > $\Delta\nu_{1/2}$(exp).

Unlike the broad IVCT band in the NIR region for the tppz bridged 8^n or 9^n (Figures 4.6, 4.7), the structurally characterized valence delocalized mixed-valent complex {[PPN][Cl$_3$Ru$^{II.5}$(μ-tppz)Ru$^{II.5}$Cl$_3$]}{[PPN][10]} (PPN$^+$ = μ-nitrido-bis(triphenylphosphane) cation) with a very large K_c value of 10^{12} displayed three IT bands (see later) in the NIR to IR region at 1800 nm (ε / M^{-1} cm^{-1} = 3500),

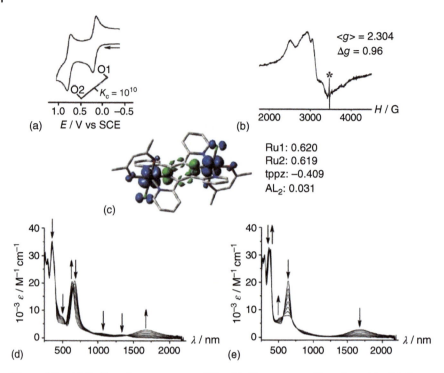

Figure 4.7 (a) Cyclic voltammograms of **8b** with AL$_2$ as an ancillary ligand in CH$_3$CN. (b) EPR, in CH$_3$CN at 4 K (*instrumental signal). (c) Mulliken spin density plot. UV–vis–NIR spectral changes in CH$_3$CN for (d) **8b** → **8b**$^+$ and (e) **8b**$^+$ → **8b**$^{2+}$.

3030 nm (ε / M^{-1} cm^{-1} = 1200), and 4550 nm (ε / M^{-1} cm^{-1} = 250) (Figure 4.6 and Table 4.1). The crystal structure of mixed-valent [PPN][**10**] showed two unequal Ru^{II}-N (pyrazine and tppz) distances of 1.940(4) (Å) and 1.912(4) (Å), corresponding to localized Ru^{II} and Ru^{III} ions, respectively, presumably because of the impact of non-planarity of the tppz bridge [58].

In certain instances such as **8**$^{3+}$, the paramagnetic Ru^{II}Ru^{III} mixed-valent state (S = 1/2) fails to show the expected EPR resonance even at 4 K (Table 4.1) because of the impact of fast relaxation process facilitated by the high spin–orbit coupling constant of Ru^{III}, $\lambda \approx 1000$ cm^{-1} [3].

4.2.2 Other Bridging Ligands

The increasing popularity of tetrazines has been used earlier in systems with various purposes [60–63].

Diruthenium(II) complexes **11**^{n} and **12**^{n} involving tetrazine-derived bis-bidentate bridging ligands, 3,6-bis-(2-pyridyl)-1,2,4,5-tetrazine (bptz) and 3,6-bis-(3,5-dimethylpyrazolyl)-1,2,4,5-tetrazine (bpytz), respectively, were developed in combination with different ancillary ligands (Figure 4.9 and Table 4.2) to access the bridge-mediated intramolecular electronic coupling of the mixed-valent Ru^{II}Ru^{III} state [64–68]. The low-energy LUMO (lowest unoccupied molecular orbital) and

4.2 Ru^{II}Ru^{III} Mixed-Valent Systems

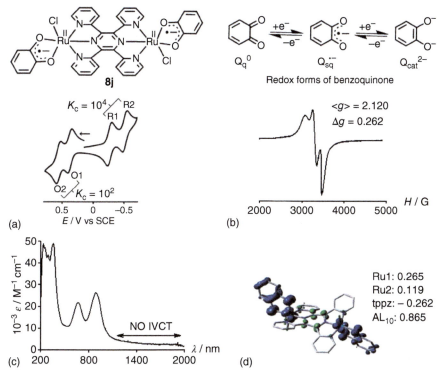

Figure 4.8 (a) Cyclic voltammograms of **8j** in CH$_3$CN. (b) EPR of **8j**$^-$ in CH$_3$CN at 4 K. (c) UV–vis–NIR spectrum of **8j**$^-$ in CH$_3$CN. (d) Mulliken spin density plot of **8j**$^-$. (e) Electronic forms of **8j**^{n}.

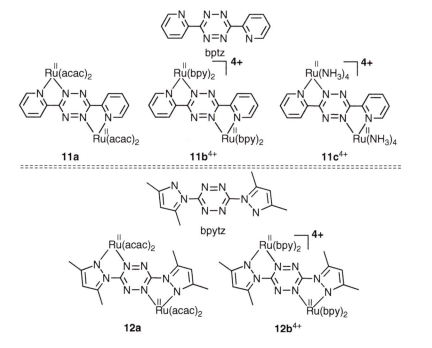

Figure 4.9 Diruthenium complexes bridged by tetrazine-derived bptz and bpytz.

Table 4.2 Varying K_c and IVCT as a function of ancillary ligands.

Complex	K_c	ν_{IVCT}/nm[a] (ε/M^{-1}cm^{-1})	$\Delta\nu_{1/2}$/cm^{-1} (cal/exp)	$<g>/\Delta g$	Class	References
11a$^+$	10^{13}	1238(20)	4320/2300	2.17	III	[64]
11b^{5+}	10^8	1483(2800)	3950/1000	c)	III	[65]
11c^{5+}	10^{15}	1433(500)	4015/2000	2.477/0.894	III	[66]
12a$^+$	10^{13}	b)	—	2.150/0.624	III	[67]
12b^{5+}	10^7	1534(1800)	3880/650	c)	III	[68]

a) CH_3CN.
b) Not reported.
c) Not resolved even at 4 K.

high coefficient of the LUMO on the coordinated N atoms of the tetrazine (tz) unit of bptz or bpytz indeed allows for effective overlap between the metal and ligand orbitals, as reflected in high to very high K_c values (10^7–10^{15}) of the intermediate mixed-valent Ru^{II}Ru^{III} states (Table 4.2). The intensity of the IVCT band in the mixed-valent states in **11**^{n} and **12**^{n} (Figure 4.9 and Table 4.2), however, varied significantly as a function of the ancillary ligands, following the order (ε/M^{-1}cm^{-1}) bpy, 2800 > NH$_3$, 500 ≫ acac, 20 and bpy, 1800 ≫ acac ≈ 0 for the bptz and bpytz series (Figure 4.10), respectively. The observed very weak to even undetectable IVCT bands despite appreciably large K_c values implies independent features of the Coulombic controlled electrochemical coupling in terms of K_c and the

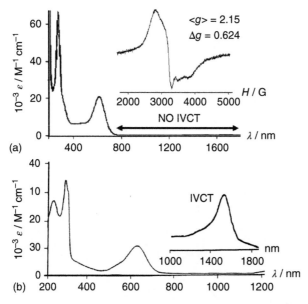

Figure 4.10 (a) UV–vis–NIR spectrum of **12a**$^+$ in CH_3CN. Inset shows the EPR of **12a**$^+$ in CH_3CN at 4 K. (b) UV–vis–NIR spectrum of **12b**$^{5+}$ in CH_3CN.

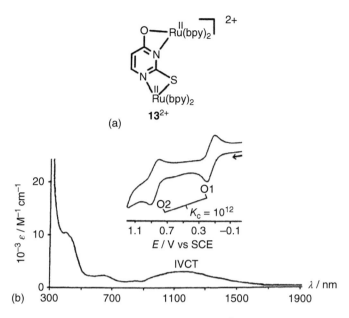

Figure 4.11 (a) Cyclic voltammograms of **13**$^{2+}$ in CH$_3$CN. (b) UV–vis–NIR electronic spectrum of **13**$^{3+}$ in CH$_3$CN.

orbital-controlled IVCT band intensity. Nevertheless, analysis of the $\Delta v_{1/2}$ values of the IVCT bands, i.e. $\Delta v_{1/2}$(cal) > $\Delta v_{1/2}$(exp) (Table 4.2), according to Hush treatment suggested a strongly coupled class III mixed-valent situation in both the cases.

The deprotonated 2-thiouracil can act as a bis-bidentate dianionic bridging ligand in **13**$^{2+}$, which is asymmetrically linked to two {Ru(bpy)$_2$}$^{2+}$ units through the [N,S]$^-$ and [N,O]$^-$ donor sites. The successive two reversible oxidative couples (O1 and O2) of **13**$^{2+}$ with a large separation in potential of 700 mV (Figure 4.11) led to the comproportionation constant (K_c) value of 10^{12} of the asymmetrical intermediate mixed-valent Ru^{II}Ru^{III} state in **13**$^{3+}$, extending the immediate impression of a strongly coupled (electrochemical) class III system as per Robin and Day classification. The mixed-valent **13**$^{3+}$ however exhibited a moderately intense but broad IVCT band at 1170 nm (ε/M^{-1}cm^{-1}: 2800) with $\Delta v_{1/2}$(cal): 4443 cm^{-1} ≈$\Delta v_{1/2}$(exp): 3060 cm^{-1}, corresponding to a weakly coupled valence localized class II mixed-valent state [69]. Hence, the large separation of the oxidation potentials (O1 and O2 in Figure 4.11) or high K_c value of 10^{12} could be mainly the consequence of donor center asymmetry in **13**$^{2+}$, i.e. RuN$_5$S versus RuN$_5$O in addition to a minor contribution from the bridge-mediated moderate electronic coupling.

The one-electron oxidation ($E°_{298}$: 0.28 V vs. SCE) of the structurally characterized oxido/(Pz)$_2$ (Pz = pyrazolate) tri-bridged diruthenium(II) complex **14** allowed the electrochemical generation of the EPR-active mixed-valent Ru^{II}Ru^{III} state in **14**$^+$ ($S = 1/2$) (Figure 4.12). The primarily metal-based EPR spectrum (<g>/Δg: 2.10/0.30) and uniform spin distribution in the Ru-O-Ru (0.324/0.317/0.324) domain of **14**$^+$ justifies a mixed electronic structural form of mixed-valent {Ru^{II}-O^{2-}/(Pz)$_2$-Ru^{III}} and radical-bridged isovalent {Ru^{II}-O$^{\bullet-}$/(Pz)$_2$-Ru^{II}} as the major and minor

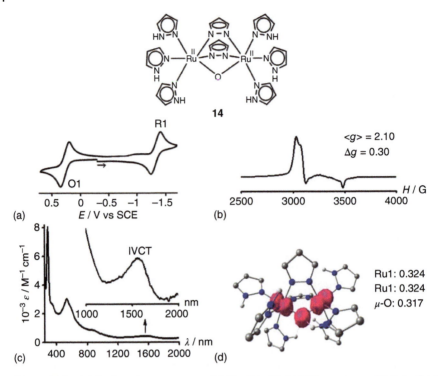

Figure 4.12 (a) Cyclic voltammograms of **14** in CH_3CN. (b) EPR spectrum of **14⁺** in CH_3CN (77 K). (c) UV–vis–NIR spectroelectrochemistry of **14⁺** in CH_3CN. (d) Mulliken spin density plot of **14⁺**.

contributions, respectively [38]. The analysis of the weak low-energy IVCT band at 1580 nm ($\varepsilon/M^{-1}cm^{-1}$: 450) of **14⁺** via the Hush formula ($\Delta v_{1/2}$(cal): 3823 $cm^{-1} \gg \Delta v_{1/2}$(exp): 1160 cm^{-1}) confirms a delocalized class III feature.

The unique oxido-bridged diruthenium complex [trpy)(bpy)Ru(O)Ru(bpy)(trpy)]$^{n+}$ (**15^n**) could be isolated and structurally characterized in three redox forms including isovalent $Ru^{III}Ru^{III}$ (**15⁴⁺**) and two mixed-valent $Ru^{III}Ru^{II}$ (**15³⁺**) and $Ru^{III}Ru^{IV}$ (**15⁵⁺**) states (Figure 4.13 and see Section 4.3). The comparative account of

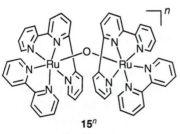

Figure 4.13 Oxido-bridged diruthenium complexes **15^n**.

15^n
$Ru^{II}Ru^{III}$: $n = 3$
$Ru^{III}Ru^{III}$: $n = 4$
$Ru^{III}Ru^{IV}$: $n = 5$

Table 4.3 Selected bond distances and bond angles of 15^n.

$$Ru1 \overset{O}{\frown} Ru2$$

Complex	Ru1-O/ Ru2-O (Å)	Ru-O-Ru [deg]
15^{3+}	2.103(2)/2.101(2)	149.55(12)
15^{4+}	1.8827(5)/1.8827(5)	164.13(15)
15^{5+}	1.844(2)/1.846(2)	170.64(15)

the structural features of 15^n with special reference to the Ru-O bond distances and Ru-O-Ru bond angles is attributed to a distinct variation of the metal redox states, the former decreasing with increasing n, i.e. $15^{5+} < 15^{4+} < 15^{3+}$, and the reverse order followed for the latter, $15^{5+} > 15^{4+} > 15^{3+}$ (Table 4.3). Further, the close Ru-O bond distances in the mixed-valent 15^{3+} ($Ru^{III}Ru^{II}$) or 15^{5+} ($Ru^{III}Ru^{IV}$) as in the case of corresponding isovalent 15^{4+} ($Ru^{III}Ru^{III}$) support a valence delocalized mixed-valent state in each case, as also suggested by the electrochemical and EPR signatures in solution.

4.3 $Ru^{III}Ru^{IV}$ Mixed-Valent Systems

Although extensive investigations were made in understanding the intricate fundamental issues involving ligand-bridged mixed-valent $Ru^{II}Ru^{III}$ states, the corresponding mixed-valent $Ru^{III}Ru^{IV}$ combination was limited to fewer recent reports [36–40, 70–72]. Frequently, the $Ru^{III}Ru^{IV}$ state was generated as a short-lived intermediate via electrochemical oxidation of stable ligand-bridged isovalent $Ru^{III}Ru^{III}$ complexes. However, a few structurally characterized oxido-bridged $Ru^{III}Ru^{IV}$ complexes were also reported in recent years.

The electrochemical one-electron oxidation of the *s-trans* configurated deprotonated glyoxalbis(2-hydroxyanil) (H_2gbha)-bridged isovalent [{(acac)$_2$Ru^{III}}$_2$ (μ-gbha)] (**16**, $S = 1$) (Figure 4.14) generated the mixed-valent species [(acac)$_2$Ru^{IV} (μ-gbha)Ru^{III}(acac)$_2$]$^+$ (**16$^+$**, $S = 1/2$). The mixed-valent **16$^+$** ($K_c = 10^6$) displayed a moderately intense broad IVCT band at 1800 nm ($\varepsilon = 2500$ M^{-1} cm^{-1}, in CH$_2$Cl$_2$) with $\Delta v_{1/2}$(exp): 2600 cm^{-1} < $\Delta v_{1/2}$(cal): 3580 cm^{-1} (Figure 4.14) and a Ru^{III}-based weak EPR spectrum (<g>/Δg: 2.09/0.41) corresponding to a "borderline" class III with a tendency toward a class II situation [70].

On the other hand, the mixed-valent $Ru^{III}Ru^{IV}$ state in C_n-bridged **17$^+$** [71] exhibited a very intense but broad IVCT band centered at 1730 nm ($\varepsilon = 25\,000$ M^{-1} cm^{-1}, in CH$_2$Cl$_2$) with $\Delta v_{1/2}$(exp): 3400 cm^{-1} ≈ $\Delta v_{1/2}$(cal): 3654 cm^{-1} in spite of a low K_c of 10^3. However, the intensity enhancement of the cumulene stretching band at 2078 cm^{-1} with the concomitant decreases of the alkyne stretching band at 2134 cm^{-1} on moving from isovalent **17** (Ru^{III}Ru^{III}) to mixed-valent Ru^{III}Ru^{IV} **17$^+$**

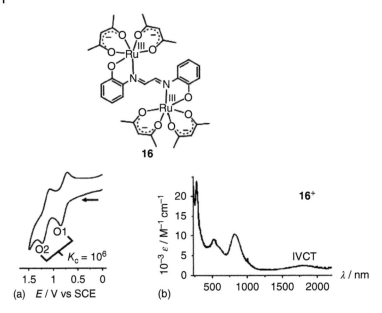

Figure 4.14 (a) Cyclic voltammograms of **16** in CH_2Cl_2. (b) UV–vis–NIR spectrum of electrogenerated **16⁺** in CH_2Cl_2.

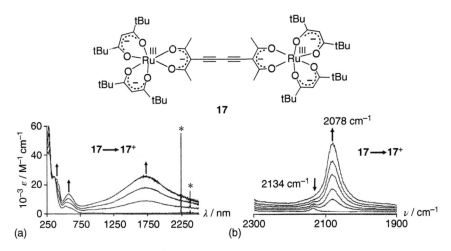

Figure 4.15 (a) UV–vis–NIR spectroelectrochemistry of **17** → **17⁺** in CH_2Cl_2 at 233 K. (b) IR spectroelectrochemistry of **17** → **17⁺** in CH_2Cl_2 at 233 K (*instrumental signal).

(Figure 4.15) revealed its valence delocalization (class III behavior) in agreement with the strong IVCT transition.

The dianionicoxamidato (2⁻)-bridged diruthenium(III) complex **18** ($S = 1$) represents a remarkable system that allows the recognition of both $Ru^{III}Ru^{IV}$ (**18⁺**,

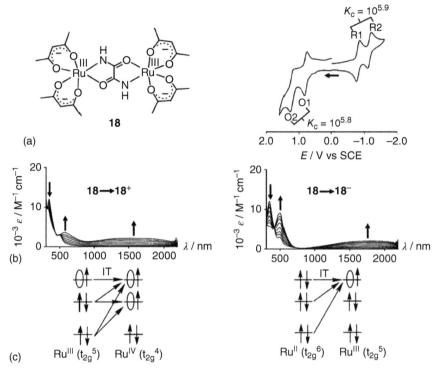

Figure 4.16 (a) Cyclic voltammograms of **18** in CH$_3$CN. (b) UV–vis–NIR spectroelectrochemistry in CH$_3$CN. (c) IT transitions in **18$^+$** (left) and **18$^-$** (right).

$S = 3/2$) and Ru^{III}Ru^{II} (**18$^-$**, $S = 1/2$) mixed-valent states in a selective molecular framework through oxidative and reductive approaches, respectively [72]. The successive metal-based two one-electron oxidation (O1 and O2) as well as reduction (R1 and R2) processes in CH$_3$CN resulted in similar K_c values of $10^{5.8}$ and $10^{5.9}$ for the intermediates **18$^+$** and **18$^-$**, respectively. Both the mixed-valent states **18$^+$** and **18$^-$** exhibited moderately intense IVCT bands at 1500 nm ($\varepsilon = 2200$ M^{-1} cm^{-1}, in CH$_3$CN) and 1800 nm ($\varepsilon = 2100$ M^{-1} cm^{-1}, in CH$_3$CN) (Figure 4.16). The IVCT bandwidths, $\Delta\nu_{1/2}$(exp)/$\Delta\nu_{1/2}$(calc): 4000/3750 cm^{-1}, of **18$^-$** were suggestive of a class II mixed-valent system. The much broader IVCT band for **18$^+$** ranging from the visible to the NIR region as compared to **18$^-$** reflects the involvement of several IT transitions (5: three holes) in the former as compared to the latter (3: one hole) (Figure 4.16) [58].

Structurally authenticated mono-oxido-bridged mixed-valent diruthenium (III/IV) complexes such as **[19]**PF$_6$ or **[20]**PF$_6$ with Ru-O-Ru angles of ≈167° were reported in combination with the ethylbis(2-pyridylmethyl)amine (ebpma) ancillary ligand [36]. Although the EPR resonance was not observed for **19$^+$** ($S = 1/2$, μ_{eff}: 1.67 BM), the frozen solution of **20$^+$** ($S = 1/2$, μ_{eff}: 1.65 BM) displayed

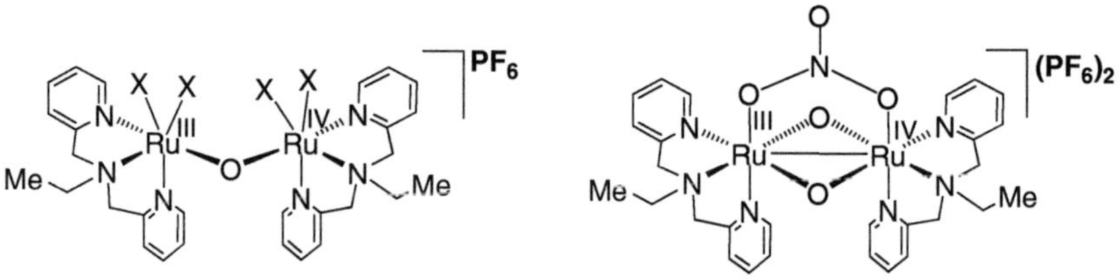

(a) [19]PF$_6$ (X=Cl), [20]PF$_6$ (X=Br)   (b)   [21](PF$_6$)$_2$

Figure 4.17 (a) Oxido-bridged and (b) nitrato-capped doubly oxido-bridged diruthenium (III)/(IV) complexes.

a broad and weak EPR signal at $g = 2.54$ in acetone/toluene at 77 K. The Ru^{III}Ru^{IV} mixed-valent states of **19$^+$** or **20$^+$** exhibit large K_c values (10^{20} or 10^{19}) and IVCT bands at 1189 and 1196 nm in CH$_3$CN with $\Delta v_{1/2}$(cal)/$\Delta v_{1/2}$(exp) of 4408/3796 cm^{-1} or 4395/3714 cm^{-1}, respectively, corresponding to a class III situation. Unlike **19$^+$** or **20$^+$**, the nitrato-capped doubly oxido-bridged mixed-valent diruthenium(III/IV) complex [**21**](PF$_6$)$_2$ (average Ru-O-Ru angle of $\approx$78°) failed to display the expected IVCT band in the NIR region [37]. However, in agreement with the Robin and Day class III description, the mixed-valent **21^{2+}** ($S = 1/2$, μ_{eff}: 2.17 BM) (Figure 4.17) exhibited a Ru^{III}-based anisotropic EPR signal with $<g>/\Delta g = 2.09/0.17$ at 77 K in acetone/toluene glass and a large K_c of 10^{24}. Similar situation of large K_c values but without detectable IVCT transitions were reported for analogous ligand-bridged Ru^{II}Ru^{III} mixed-valent states (see the preceding section).

Oxido/carboxylato- and oxido/pyrazolato-bridged diastereomeric (*meso*:$\Delta\Lambda$ and *rac*: $\Delta\Delta/\Lambda\Lambda$) mixed-valent Ru^{III}Ru^{IV}{Ru^{III}: t_{2g}^5 ($S = 1/2$), Ru^{IV}: t_{2g}^4 ($S = 0$)} complexes **22** and **23**, respectively (Figure 4.18), were structurally, spectroscopically, and electrochemically characterized, demonstrating similar features irrespective of the diastereomeric identities [39, 40]. Oxido/carboxylato- and oxido/pyrazolato-bridged **22** and **23**, respectively, displayed large K_c values (10^{19}/10^{24}) and Ru^{III}-based rhombic EPR signals (Figure 4.18). However, in contrast to **23**, no IVCT band was detected for **22**. The very weak solvent-independent IVCT band of **23** at 1440 nm ($\varepsilon = 250$ M^{-1} cm^{-1}, in CH$_3$CN) in conjunction with the high K_c value (10^{24}) and $\Delta v_{1/2}$(cal): 3944 cm^{-1} > $\Delta v_{1/2}$(exp): 2839 cm^{-1} suggests a class III mixed-valence state. In the absence of a detectable IVCT band for **22**, there is a large K_c of 10^{19}, implying a class III situation as per the Robin and Day classification.

Apart from the aforestated well-defined dimeric mixed-valent Ru^{III}Ru^{IV} complexes ([**19**]PF$_6$, [**20**]PF$_6$, [**21**](PF$_6$)$_2$, **22**, and **23**), the occurrence of the Ru^{III}Ru^{IV} mixed-valent state was also structurally authenticated in the μ_3-oxido/ μ-acetate-bridged tetraruthenium (Ru^{III}Ru^{III}Ru^{III}Ru^{IV}) setup (**24**, $S = 1/2$) involving a "butterfly" {M$_4$(μ_3–O)$_2$} motif (Figure 4.19) with a metal–metal bond (Ru2–Ru3: 2.5187(6) Å) [38]. A metal-based anisotropic EPR signal in CH$_3$CN-toluene at 100 K, a high K_c value of 1.8 × 10^{16}, the Mulliken spin density distribution of Ru1: 0.758, Ru2: −0.295, Ru3: −0.319, Ru4: 0.761, O1: −0.025, and O2: −0.014 along with a broad IVCT band at 1350 nm ($\varepsilon = 880$ M^{-1}cm^{-1}, in CH$_3$CN; $\Delta v_{1/2}$ (cal/exp): 4136/2865 cm^{-1}) (Figure 4.19) collectively suggest a class III feature.

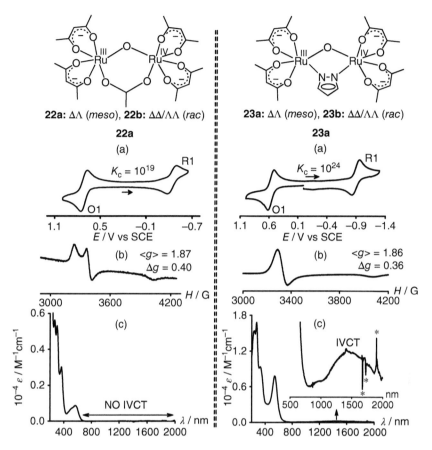

Figure 4.18 (a) Cyclic voltammograms in CH$_3$CN, (b) EPR spectra in CH$_3$CN or CH$_2$Cl$_2$ at 100 K, and (c) UV–vis–NIR spectra in CH$_3$CN of 22a (left) and 23a (right) (*instrumental signal).

4.4 Ru^{II}Ru^{I} and Ru^{I}Ru^{0} Mixed-Valent Systems

Although a large volume of diruthenium frameworks with Ru^{II}Ru^{III} and a few Ru^{III}Ru^{IV} mixed-valent states were explored over the past decades (see the preceding sections), the corresponding unconventional Ru^{II}Ru^{I} (d^{6}d^7) or Ru^{I}Ru^{0} (d^{7}d^8) mixed-valent configurations were largely missing. However, Ru^{II}Ru^{I} and Ru^{I}Ru^{0} states were recognized [73] through a complex electrochemical (E,EC,EC,E) sequence (E = electron transfer and EC = electron transfer followed by chemical process) and spectroelectrochemistry (UV–vis–NIR/EPR) of selectively designed abpy (2,2′-azobispyridine) and bpip (bis(1-phenyliminoethyl)pyrazine)-bridged isovalent precursors, [25]PF$_6$ and [26](PF$_6$)$_2$, respectively (Figure 4.20). The mixed-valent [(Cym)$_2$Ru^{II}(μ-abpy^{2-})Ru^{I}(Cym)$_2$]$^+$(25$^+$) or [(Cym)$_2$Ru^{I}(μ-bpip)Ru^{0}(Cym)$_2$]$^+$(26$^+$) with small K_c < 10$^{2.5}$ exhibited moderately intense narrow IVCT bands at ≈ 1500 nm (ε ≈ 1200 M^{-1} cm^{-1}, CH$_3$CN; $\Delta\nu_{1/2}$(exp/cal): 700 cm^{-1}/≈4000 cm^{-1}) and metal-based

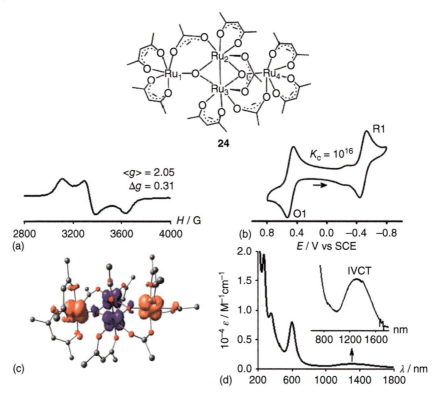

Figure 4.19 (a) EPR spectrum in CH$_3$CN at 4 K. (b) Cyclic voltammograms in CH$_3$CN and (c) Mulliken spin density plot. (d) UV–vis–NIR spectrum in CH$_3$CN. The inset shows the enlarged IVCT band.

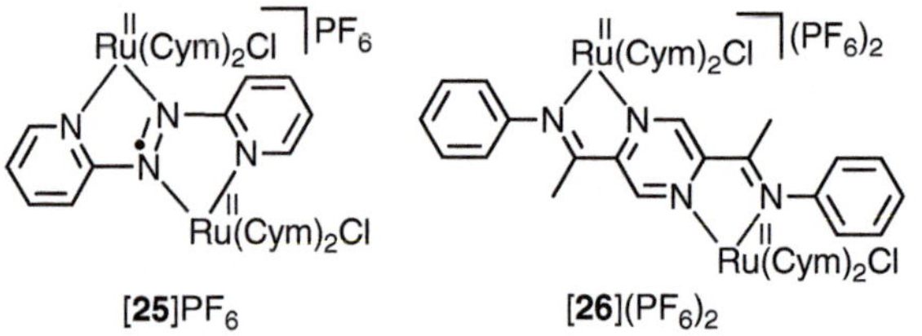

Figure 4.20 Abpy- and bpip-bridged diruthenium complexes.

EPR spectra ($g_1 = 2.155/2.210$, $g_2 = 1.988/2.027$, and $g_3 = 1.958/1.958$), corresponding to a valence-averaged system.

The Ru^I containing intermediates [73] are just an example for the special role of the azo/hydrazido ligand redox system for potential mixed valency in ligand (L^{n-})-bridged diruthenium complexes [74]. In fact, one of the earliest examples for the ambivalence Ru^{III}(μ-L^{2-})Ru^{II} ↔ Ru^{II}(μ-L$^{•-}$)Ru^{II} has concerned the diruthenium compounds of azodicarboxylates/dicarbonylhydrazides [75], followed by further related studies [76, 77]. Diamagnetic analogues Ru^{III}(μ-L)Ru^{II} ↔ Ru$^{2.5}$(μ-L$^{•-}$)Ru$^{2.5}$ have also been studied and discussed [78, 79].

Depending on energy matches, there are several other diruthenium complexes of potential non-innocent bridging ligands that have most of the redox activity centered on the bridge. Typical such bridges involve quinonoid ligands [80–84] that are relevant for dyes or biorelated systems. Thus, dinuclear coordination compounds of indigoid and anthraquinone non-innocent bridges have been reviewed recently [85, 86], but metal-based mixed valency was found to occur rarely in such cases.

While dinuclear systems constitute the prototypes of mixed-valent species [1–8, 19–22], there are also symmetrically or asymmetrically bridged tri- and tetranuclear coordination compounds [87–91]. For instance, dimer of dimer arrangement has been established for tetraruthenium compounds of conjugated tetracyano bridges [92]; however, the facile electron transfer of "TCNX" molecules may favor ligand-based instead of metal-based redox activity.

4.5 Conclusion and Outlook

This presentation of mixed-valent aspects of various bridged $Ru^n Ru^{n-1}$ systems involving a wide variety of bridging and ancillary ligands has highlighted the following salient points:

- In addition to the properties of the bridge, the electronic nature of the ancillary ligands in the complex frameworks contributes significantly to the coupling process by modulating the overall electron density of the participating metal ions.
- The divergence between the electrochemically and spectroscopically derived K_c values and IVCT energies with reference to intermetallic electronic "coupling" in the mixed-valent state could be addressed beyond the well-recognized Robin and Day classification of class I, class II, and class III, e.g. in terms of class II–III hybrids.
- The role of fast vibrational spectroscopy (e.g. IR) in differentiating moderately coupled versus strongly coupled mixed-valent states can be employed.
- Appropriate care needs to be taken in distinguishing a mixed-valent state from the radical-derived isovalent scenario, particularly for the molecular frameworks containing redox-active ligands.
- Difficulties may arise with asymmetric molecular setups as well as with the inherent limitation of spectroscopic techniques with special reference to the timescale concerned.

Strongly coupled mixed-valent systems with intense IVCT bands in the low-energy NIR region may be considered as attractive candidates for a design of near-infrared dye-based devices [93, 94] as well as for fabricating molecular electronics such as sensors [95, 96], wires [23, 97, 98], and telecommunication signaling materials [99], besides their potential as water oxidation catalysts [100–102]. Thus, further exploration in this direction with still more challenging molecular setups are envisaged in order to enhance the fundamental understanding of inner sphere electron transfer as well as expand the scope of future applications.

Acknowledgment

Financial supports received from SERB (J.C. Bose Fellowship, G.K.L.), UGC (fellowship to S.D. and S.B.), and the Land Baden-Württemberg (to W.K.), Germany, are gratefully acknowledged. The contributions of all the coauthors in the reference articles are also gratefully acknowledged.

References

1 Demadis, K.D., Hartshorn, C.M., and Meyer, T.J. (2001). The localized-to-delocalized transition in mixed-valence chemistry. *Chem. Rev.* 101: 2655–2685.
2 Ward, M.D. (1995). Metal-metal interactions in binuclear complexes exhibiting mixed valency; molecular wires and switches. *Chem. Soc. Rev.* 24: 121–134.
3 Kaim, W. and Lahiri, G.K. (2007). Unconventional mixed-valent complexes of ruthenium and osmium. *Angew. Chem. Int. Ed.* 46: 1778–1796.
4 Hazari, A.S., Indra, A., and Lahiri, G.K. (2018). Mixed valency in ligand-bridged diruthenium frameworks: divergences and perspectives. *RSC Adv.* 8: 28895–28908.
5 Brunschwig, B.S., Creutz, C., and Sutin, N. (2002). Optical transitions of symmetrical mixed-valence systems in the class II–III transition regime. *Chem. Soc. Rev.* 31: 168–184.
6 Richardson, D.E. and Taube, H. (1984). Mixed-valence molecules: electronic delocalization and stabilization. *Coord. Chem. Rev.* 60: 107–129.
7 Kaim, W. and Sarkar, B. (2007). Mixed valency in ruthenium complexes-coordinative aspects. *Coord. Chem. Rev.* 251: 584–594.
8 D'Alessandro, D.M. and Keene, F.R. (2006). Current trends and future challenges in the experimental, theoretical and computational analysis of intervalence charge transfer (IVCT) transitions. *Chem. Soc. Rev.* 35: 424–440.
9 Roelofs, T.A., Liang, W., Latimer, M.J. et al. (1996). Oxidation states of the manganese cluster during the flash-induced S-state cycle of the photosynthetic oxygen-evolving complex. *Proc. Natl. Acad. Sci.* 93: 3335–3340.
10 Klauss, A., Haumann, M., and Dau, H. (2012). Alternating electron and proton transfer steps in photosynthetic water oxidation. *Proc. Natl. Acad. Sci.* 109: 16035–16040.
11 Krewald, V., Retegan, M., Cox, N. et al. (2015). Metal oxidation states in biological water splitting. *Chem. Sci.* 6: 1676–1695.
12 Boelens, R. and Wever, R. (1980). Redox reactions in mixed-valence cytochrome c oxidase. *FEBS Lett.* 116: 223–226.
13 Proshlyakov, D.A., Pressler, M.A., and Babcock, G.T. (1998). Dioxygen activation and bond cleavage by mixed-valence cytochrome c oxidase. *Proc. Natl. Acad. Sci. U.S.A.* 95: 8020–8025.
14 Subramanian, S., Duin, E.C., Fawcett, S.E.J. et al. (2015). Spectroscopic and redox studies of valence-delocalized $[Fe_2S_2]^+$ centers in thioredoxin-like ferredoxins. *J. Am. Chem. Soc.* 137: 4567–4580.

15 Beinert, H., Holm, R.H., and Münck, E. (1997). Iron-sulfur clusters: nature's modular, multipurpose structures. *Science* 277: 653–659.

16 Kaim, W., Bruns, W., Poppe, J., and Kasack, V. (1993). Spectroscopy of mixed-valent states in dinuclear ions and metalloproteins. *J. Mol. Struct.* 292: 221–228.

17 Buser, H.J., Schwarzenbach, D., Petter, W., and Ludi, A. (1977). The crystal structure of prussian blue: $Fe_4[Fe(CN)_6]_3 \cdot xH_2O$. *Inorg. Chem.* 16: 2704–2710.

18 Herren, F., Fischer, P., Ludi, A., and Halg, W. (1980). Neutron diffraction study of prussian blue, $Fe_4[Fe(CN)_6]_3 \cdot xH_2O$. Location of water molecules and long-range magnetic order. *Inorg. Chem.* 19: 956–959.

19 Creutz, C. and Taube, H. (1969). A direct approach to measuring the Franck–Condon barrier to electron transfer between metal ions. *J. Am. Chem. Soc.* 91: 3988–3989.

20 Creutz, C. and Taube, H. (1973). Binuclear complexes of ruthenium ammines. *J. Am. Chem. Soc.* 95: 1086–1094.

21 Zwickel, A.M. and Creutz, C. (1971). Charge-transfer spectra of ruthenium (II) complexes. *J. Am. Chem. Soc.* 95: 1086–1094.

22 Kaim, W. (1983). The versatile chemistry of 1,4-diazines: organic, inorganic and biochemical aspects (Review). *Angew. Chem. Int. Ed.* 22: 171–190.

23 Burgun, A., Ellis, B.G., Roisnel, T. et al. (2014). From molecular wires to molecular resistors: TCNE, a class-III/class-II mixed-valence chemical switch. *Organometallics* 33: 4209–4219.

24 Braun-Sand, S.B. and Wiest, O. (2003). Theoretical studies of mixed-valence transition metal complexes for molecular computing. *J. Phys. Chem. A* 107: 285–291.

25 Lent, C.S., Isaksen, B., and Lieberman, M. (2003). Molecular quantum-dot cellular automata. *J. Am. Chem. Soc.* 125: 1056–1063.

26 Robin, M.B. and Day, P. (1967). Mixed valence chemistry: a survey and classification. *Adv. Inorg. Chem. Radiochem.* 10: 247–422.

27 McCleverty, J.A. and Ward, M.D. (1998). The role of bridging ligands in controlling electronic and magnetic properties in polynuclear complexes. *Acc. Chem. Res.* 31: 842–851.

28 Astruc, D. (1997). From organotransition-metal chemistry toward molecular electronics: electronic communication between ligand-bridged metals. *Acc. Chem. Res.* 30: 383–391.

29 Ito, T., Hamaguchi, T., Nagino, H. et al. (1999). Electron transfer on the infrared vibrational time scale in the mixed valence state of 1,4-pyrazine- and 4,4′-bipyridine-bridged ruthenium cluster complexes. *J. Am. Chem. Soc.* 121: 4625–4632.

30 Ito, T., Hamaguchi, T., Nagino, H. et al. (1999). Effects of rapid intramolecular electron transfer on vibrational spectra. *Science* 277: 660–663.

31 Londergan, C.H., Salsman, J.C., Ronco, S. et al. (2002). Solvent dynamical control of electron-transfer rates in mixed-valence complexes observed by infrared spectral line shape coalescence. *J. Am. Chem. Soc.* 124: 6236–6237.

32 Salsman, J.C., Kubiak, C.P., and Ito, T. (2005). Mixed valence isomers. *J. Am. Chem. Soc.* 127: 2382–2383.

33 Hage, R., Haasnoot, J.G., Nieuwenhuis, H.A. et al. (1990). Synthesis, X-ray structure, and spectroscopic and electrochemical properties of novel heteronuclear ruthenium-osmium complexes with an asymmetric triazolate bridge. *J. Am. Chem. Soc.* 112: 9245–9251.

34 Halpin, Y., Dini, D., Ahmed, H.M.Y. et al. (2010). Excited state localization and internuclear interactions in asymmetric ruthenium(II) and osmium(II) bpy/tpy based dinuclear compounds. *Inorg. Chem.* 49: 2799–2807.

35 Hush, N.S. (1967). Intervalence-transfer absorption. Part 2. Theoretical considerations and spectroscopic data. *Prog. Inorg. Chem.* 8: 391–444.

36 Suzuki, T., Matsuya, K., Kawamoto, T., and Nagao, H. (2014). Synthesis and structures of mixed-valence oxido-bridged diruthenium complexes bearing ethylbis(2-pyridyl-methyl)amine. *Eur. J. Inorg. Chem.* 722-727.

37 Suzuki, T., Suzuki, Y., Kawamoto, T. et al. (2016). Dinuclear ruthenium(III)–ruthenium(IV) complexes, having a doubly oxido-bridged and acetato- or nitrato-capped framework. *Inorg. Chem.* 55: 6830–6832.

38 Bera, S.K. and Lahiri, G.K. (2020). Unprecedented metal–metal bonded {$Ru_4(\mu_3\text{-O})_2$} butterfly core in oxido-carboxylato bridged mixed valence cluster-structural elucidation and electronic forms in accessible redox states. *Dalton Trans.* 49: 13573–13581.

39 Bera, S.K. and Lahiri, G.K. (2021). Structural and electronic forms of doubly oxido/Pz and triply oxido/(Pz)$_2$ bridged mixed valent and isovalent diruthenium complexes (Pz = pyrazolate). *Dalton Trans.* 50: 17653–17664.

40 Bera, S.K., Hazari, A.S., and Lahiri, G.K. (2019). Mixed-valent $Ru^{III}Ru^{IV}$ configuration in an oxido–carboxylato-bridged diastereomeric pair. *Inorg. Chem.* 58: 12538–12541.

41 Das, A., Scherer, T., Maji, S. et al. (2011). Reductive approach to mixed valency ($n = 1^-$) in the pyrazine ligand-bridged [(acac)$_2$Ru(μ-L^{2-})Ru(acac)$_2$]^n (L^{2-} = 2,5-Pyrazinedicarboxylate) through experiment and theory. *Inorg. Chem.* 50: 7040–7049.

42 Sedney, D. and Ludi, A. (1981). Synthesis and characterization of a series of binuclear ruthenium complexes bridged by 2,5-pyrazine dicarboxylate. *Inorg. Chim. Acta* 47: 153–158.

43 Dürr, M., Klein, J., Kahnt, A. et al. (2017). Redox behavior of a dinuclear ruthenium(II) complex bearing an uncommon bridging ligand: insights from high-pressure electrochemistry. *Inorg. Chem.* 56: 14912–14925.

44 Brady, I., Leane, D., Hughes, H.P. et al. (2004). Electronic properties of Ru(II) complexes bound to a bisphenolate bridge with low lying π^* orbitals. *Dalton Trans.* 334-341.

45 Maji, S., Sarkar, B., Mobin, S.M. et al. (2007). Non-innocent behaviour of ancillary and bridging ligands in homovalent and mixed-valent ruthenium complexes [A$_2$Ru(μ-L)RuA$_2$]^n, A = 2,4-pentanedionato or 2-phenylazopyridine, L^{2-}= 2,5-bis(2-oxidophenyl)pyrazine. *Dalton Trans.* 56: 2411–2418.

46 Mondal, P., Agarwala, H., Jana, R.D. et al. (2014). Sensitivity of a strained C–C single bond to charge transfer: redox activity in mononuclear and dinuclear ruthenium complexes of bis(arylimino)acenaphthene (BIAN) ligands. *Inorg. Chem.* 53: 7389–7403.

47 Kundu, T., Schweinfurth, D., Sarkar, B. et al. (2012). Strong metal–metal coupling in mixed-valent intermediates [Cl(L)Ru(μ-tppz)Ru(L)Cl]$^+$, L = β-diketonato ligands, tppz = 2,3,5,6-tetrakis(2-pyridyl)pyrazine. *Dalton Trans.* 41: 13429–13440.

48 Chanda, N., Sarkar, B., Fiedler, J. et al. (2003). Synthesis and mixed valence aspects of [{(L)ClRu}$_2$(μ-tppz)]$^{n+}$ incorporating 2, 2'-dipyridylamine (L) as ancillary and 2,3,5,6-tetrakis(2-pyridyl)pyrazine (tppz) as bridging ligand. *Dalton Trans.* 3550–3555.

49 Kundu, T., Sarkar, B., Mondal, T.K. et al. (2010). Carboxylate tolerance of the redox-active platform [Ru(μ-tppz)Ru]^n, where tppz = 2,3,5,6-tetrakis(2-pyridyl) pyrazine, in the electron-transfer series [(L)ClRu(μ-tppz)RuCl(L)]^n, n = 2+, +, 0, −, 2−, with 2-picolinato, quinaldato, and 8-quinolinecarboxylato ligands (L$^−$). *Inorg. Chem.* 49: 6565–6574.

50 Chanda, N., Laye, R.H., Chakraborty, S. et al. (2002). Dinuclear ruthenium(II) complexes [{(L)ClRu^{II}}$_2$(μ-tppz)]$^{2+}$ (L=an arylazopyridine ligand) incorporating tetrakis(2-pyridyl)- pyrazine (tppz) bridging ligand: synthesis, structure and spectroelectrochemical properties. *Dalton Trans.* 3496-3504.

51 Hartshorn, C.M., Daire, N., Tondreau, V. et al. (1999). Synthesis and characterization of dinuclear ruthenium complexes with tetra-2-pyridylpyrazine as a bridge. *Inorg. Chem.* 38: 3200–3206.

52 Koley, M., Sarkar, B., Ghumaan, S. et al. (2007). Probing mixed valence in a new tppz-bridged diruthenium(III,II) complex {(μ-tppz)[Ru(bik)Cl]$_2$}$^{3+}$ (tppz= 2,3,5,6-tetrakis(2-pyridyl)pyrazine, bik =2,2'-bis(1-methylimidazolyl)ketone): EPR Silence, intervalence absorption, and v_{CO} line broadening. *Inorg. Chem.* 46: 3736–3742.

53 Chanda, N., Sarkar, B., Kar, S. et al. (2004). Mixed valence aspects of diruthenium complexes [{(L)ClRu}$_2$(μ-tppz)]$^{n+}$ incorporating 2-(2-pyridyl)azoles (L) as ancillary functions and 2,3,5,6-tetrakis(2-pyridyl)pyrazine (tppz) as bis-tridentate bridging ligand. *Inorg. Chem.* 43: 5128–5133.

54 Kundu, T., Sarkar, B., Mondal, T.K. et al. (2011). Redox-rich spin–spin-coupled semiquinone ruthenium dimers with intense near-IR absorption. *Inorg. Chem.* 50: 4753–4763.

55 Dattelbaum, D.M., Hartshorn, C.M., and Meyer, T.J. (2002). Direct measurement of excited-state intervalence transfer in [(tpy)Ru^{III}(tppz$^{•−}$)Ru^{II}(tpy)]$^{4+}$ by time-resolved near-infrared spectroscopy. *J. Am. Chem. Soc.* 124: 4938–4939.

56 Wadman, S.H., Havenith, R.W.A., Hartl, F. et al. (2009). Redox chemistry and electronic properties of 2,3,5,6-tetrakis(2-pyridyl)pyrazine-bridged diruthenium complexes controlled by N,C,N'-biscyclometalated ligands. *Inorg. Chem.* 48: 5685–5696.

57 Nagashima, T., Nakabayashi, T., Suzuki, T. et al. (2014). Tuning of metal–metal interactions in mixed-valence states of cyclo metalated dinuclear ruthenium and

osmium complexes bearing tetrapyridylpyrazine or -benzene. *Organometallics* 33: 4893–4904.

58 Rocha, R.C., Rein, F.N., Jude, H. et al. (2008). Observation of three intervalence-transfer bands for a class II–III mixed-valence complex of ruthenium. *Angew. Chem. Int. Ed.* 47: 503–506.

59 Miller, J. and Min, K.S. (2009). Oxidation leading to reduction: redox-induced electron transfer (RIET). *Angew. Chem. Int. Ed.* 48: 262–272.

60 Kaim, W. (2002). The coordination chemistry of 1,2,4,5-tetrazines. *Coord. Chem. Rev.* 230: 127–139.

61 Miomandre, F. and Audebert, P. (2020). 1,2,4,5-Tetrazines: an intriguing heterocycles family with outstanding characteristics in the field of luminescence and electrochemistry. *J. Photochem. Photobiol., C* 44: 100372.

62 Lipunova, G.N., Nosova, E.V., Zyryanov, G.V. et al. (2021). 1,2,4,5-Tetrazine derivatives as components and precursors of photo- and electroactive materials. *Org. Chem. Front.* 8: 5182–5205.

63 Vergara, M.M., García Posse, M.E., Fagalde, F. et al. (2010). Mixed-valency with cyanide as terminal ligands: Diruthenium (III,II) complexes with the 3,6-bis(2-pyridyl)-1,2,4,5-tetrazine bridge and variable co-ligands (CN^- vs bpy or NH_3). *Inorg. Chim. Acta* 363: 163–167.

64 Chellamma, S. and Lieberman, M. (2001). Synthesis and properties of $[Ru_2(acac)_4(bptz)]^{n+}$ ($n = 0, 1$) and crystal structure of $[Ru_2(acac)_4(bptz)]$. *Inorg. Chem.* 40: 3177–3180.

65 Gordon, K.C., Burrell, A.K., Simpson, T.J. et al. (2002). Probing the nature of the redox products and lowest excited state of $[(bpy)_2Ru(\mu\text{-}bptz)Ru(bpy)_2]^{4+}$: a resonance raman study. *Eur. J. Inorg. Chem.* 554–563.

66 Poppe, J., Moscherosch, M., and Kaim, W. (1993). An unusually weak intervalence transition in a very stable bis-chelate analogue of the mixed-valent Creutz-Taube ion. UV/vis/near-IR and EPR spectroelectrochemistry of $[(NH_3)_4Ru(\mu\text{-}bptz)Ru(NH_3)_4]^{n+}$ (bptz = 3,6-Bis(2-pyridyl)-1,2,4,5-tetrazine; $n = 3–5$). *Inorg. Chem.* 32: 2640–2643.

67 Patra, S., Sarkar, B., Ghumaan, S. et al. (2004). Isovalent and mixed-valent diruthenium complexes $[(acac)_2Ru^{II}(\mu\text{-}bpytz)Ru^{II}(acac)_2]$ and $[(acac)_2Ru^{II}(\mu\text{-}bpytz)Ru^{III}(acac)_2](ClO_4)$ (acac =acetylacetonate and bpytz= 3,6-bis(3,5-dimethylpyrazolyl)-1,2,4,5-tetrazine): synthesis, spectroelectrochemical, and EPR Investigation. *Inorg. Chem.* 43: 6108–6113.

68 Sarkar, B., Laye, R.H., Mondal, B. et al. (2002). Synthesis, structure and spectroelectrochemical properties of a dinuclear ruthenium complex exhibiting a strong electronic interaction across a 1,2,4,5-tetrazine bridging ligand. *J. Chem. Soc., Dalton. Trans.* 2097–2101.

69 Chakraborty, S., Laye, R.H., Munshi, P. et al. (2002). Dinuclear bis(bipyridine) ruthenium(II) complexes $[(bpy)_2Ru\{L\}^{2-}Ru^{II}(bpy)_2]^{2+}$ incorporating thiouracil-based dianionic asymmetric bridging ligands: synthesis, structure, redox and spectroelectrochemical properties. *Dalton Trans.* 2348–2353.

70 Kar, S., Sarkar, B., Ghumaan, S. et al. (2005). A new coordination mode of the photometric reagent glyoxalbis(2-hydroxyanil) (H_2gbha): Bis-bidentate bridging

by gbha^{2-} in the redox series {(μ-gbha)[Ru(acac)$_2$]$_2$}^{n} (n = −2, −1, 0, +1, +2), including a radical-bridged diruthenium(III) and a Ru^{III}/Ru^{IV} intermediate. *Inorg. Chem.* 44: 8715–8722.

71 Hoshino, Y., Higuchi, S., Fiedler, J. et al. (2003). Long-range electronic coupling in various oxidation states of a C$_4$-linked tris(β-diketonato)ruthenium dimer. *Angew. Chem. Int. Ed.* 42: 674–677.

72 Agarwala, H., Scherer, T., Maji, S. et al. (2012). Correspondence of Ru^{III}Ru^{II} and Ru^{IV}Ru^{III} mixed valent states in a small dinuclear complex. *Chem. Eur. J.* 18: 5667–5675.

73 Sarkar, B., Kaim, W., Fiedler, J., and Duboc, C. (2004). Molecule-bridged mixed-valent intermediates involving the Ru^I oxidation state. *J. Am. Chem. Soc.* 126: 14706–14707.

74 Mondal, S., Schwederski, B., Frey, W. et al. (2018). At the borderline between metal-metal mixed valency and a radical bridge situation: Four charge states of a diruthenium complex with a redox-active bis(*mer*-tridentate) ligand. *Inorg. Chem.* 57: 3983–3992.

75 Kasack, V., Kaim, W., Binder, H. et al. (1995). When is an odd-electron dinuclear complex a mixed-valent species? Tuning of ligand-to-metal spin shifts in diruthenium complexes of noninnocent bridging ligands OC(R)NNC(R)O. *Inorg. Chem.* 34: 1924–1933.

76 Jana, R., Sarkar, B., Bubrin, D. et al. (2010). Structure, electrochemistry and spectroscopy of a new diacylhydrazido-bridged diruthenium complex with a strongly near-infrared absorbing Ru^{III}Ru^{II} intermediate. *Inorg. Chem. Commun.* 13: 1160–1162.

77 Roy, S., Sarkar, B., Imrich, H.-G. et al. (2012). Charged but found "not guilty": Innocence of the suspect bridging ligands [RO(O)CNNC(O)OR]$^{2-}$ = L^{2-} in [(acac)$_2$Ru(μ-L)Ru(acac)$_2$]^n, n = +,0,−,2−. *Inorg. Chem.* 51: 9273–9281.

78 Sarkar, B., Patra, S., Fiedler, J. et al. (2005). Theoretical and experimental evidence for a new kind of spin-coupled singlet species: Isomeric mixed-valent complexes bridged by a radical anion ligand. *Angew. Chem. Int. Ed.* 44: 5655–5658.

79 Sarkar, B., Patra, S., Fiedler, J. et al. (2008). Mixed-valent metals bridged by a radical ligand: Fact or fiction based on structure-oxidation state correlations. *J. Am. Chem. Soc.* 130: 3532–3542.

80 Ernst, S., Hänel, P., Jordanov, J. et al. (1989). Stable binuclear o- and p-semiquinone complexes of [Ru(bpy)$_2$]$^{2+}$, Radical ion versus mixed valence dimer formulation. *J. Am. Chem. Soc.* 111: 1733–1738.

81 McCleverty, J.A. and Ward, M.D. (2002). Non-innocent behavior in mononuclear and polynuclear complexes: consequences for redox and electronic spectroscopic properties. *J. Chem. Soc., Dalton Trans.* 275–288.

82 Dei, A., Gatteschi, D., Sangregorio, C., and Sorace, L. (2004). Quinonoid metal complexes: toward molecular switches. *Acc. Chem. Res.* 237: 827–835.

83 Ansari, M.A., Beyer, K., Schwederski, B. et al. (2018). Diruthenium complexes of *p*-benzoquinone-imidazole hybrid ligands: innocent or non-innocent behavior of the quinone moiety. *Chem. Asian J.* 13: 2947–2955.

84 Das, A.K., Sarkar, B., Fiedler, J. et al. (2009). A five-center redox system: molecular coupling of two non-innocent imino-*o*-benzoquinonato-ruthenium functions through a π acceptor bridge. *J. Am. Chem. Soc.* 131: 8895–8902.

85 Kaim, W. and Lahiri, G.K. (2019). The coordination potential of indigo, anthraquinone and related redox-active dyes. *Coord. Chem. Rev.* 393: 1–8.

86 Kumari, M., Bera, S.K., Blickle, S. et al. (2021). The indigo isomer epindolidione as redox-active bridging ligand for diruthenium complexes. *Chem. Eur. J.* 27: 5461–5469.

87 Roy, S., Sarkar, B., Duboc, C. et al. (2009). Heterohexanuclear (Cu_3Fe_3) complexes of substituted hexaazatrinaphthylene (HATN) ligands: Twofold BF_4^- association in the solid and stepwise oxidation (3e) or reduction (2e) to spectroelectrochemically characterized species. *Chem. Eur. J.* 15: 6932–6939.

88 Grange, C.S., Meijer, A.J.H.M., and Ward, M.D. (2010). Trinuclear ruthenium dioxolene complexes based on the bridging ligand hexahydroxytriphenylene: electrochemistry, spectroscopy, and near-infrared electrochromic behaviour associated with a reversible seven-membered redox chain. *Dalton Trans.* 39: 200–211.

89 Ansari, M.A., Mandal, A., Beyer, K. et al. (2017). Non-innocence and mixed valency in tri- and tetranuclear ruthenium complexes of a heteroquinone bridging ligand. *Dalton Trans.* 46: 15589–15598.

90 Moscherosch, M., Waldhör, E., Binder, H. et al. (1995). Tetranuclear pentaammineruthenium complexes bridged by π-conjugated tetracyano ligands related to TCNE synthesis and spectroscopy of different oxidation states. *Inorg. Chem.* 34: 4326–4335.

91 Jana, R., Schwederski, B., Fiedler, J., and Kaim, W. (2012). Facilitated reduction and oxidation of $\{[Ru(NH_3)]_4(\mu_4\text{-TCNX})\}^{8+}$ by changing from TCNX = TCNQ to $TCNQF_4$. *Polyhedron* 44: 174–178.

92 Záliš, S., Sarkar, B., Duboc, C., and Kaim, W. (2009). Evidence for the dimer-of-(mixed-valent dimers) configuration in tetranuclear $\{(\mu_4\text{-TCNX})[Ru(NH_3)_5]_4\}^{8+}$, TCNX = TCNE and TCNQ, from DFT calculations. *Chem. Monthly* 140: 765–773.

93 Bai, Y., Zhang, J., Wen, D. et al. (2019). Fabrication of remote controllable devices with multistage responsiveness based on a NIR light induced shape memory ionomer containing various bridge ions. *J. Mater. Chem. A.* 7: 20723–20732.

94 Zhang, K., Liu, Y., Hao, Z. et al. (2020). A feasible approach to obtain near-infrared (NIR) emission from binuclear platinum(II) complexes containing centrosymmetric isoquinoline ligand in PLEDs. *Org. Electron.* 87, 105902: 1–8.

95 Nijhuis, C.A., Ravoo, B.J., Huskens, J., and Reinhoudt, D.N. (2007). Electrochemically controlled supramolecular systems. *Coord. Chem. Rev.* 251: 1761–1780.

96 Hutter, L.H., Muller, B.J., Kore, K. et al. (2014). Robust optical oxygen sensors based on polymer bound NIR-emitting platinum(II)– benzoporphyrins. *J. Mater. Chem. C* 2: 7589–7598.

97 Rocha, R.C. and Toma, H.E. (2002). Benzotriazolate-bridged ruthenium dinuclear and trinuclear complexes. *Polyhedron* 21: 2089–2098.
98 Joachim, C., Gimzewski, J.K., and Aviram, A. (2000). Electronics using hybrid-molecular and mono-molecular devices. *Nature* 408: 541–548. (and references cited therein).
99 LeClair, G. and Wang, Z.Y. (2009). Optical attenuation at the 1,550-nm wavelength in a reflective mode using electrochromic ruthenium complex film. *J. Solid State Electrochem.* 13: 365–369.
100 Bozoglian, F., Romain, S., Ertem, M.Z. et al. (2009). The Ru-Hbpp water oxidation catalyst. *J. Am. Chem. Soc.* 131: 15176–15187.
101 Zhang, P., Hou, X., Liu, L. et al. (2015). Two-dimensional π-conjugated metal bis(dithiolene) complex nanosheets as selective catalysts for oxygen reduction reaction. *J. Phys. Chem. C* 119: 28028–28037.
102 Liu, Y., Xiang, R., Du, X. et al. (2014). An efficient oxygen evolving catalyst based on a μ-O diiron coordination complex. *Chem. Commun.* 50: 12779–12782.

5

Electronic Communication in Mixed-Valence (MV) Ethynyl, Butadiynediyl, and Polyynediyl Complexes of Iron, Ruthenium, and Other Late Transition Metals

Sheng Hua Liu[1], Ya-Ping Ou[2], and František Hartl[3]

[1]*Central China Normal University, National Key Laboratory of Green Pesticide, College of Chemistry, No. 152 Luoyu Road, Wuhan 430079, China*
[2]*Hengyang Normal University, College of Chemistry and Material Science, Key Laboratory of Functional Metal-Organic Compounds of Hunan Province, No. 165 Huangbai Road, Hengyang, Hunan 421008, China*
[3]*University of Reading, Department of Chemistry, Whiteknights, Reading RG6 6DX, UK*

5.1 Introduction

Organometallic mixed-valence (MV) complexes are commonly explored to investigate electronic communication between two redox centers in different oxidation states. The organometallic centers (M_1 and M_2) are linked by unsaturated π-conjugated ligands [1–4]. The inequivalent sites with an intermediate electronic interaction are readily reversed by an intervalence charge transfer (IVCT) in transition-metal-containing chromophores, and there is a medium barrier to this interconversion. One extreme to this most common weakly or moderately strongly coupled asymmetric (Class II) situation are MV complexes with distinct oxidation states but very limited electronic interaction and interconversion (Class I). On the other hand, in the scarcely seen fully delocalized, symmetrical MV regime (Class III), the oxidation states of very strongly interacting redox centers are equivalent. The degree of MV delocalization is tunable by ligand substitution. Therefore, in classical MV models, conjugated ligands bridging in dinuclear or polynuclear organometallic complexes play a very important role in understanding the fundamental electron transfer (ET) processes between the redox-active sites, which may also be replaced by electrode surfaces [5–8].

Recently, many organometallic MV models [L_nM–bridge–ML_n], with M stabilized by ancillary ligands serving as redox-active termini, have been prepared, and their electrochemistry and electron transfer properties were investigated [1–3, 9–14]. Common redox-active end groups, such as ferrocenyl [15–17], M(dppe)Cp* or M(dppe)$_2$Cl (M = Ru, Fe, Os, and Mo; dppe:1,2-bis(diphenylphosphino)ethane; and Cp* = pentamethylcyclopentadienyl) [18–23], and RuCl(CO) (P-ligand) (P-ligand = (P^iPr_3)$_2$ or (PMe$_3$)$_3$) [24–27], cyclometalated Ru [9, 28, 29], and some other redox-active metal termini [1–3, 30–33], were introduced to control and tune the redox behavior and improve the electrochemical reversibility. On that basis,

Mixed-Valence Systems: Fundamentals, Synthesis, Electron Transfer, and Applications, First Edition.
Edited by Yu-Wu Zhong, Chun Y. Liu, and Jeffrey R. Reimers.
© 2023 WILEY-VCH GmbH. Published 2023 by WILEY-VCH GmbH.

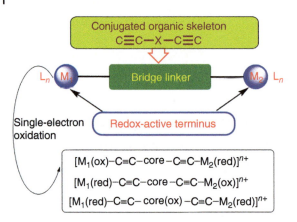

Figure 5.1 Dinuclear metal–ethynyl complexes [L$_n$M$_1$–C≡C–X–C≡C–M$_2$L$_n$], with X serving as the bridge core, and their single-electron oxidation states.

different carbon-rich bridging ligands, including polyene, polyyne, condensed aromatic hydrocarbons, and heteroaromatic ring units, have been applied to promote and mediate electronic communication between two redox-active metal–ethynyl centers [1, 2, 9, 10].

Rigid wire-like metal diynediyl complexes, with the metal center bound to the ethynyl group, M—C≡C [34–37], feature extensively π-conjugated systems along the molecular backbones, in which the bridge is made up of ethynyl (—C≡C) linkers and the bridge core X that may also be redox-active itself (oligothiophenes, oligothienoacenes, naphthalene, anthracene, etc.). These complexes provide unique advantages in straightforward syntheses, facile functionalization, and oxidative/thermal/photochemical stability. The MV systems [M$_1$—C≡C—core—C≡C—M$_2$]$^{n+}$ may thus include three different localized-valence (charge) isomers (valence tautomers), namely, [M$_1$(ox)—C≡C—core—C≡C—M$_2$(red)]$^{n+}$, [M$_1$(red)—C≡C—core—C≡C—M$_2$(ox)]$^{n+}$, and [M$_1$(red)—C≡C—core(ox)—C≡C—M$_2$(red)]$^{n+}$. They have been paid increasing attention in forming novel molecular wires with a long-range charge transfer performance (Figure 5.1) [10, 11, 37–41]. Therefore, in this chapter, we mainly focus on introducing the syntheses and structural properties of a series of homodinuclear metal–ethynyl complexes and the description of their mixed-valence states and electron transfer phenomena defined by the nature of the metal–ethynyl termini and the core of the bridging ligand. Accordingly, the content is divided into three main sections: (i) iron–ethynyl complexes, (ii) ruthenium–ethynyl complexes, and (iii) other transition metal–ethynyl complexes.

5.2 Iron–Ethynyl Complexes

For a given bridge, the characteristics of organometallic molecular wires strongly depend on the electronic properties of the metal-based end group and its electron exchange ability (interconversion) with its terminal counterpart through the bridge (the conjugated linkers and the bridge core, if any). The electron-rich (dppe)Cp*Fe moiety has served as a classical redox-active end group because of its convenient preparation developed by the research group of Lapinte in the past two decades [10, 41]. Lapinte and coworkers have synthesized a range of organometallic wires containing the (dppe)Cp*Fe–ethynyl electrophore and thoroughly investigated their structural, electronic, and magnetic coupling properties by X-ray crystallography, electrochemistry, IR, NMR, EPR, and ^{57}Fe Mössbauer spectroscopies, as well as theoretical calculations [42–45]. Recently, some novel and multifunctional dinuclear iron–ethynyl complexes have also been presented by other groups [46–48].

5.2.1 Dinuclear Iron–Ethynyl Complexes with Butadiynediyl Bridge

1,3-Butadiyn-1,4-diyl (—C≡C—C≡C—) and its derivatives with condensed polycyclic aromatic hydrocarbons in the core position (Figure 5.1) have often been used as bridging ligands to link the CpFe– or Cp*Fe–ethynyl redox-active termini [10, 41–45]. First, the Lapinte group has been very active in this area. The primary dinuclear iron complex with the full-carbon bridge and absent core is [{FeCp*(dppe)}$_2$(μ-C≡C—C≡C)] (complex **1**, Figure 5.2) [41, 49]. Its one-electron oxidation generates a radical cation with the spin density localized on the Fe centers (MV intermediates **1a** ↔ **1b**). Near-IR (NIR) absorption spectroscopy of **1a** ↔ **1b** reveals a very strong electronic coupling between the metallic termini characteristic of a fully delocalized Robin–Day Class III compound [41]. Further one-electron oxidation affords biradical **1**$^{2+}$ in a spin-singlet state (**1c**) with stabilizing *anti*-ferromagnetic coupling and a spin-triplet (diradical) state (**1d**). Experimental results have indicated that the energy gap between the two spin isomers **1c** and **1d** is sufficiently small ($\Delta_{GST} = -18.2\,\text{cm}^{-1}$) for both states to be

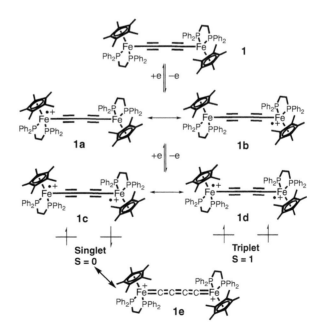

Figure 5.2 One-electron and two-electron oxidation of complex **1**. Source: Adapted from Halet and Lapinte [10].

populated. A small contribution from the cumulenic resonance structure **1e** further stabilizes the dicationic spin-singlet state.

5.2.2 Dinuclear Iron–Ethynyl Complexes with Diynediyl, Polycyclic Aromatic Hydrocarbons and Heterocycles in the C_4 Bridge Core

At a later date, numerous dinuclear iron–ethynyl complexes with different conjugated bridge cores (Figure 5.1) were synthesized by Lapinte and coworkers. These cores usually include conjugated ethynylene, condensed aromatic hydrocarbons, and aromatic heterocycles (Figure 5.3) [10, 41]. Singly oxidized species 2^+–11^+ are prepared by one-electron oxidation of neutral precursors **2–11**. These MV complexes were isolated, fully characterized by various spectroscopic techniques, and found to exhibit on average dominant characteristics of metal-localized oxidation. The key parameters (v_{max}, $\Delta v_{1/2}$, ε_{max}, and H_{ab}) of 1^+–9^+ are summarized in Table 5.1 [10]. Among them, 1^+, 2^+, 4^+, and 5^+ proved to belong to the Class III MV system [49–52], while monocation 3^+ was recognized as a borderline Class II/III complex and 9^+ as a weakly coupled Class II MV complex [53]. From the H_{ab} data given in Table 5.1, it can be stated that the electronic couplings are very strong for these Class III MV compounds, but the H_{ab} parameter decreases with the number of carbon atoms in the carbon chain. In addition, for the isomeric diethynyl–pyridine-bridged complexes [2,6-{Cp(dppe)Fe—C≡C—}$_2$(μ-NC$_5$H$_3$)] (**7**)

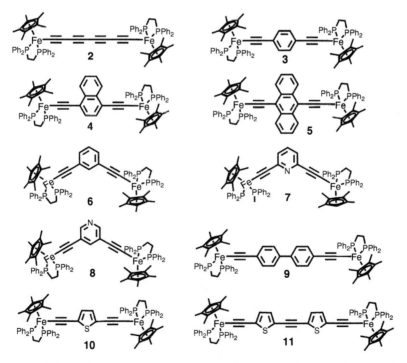

Figure 5.3 Different conjugated bridge cores in dinuclear Cp*(dppe)Fe–ethynyl complexes.

Table 5.1 Low-energy component of the IVCT band for MV dinuclear Cp*(dppe)Fe–ethynyl complexes and electronic couplings.

Compound	Class	Solvent	v_{max} (cm^{-1})	$\Delta v_{1/2}$ (cm^{-1})	ε_{max} (M^{-1} cm^{-1})	d_{ab} (Å)°	H_{ab} (cm^{-1})
1$^+$	III	CH$_3$CN	7704	3.26 × 10^3	1.3 × 10^4	7.436	3850[a]
2$^+$	III	CH$_3$CN	5157	1.30 × 10^3	3.1 × 10^4	12.655	2580[a]
3$^+$	III/II	CH$_2$Cl$_2$	4000	2.12 × 10^3	1.3 × 10^4	11.6	2000[a]
4$^+$	III	CH$_2$Cl$_2$	4280	1.43 × 10^3	2.1 × 10^4	11.7	2140[a]
5$^+$	III	CH$_3$CN	4340	1.66 × 10^3	3.9 × 10^3	11.9	2170[a]
6$^+$	II	CH$_2$Cl$_2$	5540	3.7 × 10^3	3.1 × 10^2	10.2	161[b]
7$^+$	II	CH$_2$Cl$_2$	4050	3.6 × 10^3	2.4 × 10^2	9.7	126[b]
8$^+$	II	CH$_2$Cl$_2$	5450	3.6 × 10^3	1.8 × 10^2	10.1	121[b]
9$^+$	II	CH$_2$Cl$_2$	6250	3.75 × 10^3	5.6 × 10^2	16.1	145[b]
12$^+$	II	CH$_2$Cl$_2$	4325	3.3 × 10^3	1.1 × 10^2	8.5	110[b]
13$^+$	II	CH$_2$Cl$_2$	6990	4.1 × 10^3	0.2 × 10^2	10	50[b]
14$^+$	II	CH$_2$Cl$_2$	6400	3.7 × 10^3	1.0 × 10^2	11.1	285[b]
15$^+$	II	CH$_2$Cl$_2$	6700	3.8 × 10^3	2.6 × 10^2	12.8	130
16$^+$	II	CH$_2$Cl$_2$	8500	4.5 × 10^3	0.6 × 10^2	14.5	64

a) Calculated from $H_{ab} = v_{max}/2$.
b) Calculated from $H_{ab} = (2.06 \times 10^{-2}/d_{ab})(\varepsilon_{max} v_{max} \Delta v_{1/2})^{1/2}$.
Source: Halet and Lapinte [10]/Reproduced with permission of Elsevier.

and [3,5-{Cp(dppe)Fe—C≡C–}$_2$(μ-NC$_5$H$_3$)] (**8**) [54], as compared with complex **6** of a similar shape [55, 56], the position of the nitrogen heteroatom in the core ring plays a very important role in determining the electronic and magnetic couplings in the respective mono-oxidized and di-oxidized species. With Py-N in the long branch, **8^{2+}** exhibits a stronger ferromagnetic interaction between the two Fe(III) spin carriers than that encountered in **7^{2+}** with Py-N in the short branch. While the MV properties of the two isomeric compounds, **7$^+$** and **8$^+$**, feature rather large difference and complexity, they both belong to the Robin–Day Class II group, as evidenced by MR-CI results and confirmed experimentally by Mössbauer spectroscopy. For the diethynyl–thiophene-bridged complex **10**, the NIR spectrum of singly oxidized MV compound **10$^+$** presents a strong absorption band while parent **10** and its dication are completely silent in this range. Investigation of the solvent dependence of the IVCT band shows a small increase of v_{max} with the polarity of the solvent. This effect comparable with those found for **2$^+$** confirms the Class III MV characteristic of **10$^+$**. Although the Fe–Fe distance in complex **11** is significantly longer compared to that in **9**, radical cations **11$^+$** correspond to Robin–Day Class II classification with sizable electronic coupling parameters (**11$^+$**, $H_{ab} = 262$ cm^{-1}, and $d_{Fe-Fe} = 17.7$ Å), which is ascribed to the electron-rich characteristic of the thiophene rings in the bridge of complex **11** [57].

5.2.3 Dinuclear Iron–Ethynyl Complexes with Non-conjugated C_4 Bridge Core

In addition to the above conjugated bridging ligands, Lapinte and coworkers also reported other segments in the bridge core, such as saturated structures and redox-active inorganic or organic units [58–61]. For example, in 2004, a series of dinuclear iron–ethynyl complexes containing saturated oligomethylene groups inserted into the carbon-rich bridge [X = $(CH_2)_n$, complex **12** ($n = 3$) and **13** ($n = 4$)] was constructed (Figure 5.4a) [58]. In this case, cyclic voltammetry (CV) and IR, Mössbauer, EPR, and UV–vis spectroscopies provided evidence for a weak through-bridge electronic interaction between the terminal iron sites. The key parameters (v_{max}, $\Delta v_{1/2}$, ε_{max}, and H_{ab}) of MV radical cations **12$^+$** and **13$^+$** in Table 5.1 clearly document that a through-bridge electron transfer still occurs. Subsequently, saturated polysilane backbones featuring unique electronic properties were incorporated into the carbon-rich bridges between the iron centers, and the corresponding complexes **14–16** in three different oxidation states were characterized (Figure 5.4a) [59]. The results of electrochemical studies have proven that only **14** shows a considerable half-wave potential separation, which suggested low thermal stability of the other two singly oxidized species, **15$^+$** and **16$^+$**, toward disproportionation. However, the UV–vis–NIR spectroelectrochemical detection of **14$^+$**–**16$^+$** led to the observation of IVCT absorption bands in the NIR region, which proved the existence of photo-driven long-range Fe (II)-to-Fe (III) electron transfer processes (the Fe–Fe distances vary between 11.0 and 14.5 Å) through the hybrid carbon–silicon bridges. Therefore, the degree of charge transfer largely depends on the distance between the redox-active terminal groups. The electronic coupling parameters H_{ab}, derived from the IVCT bands, decrease on going from **14$^+$** to **16$^+$** (Table 5.1). Comparison of the H_{ab} values for **12$^+$** and **13$^+$** with those for **14$^+$**–**16$^+$** reveals that the electronic coupling becomes roughly two times smaller for each additional Si—Si bond and three times smaller for the addition of an extra C—C single bond. This result suggests that the $Si(CH_3)$—$Si(CH_3)_2$ units are slightly better than the CH_2—CH_2 units as elemental molecular conductors.

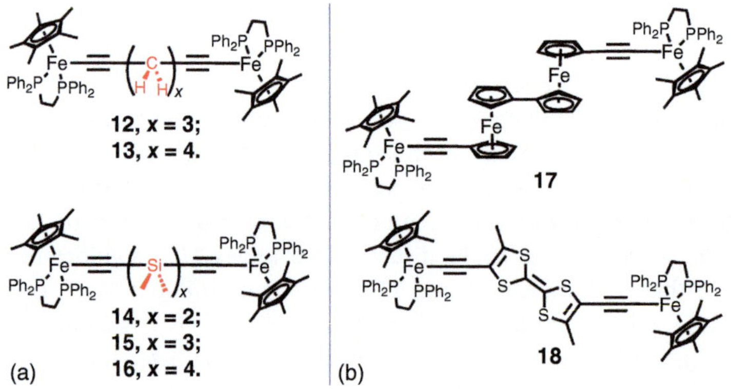

Figure 5.4 Dinuclear iron–ethynyl complexes with non-conjugated C_4 bridge cores.

To enhance the electronic communication between the iron–ethynyl groups and facilitate the long-distance electron transfer, Lapinte and coworkers incorporated redox-active units into the core position in the C_4 bridge chain. With this aim, [{FeCp*(dppe)}$_2$(μ-C≡C—biferrocenyl—C≡C)] (complex **17**) was reported in 2010 [60]. Experimental results have indicated that MV complex **17$^+$** complies with a Robin–Day Class II assignment. In addition, the incorporation of organic functional groups, such as tetrathiafulvalene (TTF), into the bridge core (complex **18**) has also supported and tuned electron transfer capabilities, optical activity, and magnetic properties [61].

5.2.4 Functionalized Dinuclear Iron–Ethynyl Complexes

Diverse functionalized dinuclear iron–ethynyl complexes have also been introduced by fellow research groups. For instance, in 2013, Berke and coworkers innovatively constructed a series of novel iron–ethynyl complexes **19–22** with buta-1,3-diyn-1-yl termini from the precursors *trans*-[Fe(depe)$_2$I$_2$] (depe = 1,2-bis(diethylphosphino)ethane) and **19**, as shown in Figure 5.5 [62]. These dinuclear complexes can further be utilized to build up linear long-distance polynuclear molecular wires through a series of reactions. Electrochemical studies indicate that the MV states of these dinuclear complexes are stable. Complexes **20, 21**, and **22** display three reversible oxidation waves, one being significantly positively shifted. The two close-lying anodic waves have been attributed to the oxidation of the metal centers, while the remaining shifted one is due to the oxidation of the bridging 1,3-butadiyn-1,4-diyl ligand. In other respect, molecular wires with the terminal —C≡C—C≡C—SnMe$_3$ group (complex **22**) also successfully converted to molecular junctions with Au–C σ-bonds through an in situ transmetallation reaction [63]. In 2006, Akita and coworkers reported a redox-active dinuclear iron–ethynyl complex with the dithienylethene (DTE) core, [{FeCp*(dppe)}$_2$(μ-C≡C—DTE—C≡C)] (**23**) [64], which shows a photochromic behavior, switching ON and OFF the electronic communication between the two metal centers. An IVCT band at 5300 cm^{-1} was observed in the NIR absorption spectrum of the closed isomer **23-C$^+$**, in contrast with the open isomer **23-O$^+$** that is silent in this region. The communication performance between the two iron centers through the DTE bridge has been characterized by the electronic coupling H_{ab} parameter that dropped from 365 cm^{-1}

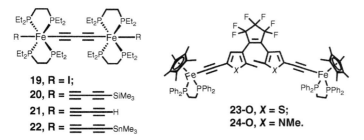

Figure 5.5 Functionalized dinuclear iron–ethynyl complexes.

in **23-C⁺** to 0 cm⁻¹ in **23-O⁺**. The remarkable switching behavior has its origin in the differently conjugated systems, that is, cross-conjugated in open isomer **23-O⁺** and fully conjugated in closed isomer **23-C⁺**. In 2015, the team went on to construct the dinuclear iron–ethynyl complexes with dipyrrolylethene in the bridge core (complex **24-O**) [47]. However, complex **24-O** showed no photochromic behavior and only a weak electronic interaction between the metal centers. Further discussions on dimetallic complexes with a switchable bridge are available in Chapter 12 of this book.

5.3 Ruthenium–Ethynyl Complexes

Dinuclear ruthenium–ethynyl complexes based on a few classical types of redox-active metallic termini have often been explored to test diverse types of bridging ligands in symmetric MV systems. The degree of electronic communication between two metal centers depends on several key factors, including the nature of the molecular bridge (linkers and core), ancillary ligands, and intrinsic properties of the transition metals, which have been investigated by several groups [1, 11, 18, 22, 38, 65–69]. Based on the molecular nature and the type of the ruthenium-centered end groups, we have divided the series of dinuclear ruthenium–ethynyl complexes into four families described in detail in the following sections: (i) dinuclear ruthenium–ethynyl complexes with Cp′(L₂)Ru-based termini [Cp′ = Cp, Cp*; L₂ = dppe or (PPh₃)₂], (ii) dinuclear ruthenium–ethynyl complexes with Ru(dppe)₂X-based termini, (iii) ruthenium–ethynyl complexes with alternating polyyndiyl and capped Ru–Ru units, and (iv) ruthenium–ethynyl complexes with other ruthenium–ethynyl termini and core units.

5.3.1 Dinuclear Ruthenium–Ethynyl Complexes with Cp′(L₂)Ru-Based Termini

In recent years, the groups of Bruce, Low, Hartl, and Liu have focused on developing different models of conjugated molecular wires bridging Cp′(L₂)Ru-based redox-active termini and investigated their electronic coupling properties. The two simplest diruthenium complexes with buta-1,3-diyn-1,4-diyl [{Ru(PP)Cp′}₂(μ-C≡C—C≡C)] [Cp′ = Cp, PP = (PPh₃)₂ (**25**); Cp′ = Cp*, PP = dppe (**26**)] were synthesized by Bruce et al. [70, 71]. They are readily oxidized in four consecutive one-electron steps. The oxidized members of the series were characterized by various molecular spectroscopic, structural, and computational methods. Each of the monocations **25⁺** and **26⁺** give rise to an IVCT band in the NIR region of the spectrum. Their position of v_{max} was found to be virtually independent of solvent, the widths of the IVCT bands at half-height ($\Delta v_{1/2}$) are considerably narrower than the calculated value, and **25⁺** and **26⁺** correspond to Robin–Day Class III classification with quite sizable electronic coupling parameters (**25⁺**, H_{ab} = 0.71 eV, **26⁺**, H_{ab} = 0.63 eV). Subsequently, the groups of Low and Liu gradually focused on changing the properties of the bridging ligand, such as the degree of conjugation, coplanarity, and length, and investigated the impact of the bridge modification on the electron transfer between the redox-active metal centers (Figure 5.6). Conjugated and coplanar

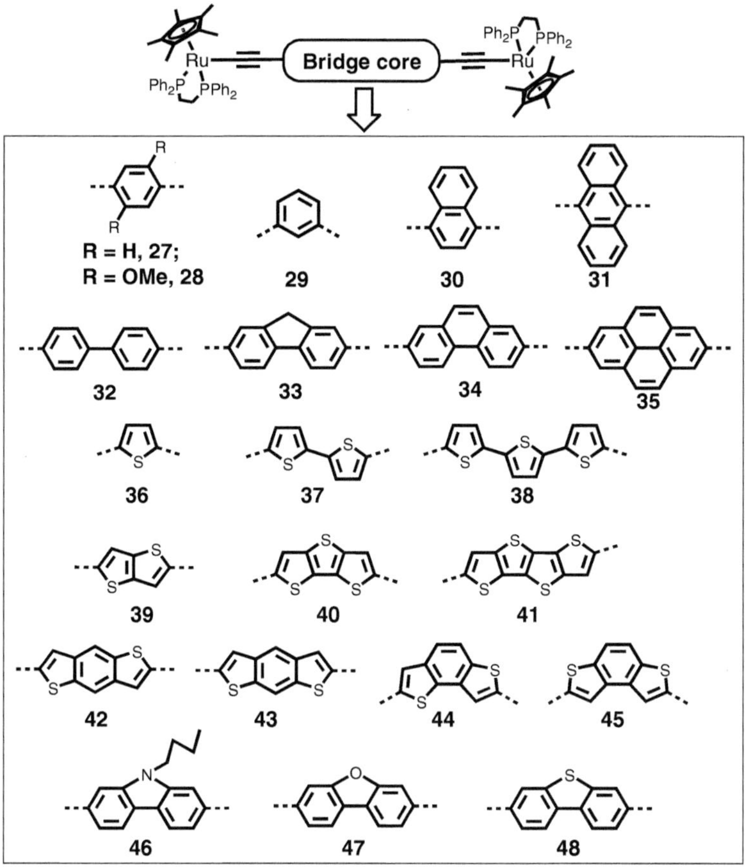

Figure 5.6 Conjugated aromatic rings and heterocycles in the core of the dinuclear RuCp*(dppe)–ethynyl complexes.

skeletons have been selected preferentially as bridging ligands, such as polycyclic aromatic hydrocarbons (**27–31** and **34–35**) [72, 73], oligophenylene (**32** and **33**) [73], oligothiophene (**36–38**) [74], oligothienoacene (**39–41**) [66], benzodithiophene (**42–45**) [75], dibenzoheterocycle (**46–48**) [76], and so on. Electrochemical and spectroscopic investigations of these complexes have revealed that the increasing conjugation along the bridge chain and its electron-rich nature are beneficial for a stronger electronic communication between two redox-active ruthenium centers while maintaining the same distance between them.

These one-electron-oxidized diruthenium MV species exhibit a characteristic intense and broad absorption band in the near-IR region, which has frequently been assigned the IVCT characteristic. However, recent EPR experiments and DFT calculations of the spin density distribution in the cationic complexes have revealed that the unpaired electron/hole is delocalized over both end-capping Ru centers and the bridging ligand. The established strong participation of the conjugated bridging ligand in the oxidation of the parent complex applies for a majority of the studied dinuclear ruthenium–ethynyl complexes. For example, the diethynylaromatic ligand-bridged diruthenium complexes **27** and **31**, reported by Low and coworkers [72], exhibit two consecutive oxidations to the corresponding monocations at $E_{1/2}$ = +0.01 V and −0.17 V vs. ferrocenium/ferrocene, respectively. The negative potential shift going from the 1,4-phenylene to 9,10-anthrylene bridge core is attributed the more thermodynamically favored larger size of the aromatic π-system of the 9,10-anthrylene ligand core, providing evidence for its strong involvement in the oxidation. Indeed, the spin density in **27$^+$** and **31$^+$** on the ligand core is high, increasing from 25% for 1,4-phenylene to 58% for 9,10-anthrylene, while the Ru contribution decreases from 34% to 18% in the same direction. The C_a of both ethynyl linkers also bears significant spin density, 24% and 30%, respectively. The second oxidation step, generating the corresponding dications, is separated in both cases by c. 300 mV. A relatively intense electronic absorption in the monocationic states is observed in the NIR region, which consists of a sum of three deconvoluted Gaussian-shaped sub-bands under the absorption envelope. Two of them have been assigned on the grounds of theoretical calculations an ML(bridge$^+$)CT characteristic, converting the ground-state cationic complex with the largely oxidized aromatic bridge core to an excited state with a MV characteristic. These MLCT sub-bands shift to a higher energy for the larger 9,10-anthrylene core as the bridge-based monocationic state becomes stabilized. The third NIR transition has an IVCT characteristic, corresponding to a MV ground state with the metallic centers weakly coupled through the orbitals of the diethynylaromatic bridge. The interconversion of the thermally populated bridge-based (delocalized) and MV (localized) ground states in **27$^+$** and **31$^+$** is facilitated by the rotation of the diethynylaromatic ligands, which causes decoupling of the bridge core from the Ru centers in the MV state because of the diminished overlap of the bridge π-orbitals and metal d_π orbitals of appropriate symmetry in a certain molecular conformation. The coexistence and rapid interconversion of the ground states with the positive charge localized largely on the bridge core, and one of the metal centers is also evidenced by the IR spectroelectrochemical data in the ν(C≡C) region.

In general, the bridge-localized radical characteristics often correspond to a dominant non-innocent redox behavior [72–75, 77]. That is to say, the oxidation pattern delocalized over the molecular backbone and the diverse nature of the redox-derived products are then reflected in the complexity of low-energy electronic transitions in the singly oxidized specie with multiple characteristics. For example, **29$^+$** features broad NIR absorptions that can be deconvoluted into several Gaussian-shaped IVCT sub-bands, while mixing up with other types of electronic excitation, such as MLCT or intraligand π–π* electronic absorptions, belonging to different conformers in the solution. In addition, the NIR band envelope of this class of MV complexes often obtains partial contributions from resolved vibrational progression and additional optical excitations gaining intensity from the low symmetry at the Ru center [78] and spin–orbit coupling (despite the low spin–orbit coupling constant of Ru). However, for strongly coupled MV systems, the above factors have a rather small effect on the NIR absorption. Low, Kaupp, and coworkers have revealed that the existence of conformationally unrestricted rotamers (e.g., in

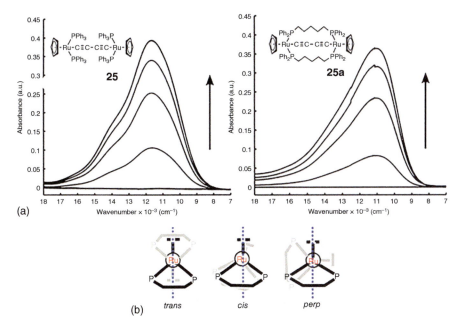

Figure 5.7 (a) NIR spectra recorded on electrochemical oxidation of **25** to **25⁺** (left) and **25a** to **25a⁺** (right) in dichloromethane/Bu₄NPF₆. Source: Parthey et al. [78]/Reproduced with permission from John Wiley & Sons. (b) Schematic representation of optimized rotamers of **33⁺** and **34⁺** with the different relative orientation of the two ruthenium centers (black and gray) and the plane of the conjugated diethynyl polyaromatic bridge (blue dashed lines). Source: Zhang et al. [73]/Reproduced with permission from American Chemical Society.

25 compared to constrained **25a**) is indeed a crucial variable that controls the rise of an IVCT absorption and participation of different electrophores (Figure 5.7a). The explanation is based on different relative conformations of the bridge backbone and the metallic redox centers encountered (with a support from molecular modeling) in this type of complexes (Figure 5.7b) [72–75, 78, 79].

The series of diruthenium diethynyl complexes with the bridge core consisting of oligothiophenes (**36–38**) and α,β-fused oligothienoacenes (**36, 39–41**) studied by Liu and coworkers [66, 74] represent a situation where the electrochemical oxidation to the corresponding monocations and dications is localized predominantly on the heteroaromatic bridge core, with a strong participation of the ethynyl linkers and a smaller contribution from the metallic termini. Going from the diethynyl thiophene bridge in **36⁺** to the diethynyl tetrathienoacene bridge in **41⁺**, the calculated spin density on the bridge core increases from 36% to 56% and decreases both on the diethynyl linkers from 36% to 24% and the Ru centers from 28% to 20%. The increasing participation of the longer oligothienoacene bridge core in the initial one-electron oxidation is reflected in a gradual moderate shift of the electrode potential to less negative values, viz. from −0.39 V for **36** (thiophene) to −0.18 V for **41** (tetrathienoacene) vs. ferrocenium/ferrocene. The electrode potential of the second oxidation generating the corresponding dications remains almost unaffected by the core length. Overall, the $\Delta E_{1/2}$ values decrease along the series from 320 mV for **36** to 135 mV for **41**. These trends reflect the stabilizing effect of the spin delocalization on the cationic complexes and the decreasing electronic communication between the ruthenium centers through the molecular orbitals of the longer diethynyl oligothienoacene bridge. The differences are even more pronounced along the less conjugated diethynyl oligothiophene series (**36–38**) [74]. The cationic complexes **36**$^+$, **39**$^+$–**41**$^+$, differently from the above diruthenium diethynylaromatic series represented by **27**$^+$ and **31**$^+$, cannot be assigned as valence-trapped MV species. Their electronic absorption spectra in the NIR region exhibit a strong asymmetric band that has been attributed to a β-HOSO → β-LUSO excitation having mixed π–π*(intrabridge)/ML(bridge)CT characteristics, in line with the dominant oligothienoacene bridge core oxidation (the bridge-based ground state) also supported by the EPR data. A very similar α-HOSO → α-LUSO excitation is blue-shifted to the visible spectral region. No evidence has been obtained for the IVCT absorption in the alternative MV ground state. Also, the IR spectra show no ν(C≡C) absorption that could be assigned to the MV ground state of the cationic complexes. Complex **36**$^+$ exhibits only a single IR ν_{as}(C≡C) band. Notably, the ν_s(C≡C) mode gains significant intensity and becomes increasingly separated from ν_{as}(C≡C) (=$\Delta\tilde{\nu}$) with the increasing wavelength (λ_{max}) of the NIR absorption with the significant ML(bridge)CT component. This trend is best illustrated for complex **39**$^+$ with *anti*-dithienoacene (thieno[3,2-*b*]thiophene) in the bridge core and closely related cations **39a**$^+$ with *syn*-dithienoacene (thieno[2,3-*b*]thiophene) and **39b**$^+$ with *syn*-dimethyldithienoacene (3,4-dimethylthieno[2,3-*b*]thiophene): **39**$^+$ (λ_{max} = 1569 nm; $\Delta\tilde{\nu}$ = 26 cm^{-1}), **39a**$^+$ (λ_{max} = 2500 nm; $\Delta\tilde{\nu}$ = 69 cm^{-1}), and **39b**$^+$ (λ_{max} = 2614 nm; $\Delta\tilde{\nu}$ = 91 cm^{-1}). The remaining diethynyl oligothienacene complexes also follow this trend: **36**$^+$ (λ_{max} = 1390 nm; $\Delta\tilde{\nu}$ < 10 cm^{-1}), **40**$^+$ (λ_{max} = 1800 nm; $\Delta\tilde{\nu}$ = 42 cm^{-1}), **41**$^+$ (λ_{max} = 2028 nm; $\Delta\tilde{\nu}$ = 55 cm^{-1}). This behavior has been attributed to a strong vibronic coupling of the ν_s(C≡C) mode of the oxidized —[C≡C—core—C≡C]$^+$— ↔ =[C=C=core=C=C]$^+$= bridge to the low-lying π–π*(intrabridge)/ML(bridge)CT electronic transition, which increases with the reduced energy gap between the two processes. This phenomenon, with potential applicability in molecular electronic devices, appears to be general in nature for the diruthenium diethynyl(core$^+$)-bridged complexes, as revealed, for example, by the literature data for the series **27**$^+$–**31**$^+$ [72] and **42**$^+$–**45**$^+$ [75].

To tune the electronic properties of diruthenium-based [{RuCp*(dppe)}$_2$(μ-C≡C—core—C≡C)]$^{n+}$ systems, some new types of core units have been introduced. Low and coworkers prepared complexes **49** and **50** [80], bridged by 10- and 12-vertex diethynyl-*p*-carboranes, respectively (Figure 5.8). Cyclovoltammetric studies have proven a significant separation of the two largely metal-based one-electron anodic waves in these dinuclear complexes. The UV–vis–NIR spectroscopic analysis has revealed that both the 10- and 12-vertex cages permit electronic communication between the two ruthenium centers over a distance of c. 12.5 Å, with the 10-vertex cage leading to a slightly higher coupling parameter (H_{ab} = 260 cm^{-1} (**49**$^+$) and 140 cm^{-1} (**50**$^+$)). Monocations **49**$^+$ and **50**$^+$ have been regarded as valence-localized (Class II) MV organometallic compounds. In 2012, Liu and coworkers incorporated

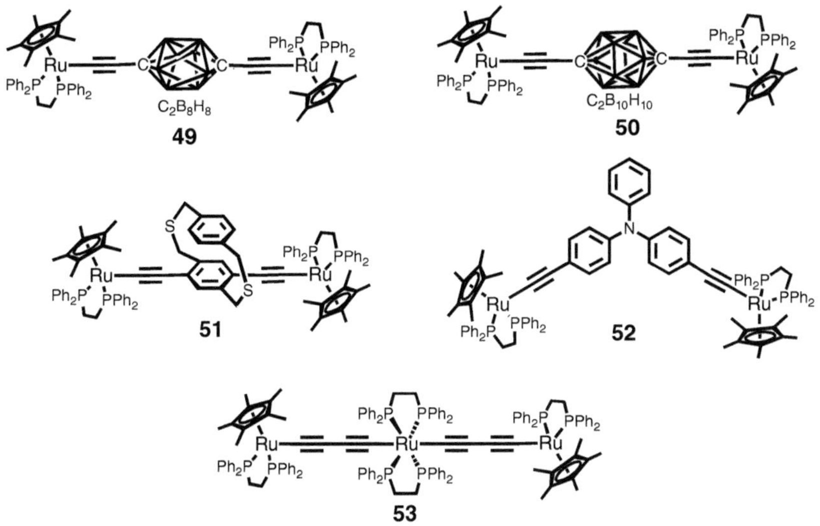

Figure 5.8 Several new types of bridging units used in organometallic molecular wires.

dithia[3.3]paracyclophane as the bridge core unit in complex **51**, featuring a transannular π–π interaction associated with a high degree of structural rigidity, and explored its effect on the electronic communication through the bridge backbone [81]. In addition, the team implemented electron-rich triarylamines (TAA) into the bridge core. Non-linear homodinuclear complex **52** exhibits a significantly delocalized characteristic and a marked electronic communication between the ruthenium centers through the diethynyl–TAA bridge [22]. On the other hand, the group of Bruce developed a synthetic route toward long-distance molecular wires by incorporating metal units into polyynediyl linkers to solve the problem of poor chemical stability of the oxidized species and a tendency to decrease the electronic interaction between the termini. Therefore, they synthesized symmetric bis(metalla diynediyl)ruthenium(II) complex **53** [82] containing Ru(dppe)$_2$ as the central linking group. DFT calculations have shown its HOMO delocalized over the Ru—C$_4$—Ru—C$_4$—Ru chain, suggesting that there is an electronic interaction between the terminal RuCp*(dppe) groups through the C$_4$-chains and the central Ru(dppe)$_2$ unit. This strategy opens a new avenue to the construction of molecular wires with long-range charge transfer properties.

5.3.2 Dinuclear Ruthenium–Ethynyl Complexes with Ru(dppe)$_2$X-Based Termini

Classical diruthenium complex **54** with the Ru(dppe)$_2$Cl termini was first reported by Dixneuf and coworkers [83]. They obtained it in 58% yield by the reaction of 1,4-diethynylbenzene with 2 equiv cis-[RuCl$_2$(dppe)$_2$] in the presence of NaPF$_6$ in THF and subsequent triethylamine (NEt$_3$) treatment (Figure 5.9, Route I). Alternatively, complex **55** was prepared from 1,4-diethynyl-2,5-dimethoxybenzene and [RuCl(dppe)$_2$](OTf) (Figure 5.9, Route II) [84]. **54$^+$** and **55$^+$** present rather

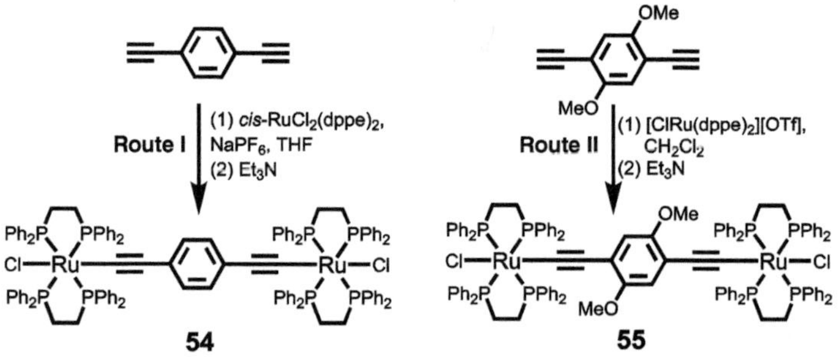

Figure 5.9 Two synthetic routes for the Ru(dppe)$_2$Cl-terminated dinuclear Ru–ethynyl complexes.

sharp and intense low-energy absorption bands assigned as SOMO-n → SOMO transitions within an extended open-shell organometallic chromophore involving some charge transfer from the metal end groups to the central arene part of the bridge. Therefore, both complexes **54** and **55** are a token of the strong participation of the bridge in the overall oxidation process.

Based on these synthetic methods, conjugated bridge chains of different lengths have been incorporated into such systems (**56**, **57**, and **59**) (Figure 5.10) [67, 85]. Electrochemical studies show two one-electron reversible oxidation steps for **56**. Complex **57** only displays one reversible oxidation wave. UV–vis–NIR absorption spectra show a long-wavelength band for **56**$^+$. The authors have claimed that this band cannot be attributed to an IVCT process as the observed intensities are too high for such weakly coupled systems (as assumed from electrochemistry); although, the

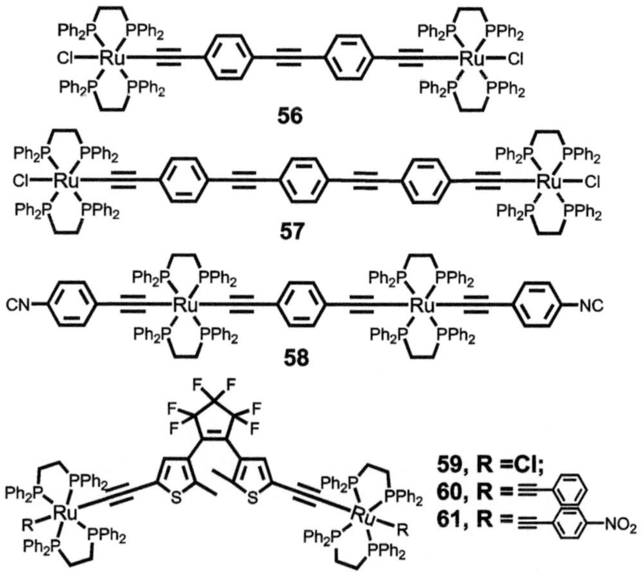

Figure 5.10 Dinuclear homometallic complexes with Ru(dppe)$_2$R groups as redox-active termini.

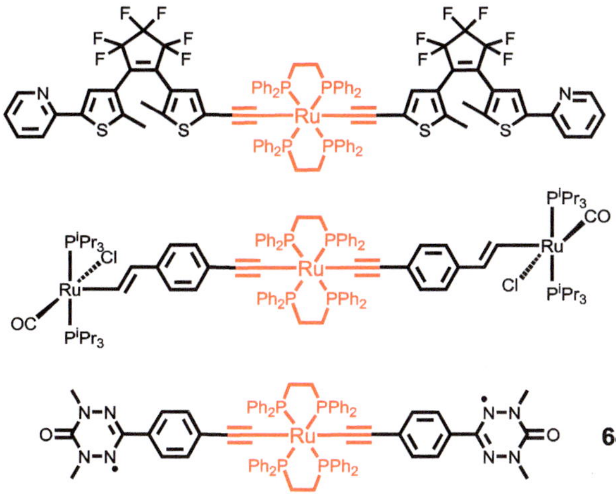

Figure 5.11 Bis(ethynyl)Ru(dppe)$_2$ central coupling units in multi-responsive molecular switches and long-distance molecular wires.

calculated and observed bandwidths are in better agreement. Oxidized **57$^+$** does not show such a long-wavelength band. This supports the results obtained from the CV studies. There is no apparent electronic communication between both ruthenium centers, and they are both oxidized independently at the same potential. To increase the length of the conjugated molecular chain, the terminal chloride ligand at the ruthenium centers was further substituted by ethynyl groups, for example, in complexes **58** and **60–61** [86, 87]. Complex **58**, end capped with isocyanide (—N≡C) surface-anchoring groups, can form self-assembled monolayers (SAMs) on Au substrates, which were examined for electronic communication along the molecular backbone [87]. In addition, the precursor cis-[Ru(dppe)$_2$Cl$_2$] can substitute both chloride ligands to form an intermediate in the construction of multi-functional complexes with the Ru(dppe)$_2$ moiety in the bridge core. Recently, the preparation and properties of several novel multi-responsive molecular switches [88] (such as complex **62**) and redox-active linear oligonuclear molecular wires [84, 89] (such as complexes **63** and **64**) have been reported by the groups of Chen and Rigaut (Figure 5.11).

5.3.3 Ruthenium–Ethynyl Complexes with Alternating Polyyndiyl and Capped Ru–Ru Units

Bimetallic compounds containing a polyyndiyl bridge have fascinated communities of organometallic and materials chemists for decades. Ren et al. first reported a carbon-rich polyynediyl molecular wire with incorporated capped redox-active diruthenium units [Ru$_2$(ap)$_4$] (ap = 2-anilinopyridinate) (**67**) [90], which represents a new type of a conjugated molecular wire serving as a model for studies of electron transfer properties. In the same year, Lehn and coworkers reported a similar compound with incorporated redox-active capped diruthenium units [Ru$_2$(DPhF)$_4$]

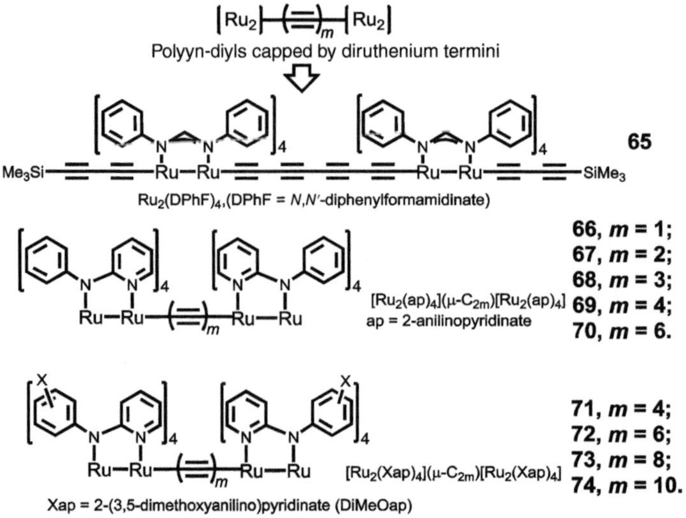

Figure 5.12 Diruthenium–polyynediyl–diruthenium molecular wires.

(DPhF = N,N'-diphenylformamidinate) (**65**) [91]. A series of related diruthenium polyynediyl compounds was investigated extensively by Ren and coworkers. The most illustrative types are [Ru$_2$(ap)$_4$](μ-C$_{2m}$)[Ru$_2$(ap)$_4$] (m = 1–4 and 6) (**66–70**) [91, 92] and symmetric [{Ru$_2$(Xap)$_4$}$_2$(μ-C$_{2m}$)] (m = 4, 6, 8, and 10) (**71–74**) [93] (Figure 5.12). The electron transfer characteristics between the capped Ru–Ru termini linked by the polyynediyl chains were explored by cyclovoltammetric and spectroelectrochemical measurements that have reflected a structure–property relationship at longer distances. The comparison of all CV plots and electrochemical potential data, determined from differential pulse voltammetry (DPV) measurements, indicated that the half-wave potential difference ($\Delta E_{1/2}$) for the two sequential one-electron oxidation gradually decreases as the polyyn-diyl elongates and the degree of the electronic coupling between the two dinuclear Ru–Ru centers decays with the increasing distance.

In addition, the capped diruthenium units and ferrocenyl groups were combined as end groups or core units to form diverse complexes with three (di)iron centers connected by polyynediyl chains (**75–80**) (Figure 5.13) [94–96]. The electronic interaction between the diruthenium and ferrocenyl centers has illustrated that the redox-active transition metal units can be utilized to enhance the electronic coupling across a carbon-rich backbone.

5.3.4 Ruthenium–Ethynyl Complexes with Other Ruthenium–Ethynyl Termini and Core Units

Recently, some novel dinuclear ruthenium–ethynyl complexes with diverse ancillary ligands have been reported by the Chen group (Figure 5.14). Characteristic rod-like dinuclear ruthenium polyynediyl complexes terminated with redox-active

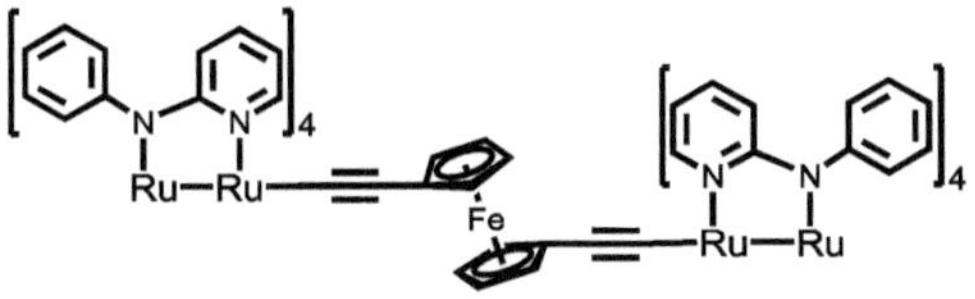

75

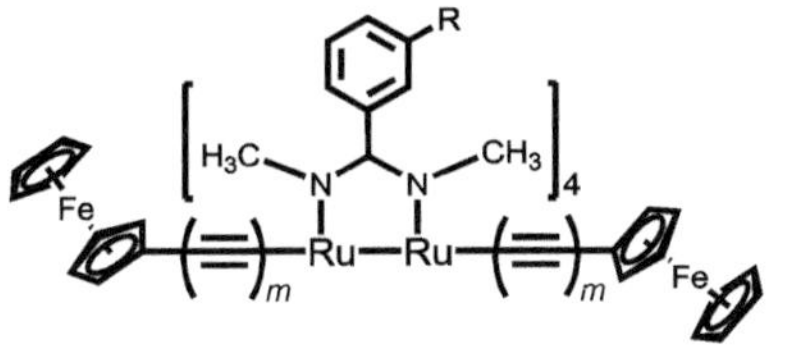

76, *m* = 1; 77, *m* = 2; 78, *m* = 3; 79, *m* = 4.

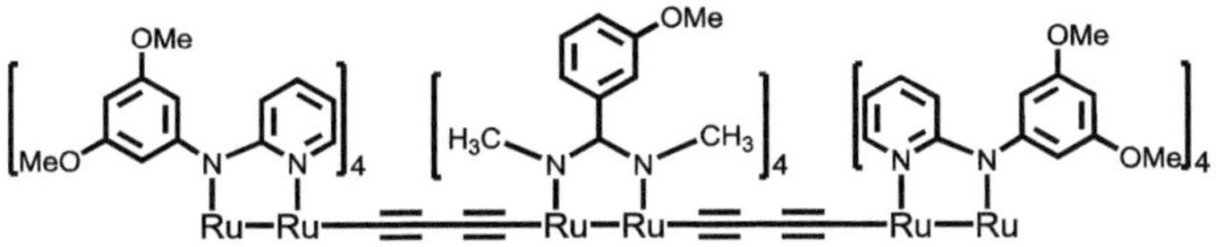

80

Figure 5.13 Organometallic molecular wires constructed from capped diruthenium core units connected to capped diruthenium and ferrocenyl terminals by polyynediyl chains.

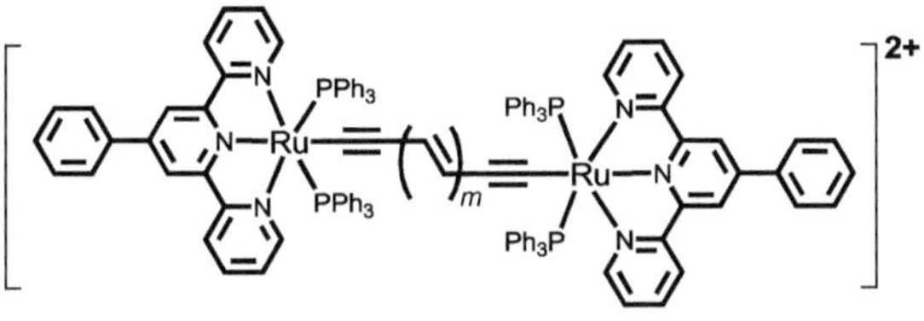

81, *m* = 0; 82, *m* = 1; 83, *m* = 2; 84, *m* = 3

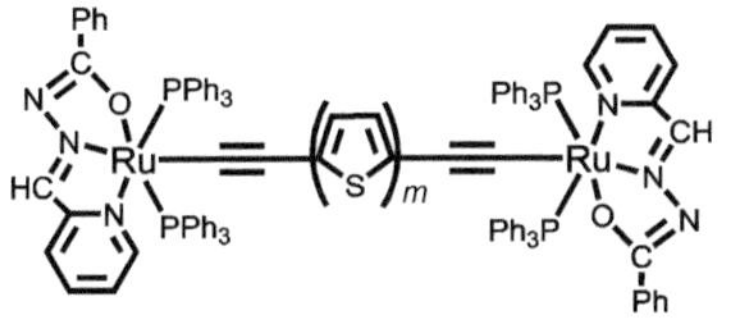

85, *m* = 1; 86, *m* = 2; 87, *m* = 3

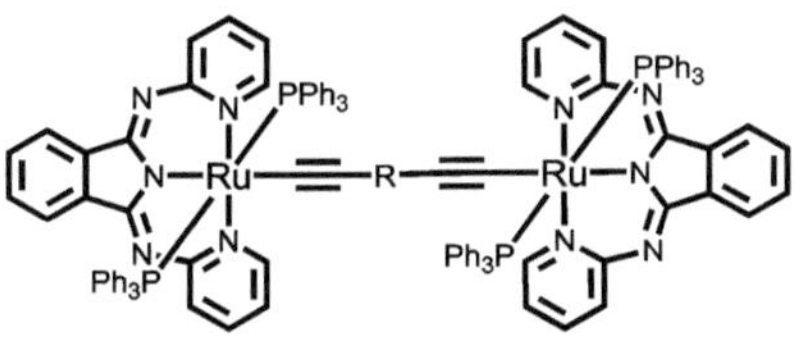

88 R = —

89 R =

90 R =

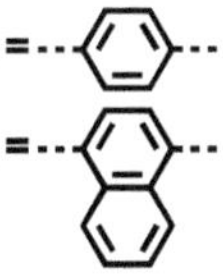

R = **91**

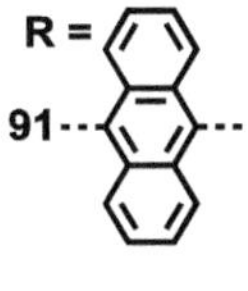

Figure 5.14 A range of dinuclear Ru–ethynyl complexes with a variety of auxiliary ligands and conjugated bridge cores.

organometallic groups $[Ru(bph)(PPh_3)_2]^+$ (bph = *N*-(benzoyl)-*N′*-(picolinylidene)-hydrazine) (**81–84**) [97], $[Ru(Phtpy)(PPh_3)_2]^{2+}$ (Phtpy = 4′-phenyl-2,2′:6′,2″-terpyridine) (**85–87**) [98], or $[Ru(BPI)(PPh_3)_2]$ BPI = 1,3-bis(2-pyridylimino) isoindoline (**89–91**) [68] were synthesized and investigated by UV–vis–NIR absorption spectroscopies and cyclic and differential pulse voltammetric methods. The visible–NIR absorption spectral studies of the mixed-valence complexes **81**$^+$–**83**$^+$ have demonstrated that electronic delocalization along the molecular rods attenuates dramatically with the increase of the ethenyl number. Complex **81**$^+$ is characteristic of a Class III MV system with strong electronic delocalization. The IVCT absorption and the redox properties of complex **82$^+$** reveal that its nature lies between an electronically delocalized and valence-trapped system. By contrast, **83$^+$** shows a typical Robin–Day Class II MV behavior. In addition, auxiliary ligands have some influence on the electronic communication. The molecular wires containing electron-rich auxiliary ligands such as bph or BPI exhibit a more efficient electronic communication compared to congeners with an identical bridge and electron-deficient auxiliary ligands such as Phtpy. Notably, for complexes **89–91**, the gradual increase in size of the π-conjugated system of the aromatic core favors the electronic delocalization along the Ru—C≡C—R—C≡C—Ru backbone, which is attributed to an increasing participation of the chain core in the delocalization of the spin density in the singly oxidized state.

5.4 Other Transition Metal–Ethynyl Complexes

Transition-metal-terminated polyynediyl compounds, [M]—C$_{2m}$—[M], serve as convenient organometallic MV models that provide an alternative way to investigate their ET and charge (de)localization properties. The dinuclear iron- and ruthenium–ethynyl/polyynediyl complexes with a variety of bridge cores and ancillary ligands have been introduced in the preceding two key sections. Other late transition metals from Groups 6–9 in homodimetallic ethynyl complexes in the MV states are reviewed in this section. They are sub-divided into three categories: (i) dinuclear Group 6 (Cr and Mo) metal–ethynyl complexes, (ii) dinuclear Group 7 (Mn, Re) metal–polyynediyl complexes, and (iii) dinuclear Group 8 (Os) and Group 9 (Co) metal–polyyndiyl complexes.

5.4.1 Dinuclear Group 6 (Cr and Mo) Metal–Ethynyl Complexes

The Group 6 metal triad bound to ethynyl, especially Cr and Mo, have also been adopted as redox-active groups in the terminal or bridge core positions to explore the electronic communication properties along the linear backbone [99, 100] (Figure 5.15). In 2008, Berben and Kozimor reported an interesting, unusually stable complex containing two Cr(I) centers bridged by a neutral dinitrogen moiety, viz. [*trans,trans*-{(Me$_3$SiC≡C)(dmpe)$_2$Cr}$_2$(μ-N$_2$)] (**92**) [99]. Its cyclic voltammograms displayed two separated one-electron oxidation steps with a comproportionation constant of $K_c = 10^{4.5}$. The mixed-valence state **92$^+$** has been regarded as a Robin–Day Class II compound. In 2012, Berke and coworkers described mono- and dinuclear chromium–ethynyl complexes [100]. Depicted examples **93–95** contain the butadiynediyl bridge lacking the bridge core X. However, electrochemical responses of these complexes only displayed a single reversible two-electron oxidation wave, which complies with a less common Class I behavior based on the Robin–Day classification.

Regarding the Mo center, Low and coworkers studied thoroughly a series of dinuclear molybdenum–ethynyl complexes **96–98** [101–103]. Spectroscopic

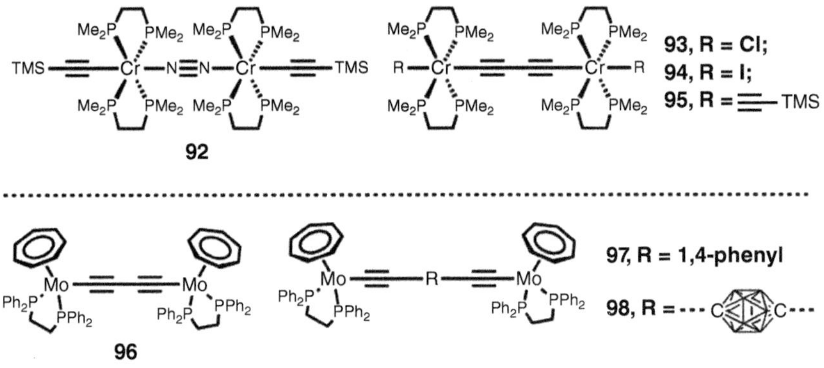

Figure 5.15 Dinuclear Cr- and Mo–ethynyl/butadiynediyl complexes.

investigations of butadiynediyl-bridged [{Mo(dppe)(η^7-C$_7$H$_7$)}$_2$(μ-C≡C—C≡C)] MV complex **96$^+$** by IR, Raman, UV–vis–NIR, and EPR spectroscopies revealed properties characteristic of a weakly coupled d^5/d^6 MV complex with a localized electronic structure. In addition, 1,12-bis(ethynyl)-1,12-carborane as the bridging ligand was used to build up dinuclear molybdenum–ethynyl complex **98** that exhibited well-defined molybdenum-based oxidation. The corresponding cation **98$^+$** has been classified as a weakly coupled Robin–Day Class II MV system.

5.4.2 Dinuclear Group 7 (Mn and Re) Metal–Polyynediyl Complexes

The nature of the transition metal center and its significant impact on the redox chemistry play a significant role in electron transfer processes and determine the properties of molecular wire models. Therefore, Berke and coworkers focused on building up dinuclear Mn(μ-C≡C)$_2$Mn complexes bearing acetylide end groups. Notably, the C–C coupling of the end groups presents a potential for building long-distance oligonuclear molecular wires [104, 105]. Namely, in 2005, they reported the syntheses of Mn(II)/Mn(III) MV complexes [{Mn-(dmpe)$_2$(C≡CR)}$_2$(μ-C$_4$)](PF$_6$) (R = SiEt$_3$, **99$^+$**; R = Si^{i}Pr$_3$, **100$^+$**; and R = Si(^{t}Bu)Me$_2$, **101$^+$**) from the corresponding mononuclear Mn(III) precursors [Mn(dmpe)$_2$(C≡CR)(C≡CH)][PF$_6$] treated with 1,8-diazabicyclo[5.4.0]undec-7-ene (DBU) [105] (Figure 5.16). The corresponding neutral dinuclear Mn(II)/Mn(II) compounds **99–101** were prepared by one-electron reduction of the precursor cationic complexes with permethylated cobaltocene. On the other hand, stable dications **99^{2+}–101^{2+}** were obtained by oxidation with ferrocenium. The cyclic voltammograms of **99–101** exhibit two reversible anodic waves separated by c. 560 mV, which corresponds to large $K_c \sim 4 \times 10^9$. Single-crystal X-ray diffraction has revealed two equivalent Mn centers for complex **99$^+$** (Mn–C = 1.768 Å), suggesting a strong electronic communication between them.

For the other accessible Group 7 metal, rhenium, Gladysz and coworkers synthesized the metal-terminated carbon chains [{ReCp*(NO)(PPh$_3$)}$_2$(μ-C≡C)$_m$, $m \leq 10$] (**102**) (Figure 5.16). The Re(μ-C≡C)$_2$Re species showed two separate one-electron

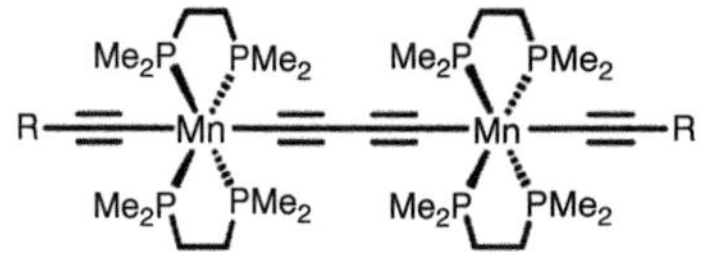

99, R = SiEt$_3$;
100, R = Si^{i}Pr$_3$;
101, R = Si^t(Bu)(Me)$_2$

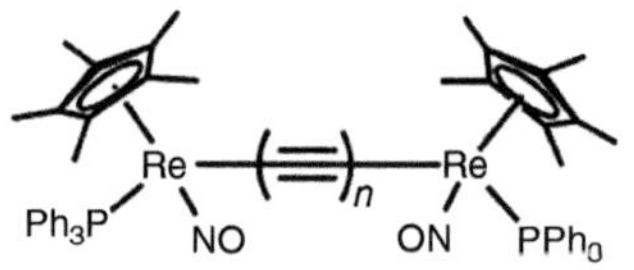

n = 2, 3, 4, 5, 6, 8, 10

102

Figure 5.16 Dinuclear manganese- and rhenium–polyynediyl complexes.

oxidations in their cyclic voltammetric responses [106–109]. For the longer chains, viz. Re(μ-C≡C)$_3$Re and Re(μ-C≡C)$_4$Re, the oxidations become increasingly irreversible. The presence of merely a single, presumably two-electron oxidation wave for the Re(μ-C≡C)$_{10}$Re chain indicates that for this chain length, the rhenium centers are effectively isolated and do not electronically communicate through the conjugated bridge.

5.4.3 Dinuclear Group 8 (Os) and Group 9 (Co) Metal–Polyyndiyl Complexes

Osmium that completes the Group 8 triad after iron and ruthenium introduced herein separately can also form the corresponding ethynyl and polyyndiyl complexes (Figure 5.17a). Bruce et al. described dinuclear osmium–ethynyl complex **103** bearing the same ancillary ligands, OsCp*(dppe)$^+$ [110], to conduct a direct comparison with the analogous Fe and Ru complexes. Osmium complex **103** is readily oxidized at modest potentials to afford a range of mono-, di-, tri-, and tetra-cationic derivatives, whose spectroscopic properties are generally similar to those described for the ruthenium analogs. The first-row transition metal–butadiynediyl complex exhibits largely Fe-centered redox events, while the heavier second- (Ru) and third-row (Os) systems feature more carbon characteristic of the redox frontier orbitals. In addition, [{Os(PPh$_3$)$_2$Cp}$_2${μ-(C≡C)$_{m+1}$}] (m = 1, 2, and 3) complexes (**104–106**), featuring different polyynediyl bridge lengths, were also synthesized by this research group [111]. The comparison of the redox properties of [{M(PPh$_3$)$_2$Cp}$_2${μ-(C≡C)$_{m+1}$}] (M = Ru and Os) shows that the oxidation potentials of the osmium complexes are, as expected, invariably lower (by 160–640 mV) than those of the Ru analogs, in spite of their identical ligand environments. In addition, the dinuclear osmium complexes exhibit a smaller oxidative potential separation ($\Delta E_{1/2}$) for the same metal–metal distances.

A range of mononuclear alkynyl metal complexes containing a Group 9 metal center (Co, Rh, or Ir) exhibit unique optical and luminescence properties [112–114]. A recent study has documented that dinuclear cobalt complexes differ from other Group 9 organometallic oligoynes and polyynes in their electronic coupling properties. For example, Ren and coworkers reported a family of polyynediyl-bridged, (μ-C$_{2m}$) (m = 2–4), dinuclear complexes with Co^{III}(cyclam)Cl (cyclam = 1,4,8,11-tetraazacycloctetradecane) termini (**107**) [115]. The cyclovoltammetric analysis of

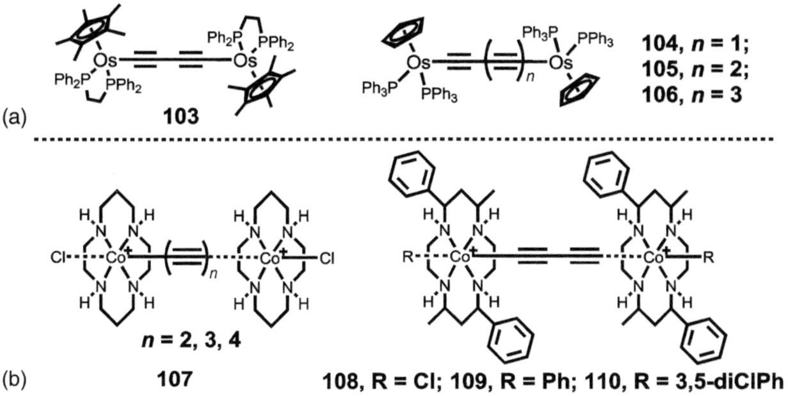

Figure 5.17 (a) Dinuclear osmium-polyynediyl complexes. (b) Dinuclear cobalt–polyynediyl complexes.

this series of complexes has revealed a weak Co…Co interaction through the bridge, which is attenuated by the polyyne chain length. Later, the same group selected MPC (MPC = 5,12-dimethyl-7,14-diphenyl-1,4,8,11-tetraazacyclotetradecane) as the ligand that can weaken the ligand field around the equatorial plane, allowing for a stronger axial coordination, and described the syntheses and characterization of a series of butadiynediyl-bridged Co^{III}(MPC) complexes (Figure 5.17b) [116], terminated with chloride (**108**), phenylacetylide (**109**), or 3,5-dichlorophenylacetylide (**110**) ligands. Cyclovoltammetric analysis, using NBu_4BArF (NBu_4BArF = tetrabutylammonium tetrakis[3,5-bis(trifluoromethyl)phenyl]borate) as the supporting electrolyte, resulted in two well-separated reversible cathodic waves ($Co^{III/II}$), with up to 160 mV separation for **109**. The results indicate that changing the ancillary ligand structures can effectively improve the redox reversibility and electronic coupling between the Co^{III}–ethynyl units.

5.5 Concluding Remarks and Outlook

In this chapter, we have outlined a range of symmetrical oligonuclear Group 6–9 metal complexes with redox-active metal–ethynyl termini linked directly to each other or to a core unit of the bridge represented by diverse conjugated, saturated, or ligated metallic groups. The transition metals in the terminal parts of the molecular chains include chromium, molybdenum, manganese, rhenium, iron, ruthenium, osmium, and cobalt in diverse oxidation states, forming redox series. Their electronic coupling properties in the MV states have been assessed. The degree of electronic communication between the metal–ethynyl centers in the MV systems can be evaluated on the grounds of electrochemical data, IR, UV–vis–NIR spectroelectrochemical monitoring, and EPR spectra in combination with theoretical calculations. The MV characteristics strongly depend on the structural properties of the bridging ligands, such as the length and coplanarity, and the degree of their conjugation. The interplay between the bridge properties and the nature of the linked metal centers, the stability and localization of the oxidation states, and the strength of the coupling between the metal centers and the ancillary ligands also effectively affects the electron transfer processes. Special attention needs to be paid to the analysis of electronic absorption of the open-shell bimetallic complexes in the NIR and short-wave infrared (SWIR) region. Apart from the characteristic IVCT absorption attributed to the thermal population of MV states (Class II and III, and the borderline between them), the coexistence of bridge-localized oxidation (consistent with a high degree of C≡C–core–C≡C π-characteristic in the frontier orbitals and a redox non-innocent characteristic) introduces low-lying $\pi-\pi^*$ (intra-bridge) or MLCT electronic transitions in this region. Their simultaneous population may be detected in configurational rotamers where the plane of an diethynyl aromatic core of the bridge varies its orientation with respect to the metal d-orbitals of appropriate π-symmetry. To sum up, the family of homodinuclear metal–ethynyl complexes offers a great potential in exploring thermal or photo-induced intramolecular electron transfer properties. However, despite the rigid structure of the metal–ethynyl units, the synthesis of extended MV molecular wire models with remote charge transfer ability remains a challenge.

Acknowledgment

Financial support from the National Natural Science Foundation of China (nos. 21772054 and 21602049) and the Natural Science Foundation of Hunan Province, China (nos. 2017JJ3004 and 2021JJ30058). F.H. also thanks the University of Reading for continued support of the spin-out spectroelectrochemistry reading project and the CCNU Wuhan for his participation in the Global Talent Plan 111 Project (B17019).

References

1 Aguirre-Etcheverry, P. and O'Hare, D. (2010). Electronic communication through unsaturated hydrocarbon bridges in homobimetallic organometallic complexes. *Chem. Rev.* 110: 4839–4864.
2 Ceccon, A., Santi, S., Orian, L., and Bisello, A. (2004). Electronic communication in heterobinuclear organometallic complexes through unsaturated hydrocarbon bridges. *Coord. Chem. Rev.* 248: 683–724.
3 Kaim, W. and Lahiri, G.K. (2007). Unconventional mixed-valent complexes of ruthenium and osmium. *Angew. Chem. Int. Ed.* 46: 1778–1796.
4 Demadis, K.D., Hartshorn, C.M., and Meyer, T.J. (2001). The localized-to-delocalized transition in mixed-valence chemistry. *Chem. Rev.* 101: 2655–2685.
5 Caballero, A., Lloveras, V., Curiel, D. et al. (2007). Electroactive thiazole derivatives capped with ferrocenyl units showing charge-transfer transition and selective ion-sensing properties: a combined experimental and theoretical study. *Inorg. Chem.* 46: 825–838.

6 Pinheiro, S.O., Paulo, T.F., Abreu, D.S. et al. (2012). Mixed-valence state of symmetric diruthenium complexes: synthesis, characterization, and electron transfer investigation. *Dalton Trans.* 41: 14540–14546.

7 Qi, H., Gupta, A., Noll, B.C. et al. (2005). Dependence of field switched ordered arrays of dinuclear mixed-valence complexes on the distance between the redox centers and the size of the counterions. *J. Am. Chem. Soc.* 127: 15218–15227.

8 Tanaka, Y., Kiguchi, M., and Akita, M. (2017). Inorganic and organometallic molecular wires for single-molecule devices. *Chem. Eur. J.* 23: 4741–4749.

9 Zhong, Y.-W., Gong, Z.-L., Shao, J.-Y., and Yao, J. (2016). Electronic coupling in cyclometalated ruthenium complexes. *Coord. Chem. Rev.* 312: 22–40.

10 Halet, J.-F. and Lapinte, C. (2013). Charge delocalization vs localization in carbon-rich iron mixed-valence complexes: a subtle interplay between the carbon spacer and the (dppe)Cp*Fe organometallic electrophore. *Coord. Chem. Rev.* 257: 1584–1613.

11 Tanaka, Y. and Akita, M. (2019). Organometallic radicals of iron and ruthenium: similarities and dissimilarities of radical reactivity and charge delocalization. *Coord. Chem. Rev.* 388: 334–342.

12 Cheng, T., Shen, D.X., Meng, M. et al. (2019). Efficient electron transfer across hydrogen bond interfaces by proton-coupled and -uncoupled pathways. *Nat. Commun.* 10: 1531–1540.

13 Linseis, M., Záliš, S., Zabel, M., and Winter, R.F. (2012). Ruthenium stilbenyl and diruthenium distyrylethene complexes: aspects of electron delocalization and electrocatalyzed isomerization of the Z-Isomer. *J. Am. Chem. Soc.* 134: 16671–16692.

14 D'Alessandro, D.M., Dinolfo, P.H., Davies, M.S. et al. (2006). Underlying spin–orbit coupling structure of intervalence charge transfer bands in dinuclear polypyridyl complexes of ruthenium and osmium. *Inorg. Chem.* 45: 3261–3274.

15 Hildebrandt, A. and Lang, H. (2013). (Multi)ferrocenyl five-membered heterocycles: excellent connecting units for electron transfer studies. *Organometallics* 32: 5640–5653.

16 Zhang, M.-X., Zhang, J., Yin, J. et al. (2018). Anodic electrochemistry of mono- and dinuclear aminophenylferrocene and diphenylaminoferrocene complexes. *Dalton Trans.* 47: 6112–6123.

17 Rossi, S., Bisello, A., Cardena, R. et al. (2017). Benzodithiophene and benzotrithiophene as π cores for two and three-blade propeller-shaped ferrocenyl-based conjugated systems. *Eur. J. Org. Chem.* 2017: 5966–5974.

18 Roy, S.S., Sil, A., Giri, D. et al. (2018). Diruthenium(II)-capped oligothienylethynyl bridged highly soluble organometallic wires exhibiting long-range electronic coupling. *Dalton Trans.* 47: 14304–14317.

19 Makhoul, R., Sahnoune, H., Dorcet, V. et al. (2015). 1,2-Diethynylbenzene-bridged [Cp*(dppe)Fe]$^{n+}$ units: effect of steric hindrance on the chemical and physical properties. *Organometallics* 34: 3314–3326.

20 Burgun, A., Gendron, F., Schauer, P.A. et al. (2013). Straightforward access to tetrametallic complexes with a square array by oxidative dimerization of organometallic wires. *Organometallics* 32: 5015–5025.

21 Bruce, M.I., Costuas, K., Davin, T. et al. (2007). Syntheses, structures and redox properties of some complexes containing the Os(dppe)Cp* fragment, including [{Os(dppe)Cp*}2(1–C≡CC≡C)]. *Dalton Trans.* 2007: 5387–5399.

22 Zhang, J., Guo, S.-Z., Dong, Y.B. et al. (2017). Multistep oxidation of diethynyl oligophenylamine-bridged diruthenium and diiron complexes. *Inorg. Chem.* 56: 1001–1015.

23 Gauthier, N., Olivier, C., Rigaut, S. et al. (2008). Intramolecular optical electron transfer in mixed-valent dinuclear iron-ruthenium complexes featuring a 1,4-diethynylaryl spacer. *Organometallics* 27: 1063–1072.

24 Liu, S.H., Hu, Q.Y., Xue, P. et al. (2005). Synthesis and characterization of $C_{10}H_{10}$-bridged bimetallic ruthenium complexes. *Organometallics* 24: 769–772.

25 Ou, Y.-P., Jiang, C., Wu, D. et al. (2011). Synthesis, characterization, and properties of anthracene-bridged bimetallic ruthenium vinyl complexes [RuCl(CO)(PMe$_3$)$_3$]$_2$(μ-CH=CH–anthracene-CH=CH). *Organometallics* 30: 5763–5770.

26 Rotthowe, N., Zwicker, J., and Winter, R.F. (2019). Influence of quinoidal distortion on the electronic properties of oxidized divinylarylene-bridged diruthenium complexes. *Organometallics* 38: 2782–2799.

27 Pevny, F., Winter, R.F., Sarkar, B., and Záliš, S. (2010). How to elucidate and control the redox sequence in vinylbenzoate and vinylpyridine bridged diruthenium complexes. *Dalton Trans.* 39: 8000–8011.

28 Wang, H., Shao, J.-Y., Duan, R. et al. (2021). Synthesis and electronic coupling studies of cyclometalated diruthenium complexes bridged by 3,3′,5,5′-tetrakis(benzimidazol-2-yl)-biphenyl. *Dalton Trans.* 50: 4219–4230.

29 Yao, C.-J., Zhong, Y.-W., and Yao, J. (2011). Charge delocalization in a cyclometalated bisruthenium complex bridged by a noninnocent 1.2.4.5-tetra(2-pyridyl)benzene ligand. *J. Am. Chem. Soc.* 133: 15697–15706.

30 Zhang, H.L., Zhu, G.Y., Wang, G. et al. (2015). Electronic coupling in [Mo2]–bridge–[Mo2] systems with twisted bridges. *Inorg. Chem.* 54: 11314–11322.

31 Ansari, M.A., Mandal, A., Paretzki, A. et al. (2016). 1,5-Diamido-9,10-anthraquinone, a centrosymmetric redox-active bridge with two coupled β-ketiminato chelate functions: symmetric and asymmetric diruthenium complexes. *Inorg. Chem.* 55: 5655–5670.

32 Pagels, N., Albrecht, O., Görlitz, D. et al. (2011). Electronic coupling through intramolecular π–π interactions in biscobaltocenes: a structural, spectroscopic, and magnetic study. *Chem. Eur. J.* 17: 4166–4176.

33 Huang, J., Lin, R., Wu, L. et al. (2010). Synthesis, characterization, and electrochemical properties of bisosmabenzenes bridged by diisocyanides. *Organometallics* 29: 2916–2925.

34 Cao, Z., Forrest, W.P., Gao, Y. et al. (2012). *trans*-[Fe(cyclam)(C$_2$R)$_2$]$^+$: a new family of iron(III) bis-alkynyl compounds. *Organometallics* 31: 6199–6206.

35 West, P.J., Cifuentes, M.P., Schwich, T. et al. (2012). Syntheses and spectroscopic, structural, electrochemical, spectroelectrochemical, and

theoretical studies of osmium(II) mono- and bis-alkynyl complexes. *Inorg. Chem.* 51: 10495–10502.

36 Natoli, S.N., Zeller, M., and Ren, T. (2017). An aerobic synthetic approach toward bis-alkynyl cobalt(III) compounds. *Inorg. Chem.* 56: 10021–10031.

37 Costuas, K. and Rigaut, S. (2011). Polynuclear carbon-rich organometallic complexes: clarification of the role of the bridging ligand in the redox properties. *Dalton Trans.* 40: 5643–5658.

38 Low, P.J. (2013). Twists and turns: studies of the complexes and properties of bimetallic complexes featuring phenylene ethynylene and related bridging ligands. *Coord. Chem. Rev.* 257: 1507–1532.

39 Haque, A., Al-Balushi, R.A., Al-Busaidi, I.J. et al. (2018). Rise of conjugated poly-ynes and poly(Metalla-ynes): from design through synthesis to structure–property relationships and applications. *Chem. Rev.* 118: 8474–8597.

40 Gendron, F., Groizard, T., Guennic, B.L., and Halet, J.-F. (2020). Electronic properties of poly-yne carbon chains and derivatives with transition metal end-groups. *Eur. J. Inorg. Chem.* 2020: 667–681.

41 Lapinte, C. (2008). Magnetic perturbation of the redox potentials of localized and delocalized mixed-valence complexes. *J. Organomet. Chem.* 693: 793–801.

42 Makhoul, R., Gluyas, J.B.G., Vincent, K.B. et al. (2018). Redox properties of ferrocenyl ene-diynyl-bridged Cp*(dppe)M–C≡C–1,4-(C6H4) complexes. *Organometallics* 37: 4156–4171.

43 Burgun, A., Ellis, B.G., Roisnel, T. et al. (2014). From molecular wires to molecular resistors: TCNE, a class-III/class-II mixed-valence chemical switch. *Organometallics* 33: 4209–4219.

44 Burgun, A., Gendron, F., Sumby, C.J. et al. (2014). Hexatriynediyl chain spanning two Cp*(dppe)M termini (M = Fe, Ru): evidence for the dependence of electronic and magnetic couplings on the relative orientation of the termini. *Organometallics* 33: 4209–4219.

45 Makhoul, R., Hamon, P., Roisnel, T. et al. (2020). A tetrairon dication featuring tetraethynylbenzene bridging ligand: a molecular prototype of quantum dot cellular automata. *Chem. Eur. J.* 26: 8368–8371.

46 Malvolti, F., Rouxel, C., Triadon, A. et al. (2015). 2,7-Fluorenediyl-bridged complexes containing electroactive "Fe(η^5-C$_5$Me$_5$)(κ^2-dppe)C≡C–" end groups: molecular wires and remarkable nonlinear electrochromes. *Organometallics* 34: 5418–5437.

47 Tanaka, Y., Ishisaka, T., Koike, T., and Akita, M. (2015). Synthesis and properties of diiron complexes with heteroaromatic linkers: an approach for modulation of organometallic molecular wire. *Polyhedron* 86: 105–110.

48 Tohm, A., Grelaud, G., Argouarch, G. et al. (2013). Redox-induced reversible P-P bond formation to generate an organometallic $\sigma^4\lambda^4$-1,2-biphosphane dication. *Angew. Chem. Int. Ed.* 52: 4445–4448.

49 Narvor, N.L., Toupet, L., and Lapinte, C. (1995). Elemental carbon chain bridging two iron centers: syntheses and spectroscopic properties of [Cp*(dppe)Fe–C$_4$–FeCp*(dppe)]$^{n+}$·n[PF$_6$]$^-$. X-ray crystal structure of the mixed valence complex (n = 1). *J. Am. Chem. Soc.* 117: 7129–7138.

50 Coat, F. and Lapinte, C. (1996). Molecular wire consisting of a C8 chain of elemental carbon bridging two metal centers: synthesis and characterization of [{Fe(η^5-C$_5$Me$_5$)(dppe)}$_2$(μ-C8)]. *Organometallics* 15: 477–479.

51 Stang, S.L., Paul, F., and Lapinte, C. (2000). Molecular wires: synthesis and properties of the new mixed-valence complex [Cp*(dppe)Fe–C≡C–X–C≡C–Fe(dppe)Cp*][PF$_6$] (X = 2,5 –C$_4$H$_2$S) and comparison of its properties with those of the related all-carbon-bridged complex (X = –C4–). *Organometallics* 19: 1035–1043.

52 Montigny, F., Argouarch, G., Costuas, K. et al. (2005). Electron transfer and electron exchange between [Cp*(dppe)Fe]$^{n+}$ (n = 0, 1) building blocks mediated by the 9,10-bis(ethynyl)anthracene bridge. *Organometallics* 24: 4558–4572.

53 Ghazala, S.I., Paul, F., Toupet, L. et al. (2006). Di-organoiron mixed valent complexes featuring "(η^2-dppe)(η^5-C$_5$Me$_5$)Fe" endgroups: smooth class-III to class-II transition induced by successive insertion of 1,4-phenylene units in a butadiyne-diyl bridge. *J. Am. Chem. Soc.* 128: 2463–2476.

54 Costuas, K., Cador, O., Justaud, F. et al. (2011). 3,5-Bis(ethynyl)pyridine and 2,6-bis(ethynyl)pyridine spanning two Fe(Cp*)(dppe) units: role of the nitrogen atom on the electronic and magnetic couplings. *Inorg. Chem.* 50: 12601–12622.

55 Weyland, T., Costuas, K., Toupet, L. et al. (2000). Organometallic mixed-valence systems. two-center and three-center compounds with meta connections around a central phenylene ring. *Organometallics* 19: 4228–4239.

56 Lu, Y., Quardokus, R., Lent, C.S. et al. (2010). Charge localization in isolated mixed-valence complexes: an STM and theoretical study. *J. Am. Chem. Soc.* 132: 13519–13524.

57 Roué, S., Sahnoune, H., Toupet, L. et al. (2016). Double insertion of thiophene rings in polyynediyl chains to stabilize nanoscaled molecular wires with [Cp*(dppe)Fe] termini. *Organometallics* 35: 2057–2070.

58 Roué, S., Lapinte, C., and Bataille, T. (2004). Organometallic mixed-valence systems. electronic coupling through an alkyndiyl bridge incorporating methylene groups. *Organometallics* 23: 2558–2567.

59 Hamon, P., Justaud, F., Cador, O. et al. (2008). Redox-active organometallics: magnetic and electronic couplings through carbon-silicon hybrid molecular connectors. *J. Am. Chem. Soc.* 130: 17372–17383.

60 Lohan, M., Justaud, F., Lang, H., and Lapinte, C. (2012). Synthesis, spectroelectrochemical, and EPR spectroscopic studies of mixed bis(alkynyl) biferrocenes of the type (LnMC≡C)(LnM'C≡C)bfc. *Organometallics* 31: 3565–3574.

61 Makhoul, R., Kumamoto, Y., Miyazaki, A. et al. (2014). Synthesis and properties of a mixed-valence compound with single-step tunneling and multiple-step hopping behavior. *Eur. J. Inorg. Chem.* 2014: 3899–3911.

62 Lissel, F., Fox, T., Blacque, O. et al. (2013). Stepwise construction of an iron-substituted rigid-rod molecular wire: targeting a tetraferra–tetracosa-decayne. *J. Am. Chem. Soc.* 135: 4051–4060.

63 Lissel, F., Schwarz, F., Blacque, O. et al. (2014). Organometallic single-molecule electronics: tuning electron transport through X(diphosphine)$_2$FeC$_4$Fe

(diphosphine)$_2$X building blocks by varying the Fe–X–Au anchoring scheme from coordinative to covalent. *J. Am. Chem. Soc.* 136: 14560–14569.

64 Tanaka, Y., Inagaki, A., and Akita, M. (2007). A photoswitchable molecular wire with the dithienylethene (DTE) linker, (dppe)(μ^5-C$_5$Me$_5$)Fe–C≡C–DTE–C≡C–Fe(μ^5-C$_5$Me$_5$)(dppe). *Chem. Commun.* 2007: 1169–1171.

65 Gückel, S., Safari, P., Mohammad, S. et al. (2021). Iron versus ruthenium: evidence for the distinct differences in the electronic structures of hexa-1,3,5-triyn-1,6-diyl-bridged complexes [Cp*(dppe)M{μ-(C≡C)$_3$}{M(dppe)Cp*}]$^+$ (M = Fe, Ru). *Organometallics* 40: 346–357.

66 Ou, Y.-P., Zhang, J., Zhang, M.-X. et al. (2017). Bonding and electronic properties of linear diethynyl oligothienoacene-bridged diruthenium complexes and their oxidized forms. *Inorg. Chem.* 56: 11074–11086.

67 Klein, A., Lavastre, O., and Fiedler, J. (2006). Role of the bridging arylethynyl ligand in bi- and trinuclear ruthenium and iron complexes. *Organometallics* 25: 635–643.

68 Zhang, D.-B., Wang, J.-Y., Wen, H.-M., and Chen, Z.-N. (2014). Electrochemical, spectroscopic, and theoretical studies on diethynyl ligand bridged ruthenium complexes with 1,3-bis(2-pyridylimino)isoindolate. *Organometallics* 33: 4738–4746.

69 Forrest, W.P., Choudhuri, M.M.R., Kilyanek, S.M. et al. (2015). Synthesis and electronic structure of Ru$_2$(Xap)$_4$(Y-gem-DEE) type compounds: effect of cross-conjugation. *Inorg. Chem.* 54: 7645–7652.

70 Bruce, M.I., Low, P.J., Costuas, K. et al. (2002). Oxidation chemistry of metal-bonded C4 chains: a combined chemical, spectroelectrochemical, and computational study. *J. Am. Chem. Soc.* 122: 1949–1962.

71 Bruce, M.I., Ellis, B.G., Low, P.J. et al. (2003). Syntheses, structures, and spectro-electrochemistry of {Cp*(PP)Ru}C≡CC≡C{Ru(PP)Cp*} (PP = dppm, dppe) and their mono- and dications. *Organometallics* 22: 3184–3198.

72 Fox, M.A., Guennic, B.L., Roberts, R.L. et al. (2011). Simultaneous bridge-localized and mixed-valence character in diruthenium radical cations featuring diethynylaromatic bridging ligands. *J. Am. Chem. Soc.* 133: 18433–18446.

73 Zhang, J., Zhang, M.-X., Sun, C.-F. et al. (2015). Diruthenium complexes with bridging diethynyl polyaromatic ligands: synthesis, spectroelectrochemistry, and theoretical calculations. *Organometallics* 34: 3967–3978.

74 Ou, Y.-P., Xia, J., Zhang, J. et al. (2013). Experimental and theoretical studies of charge delocalization in biruthenium-alkynyl complexes bridged by thiophenes. *Chem. Asian J.* 8: 1–12.

75 Ou, Y.-P., Zhang, J., Zhang, F. et al. (2016). Bonding and electronic properties of linear diethynyl oligothienoacene-bridged diruthenium complexes and their oxidized forms. *Dalton Trans.* 45: 6503–6516.

76 Zhang, J., Ou, Y.-P., Xu, M. et al. (2014). Synthesis and characterization of dibenzoheterocycle-bridged dinuclear ruthenium alkynyl and vinyl complexes. *Eur. J. Inorg. Chem.* 2014: 2941–2951.

77 Fox, M.A., Farmer, J.-D., Roberts, R.L. et al. (2009). Noninnocent ligand behavior in diruthenium complexes containing a 1,3-diethynylbenzene bridge. *Organometallics* 28: 5266–5269.

78 Parthey, M., Gluyas, J.B.G., Schauer, P.A. et al. (2013). Refining the interpretation of near-infrared band shapes in a polyynediyl molecular wire. *Chem. Eur. J.* 19: 9780–9784.

79 Parthey, M., Gluyas, J.B.G., Fox, M.A. et al. (2014). Mixed-valence ruthenium complexes rotating through a conformational robin-day continuum. *Chem. Eur. J.* 20: 6895–6908.

80 Fox, M.A., Roberts, R.L., Baines, T.E. et al. (2008). Ruthenium complexes of C, C′-bis(ethynyl)carboranes: an investigation of electronic interactions mediated by spherical pseudo-aromatic spacers. *J. Am. Chem. Soc.* 130: 3566–3578.

81 Xia, J.-L., Man, W.-Y., Zhu, X. et al. (2012). Synthesis and characterization of dithia[3.3]paracyclophane-bridged binuclear ruthenium vinyl and alkynyl complexes. *Organometallics* 31: 5321–5333.

82 Bruce, M.I., Guennic, B., Scoleri, N. et al. (2012). Extending metal-capped polyynediyl molecular wires by insertion of inorganic metal units. *Organometallics* 31: 4701–4706.

83 Lavastre, O., Plass, J., Bachmann, P. et al. (1997). Ruthenium or ferrocenyl homobimetallic and RuPdRu and FePdFe heterotrimetallic complexes connected by unsaturated, carbon-rich –C≡CC6H4C≡C– Bridges. *Organometallics* 16: 184–189.

84 Wuttke, E., Pevny, F., Hervault, Y.-M. et al. (2012). Fully delocalized (ethynyl)(vinyl)phenylene bridged triruthenium complexes in up to five different oxidation states. *Inorg. Chem.* 51: 1902–1915.

85 Olivier, C., Kim, B., Touchard, D., and Rigaut, S. (2008). Redox-active molecular wires incorporating ruthenium (II) σ-arylacetylide complexes for molecular electronics. *Organometallics* 27: 509–518.

86 Liu, Y., Ndiaye, C.M., Lagrost, C. et al. (2014). Diarylethene-containing carbon-rich ruthenium organometallics: tuning of electrochromism. *Inorg. Chem.* 53: 8172–8188.

87 Kim, B., Beebe, J.M., Olivier, C. et al. (2007). Temperature and length dependence of charge transport in redox-active molecular wires incorporating ruthenium(II) bis(σ-arylacetylide) complexes. *J. Phys. Chem. C* 111: 7521–7526.

88 Li, B., Wang, J.-Y., Wen, H.-M. et al. (2012). Redox-modulated stepwise photochromism in a ruthenium complex with dual dithienylethene-acetylides. *J. Am. Chem. Soc.* 134: 16059–16067.

89 Piazza, E.D., Merhi, A., Norel, L. et al. (2015). Ruthenium carbon-rich complexes as redox switchable metal coupling units. *Inorg. Chem.* 54: 6347–6355.

90 Ren, T., Zou, G., and Alvarez, J.-C. (2000). Facile electronic communication between bimetallic termini bridged by elemental carbon chains. *Chem. Commun.* 2000: 1197–1198.

91 Wong, K.-T., Lehn, J.-M., Peng, S.-M., and Lee, G.-H. (2000). Nanoscale molecular organometallo-wires containing diruthenium cores. *Chem. Commun.* 2000: 2259–2260.

92 Xu, G.-L., Zou, G., Ni, Y.-H. et al. (2003). Polyyn-diyls capped by diruthenium termini: a new family of carbon-rich organometallic compounds and distance-dependent electronic coupling therein. *J. Am. Chem. Soc.* 125: 10057–10065.

93 Cao, Z., Xi, B., Jodoin, D.S. et al. (2014). Diruthenium–polyyn-diyl–diruthenium wires: electronic coupling in the long-distance regime. *J. Am. Chem. Soc.* 136: 12174–12183.

94 Fan, Y., Liu, I.P.-C., Fanwick, P.E., and Ren, T. (2009). Dimer of diruthenium compound bridged by 1,10-diethynylferrocene: ferrocene as a weak mediator for electronic coupling. *Organometallics* 28: 3959–3962.

95 Xi, B., Xu, G.-L., Fanwick, P.E., and Ren, T. (2009). Diruthenium complexes of axial ferrocenyl-polyynyl ligands: the cases of C6Fc and C8Fc. *Organometallics* 28: 2338–2341.

96 Ying, J.-W., Liu, I.P.-C., Xi, B. et al. (2010). Linear trimer of diruthenium linked by butadiyn-diyl units: a unique electronic wire. *Angew. Chem. Int. Ed.* 49: 954–957.

97 Gao, L.-B., Liu, S.H., Zhang, L.Y. et al. (2006). Preparation, characterization, redox properties, and UV–Vis–NIR spectra of binuclear ruthenium complexes [{(Phtpy)(PPh$_3$)$_2$Ru}$_2${C≡C–(CH=CH)$_m$–C≡C}]$^{n+}$ (Phtpy = 4′-phenyl-2,2′:6′,2″-terpyridine). *Organometallics* 25: 506–512.

98 Gao, L.-B., Kan, J., Fan, Y. et al. (2007). Wirelike dinuclear ruthenium complexes connected by bis(ethynyl)oligothiophene. *Inorg. Chem.* 46: 5651–5664.

99 Berben, L.A. and Kozimor, S.A. (2008). Dinitrogen and acetylide complexes of low-valent chromium. *Inorg. Chem.* 47: 4639–4647.

100 Egler-Lucas, C., Blacque, O., Venkatesan, K. et al. (2012). Dinuclear and mononuclear chromium acetylide complexes. *Eur. J. Inorg. Chem.* 2012: 1536–1545.

101 Brown, N.J., Lancashire, H.N., Fox, M.A. et al. (2011). Molybdenum complexes of C,C-bis(ethynyl)carboranes: design, synthesis, and study of a weakly coupled mixed-valence compound. *Organometallics* 30: 884–894.

102 Fitzgerald, E.C., Brown, N.J., Edge, R. et al. (2012). Orbital symmetry control of electronic coupling in a symmetrical, all-carbon-bridged "mixed valence" compound: synthesis, spectroscopy, and electronic structure of [{Mo(dppe)(η-C$_7$H$_7$)}$_2$(μ-C$_4$)]$^{n+}$ (n = 0, 1, or 2). *Organometallics* 31: 157–169.

103 Ladjarafi, A., Costuas, K., Meghezzi, H., and Halet, J.-F. (2015). Electronic structure of modelized vs. real carbon-chain containing organometallic dinuclear complexes: similarities and differences. *J. Mol. Model.* 21: 71–80.

104 Kheradmandan, S., Heinze, K., Schmalle, H.W., and Berke, H. (1999). Electronic communication in C$_4$-bridged binuclear complexes with paramagnetic bisphosphane manganese end groups. *Angew. Chem. Int. Ed.* 38: 2270–2273.

105 Venkatesan, K., Fox, T., Schmalle, H.W., and Berke, H. (2005). Synthesis and characterization of redox-active C$_4$-bridged rigid-rod complexes with acetylide-substituted manganese end groups. *Organometallics* 24: 2834–2847.

106 Dembinski, R., Bartik, T., Bartik, B. et al. (2000). Toward metal-capped one-dimensional carbon allotropes: wirelike C_6–C_{20} polyynediyl chains that span two redox-active (η^5-C_5Me_5)Re(NO)(PPh$_3$) endgroups. *J. Am. Chem. Soc.* 122: 810–822.

107 Meyer, W.E., Amoroso, A.J., Horn, C.R. et al. (2001). Synthesis and oxidation of dirhenium C4, C6, and C8 complexes of the formula (η^5-C_5Me_5)Re(NO)(PR$_3$)(C≡C)n(R$_3$P)(ON)Re(η^5-C_5Me_5) (R = 4-C_6H_4R', c-C_6H_{11}): in search of dications and radical cations with enhanced stabilities. *Organometallics* 20: 1115–1127.

108 Jiao, H., Costuas, K., Gladysz, J.A. et al. (2003). Bonding and electronic structure in consanguineous and conjugal iron and rhenium sp carbon chain complexes [MC_4M']$^{n+}$: computational analyses of the effect of the metal. *J. Am. Chem. Soc.* 125: 9511–9522.

109 Herrmann, C., Neugebauer, J., Gladysz, J.A., and Reiher, M. (2005). Theoretical study on the spin-state energy splittings and local spin in cationic [Re]–C_n–[Re] complexes. *Inorg. Chem.* 44: 6174–6182.

110 Bruce, M.I., Costuas, K., Davin, T. et al. (2007). Syntheses, structures and redox properties of some complexes containing the Os(dppe)Cp* fragment, including [{Os(dppe)Cp*}$_2$(μ-C≡CC≡C)]. *Dalton Trans.* 2007: 5387–5399.

111 Bruce, M.I., Kramarczuk, K.A., Skelton, B.W., and White, A.H. (2010). Syntheses, structures and redox properties of {Os(PPh$_3$)$_2$Cp}$_2${μ-(C,C)$_x$} (x = 2, 3, 4): comparisons with the Ru analogues. *J. Organomet. Chem.* 695: 469–473.

112 Steffen, A., Ward, R.M., Tay, M.G. et al. (2014). Regiospecific formation and unusual optical properties of 2,5-bis(arylethynyl)rhodacyclopentadienes: a new class of luminescent organometallics. *Chem. Eur. J.* 20: 3652–3666.

113 Sieck, C., Tay, M.G., Thibault, M.H. et al. (2016). Reductive coupling of diynes at rhodium gives fluorescent rhodacyclopentadienes or phosphorescent rhodium 2,2′-biphenyl complexes. *Chem. Eur. J.* 22: 10523–10532.

114 Obara, S., Itabashi, M., Okuda, F. et al. (2006). Highly phosphorescent iridium complexes containing both tridentate bis (benzimidazolyl)-benzene or-pyridine and bidentate phenylpyridine: synthesis, photophysical properties, and theoretical study of Ir-bis (benzimidazolyl) benzene complex. *Inorg. Chem.* 45: 8907–8921.

115 Cook, T.D., Natoli, S.N., Fanwick, P.E., and Ren, T. (2015). Dimeric complexes of CoIII(cyclam) with a polyynediyl bridge. *Organometallics* 34: 686–689.

116 Mash, B.L., Yang, Y., and Ren, T. (2020). Improving redox reversibility and intermetallic coupling of Co(III) alkynyls through tuning of frontier orbitals. *Organometallics* 39: 2019–2025.

6

Electron Transfer in Mixed-Valence Ferrocenyl-Functionalized Five- and Six-Membered Heterocycles

Peter Frenzel[1] and Heinrich Lang[2]

[1] *TU Chemnitz, Institute of Chemistry, Inorganic Chemistry, Faculty of Natural Sciences, Straße der Nationen 62, D-09107 Chemnitz, Germany*
[2] *TU Chemnitz, Research Centre MAIN, Rosenbergstr. 6, D-09126 Chemnitz, Germany*

6.1 Introduction

Since the first reports on heterocyclic chemistry [1–3], great progress has augmented in this field of chemistry, and these days, almost two-thirds of all organic compounds include such subjects. Remarkable enrichment of this class of molecules results now-a-days from inorganic, metal–organic, and organometallic chemistry. The outstanding chemical and physical properties of this family of compounds fascinate many scientists as these compounds show a rich structural diversity; intriguing physical, photochemical, and electrochemical properties; and furthermore possess a rich potential of many applications. Examples include the attachment of organometallic or metal–organic fragments to heterocyclic cores as such species are excellent templates for diverse (semi)conducting polymers and electroactive materials. Especially, ferrocenyl-functionalized heterocycles are well suited for studying intramolecular electronic metal–metal interactions through the heterocyclic anti-aromatic or aromatic backbone [4–8].

Robin and Day [9–11] as well as Allen, Hush, and Marcus [12–18] established the concept of mixed valency (MV) and hence facilitated the understanding of electronic interactions. In this context, Robin and Day organized mixed-valence species into three classes depending on the magnitude of charge transfer and delocalization: class I compounds are MV species displaying no electronic coupling between the donor and acceptor units of the molecule, and hence, intramolecular charge transfer is not induced neither optically nor thermally. In class III compounds, the electrons of the redox-active terminal groups are fully delocalized, and hence, charge transfer between the redox centers is reflected as a resonance between varying mesomeric structures. Class II compounds queue up between class I and III types. While in the ground state they are charge localized (or partially delocalized), the charge-delocalized excited state allows inter-valence charge transfer (IVCT). For a discussion of this matter, see Chapters 1, 3, and 13 within this monograph.

Mixed-Valence Systems: Fundamentals, Synthesis, Electron Transfer, and Applications, First Edition.
Edited by Yu-Wu Zhong, Chun Y. Liu, and Jeffrey R. Reimers.
© 2023 WILEY-VCH GmbH. Published 2023 by WILEY-VCH GmbH.

In 1970, short after the development of the Creutz–Taube ion $\{[Ru(NH_3)_5]_2(C_4H_4N_2)\}^{5+}$ [19], Cowan and coworker reported about the electrochemical behavior of diferrocene, diferrocenylacetylene, diferrocenylbutadiyne, bis(fulvalene)diiron, and [2.2]ferrocenophane-1,13-diyne [20–23]. They were able to show that within these molecules, the distance between the redox-active iron centers and the molecular geometry affect the electron transfer properties. Since that time, many readings concerning the use of ferrocenyl [Fc, $Fe(\eta^5\text{-}C_5H_4)(\eta^5\text{-}C_5H_5)$] building blocks as redox-active organometallic components appeared in a great series of diverse families of compounds. However, these studies do not allow an easy comparison of the obtained data as, for example, they possess different geometries and differ in their measurement conditions. In addition, until now, only less spectroelectrochemical studies were carried out, resulting in a variation of the electrostatic contribution to the observable redox splitting.

Hence, in this chapter, a comparative overview of five- and six-membered antiaromatic and aromatic heterocyclic compounds featuring at least two redox-active Fc termini (directly) bonded to the heterocycle will be highlighted, and therefore, the synthesis, structure, reactivity, and (spectro)electrochemistry of such molecules including current trends in this outstanding field of science will be envisaged, allowing to show how systematic changes, for example, within the bridging systems or at the Fc groups offer insight into the electron transfer properties and mechanism. However, additional spacer units between the heterocyclic core and the organometallic sandwich termini declining or inhibiting electron transfer as well as not planar *non*-aromatic heterocycles will not be addressed here. In addition, as Fc-functionalized heterocycles gained great visibility during the past years, which is documented in several reviews e.g. [6–8, 18, 24–28], this chapter is intended to cover continuing major aspects of this family of molecules published after 2011.

6.2 Ferrocenyl-Functionalized Five-Membered Heterocycles

In this section, five-membered heterocyclic compounds featuring main group elements of Groups 13–16 from the periodic table of the elements containing at least two Fc moieties (directly) bonded to the heterocyclic core will be discussed. In general, the electrochemical measurements were carried out in anhydrous dichloromethane solutions containing $[Bu_4N][B(C_6F_5)_4]$ as supporting electrolyte under inert conditions at ambient temperature unless otherwise stated. In comparison to smaller counterions including $[PF_6]^-$ or $[Cl]^-$, the $[B(C_6F_5)_4]^-$ anion stabilizes highly charged species in solution, minimizing ion-pairing effects [29–32]. Hence, minimization of this effect leads to an increase in the observed redox potential splitting. Spectroelectrochemical measurements (UV–vis/NIR) were carried out in an optically transparent thin-layer electrode (OTTLE) cell [33]. In addition, all potentials are referenced to the FcH/[FcH]$^+$ (FcH = $Fe(\eta^5\text{-}C_5H_5)_2$) redox couple [34, 35].

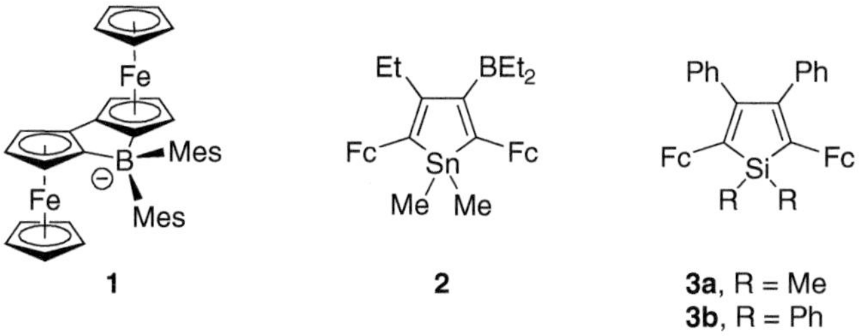

Figure 6.1 Compounds **1–3**. Source: Adapted from [36–38].

6.2.1 Five-Membered Heterocyclic Compounds with Group 13 Elements

To the best of our knowledge, Fc-substituted five-membered heterocycles with Group 13 elements have not been reported. However, in a broader sense, fused biferrocene (pS,pS)-2,2″-(μ-dimesitylboryl)-1,1″-biferrocene (**1**) can be considered as a molecule comprising a ^cC_4B ring (Figure 6.1). Compound **1** is accessible in a multi-step synthetic methodology as described by Jäkle in 2015 [36]. It is redox active as cyclic (CV) and square wave voltammetry (SWV) studies confirmed. UV–vis/NIR spectra display IVCT bands at ca. 4000 cm^{-1} (ε_{max} = 300 l mol^{-1} cm^{-1}) and 5000 cm^{-1} (ε_{max} = 250 l mol^{-1} cm^{-1}), confirming electronic coupling in the mixed-valence species [**1**]$^+$ (class II species according to the Robin–Day classification [9–11]). In addition, mono-oxidized [**1**]$^+$ was available by chemical oxidation upon addition of [AgPF$_6$] or under aerobic conditions. The structure of [**1**]$^+$ in the solid state was determined by single-crystal X-ray structure analysis, specifying that it contains distinct Fe(II)/Fe(III) sites.

6.2.2 Five-Membered Heterocyclic Compounds with Group 14 Elements

Based on 2,5-diferrocenyl-stannacyclopenta-2,4-diene (**2**) [37], we enriched this only less-developed family of compounds by 2,5-diferrocenyl-substituted siloles (Figure 6.1) [38]. 2,5-Fc$_2$-3,4-Ph$_2$-$^cC_4$SiR$_2$ (**3a**, R = Me; **3b**, R = Ph) molecules were prepared by a consecutive reductive cyclization reaction of SiR$_2$(C≡CPh)$_2$ with lithium naphthalenide followed by the introduction of the Fc ligands by applying the Negishi C,C cross-coupling protocol, whereas the silole ring was applied either as a vinyl halogenide or as a zinc organic species [38].

The electrochemical behavior of **3a,b** was studied by CV and SWV, and the nature of the redox products was investigated by *in situ* UV–vis/NIR spectroelectrochemistry (Figure 6.2). It was shown that the Fc ligands in **3a,b** undergo two sequential Fc-based redox processes, whereby their separation [$\Delta E^{0'} = \Delta E_2^{0'} - \Delta E_1^{0'}$ = 300 mV (**3a**), = 280 mV (**3b**)] is in the range of structural similar systems, i.e. 2,5-diferrocenyl-1-phenyl-1H-phosphole (280 mV) [39] and 2,5-diferrocenyl furan (290 mV) [40] (Sections 6.2.3 and 6.2.4). It was found that the more electron-rich siloles displayed a lower redox separation between the distinct Fc oxidation events.

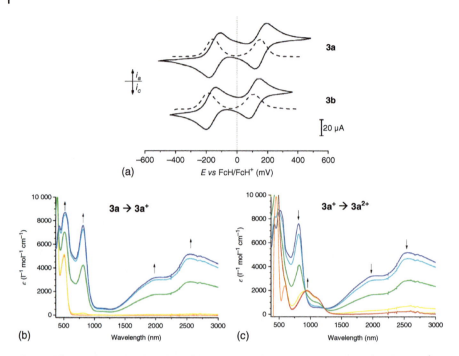

Figure 6.2 (a) CVs (solid lines) and SWVs (dotted lines) of **3a,b** in CH_2Cl_2 (1.0 mmol l^{-1}) at 25 °C (scan rate 100 mV s^{-1} and supporting electrolyte 0.1 mol l^{-1} [Bu$_4$N][B(C$_6$F$_5$)$_4$]). (b, c) UV–vis/NIR spectra of **[3a]**$^{n+}$ ($n = 0$, 1, and 2) in CH_2Cl_2 (2.0 mmol l^{-1}) at increasing potentials vs. Ag/AgCl: (b) −200 to 300 mV and (c) 300–700 mV at 25 °C (supporting electrolyte 0.1 mol l^{-1} [Bu$_4$N][B(C$_6$F$_5$)$_4$]). Arrows indicate an increase or decrease in the absorptions. Source: [38] Reproduced with permission of the American Chemical Society.

Spectroelectrochemical studies support electron charge transfer interactions among the Fc/[Fc]$^+$ termini in the mixed-valence species across the silole bridging unit. IVCT transitions were observed at 4700 cm^{-1} (**[3a]**$^+$: ε_{max} = 3150 l mol^{-1} cm^{-1}, $\Delta\nu_{1/2}$ = 2950 cm^{-1}) and 4650 cm^{-1} (**[3b]**$^+$: ε_{max} = 2270 l mol^{-1} cm^{-1}, $\Delta\nu_{1/2}$ = 3300 cm^{-1}). These values correspond well with the ones found, for example, in 2,5-diferrocenyl phospholes [39] and diferrocenyl *cis*-butadiene [41], agreeing that the C$_4$ unit at the molecule's backbone mediates the electronic coupling. Mixed-valent **[3a,b]**$^+$ are moderate to weakly coupled class II species according to the classification of Robin and Day [9–11]. Despite being more electron rich, mono-cationic **[3b]**$^+$ reveals less-intense IVCT absorptions than **[3a]**$^+$, which is most probably attributed to steric interactions. DFT calculations are consistent with the class II description of **[3a,b]**$^+$, with apparent low differences in energy between the *syn*- and *anti*-conformations, suggesting a distribution of molecular geometries and hence variation in electronic coupling in solution [38].

6.2.3 Five-Membered Heterocyclic Compounds with Group 15 Elements

In this section, Fc-functionalized heterocycles such as pyrroles, *aza*dipyrromethenes, difluoroboryl-*aza*dipyromethenes, pyrazoles, dihydropyrazoles, triazoles, maleimides, and phospholes will be discussed.

Ferrocenyl-functionalized five-membered heteroaromatics, such as Fc pyrroles, with the Fc groups acting as electron reservoir, play a significant role as model compounds for the study of electronic metal–metal interactions in, for example, molecular wire molecules or conducting polymers [4, 5, 42–44]. In general, they are suitable for the design of electroactive materials. Hence, within this section, the synthesis of a series of mainly 2,5-, 3,4-, and 2,3,4,5-Fc-substituted pyrroles will be discussed (Scheme 6.1). Also, it will be envisaged how systematic changes within the pyrrole bridging unit and the Fc influence the electron transfer properties (Table 6.1).

In principle, for the synthesis of pyrroles **4–12**, the following methodologies exist: (i) Negishi C,C coupling of the respective dibromo- or diiodo-pyrroles with an excess of FcZnCl in the presence of catalytic amounts of $[Pd(PPh_3)_4]$ [40, 45, 47–51, 53], (ii) Sonogashira C,C cross-coupling of the dibromo or diiodo pyrrole derivatives with FcC≡CH and $[PdCl_2(PPh_3)_2]$ as catalyst [52], or (iii) cycloaddition of FcC≡C—C≡CFc with aromatic amines [46]; however, the yields in the latter synthesis procedure are much lower than those within the C,C coupling reactions (Scheme 6.1). For the synthesis of (oligo)pyrroles **10–12**, multi-step preparation methods were applied, based on the Negishi C,C coupling procedure [49].

The redox properties of all Fc-functionalized pyrroles were studied by CV, SWV, LSV (linear sweep voltammetry), and *in situ* UV–vis/NIR and/or IR spectroscopy in anhydrous CH_2Cl_2 solutions using as supporting electrolyte a 0.1 mol l^{-1} $[Bu_4N][B(C_6F_5)_4]$ or a less common $[Bu_4N][PF_6]$ solution (Table 6.1).

As can be seen from Table 6.1 and Figure 6.3, well-separated reversible redox events in each of the respective di- or tetraferrocenyl-1-phenyl-1*H*-pyrroles exist. The observed $\Delta E^{0'}$ values ($\Delta E^{0'}$ = difference between the first and second (di-substituted) or the first and second, second and third, and third and fourth (tetra-substituted) oxidation) range from 160 to 600 mV. It was found that the electronic metal–metal interactions via the heterocyclic connectivity can be influenced by either the substituents at the nitrogen phenyl group and/or the Fc ligands [40, 45, 46, 48]. For example, NMR studies confirmed within the series of different Ph-substituted pyrroles that a systematic change from electron-donating (NMe_2, Et, Me, OMe, OEt, etc.) via electron-neutral (H) to electron-withdrawing (F, CO_2Et, C(O)H, C≡N, etc.) groups is characteristic and hence has a significant effect on the electronic density in the heterocyclic core [45, 46]. Furthermore, it could be shown that the electron-donating and electron-withdrawing substituents R^1 at the phenyls affect the redox events of the Fc oxidations, as a linear relationship between the σ Hammett constants and the respective $E_1^{0'}$ values as well as the separation of the

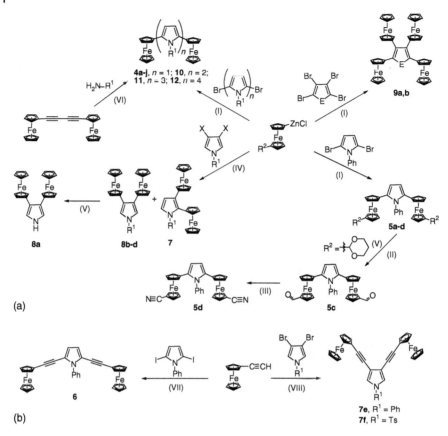

Scheme 6.1 Synthetic strategies for the preparation of Fc-substituted pyrroles **4–12** (for abbreviations, see Table 6.1) ((I) [Pd(PPh$_3$)$_4$], thf, 60 °C, 48 hours; synthesis of oligomers: [Pd(CH$_2$CMe$_2$P(^{t}Bu$_2$)(μ-Cl)]$_2$, thf, 60 °C, 12 hours; (II) 1-Me-4-SO$_3$H-C$_6$H$_4$, 20 hours, CH$_2$Cl$_2$/H$_2$O (20:1), 100 °C; (III) first: NH$_3$OH$^+$Cl$^-$, py/H$_2$O (2:1), 1 hour; second: [Cu(SO$_4$)$_2$], NEt$_3$/-CH$_2$Cl$_2$ (1:10, v/v), dicarbodiimide, 10 hours; (IV) [Pd(PPh$_3$)$_4$], 72 hours, reflux; (V) R^1 = Si(^{i}Pr)$_3$: [Bu$_4$N]F, thf, 25 °C, 10 minutes; (VI) [CuCl], 100 °C; (VII) [CuI]/[PdCl$_2$(PPh$_3$)$_2$], ^{i}Pr$_2$NH; and (VIII) [CuI], ^{i}Pr$_2$NH, thf, PPh$_3$, [PdCl$_2$(PPh$_3$)$_2$], 65 °C, 72 hours) [45–53].

redox potentials exists (Figure 6.3) [45]. The electronic influence is not restricted to the redox potentials. Likewise, $\Delta E^{0\prime}$ correlates nicely with the Hammett constant. In addition, NIR studies endorse electronic coupling in the mixed-valence species between the Fc and [Fc]$^+$ termini; the oscillator strength f of the charge transfer transition depends on the electron-donating/electron-withdrawing characteristic of the corresponding nitrogen-bonded Ph groups. It could be shown that the most electron-rich pyrrole unit realizes a high degree of intermetallic communication as the highest f values were observed in molecules featuring electron-donating groups (Table 6.1) [45]. Hence, this finding allows a direct intermetallic communication by modification of the compound's functional groups. Because of the specific adjustment of electron transfer, such species can be regarded as model compounds for single-molecule transistors. The linear relationship between $\Delta E^{0\prime}$ and f is in

Table 6.1 Selected (spectro)electrochemical data of pyrroles 4–6 and 8–12.

Compound	R^1	$E_1^{0'a}$ $(\Delta E_p)^{b}$ (mV)	$E_2^{0'a}$ $(\Delta E_p)^{b}$ (mV)	$E_3^{0'a}$ $(\Delta E_p)^{b}$ (mV)	$E_4^{0'a}$ $(\Delta E_p)^{b}$ (mV)	$\Delta E^{0'c}$ (mV)	$\nu_{max}^{d)}(\varepsilon_{max})^{e)}$ [cm^{-1}]($l\,mol^{-1}$ cm^{-1})	$\Delta\nu_{1/2}^{f)}$ (cm^{-1})	$(\Delta\nu_{1/2})_{theo}^{g)}$ (cm^{-1})	References
4a[h/i]	CH_3	−206(65)	204(65)			410	4750(3145)	2314	2312	[47]
4b[h/i]	C_6H_5	−240(68)	210(75)			450	4820(4200)	2369	3340	[45]
		−198(80)	117(85)			315[j]				[53]
4c[h/i]	N-C_6H_4-4-Me	−250(71)	205(75)			455	4825(4279)	2337	3339	[45]
4d[k]	N-C_6H_4-4-Et	55	379			324				[46]
4e[h/i]	N-C_6H_4-4-NMe_2	−305(61)	175(68)			480	5055(4678)	2358	3417	[45]
4f[h/i]	N-C_6H_4-4-OMe	−255(60)	205(69)			460	4870(4296)	2360	3354	[45]
4g[k]	N-C_6H_4-4-OEt	17	345			328				[46]
4h[h/i]	N-C_6H_4-4-CO_2Et	−190(62)	230(69)			420	4554(3445)	2438	3243	[45]
4i[k]	N-C_6H_4-4-F	89	381			292				[46]
4j[h/i]	N-C_6H_4-3-F	−210(61)	215(65)			425	4586(3584)	2413	3254	[45]

(Continued)

Table 6.1 (Continued).

Compound	R²	$E_1^{0'a}$ $(\Delta E_p)^{b)}$ (mV)	$E_2^{0'a}$ $(\Delta E_p)^{b)}$ (mV)	$E_3^{0'a}$ $(\Delta E_p)^{b)}$ (mV)	$E_4^{0'a}$ $(\Delta E_p)^{b)}$ (mV)	$\Delta E^{0'c}$ (mV)	$\nu_{max}^{d)}(\varepsilon_{max})^{e)}$ [cm⁻¹](L mol⁻¹ cm⁻¹)	$\Delta\nu_{1/2}^{f)}$ (cm⁻¹)	$(\Delta\nu_{1/2})_{theo}^{g)}$ (cm⁻¹)	References
5a[l/i]	CHO₂(CH₂)₃	−235(66)	160(77)			395	4820(5750)	3400	3340	[48]
		−160(66)	160(77)			320[m]				[48]
5b[l/i]	3,5-(CF₃)₂-C₆H₃	−80(68)	405(74)			485	4520(4400)	3190	3240	[48]
		80(65)	460(64)			380[m]				[48]
5c[l/i]	CHO	0(64)	555(91)			555	5230(4660)	2550	3480	[48]
		55(70)	455(70)			400[m]				[48]
5d[l/i]	N≡C—C≡C	260(60)	860(61)			600	5430(3630)	2440	3550	[48]
		115(58)	535(76)			420[m]				[48]
6[l/i]		121(204)				106	6625(1500)	3980	3912	[50]
	R¹/X									
8a[n]	H/–	−150(−170)	30			180 290[o)]				[51] [51]
			120							
8b[l/i]	C₆H₅/–	−150(72)	135(74)			285	6285(95)	4165	3810	[52]
8c[n]	Si[i)]Pr₃/–	−225	−50			175				[51]
		−180	120			300[o)]				[51]
8d[l/i]	N-SO₂-C₆H₄-4-Me/–	40(70)	295(66)			255	7210(65)	4115	4081	[52]
8e[l)]	C₆H₅/C≡C	105(146)				110				[52]
8f[l)]	N-SO₂-C₆H₄-4-Me/C≡C	140(132)				95				[52]

	R¹									
9a[h/i]	C$_6$H$_5$	−280(62)	51(63)	323(62)	550(61)	331/272/227	4752(4900)	2719	3313	[53]
		−226(72)	−26(72)	134(104)		200/160[j]				[53]
9b[h/i]	CH$_3$	−280(67)	−15(68)	385(71)	609(72)	265/400/224	6214(1045)	2592	3789	[47]
							6237(948)	2682	3796[p]	[47]
	n									
10[l]	2	−250(74)	70(74)	810(74)		320[q]				[49]
11[l]	3	−230(68)	−65(74)	480(74)	1080(72)	165[q]				[49]
12[l]	4	−175(90)	265(66)	725(63)		440[q]				[49]

a) $E_n^{0'}$ = Formal potential of nth Fc-related redox process.
b) ΔE_p = Difference between oxidation/reduction potentials.
c) $\Delta E^{0'}$ = Potential difference between two Fc-related redox processes.
d) ν_{max} = Energy of IVCT band.
e) ε_{max} = Extinction coefficient of IVCT band.
f) $\Delta \nu_{1/2}$ = Full width at half-height of IVCT band.
g) Values calcd. as $(\Delta \nu_{1/2})_{theo} = (2310 \cdot \nu_{max})^{1/2}$ [13].
h) CV studies: potentials vs. FcH/[FcH]$^+$, scan rate 100 mV s^{-1} at a glassy carbon electrode of 0.5 mmol l^{-1} pyrrole CH$_2$Cl$_2$ solutions containing 0.1 mol l^{-1} of [NBu$_4$][B(C$_6$F$_5$)$_4$] as the supporting electrolyte, 25 °C.
i) UV–vis/NIR studies in CH$_2$Cl$_2$ containing 0.1 mol l^{-1} [NBu$_4$][B(C$_6$F$_5$)$_4$] as the supporting electrolyte, 25 °C using an OTTLE cell [33].
j) CV studies: analog to [h] however, [NBu$_4$][PF$_6$] was used as an electrolyte.
k) CV studies: analog to [h] at a Pt disk electrode in CH$_2$Cl$_2$ containing 0.1 mol l^{-1} of [NBu$_4$][PF$_6$] as the supporting electrolyte.
l) CV studies: analog to [h] however, the analyte concentration amounts to 1.0 mmol l^{-1}.
m) CV studies: analog to [l], [NBu$_4$][PF$_6$] was used as an electrolyte.
n) CV studies: analog to [h] Pt working electrode in CH$_2$Cl$_2$ containing 0.1 mol l^{-1} of [NBu$_4$]ClO$_4$ as the supporting electrolyte, 25 °C.
o) CV studies: analog to [n] however, 0.05 mol l^{-1} [NBu$_4$][B(C$_6$F$_5$)$_4$] was used as an electrolyte.
p) Values of di-cationic species.
q) $\Delta E^{0'}$ = Potential difference between the first and second redox process.

Source: Adapted from [40, 45–53].

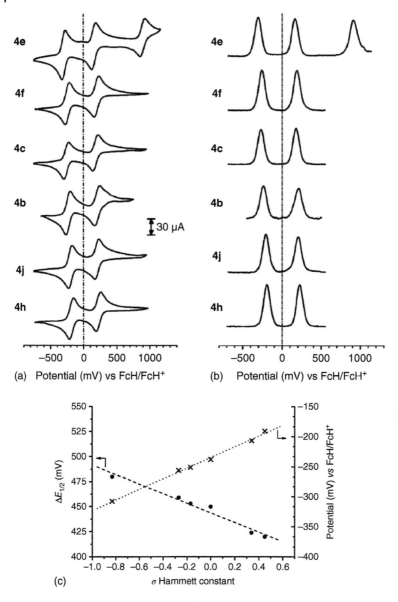

Figure 6.3 (a) CVs and (b) SWVs of **4b,c,e,f,h,j**. (c) Left ordinate: correlation of $\Delta E^{0\prime}$ values and the σ Hammett constant, linear fit ($R^2 = 0.978$, dashed line); right ordinate: correlation of $E_1^{0\prime}$ values and the σ Hammett constant, linear fit ($R^2 = 0.995$, dotted line). Source: [45] Reproduced with permission of the Royal Society of Chemistry.

consensus with the theoretical assumption of a series of compounds possessing similar geometries and hence akin electrostatic properties.

Moreover, it could be shown that different substituents at the NPh unit influence not only the electronic properties of **4a–j** but also electron-donating/electron-withdrawing groups at the Fc 1′ position (**5a–e**) [48]. Electron-withdrawing substituents R^2 at Fc shift the redox potentials anodically ($R^2 = {}^cC_4H_7O_2 > H > 3,5\text{-}(CF_3)_2\text{-}C_6H_3 > C(O)H > C\equiv N$), and the first and second redox processes are more separated when the electron-withdrawing effect of R^2 increased; for example, compound **5d** featuring C≡N functionalities possesses with 600 mV, the highest redox separation measured for Fc-substituted (hetero)aromatics so far (Table 6.1).

The influence of the increasing number of n ($n = 1, 2, 3,$ and 4) in the wire-like oligo-pyrroles **4a** and **10–12** (Scheme 6.1) on the electronic interaction between the redox-active iron centers was also studied by CV, SWV, and UV–vis/NIR spectroscopy (Table 6.1) [49]. With exception of the quarter-pyrrole **12** (the iron centers are oxidized simultaneously), all other oligomers undergo discrete Fc oxidations ($\Delta E^{0'} = 450$ mV ($n = 1$), 320 mV ($n = 2$), and 165 mV ($n = 3$)). The splitting of the Fc-based redox couples decreases with increasing oligo-pyrrole chain length. *In situ* UV–vis/NIR measurements displayed for the monomeric compound ($n = 1$) an intense IVCT absorption (Table 6.1), while the classical two-state model appears not to be applicable to the lengthier oligo-pyrroles [49]. For the quarter-pyrrole, a strong absorption is found in the near-IR region, which is arising from a photo-induced charge transfer from the pyrrole chain to the redox-active Fc termini. CT absorptions were also found in the other pyrroles ($n = 1, 2,$ and 3) when twice oxidized. Furthermore, within the series of oligo-pyrroles, this CT band is shifted bathochromically, when the chain length of the backbone is increased [49].

Tetra-ferrocenyl pyrroles **9a,b** display four electrochemically reversible one-electron transfer processes in weakly coordinating [Bu$_4$N][B(C$_6$F$_5$)$_4$], while in [Bu$_4$N][PF$_6$], only three events are characteristic with $\Delta E^{0'}$ values from 220 to 400 mV and reduction potentials between −280 and 610 mV (Table 6.1) [40, 53]. Compared to 2,5-Fc$_2$-$^cC_4H_2$NPh (**4a**), the differences in the potential of the first and second redox couple are lower, which can be explained by the increased steric demand of the Fc groups in **9a,b** and therefore the decreased tendency of the Fc cyclopentadienyls to be coplanar with the pyrrole core as evidenced by single-crystal X-ray structure analysis [40, 53]. *In situ* UV–vis/NIR studies were carried out whereby by increasing the potential (25–100 mV), cationic [**9a,b**]$^+$, [**9a,b**]$^{2+}$, [**9a,b**]$^{3+}$, and finally [**9a,b**]$^{4+}$ were obtained with increasing extinction of the NIR bands. *In situ* formed [**9a,b**][B(C$_6$F$_5$)$_4$] shows a strong band at 4752 cm^{-1} ([**9a**]$^+$) or at 6214 cm^{-1} ([**9b**]$^+$), similar to that observed for **4a** (Table 6.1), confirming electronic communication between the Fc/[Fc]$^+$ ligands in the mono-cationic species via the $^cC_4$N connectivity. Comparing tetra-ferrocenyl pyrroles **9a,b** with the respective thiophene compound (Section 6.2.4) in which the interaction between the Fc entities is mainly based on electrostatic effects, the pyrrole derivative can be classified as a weakly class II system according to Robin and Day [9–11].

Comparison of the 2,5-diferrocenyl pyrroles **4a–j** with the corresponding 3,4-isomers **8a–d** displays that the $\Delta E^{0'}$ values strongly depend on the substitution pattern (2,5-isomer: 290–480 mV; 3,4-isomer: 170–300 mV) (Table 6.1), which significantly differs from the respective thiophene species (Section 6.2.4), showing rather low differences in the redox splitting between the appropriate isomers [6, 7, 18, 40].

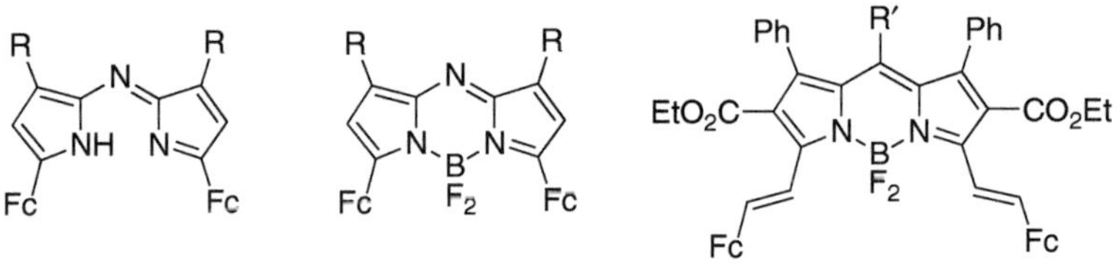

Figure 6.4 Fc-*aza*dipyrromethenes (left) and their corresponding *aza*BODIPY complexes (right). Source: Adapted from [54–56].

Spectroelectrochemical studies confirmed that compounds in which the Fc groups are directly bonded to the heterocycle are class II species according to the classification of Robin and Day [9–11], while those in which the Fc units are linked by a C≡C connectivity to pyrrole can be categorized as class I systems [50, 51].

In a series of papers, Nemykin and coworkers discussed the synthesis, redox properties, and electronic coupling of compounds **13–15** (Figure 6.4) [54–56]. Such molecules are of interest for applications in light harvesting, molecular electronics, electro- as well as photocatalysis, and, for example, redox-switchable fluorescence. The electronic structure of Fc-functionalized BODIPY's comprises the Fc-centered HOMO and the BODIPY-centered LUMO, building a platform for intense low-energy metal-to-ligand charge transfer (MLCT) transitions. Compounds **13–15** can be prepared by consecutive synthetic procedures by applying the well-established chalcone-type methodology [54–56].

All compounds were characterized by (spectro)electrochemistry. Their electronic structure, redox behavior, and UV–vis/NIR spectra were correlated with DFT and TD-DFT calculations. For example, the molecules containing two Fc entities being bonded to the pyrrolic α-position show in the CV (CH_2Cl_2, 0.05 M [Bu_4N][$B(C_6F_5)_4$]) two reversible one-electron Fc oxidations at 0–400 mV. The difference between the first and second oxidation is 340 mV (**13a**), 460 mV (**14a**), 210 mV (**15b**), and 170 mV (**15a**), which is in agreement with other Fc-functionalized BODIPY dyads [57]. While the first oxidation potentials in **15a,b** only differ by 30 mV, the two separated reduction events (first reduction potential) are shifted by 360 mV, when one goes from **15b** to **15a**, which is due to stabilization of the LUMO in **15b** by the electron-withdrawing C≡N group. The Fc units in **13a** and **14a** are electronically coupled, while in **15a,b** no prominent IVCT band in the NIR region could be seen. DFT and TD-DFT results on **13a** and **14a** are in good agreement with the obtained experimental data, signifying Fc-centered MOs in the HOMO region and chromophore π^* characteristic LUMO [56]. For tetra-ferrocenyl-substituted compounds **13b** and **14b** in which the Fc ligands are bonded to the α- and β-positions of the pyrrolic backbone, the stepwise oxidation of the four Fc groups occurs between −140 and 700 mV [55]. Comparing the difference of the first and second oxidation of **13b** (280 mV) and **14b** (160 mV) with the one of **13a** and **14a** shows that they are smaller. However, the values point to high electrochemical comproportionation constants K_c [4.65×10^4 (**13b**) and 5.08×10^2 (**14b**)]. Spectroelectrochemical studies allowed for the characterization of the mixed-valence forms, and IVCT band analysis endorses that [**13b**]⁺ and [**14b**]⁺ belong to weakly coupled class II species in the Robin–Day classification [9–11]. It should be noted that Mössbauer spectroscopy specified two discrete doublets for Fe(II) as well as Fe(III) at ambient temperature for [**13b**]$^{n+}$ and [**14b**]$^{n+}$ ($n = 1$, 2, and 3) correlating well with the spectroelectrochemical experiments. DFT and TD-DFT came to the same result as already discussed earlier with Fc-centered HOMO and chromophore-centered LUMO and prime spin localization at the Fc unit attached to the α-position of the heterocycle [55].

In 2020, the family of Fc-functionalized pyrazoles was enriched by 1-R′-3,5-Fc$_2$-^{c}C$_3$HN$_2$ (R′ = H, **16a**; Me, **16b**; and Ph, **16c**) [58]. The redox behavior of these molecules was determined by CV and SWV. All of the three compounds show two reversible one-electron redox processes that confirm the separate oxidation of the Fc groups. It can be noted that the redox behavior of **16a–c** is very similar with $E_1^{0\prime} = -30$ mV and $E_2^{0\prime} = $ c. 210 mV, evidencing only minor impact of the N-bonded H, Me, or Ph groups on the charge transfer process or the electrostatics in mixed valence [**16a–c**]⁺ [58]. In order to get a deeper insight into the spectroscopic details of the latter species, in situ UV–vis/NIR studies were accomplished. However, it appeared that for [**16a–c**]⁺, no IVCT absorptions of considerable strength could be detected. Hence, the redox separation is mainly caused by electrostatic interactions.

Also, a series of unsymmetrical Fc-substituted dihydropyrazoles was discussed toward their synthesis, structure, and electrochemical behavior. This family of compounds includes 1-phenylsulfonyl-3,5-Fc$_2$-4,5- (**17a**) [59], 3,5-Fc$_2$-1-(C(O)CH$_2$CH$_2$CO$_2$Et-4,5- (**17b**) [60], 1-*p*-pyridine acyl-3-Fc-5-(2,2-diferrocenylpropane)-4,5- (**17c**), and 1-phenyl acyl-3-Fc-5-(2,2-diferrocenylpropane)-4,5-dihydropyrazole (**17d**) [61]. Representatively, the CV and DPV as well as the electron transfer mechanism in the redox process of **17a** are discussed [59]. The CV shows two separated and reversible redox events at 522 and 664 mV; analysis of the CVs indicates a diffusion-controlled process. The first oxidation takes place on the iron center in position 3 and the second one at Fc in position 5, giving the respective mono- or di-cationic species. From DFT calculations, it was found that the iron center in position 3 is more electron-rich and hence serves as an electron donor to the iron atom in position 5.

In Figure 6.5, Fc-functionalized triazoles **18–21** are depicted, which were prepared by applying the Cu-catalyzed *click* (1,3-diploar cycloaddition) reaction (CuAAC) of terminal alkynes (ethynylferrocene or 1,4-bis(trimethylsilyl)-1,3-butadiyne) and azido ferrocene [62, 63]. Táraga and Molino used these molecules as receptors for sensing studies toward anions (ClO$_4^-$ and CF$_3$SO$_3^-$) and cations (Li⁺, Na⁺, K⁺, Ca²⁺, Mg²⁺, Ni²⁺, Cd²⁺, Zn²⁺, Pb²⁺ Hg²⁺, and Cu²⁺), and hence, the compounds have been studied by electrochemical (CV and Osteryoung square-wave voltammetry [OSWV]) and optical techniques.

In general, compounds **18a–c** show two reversible one-electron redox events at $E_1^{0\prime} = 50$ mV and $E_2^{0\prime} = 250$ mV ($\Delta E^{0\prime} = 200$ mV) for **18a**, $E_1^{0\prime} = 50$ mV and $E_2^{0\prime} = 310$ mV ($\Delta E^{0\prime} = 260$ mV) for **18b**, and $E_1^{0\prime} = 30$ mV and $E_2^{0\prime} = 240$ mV ($\Delta E^{0\prime} = 210$ mV) for **18c** (**18a** and **18c**: CH$_3$CN/CH$_2$Cl$_2$ (ratio 4 : 1, $c = 5 \times 10^{-4}$ M) and **18b**: CH$_3$CN/CH$_2$Cl$_2$ (ratio 3 : 2, $c = 5 \times 10^{-4}$ M) (Figure 6.5). In summary, it

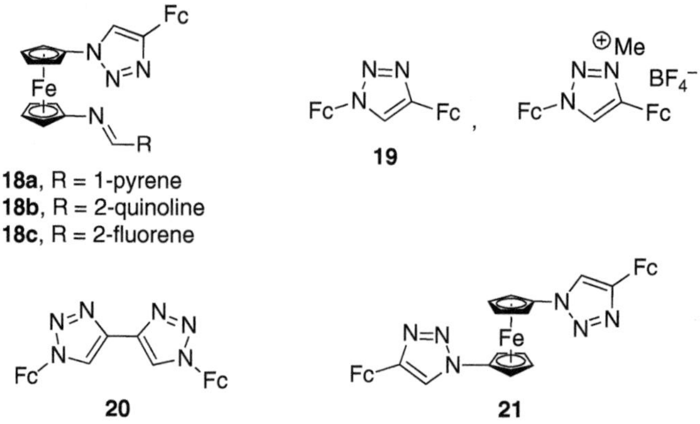

Figure 6.5 Ferrocenyl-functionalized triazoles **18–21**. Source: Adapted from [62, 63].

can be stated that **18b** performs as a colorimetric receptor for Ni^{2+} and Cd^{2+} ions, while **18c** acts as a selective fluorescent receptor for Pb^{2+} [62]. Tentative binding modes are proposed on the basis of ^{1}H NMR titration studies and DFT calculations.

Major features resulting from the sensing investigations are that **19–21** (Figure 6.5) undertake exceptional cathodic shifts of the oxidation event of the Fc/[Fc]$^+$ redox couple in the presence of anions such as F^-, AcO^-, $H_2PO_4^-$, and $HP_2O_7^{3-}$ [63]. For the use of further Fc-functionalized triazole derivatives, see Ref. [64].

In 2017, Sakar and coworkers reported on the synthesis and electrochemical (CV) as well as spectroelectrochemical (UV-vis/NIR) properties of heterotrinuclear (Fe–Cu–Fe) and heteropentanuclear (Fe–Cu–Fe–Cu–Fe) copper(I)-ferrocenyl pyridyl-triazoles **22** and **23** (Figure 6.6 (left)) [65].

As it can be seen from Figure 6.6 (middle), the Fe–Cu–Fe compound **22** owns two closely spaced one-electron oxidations at 280 and 390 mV (CH_2Cl_2, 0.1 M

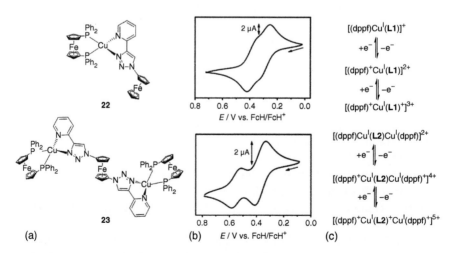

Figure 6.6 (a) Molecules **22** and **23**. (b) CVs. (c) Redox scheme for electron transfer. Source: [65] Reproduced with permission of Wiley-VCH.

[Bu$_4$N][PF$_6$]), while for Fe–Cu–Fe–Cu–Fe **23**, one two-electron and one one-electron processes are characteristic (370 and 550 mV) (changing to other solvents or electrolytes such as CH$_3$C≡N and [Bu$_4$N][B(C$_6$F$_5$)$_4$] did not have an influence on the separation between the redox waves) [65]. As compared to **22**, the oxidation potentials of **23** are positively shifted, which can be explained by the fact that the second Cu(I) ion imparts an additional positive charge to the complex. It further could be shown that the Cu(I) ions act as innocent connectivities. For the respective electron transfer steps in both complexes, see the scheme depicted in Figure 6.6 (right) [65].

In 2011 and 2016, the solvatochromism as well as the (spectro)electrochemistry (CV, SWV, and DPV; *in situ* UV–vis/NIR; and IR) of 2,5-Fc$_2$-*maleimide* (**24a**) [66] and 2,5-Fc$_2$-*N*-methylmaleimide (**24b**) [67] was reported. In these compounds, the two Fc ligands are *anti*-oriented because of their steric bulkiness. Electronic structures and solvatochromic properties of the molecules were investigated by UV–vis spectroscopy and polarized continuum model DFT and time-dependent DFT approaches. For **24a**, it was found that the calculated vertical excitation energies are consistent with the experimental data, suggesting the dominance of metal-to-ligand charge transfer bands in the visible region of the UV–vis spectrum.

Both compounds exhibit two well-resolved Fc-based redox events in the *non*-coordinating CH$_2$Cl$_2$/[Bu$_4$N][B(C$_6$F$_5$)$_4$] system (**24a**: 96 mV, 441 mV, and $\Delta E^{0'} = 345$ mV; **24b**: 50 mV, 380 mV, and $\Delta E^{0'} = 330$ mV). The difference between the first and second redox potential points to some electronic communication between the two Fc groups in the mixed-valence species, which was confirmed by the observation of IVCT bands in the UV–vis/NIR spectra. For example, homovalent neutral **24b**, as expected, does not show any absorptions between 3000 and 10 000 cm^{-1}; however, when the potential is increased in step heights of 25, 50, and 100 mV, mixed-valent [**24a**]$^+$ is produced as recognized by the appearance of a weak IVCT band at 4900 cm^{-1} [66]. Band shape analysis of the IVCT excitation consents **24b** as a weakly coupled class II system according to Robin and Day [9–11] by using the criterion of Brunschwig, Creutz, and Sutin [15, 16]. In addition, *in situ* IR studies were carried out, signifying some extent of electron delocalization.

In the following, the synthesis, reactivity, reaction chemistry, and (spectro)electrochemistry, including the influence of P-bonded bulky substituents on the electronic interactions in 2,5-Fc$_2$-substituted phospholes, are discussed.

Recently, phospholes have gained great interest because of their pronounced potential in applications such as metal-containing organic light-emitting devices (OLEDs). Phospholes are also of scientific interest as they show a degree of aromaticity depending on the substituents in positions 1, 2, and 5; bulky groups force the P atom to adopt a planar surrounding. In addition, blocking the phosphorus lone pair of electrons by, for example, oxidation (H$_2$O$_2$, S$_8$, and Se) or coordination of the P atom to organometallic building blocks prevents the conjugation around the cycle and hence suppresses aromaticity. Out of these reasons, the phosphole motif offers the possibility to adjust the electronic nature of the heterocyclic ring and hence electronic effects that may be transmitted across it, which will be discussed below.

The synthesis of phospholes **25a–e** succeeds by cyclization of FcC≡C—C≡CFc with PH$_2$Ar (Ar = Fc, Ph, 2,4,6-Me$_3$-C$_6$H$_2$, 2,4,6-Ph$_3$-C$_6$H$_2$, or 2,4,6-^{t}Bu$_3$-C$_6$H$_2$)

based on a method firstly described by Märkl and Potthast [68] (Scheme 6.2). Scheme 6.2 also includes the reaction chemistry of phospholes **26–31** toward oxidation (reaction path i) and coordination (reaction path ii) (P atom and/or π system).

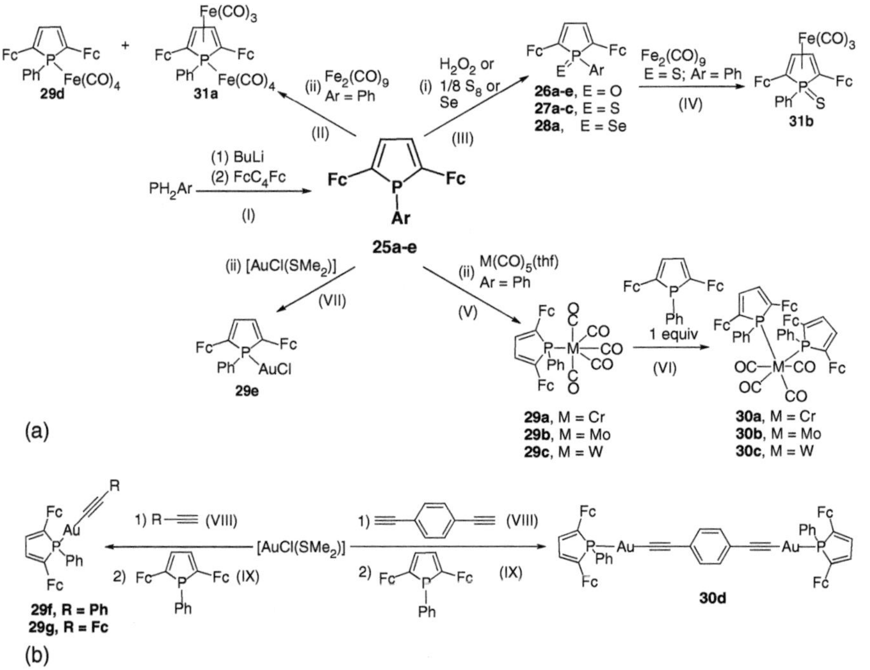

Scheme 6.2 Synthesis of phospholes **25–31**. Source: Adapted from [39, 69–71]. ((I) 1) Toluene/thf (1 : 1, v/v), 0 °C, 1 hour and 2) thf, 25 °C, 12 hours. (II) Toluene, 110 C, 8 hours. (III) CH_2Cl_2, 25 °C. (IV) Toluene, 25 °C, 8 hours. (V) thf, 25 °C, 14 hours. (VI) hv, thf, 2 hours. (VII) CH_2Cl_2, 25 °C, 14 hours. (VIII) thf/DIPA (1 : 1, v/v), [CuI], 25 °C, 30 minutes. (IX) thf, 25 °C, 1 hour.)

Phospholes **25a–e** were structurally characterized by single-crystal X-ray diffraction studies [39, 69]. It was found that an increased planarity of the pyramidal P surroundings occurs when, for example, the bulky 2,4,6-^{t}Bu$_3$-C$_6$H$_2$ group was applied (**25e**), which is in accordance with observations made elsewhere [72]. Calculations of the Bird index [73] revealed higher delocalization for phospholes with bulkier substituents. The activation enthalpy and activation entropy of the inversion at P were determined by VT-NMR spectroscopy, confirming a decrease in the activation enthalpy and coalescence temperature, with substituent enlargement at P [69]. Compared with 1-isopropyl-2-methyl-5-phenyl-phosphole [74], compounds **25a–e** show lower activation enthalpies but similar activation entropies.

CV and SWV measurements confirmed that the Fc substituents in **25–31** could be oxidized at discrete potentials with a significant separation of the individual redox events of 105–350 mV (Table 6.2 and Figure 6.7), indicating electronic interactions with substantial thermodynamic stability of the mixed-valence radical

Table 6.2 (Spectro)electrochemical data of **25–31**.

Compound	R	$E_1^{0'a)}$ $(\Delta E_p)^{b)}$ (mV)	$E_2^{0'a)}$ $(\Delta E_p)^{b)}$ (mV)	$E_3^{0'a)}$ $(\Delta E_p)^{b)}$ (mV)	$E_4^{0'a)}$ $(\Delta E_p)^{b)}$ (mV)	$E_{pa,1}/E_{pa,2}^{c)}$ (mV)	$\Delta E^{0'd)}$ (mV)	$\nu_{max}^{e)}(\varepsilon_{max})^{f)}$ $[cm^{-1}]$ $(l\,mol^{-1}\,cm^{-1})$	$\Delta\nu_{1/2}^{g)}$ (cm^{-1})	$(\Delta\nu_{1/2})_{theo}^{h)}$ (cm^{-1})	References
25a[i]	C_6H_5	−135(76)	50(62)				185				[39]
		−110(72)	170(80)				280	5000(1750)	3050	3400[j/k]	[39]
								6500(1600)	3200	3870[j]	[39]
25b[i]	Fc	−135(87)	40(72)	150(62)			175/190				[69]
		−125(76)	140(72)	450(72)			265/310	4850 (1850)	3250	3343[j/k]	[69]
25c[i]	2,4,6-Me$_3$-C$_6$H$_2$	−155(77)	55(85)				210				[69]
		−140(63)	155(74)				295	4900(2750)	2700	3365[j/k]	[69]
25d[i]	2,4,6-(C$_6$H$_5$)$_3$-C$_6$H$_2$	−185(73)	45(81)				230				[69]
		−165(75)	175(78)				340	4650(2650)	2800	3270[j/k]	[69]
25e[i]	2,4,6-^{t}Bu$_3$-C$_6$H$_2$	−200(76)	30(78)				230				[69]
		−180(70)	135(76)				315	5050(3000)	2550	3418[j/k]	[69]

(Continued)

Table 6.2 (Continued).

Compound		$E_1^{0'a)}$ $(\Delta E_p)^{b)}$ (mV)	$E_2^{0'a)}$ $(\Delta E_p)^{b)}$ (mV)	$E_3^{0'a)}$ $(\Delta E_p)^{b)}$ (mV)	$E_4^{0'a)}$ $(\Delta E_p)^{b)}$ (mV)	$E_{pa,1}/E_{pa,2}^{c)}$ (mV)	$\Delta E^{0'd)}$ (mV)	$\nu_{max}^{e)}(\varepsilon_{max})^{f)}$ [cm^{-1}] (l mol^{-1} cm^{-1})	$\Delta\nu_{1/2}^{g)}$ (cm^{-1})	$(\Delta\nu_{1/2})_{theo}^{h)}$ (cm^{-1})	References
26a[j]	(=O)C$_6$H$_5$	−20(71)	215(71)				235				[69]
26b[j]	(=O)Fc	−35(60)	235(67)	580(63)			270/345				[69]
26c[j]	(=O)-2,4,6-Me$_3$-C$_6$H$_2$	−40(64)	195(63)				235				[69]
26d[j]	(=O)-2,4,6-(C$_6$H$_5$)$_3$-C$_6$H$_2$	−80(66)	185(66)				265				[69]
26e[j]	(=O)-2,4,6-^{t}Bu$_3$-C$_6$H$_2$	−65(72)	205(76)				270				[69]
27a[j]	(=S)C$_6$H$_5$	−65(76)	100(76)				165				[39]
		−15(68)	225(74)				240	4900(1300)	4200	3360[j/k]	[39]
								(6900)	3100	3990[l]	[39]
27b[j]	(=S)Fc	−40(80)	220(81)	575(84)			260/355				[69]
27c[j]	(=S)-2,4,6-Me$_3$-C$_6$H$_2$	−45(66)	205(70)				250				[69]
28a[j]	(=Se)C$_6$H$_5$	−60(80)	100(78)				160				[39]
		−15(74)	220(80)				235	4850(1100)	4200	3350[j/k]	[39]
								(6900)	3400	3990[l]	[39]
29a[j/k]	(Cr(CO)$_5$)C$_6$H$_5$	15(86)	280(85)			1240	265	4850(1450)	3300	3344	[70]
29b[j/k]	(Mo(CO)$_5$)C$_6$H$_5$	10(78)	280(61)			1335	270	4825(1650)	3300	3337	[70]
29c[j/k]	(W(CO)$_5$)C$_6$H$_5$	20(70)	285(67)			1400	265	4850(1450)	3250	3342	[70]
29d[j]	(Fe(CO)$_4$)C$_6$H$_5$	115[m]	305[m]								[70]

	R¹										
29e[j/k]	(AuCl)C$_6$H$_5$	55(74)	315(74)					260	4800(1300)	3750	3331	[71]
29f[j/k]	(AuC≡CC$_6$H$_5$)C$_6$H$_5$	30(72)	290(80)					260	4850(1200)	3600	3346	[71]
29g[j/k]	(AuC≡CFc)C$_6$H$_5$	−35(83)	75(79)	325(77)			130/250		4800(1100)	3650	3327[n]	[71]
	ML$_n$											
30a[j/k]	Cr(CO)$_4$	−95(65)	10(86)	260(82)	410(64)	1220	105/250/150		4850(1950)	3150	3345[n]	[70]
30b[j/k]	Mo(CO)$_4$	−115(68)	15(74)	270(67)	415(65)	1325	130/255/145		4800(2000)	3150	3330[n]	[70]
30c[j/k]	W(CO)$_4$	−100(67)	20(81)	270(77)	420(66)	1365	120/250/150		4850(1800)	3250	3336[n]	[70]
30d[j/k]	AuC≡C—C$_6$H$_4$—C≡Au	50(92)	315(112)				265		4900(2700)	3200	3368[n]	[71]
	R¹											
31a[j]	Fe(CO)$_4$	110[m]	305[m]									[70]
31b[j/k]	(=S)	5(82)	195(82)				190		5050(400)	4700	3419	[70]

a) $E_n^{0'}$ = Formal potential of nth Fc-related redox process.
b) ΔE_p = Difference between oxidation/reduction potentials.
c) E_{pa} = Anodic peak potential of MI$_n$.
d) $\Delta E^{0'}$ = Potential difference between two Fc-related redox processes.
e) ν_{max} = Energy of IVCT band.
f) ε_{max} = Extinction coefficient of IVCT band.
g) $\Delta\nu_{1/2}$ = Full width at half-height of IVCT band.
h) Values calcd. as $(\Delta\nu_{1/2})_{theo} = (2310 \cdot \nu_{max})^{1/2}$ [13].
i) CV studies: potentials vs. FcH/FcH$^+$, scan rate 100 mV s^{-1} at a glassy-carbon electrode of 1.0 mmol l^{-1} phosphole CH$_2$Cl$_2$ solutions containing 0.1 mol l^{-1} of [NBu$_4$][PF$_6$] as the supporting electrolyte, 25 °C.
j) CV studies: analog to ^{i}; however, [NBu$_4$][B(C$_6$F$_5$)$_4$] was used as an electrolyte.
k) UV-vis/NIR studies in containing 0.1 mol l^{-1} [NBu$_4$][B(C$_6$F$_5$)$_4$] as the supporting electrolyte, 25 °C, using an OTTLE cell [33].
l) UV-vis/NIR studies carried out analog to^j; however, MeCN was used as the solvent.
m) E_{pa} = Anodic peak potential of Fc.
n) Values of di-cationic species.
Source: Adapted from [39, 69, 70].

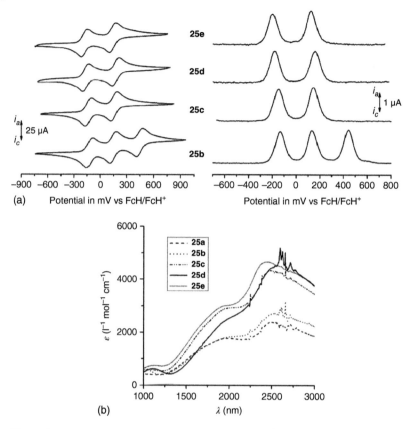

Figure 6.7 (a) CVs of **25b–e** (left, scan rate 100 mV s^{-1}) and SWVs [right, CH$_2$Cl$_2$, 1.0 mmol l^{-1}, 25 °C, supporting electrolyte [Bu$_4$N][B(C$_6$F$_5$)$_4$] (0.1 mol l^{-1}), working electrode: glassy carbon electrode (surface area 0.031 cm^{-1})]. (b) UV–vis spectra of *in situ* generated [**25a–e**]$^+$ (CH$_2$Cl$_2$ (2.0 mmol), 25 °C, [Bu$_4$N][B(C$_6$F$_5$)$_4$]). Source: [69, 70]/Reproduced with permission of the American Chemical Society and Wiley-VCH.

cations [39, 69, 70]. The respective values agree well with other species of heterocyclic-bridged diferrocenyl molecules, including pyrroles, thiophenes, furans, etc. (vide supra). The occupation of the lone pair of electrons at P (P=O, P=S, P=Se, and P–ML$_n$; Scheme 6.2) (slightly) decreases the metal–metal interaction among the ^{c}C$_4$P ring. The electronic characteristics of the compounds resemble those of a *cis*-diene structure rather than the ones of an aromatic arrangement, and the changes upon oxidation of P(III) to P(V) are therefore small. The electron-withdrawing characteristic of P(V) in mono-oxidized [**26-28**]$^+$ results in a shift of the $E_1^{0'}$ values to more positive potentials in comparison to the appropriate neutral P(III) species **25a–e** (Table 6.2). η^4-Coordination of the dienic phosphole unit to a Fe(CO)$_3$ moiety, as characteristic for **31a,b**, results in a decrease of the redox splitting in comparison to the respective uncoordinated systems [i.e. $\Delta E^{0'}$ = 190 mV for (η^{2,3,4,5})(2,5-Fc$_2$-^{c}C$_4$H$_2$P(=S)Ph))Fe(CO)$_3$ (**31b**); 240 mV for 2,5-Fc$_2$-^{c}C$_4$H$_2$P(=S)Ph (**27a**)] (Table 6.2) [69, 70].

UV–vis/NIR measurements display for the respective mono-cationic compounds' IVCT absorptions of moderate strength (Figure 6.7 and Table 6.2). In comparison to **25a** (Ar = Ph), the phospholes featuring sterically demanding substituents on P showed an increased interaction between the Fc/[Fc]$^+$ termini in the mixed-valence species. Studies with different electrolytes such as [B(C$_6$F$_5$)$_4$]$^-$ and [PF$_6$]$^-$ allowed some insights into the electrostatic contribution to the redox splitting. As a result thereof, electrostatic interactions are similar throughout the series. The highest interaction determined by spectroelectrochemistry was observed for **25e** (v_{max} = 5050 cm^{-1}, ε_{max} = 3000 l mol^{-1} cm^{-1}, and $\Delta v_{1/2}$ = 2550 cm^{-1}). Also, it could be shown that variation of the substituents Ar at P influences the electronic interactions of the Fc ligands in positions 2 and 5 of the phosphole ring (Table 6.2). All phospholes exhibit IVCT absorptions of moderate strength, confirming well to the predictions of the Hush two-state model for weakly coupled mixed-valence species [39, 69]. Compared with other Fc$_2$-substituted heterocycles, the strength of the electronic interactions in **25e** was in the same range as those of 2,5-Fc$_2$-1-Me-1*H*-pyrrole (v_{max} = 4750 cm^{-1}, ε_{max} = 3145 l mol^{-1} cm^{-1}, and $\Delta v_{1/2}$ = 2314 cm^{-1}) [40]. These conclusions were supported by DFT and TD-DFT calculations adequately modeling the observed structural and spectroscopic parameters. The theoretical studies help in assigning the various low-energy electronic transitions (LF and IVCT) and also highlight the key role of the unsaturated diene-like structure of the phosphole connectivity in promoting the Fc → [Fc]$^+$ electron transfer transition. Further planarization of the ring system and increased aromatic characteristic of the phosphole ring, as indicated by DFT-based calculations, provided an additional, and unexpected, contribution to the stabilization of the mixed-valence state [69].

6.2.4 Five-Membered Heterocyclic Compounds with Group 16 Elements

Ferrocenyl-functionalized furans, thiophenes, and oligo-thiophenes including benzothiophenes are discussed within this section. As outlined earlier, such compounds are attracting growing interest as, for example, electronic materials at the molecular scale (molecular wires). The reason is that they display (high) intramolecular charge transfer and conductivity depending on the delocalization of charge along the conjugated core relying on the degree of energy matching between the nature of the connectivity, the capping groups, and the distance between the redox-active (donor–acceptor) building blocks.

A very efficient synthesis route to di- and tetra-ferrocenyl-substituted furans of type 2,5-Fc$_2$-^{c}C$_4$H$_2$O (**32**) and 2,3,4,5-Fc$_4$-^{c}C$_4$O (**33**) was established in 2011 by applying the Negishi synthesis protocol [40, 75]. In this respect, Negishi ferrocenylation of the particular bromo-functionalized derivatives with FcZnCl in the presence of catalytic amounts of [Pd(PPh$_3$)$_4$] gave **32** and **33** in high yields. Somewhat earlier furans 2,4-Fc$_2$-3-Br-^{c}C$_4$HO (**34**) and 2,4-Fc$_2$-3-(Pd(PPh$_3$)$_2$Br)-^{c}C$_4$HO (**35**) were synthesized by basic treatment of α-bromoacetyl ferrocene and by the oxidative addition of **34** to [Pd(PPh$_3$)$_4$]; however, the compounds could only be isolated in low yields [76]. A further synthetic methodology includes the Suzuki–Miyaura *C,C* cross-coupling of α,ω'-Br$_2$-(^{c}C$_4$H$_2$O)$_n$ ($n = 1, 2, 3$, and 4) with FcB(OH)$_2$ in the presence of [PdCl$_2$(dppf)] (dppf = Fe(η^5-C$_5$H$_4$PPh$_2$)$_2$) as a catalyst to give **32** ($n = 1$), **36** ($n = 2$), **37** ($n = 3$), and **38** ($n = 4$) [77]. Compound 2,5-(FcC≡C)$_2$-^{c}C$_4$H$_2$O (**39**) was accessible by treatment of 2,5-Br$_2$-^{c}C$_4$H$_2$O with two equivalents of FcC≡CH using the Sonogashira protocol [78].

The electronic and electrochemical properties of **32–39** were investigated by CV, SWV, and *in situ* UV–vis/NIR spectroscopy and single-crystal X-ray diffractometry [40, 75–78]. (Spectro)electrochemical measurements reveal that the Fc ligands of all compounds can be individually oxidized giving two (i.e. **32**, −152 and 138 mV) or four (**33**, −237 mV, −10, 370, and 590 mV) well-separated and reversible one-electron redox events in CV and SWV with $\Delta E^{0'}$ values between 220 and 410 mV with exception of **39** for which, due to the separation of the Fc moieties from the heterocyclic core by a C≡C unit, the oxidation processes take place in a close potential range at $E^{0'} = 70$ mV with $\Delta E_p = 144$ mV, indicating a certain degree of intermetallic interaction in the molecule (IVCT transition at 7450 cm^{-1} with $\varepsilon_{max} = 565$ l mol^{-1} cm^{-1}) [40]. For the corresponding 2,4-Fc$_2$-isomers **34** and **35** with formal electrode potentials at 433 and 549 mV for **34** and 296 and 453 mV for **35**, the $\Delta E^{0'}$ values correspond to 116 and 157 mV, respectively [76].

In situ UV–vis/NIR studies on **32** and **33** confirmed electronic communication as IVCT absorptions were found at 5060 cm^{-1} ($\varepsilon_{max} = 1496$ l mol^{-1} cm^{-1}) for [**32**]$^+$, 5631 cm^{-1} ($\varepsilon_{max} = 466$ l mol^{-1} cm^{-1}) for [**33**]$^+$, and 6166 cm^{-1} ($\varepsilon_{max} = 449$ l mol^{-1} cm^{-1}) for [**33**]$^{2+}$ [40]. Similarly, [α,ω'-Fc$_2$-(^{c}C$_4$H$_2$O)$_n$]$^+$ possesses IVCT bands at 1968 nm ($\varepsilon_{max} = 1950$ l mol^{-1} cm^{-1}, $n = 1$), 1943 nm ($\varepsilon_{max} = 2190$ l mol^{-1} cm^{-1}, $n = 2$), 1808 nm ($\varepsilon_{max} = 2140$ l mol^{-1} cm^{-1}, $n = 3$), and 1939 nm ($\varepsilon_{max} = 2390$ l mol^{-1} cm^{-1}, $n = 4$) [77]. Based on the electrochemical studies and on computational tools, strong charge delocalization in the Fc oligo-furans is found. The distance decay constant was determined to be 0.066 Å^{-1}, suggesting delocalization along the furan backbone. Compared with similar Fc$_2$-substituted oligo-thiophene molecular wires, in the appropriate furan derivatives, a better energy matching between the Fc termini and the oligo-furan chain is found [77]. Computational studies indicate a slightly larger extent of delocalization in the furan-bridged species when compared with the respective thiophene analogs. This finding is in agreement with oligo-furans being more rigid and less aromatic than oligo-thiophenes (vide supra). Gidron and coworkers pointed out that the high charge delocalization in oligo-furans in combination with their strong fluorescence and their high mobility and rigidity makes them attractive materials for organic electronic applications [77].

Recently, the synthesis, solid-state structure, and electrochemical behavior (CV, DPV at a stationary Pt disc electrode, and by voltammetry at a rotating disc electrode; 1.0 mM solutions in CH$_2$Cl$_2$ containing 0.1 mM [Bu$_4$N][PF$_6$] as supporting electrolyte) of 2,5-diferrocenyl-1,3,4-oxadiazole were reported by Štěpnička, showing a broad unresolved redox event at $E^{0'} = 280$ mV consisting of two unsatisfactorily separated waves [79].

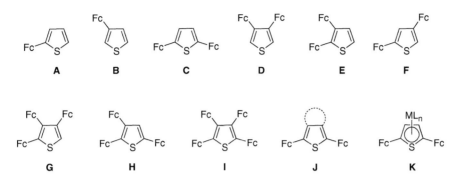

Figure 6.8 Type **A–K** Fc-thiophenes. Source: Adapted from [7, 80–88].

Oligo- and poly-thiophene-based materials play a significant role as organic conductive materials. Out of this, it is not surprising that thiophenes and oligomers thereof featuring donor–acceptor components gained great attention during the past years as they can act as model compounds for molecular wires and, for example, nanoscale electroactive materials (vide supra). Hence, in this section, the focus is directed toward ferrocenyl-functionalized thiophenes.

An overview of the mono-, di-, tri-, and tetra-ferrocenyl-substituted thiophenes (type **A–K** molecules) synthesized so far is summarized in Figure 6.8 and Table 6.3. The following synthetic methodologies exist: (i) Pd-promoted Negishi or Suzuki–Miyaura *C,C* cross-couplings [7, 80–86], (ii) the reaction of iodo-substituted thiophenes with HgFc$_2$ in the presence of [Pd(PPh$_3$)$_4$] [88], and (iii) the M(CO)$_6$-endorsed (M = Mo and W) treatment of ferrocenylalkyne with S$_8$ [87]; however, when Cr(CO)$_6$ is used as catalyst than next to 2,5-Fc$_2$-^{c}C$_4$H$_2$S, 2,6-diferrocenyldithiine was also formed [87].

All compounds were subjected to (spectro)electrochemical studies. The experimentally determined data are summarized in Table 6.3.

As it can be seen from Table 6.3, compounds **40–51** show separated reversible one-electron redox events for each of the Fc unit present. The most interesting features of the electrochemical studies are discussed in the following. In 2014, the substituent influence on charge transfer interactions in 2,5-diferrocenyl thiophenes was reported [81]. As a result thereof, it can be held that electron-withdrawing groups at Fc in combination with the electron-donating ethylenedioxy entity enforces the electronic coupling between the iron-based redox centers in the respective mixed-valence systems [81]. The compounds are class II species according to the Robin and Day classification [9–11]. From Table 6.3, it can also be generalized that Fc termini in α position of the thiophene core interact more strongly with the five-membered heterocyclic ring than those in the β position [81, 83].

Furthermore, in the series of the di-ferrocenyl isomers of the type **C–F** (Figure 6.8), the interaction between the iron centers in the mixed-valence species decreases in the series **C > E > F > D** (Table 6.3). A comparison of the peak width at half-height $\Delta v_{1/2}$ (IVCT band) and the Hush prediction $\Delta v_{1/2}$ (theor.) ([**C**]$^+$, [**D**]$^+$, [**E**]$^+$, [**F**]$^+$, [**G**]$^+$, [**H**]$^+$, [**G**]$^{2+}$, and [**H**]$^{2+}$) is consistent with the low-energy absorption of

Table 6.3 (Spectro)electrochemical data of Fc-substituted thiophenes of type **C–K** (compounds **40–51**).

Compound	$E_1{}^{0'a)}$ $(\Delta E_p)^{b)}$ (mV)	$E_2{}^{0'a)}$ $(\Delta E_p)^{b)}$ (mV)	$E_3{}^{0'a)}$ $(\Delta E_p)^{b)}$ (mV)	$E_4{}^{0'a)}$ $(\Delta E_p)^{b)}$ (mV)	$E_{pa}{}^{c)}/E_{pc}{}^{d)}$ (mV)	$\Delta E^{0'e)}$ (mV)	$\nu_{max}{}^{f)}(\varepsilon_{max})^{g)}$ (cm^{-1}) (l mol^{-1} cm^{-1})	$\Delta\nu_{1/2}{}^{h)}$ (cm^{-1})	$(\Delta\nu_{1/2})_{theo}{}^{i)}$ (cm^{-1})	References
R^1/R^2/R^3/R^4										
40a$^{j/k}$ H/H/H/H	−53(60)	195(62)				248	4994 (2129) (7020)	3843 3950	3397 4027$^{l)}$	[80] [81]
40b$^{j/k}$ C(O)H/C(O)H/H/H	235(63)	528(65)				293	4680 (3160)	3540	3290	[81]
40c$^{j/k}$ 3,5-(CF$_3$)$_2$-C$_6$H$_3$/H/H	82(64)	407(65)					6500	3380	3875$^{l)}$	[81]
40d$^{j/k}$ C≡N/C≡N/H/H	306(65)	627(70)				325	4130 (3290) (5880)	3660 3470	3090 3685$^{l)}$	[81] [81]
40e$^{j/k}$ H/(CO)$_5$W=C(OMe)/H/H	10(70)	420(70)			1135/−2095$^{m)}$	321	4710(3340) (6320)	3270 3560	3300 3821$^{l)}$	[81] [81]
40f$^{j/k}$ (CO)$_5$W=C(OMe)/(CO)$_5$W=C(OMe)/H/H	230(65)	475(70)			1135/−2095$^{m)}$	410 245	6600 4660	3220 4200	3100 3280	[82] [82]

	$R^3/R^4/R^5/R^6$								
41a[j/k]	H/H/NO$_2$/NO$_2$	274 (66)	429 (71)	−1539(71)[n)]	155	6080 (320) (9010)	4970 4490	3750 4562[l)]	[83] [83]
41b[j/k]	H/H/NMe$_2$/NMe$_2$	−168 (60)	168 (60)	1018[o)]	336	5000 (2520)	3630	3400	[83]
						(6040)	3160	3735[l)]	[83]
41c[j/k]	H/H/CH$_3$N (CH$_2$)$_2$NCH$_3$	−249(62)	169(64)	858(72)[p)]	418	4800 (3390)	3530	3330	[83]
41d[j/k]	H/H/N=CH–CH=N	−92(62)	228(62)	−1974(64)[q)]	320	4440 (3480)	3560	3568[l)]	[83]
						(5510)	3130	3200	[83]
41e[j/k]	H/H/O(CH$_2$)$_2$O	−160 (64)	153 (66)		313	4840 (3330)	3340	3571[l)]	[83]
						(5520)	3300	3440	[81]
41f[j/k]	C(O)H/C(O)H/O(CH$_2$)$_2$O	90 (65)	488 (69)		398	4280 (4600)	3090	3772[l)]	[81]
						(6160)	3520	3140	[81]
41g[j/k]	C≡N/C≡N/O(CH$_2$)$_2$O	146 (63)	582 (66)		436	4260 (4610)	3600	3496[l)]	[81]
						(5290)	3770	3140	[81]
41h[j/k]	3,5-(CF$_3$)$_2$-C$_6$H$_3$/ 3,5-(CF$_3$)$_2$-C$_6$H$_3$/ O(CH$_2$)$_2$O	−32 (68)	357 (68)		389	4200 (4600)	3590	3449[l)]	[81]
						(5150)	3310	3120	[81]
						(5140)	3490	3446[l)]	[81]
	$R^3/R^4/R^5/R^6$								
42[j/k]	H/Fc/Fc/H	−21 (68)	148 (69)		169	6098 (151)	5620	3753	[80]

(Continued)

Table 6.3 (Continued).

Compound	$E_1^{0'a)}$ $(\Delta E_p)^{b)}$ (mV)	$E_2^{0'a)}$ $(\Delta E_p)^{b)}$ (mV)	$E_3^{0'a)}$ $(\Delta E_p)^{b)}$ (mV)	$E_4^{0'a)}$ $(\Delta E_p)^{b)}$ (mV)	$E_{pa}^{c)}/E_{pc}^{d)}$ (mV)	$\Delta E^{0'e)}$ (mV)	$\nu_{max}^{f)} (\varepsilon_{max})^{g)}$ (cm^{-1}) (l mol^{-1} cm^{-1})	$\Delta \nu_{1/2}^{h)}$ (cm^{-1})	$(\Delta \nu_{1/2})_{theo}^{i)}$ (cm^{-1})	References
43$^{j/k)}$ Fc/H/Fc/H	−51 (71)	232 (73)				283	5300 (401)	5492	3499	[80]
44$^{j/k)}$ Fc/Fc/H/H	−49 (71)	195 (70)				244	5405 (70)	7761	3534	[80]
45$^{j/k)}$ Fc/Fc/Fc/H	−118 (63)	148 (61)	427 (61)			266/279	6628 (184) 7511 (244)	5736 6733	3913 4165$^{r)}$	[80] [80]
46$^{j/k)}$ Fc/H/Fc/Fc	−71 (66)	151 (62)	374 (70)			222/223	4914 (1583) 5975 (639)	4628 5895	3369 3715$^{l)}$	[80] [80]
47$^{j/k)}$ Fc/Fc/Fc/Fc	−161 (62)	58 (64)	418 (64)	604 (60)		219/360/18				[84]
$ML_n = Cr(CO)_3$ $R^5/R^3/R^4/R^6$										
48$^{j/k)}$ Fc/H/H/Fc		371$^{s)}$			540/−					[85]
49$^{j/k)}$ H/Fc/Fc/H	120$^{s)}$	492$^{s)}$			824/−					[85]
50$^{j/k)}$ H/H/Fc/Fc	119$^{s)}$	453$^{s)}$			895/−					[85]
$ML_n = [Ru(\eta^5\text{-}C_5H_5)]^+$ PF_6^-										

	$R^5/R^3/R^4/R^6$					
51a[j/k]	Fc/H/H/Fc	384			−/−1763	[86]
51b[j/k]	Fc/OCH$_2$CH$_2$O/Fc	275 (59)	360 (59)		−/−1763	85[t]	[86]
	ML_n = [Ru(η^5-C$_5$Me$_5$)]$^+$ PF$_6^-$	345 246 (63)	335 (62)		−/−1758 −/−1758	89[t]	[86] [86]
	$R^5/R^3/R^4/R^6$					
51c[j/k]	Fc/H/H/Fc	373			−/−2024		[86]
		246 (59)	344 (59)		−/−2024	98[t]	[86]
51d[j/k]	Fc/OCH$_2$CH$_2$O/Fc	319			−/−2028		[86]
		213 (59)	313 (59)		−/−2028	100[t]	[86]

a) $E_n^{o'}$ = Formal potential of nth Fc-related redox process.
b) ΔE_p = Difference between oxidation/reduction potentials.
c) E_{pa} = Anodic peak potential of ML_n or nitrogen substituents.
d) E_{pc} = Cathodic peak potential of ML_n.
e) $\Delta E^{o'}$ = Potential difference between two Fc-related redox processes.
f) ν_{max} = Energy of IVCT band.
g) ε_{max} = Extinction coefficient of IVCT band.
h) $\Delta\nu_{1/2}$ = Full width at half-height of IVCT band.
i) Values calcd. as $(\Delta\nu_{1/2})_{theo} = (2310\nu_{max})^{1/2}$ [13].
j) CV studies: potentials vs. FcH/FcH$^+$, scan rate 100 mV s^{-1} at a glassy-carbon electrode of 1.0 mmol l^{-1} thiophene CH$_2$Cl$_2$ solutions containing 0.1 mol l^{-1} of [NBu$_4$][B(C$_6$F$_5$)$_4$] as the supporting electrolyte, 25 °C.
k) UV-vis/NIR studies in CH$_2$Cl$_2$ containing 0.1 mol l^{-1}. [NBu$_4$][B(C$_6$F$_5$)$_4$] as the supporting electrolyte, 25 °C, using an OTTLE cell [33].
l) UV-vis/NIR studies analog to k), however, propylene carbonate was used as a solvent.
m) E_{pc} = Cathodic peak potential of the carbene unit.
n) $E^{o'}$ = Formal potential NO$_2$.
o) E_{pa} = Anodic peak potential of NMe$_2$.
p) $E^{o'}$ = Formal potential of amino groups.
q) $E^{o'}$ = Formal potential of imino groups.
r) Values of di-cationic species.
s) E_{pa} = Anodic peak potential of Fc/[Fc]$^+$.
t) CV studies analog to j), however, CH$_2$Cl$_2$/CH$_3$CN (ratio 1 : 1, v/v) was used, 0 °C.

Source: Adapted from [80–86].

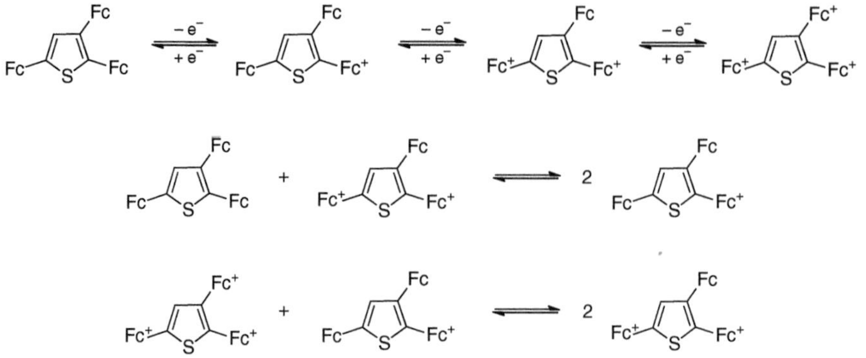

Figure 6.9 Electron transfer mechanism of 2,3,5-triferrocenyl thiophene (type **H** molecule). Source: [88]/Elsevier.

this family of molecules as class II mixed-valence systems. In contrast, during the stepwise oxidation of compounds of type **I**, however, no IVCT transition could be detected during UV–vis/NIR measurements, and hence, the *in situ* electrochemically generated ferrocenium units in the appropriate mixed-valence species communicate electrostatically (Fc/[Fc]$^+$) [84]. This differs from isostructural per-ferrocenylated furan and pyrrole compounds (Sections 6.2.3 and 6.2.4). An electron transfer mechanism of the type **H** compound based on rapid-scan time-resolved FT-IR spectroelectrochemical experiments is given in Figure 6.9 [80, 88].

In addition, on the examples of 2,5-diferrocenyl-thiophene (**40a**), -3,4-thiazol (**66**), and –*N*-(4-dimethylaminophenyl)-1*H*-pyrrole (**4e**), the influence of electrolytes with varying ion-pairing strength ([Bu$_4$N][Cl], [Bu$_4$N][PF$_6$], and [Bu$_4$N][B(C$_6$F$_5$)$_4$]) and solvents (CH$_2$Cl$_2$ (ε_r = 8.93), Me$_2$C(O) (ε_r = 20.56), MeC≡N (ε_r = 35.94), and propylene carbonate (ε_r = 64.92)) of increasing dielectric constants on the (spectro)electrochemical properties were studied in order to determine the effect of electrolyte and analyte solvation on the electron transfer interactions [29]. It was found that from the above series of electrolytes, the respective Cl$^-$-containing one showed irreversible redox processes, which is most likely initiated by a nucleophilic attack of Cl$^-$ at [Fc]$^+$ after oxidation (EC process). Hence, *in situ* UV–vis/NIR studies were carried out with [NBu$_4$][PF$_6$] and [NBu$_4$][B(C$_6$F$_5$)$_4$]. The measurements indicated that individual redox processes of the thiazol derivative **66** are barely resolved with exception of [**66**]/[B(C$_6$F$_5$)$_4$] in CH$_2$Cl$_2$. In CH$_2$Cl$_2$, a large impact on the $\Delta E^{0\prime}$ values is most prominent for strongly coupled pyrrole **4e**. In more polar solvents, the ion-dependent change in the redox splitting did not influence the coupling strength of the appropriate thiophene and pyrrole systems. Increase of ε_r resulted in a decrease of the redox splittings $\Delta E^{0\prime}$. The formation of [Fc]$^+\cdots$[PF$_6$]$^-$ pairs compensates part of the repulsive electrostatic stabilization energies by the formation of attractive electrostatic forces, reducing the overall stability of the mixed-valence compound. By increasing the solvent polarity, the formation of ion pairs becomes increasingly unlikely. The higher relative permittivity reduces the electrostatic forces. Hence, the $\Delta E^{0\prime}$ values are little for polar solvents and electrolytes [PF$_6^-$] or

[B(C$_6$F$_5$)$_4$$^-$], showing no large impact in the redox properties. As ion pair interactions with the respective [PF$_6$$^-$] anion is directing the electrochemistry for studies carried out in CH$_2$Cl$_2$, the [B(C$_6$F$_5$)$_4$$^-$] anion displays a much lower tendency to create ion pairs with the analyte. Henceforth, the electrochemical features are electrolyte dependent in CH$_2$Cl$_2$. *In situ* UV–vis/NIR studies validated the trends found in the corresponding electrochemical measurements. More polar solvents, for example, propylene carbonate, revealed a nearly electrolyte independency of the spectroscopic parameter, while solvatochromic- and ion-related changes in the spectroscopic properties are greatest pronounced for weakly coupled species. They decrease with an increase in the electron transfer coupling strength. The formation of ion pairs as well as solvent coordination to [Fc]$^+$ in the mixed-valence species resulted in a decrease of the coupling strength of the mixed-valence systems. This effect is best found in weakly coupled 2,5-diferrocenyl-3,4-thiazol. It decreases in its extent the more delocalized the ground-state of the corresponding mixed-valence system is.

The use of type **C** and **G** molecules (Figure 6.8) in an electrochemical nanogap transducer was reported in 2016 [89]. Within these studies, a new electrochemical methodology to determine the diffusive properties as a function of the oxidation state of complex electrochemically active species in a stochastic amperometric measurement was introduced. The authors indicated that by stochastic analysis, they could determine, as a function of the oxidation states of a particular redox couple, the effective diffusion coefficient and the Faradaic current produced per molecule, all in a simple experiment requiring only a mesoscopic amount of molecules in a femtoliter compartment. As a result thereof, diffusive transport is reduced for higher oxidation states and that analytes afford very high currents per molecule (15 fA) [89].

For the (ferrocenyl)thiophenes modified by tungsten Fischer carbenes (**40e,f**, Table 6.3), characteristic electrode reactions were observed for the carbene reduction and the W(CO)$_5$ oxidation [82]. The Fc groups give rise to reversible one-electron redox processes. UV–vis/NIR studies reveal metal–metal charge transfer transitions between the W(CO)$_5$ building block and the Fc substituents. The compounds could be classified as class II species according to Robin and Day [9–11].

Recently, the synthesis, structure, and (spectro)electrochemical behavior of a series of Fc-functionalized η^5-thiophene-Cr(CO)$_3$ half-sandwich (**48–50**) [85] and [Ru(η^5-ferrocenylthiophene)(η^5-C$_5$R$_5$)]$^+$ (R = H (**51a,b**) and Me (**51c,d**)) [86] sandwich compounds have also been discussed. Electrochemical studies confirmed that each Fc group could be oxidized separately with higher ΔE_{pa} values for the Fc units of the Cr(CO)$_3$ compounds (ca. 350 mV) than for the *non*-complexed species (ca. 260 mV) [85]. The potentials of the oxidation processes are shifted anodically by ca. 140 mV, confirming an electron-withdrawing characteristic of the Cr(CO)$_3$ entity. In addition, within the Fc oxidation of the compounds, the respective half-sandwich molecules decompose to give the corresponding *non*-complexed Fc-thiophenes [85]. The same goes for the respective Ru–Fc–thiophene sandwich compounds [86].

In 2012, the preparation, solid-state structure, and electrochemistry of atropisomeric 3,3′,4,4′,5,5′-hexaferrocenyl-2,2′-bithiophene (**52**) were reported [90]. Because of steric reasons, the Fc groups are out of plane tilted of the bithiophene

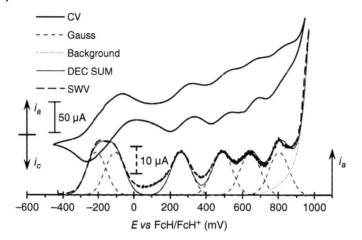

Figure 6.10 CV (scan rate: 200 mV), SWV in anisole at 100 °C (supporting electrolyte [N^{n}Bu$_4$][B(C$_6$F$_5$)$_4$] (0.1 mol l^{-1})), and deconvolution of SWV of **52**. Source: [90] Reproduced with permission of the American Chemical Society.

backbone. Because of the low solubility of the compound and the oligo-cationic species thereof, the electrochemical measurements had to be carried out in anisole at 100 °C. Five redox processes were observed at $E^{0'}$ = −165, 261, 491, 651, and 800 mV. The first wave shows a peak separation of 156 mV, indicating that this process corresponds to two redox events close together as it could be shown by deconvolution (Figure 6.10). UV–vis/NIR studies confirm that upon stepwise oxidation, the Fc and electrochemically generated [Fc]$^+$ units communicate electrostatically with each other, emphasizing that the positive charge is mainly located on the ferrocenium entities in the mixed-valence species [90].

In addition, bithiophenes, such as 3,3′-diferrocenyl-2,2′- (**53**), 3,4′-diferrocenyl-2,2′- (**54**), 4,4′-diferrocenyl-2,2′- (**55**), 5,5′-diferrocenyl-2,2′- (**56**), and 3,3′,4,4′-tetraferrocenyl-2,2′-bithiophene (**57**), were synthesized and characterized by CV, SWV, and *in situ* UV–vis/NIR studies (in CH$_2$Cl$_2$, the supporting electrolyte is [Bu$_4$N][B(3,5-(CF$_3$)$_2$-C$_6$H$_3$)$_4$]) [6]. Within this series of compounds, the Fc groups could reversibly be oxidized whereby the difference between the two redox events varies with the substitution pattern at the bithiophenes. 3,3′-Isomer **53** shows with 229 mV the highest value, as the 3,3′-substitution leads to the shortest distance between the Fc groups and hence to the strongest electrostatic interaction. In contrast, the 4,4′- (**55**) and 5,5′-isomer (**56**) exhibit $\Delta E_{1/2}$ values of 95 and 130 mV and the one in unsymmetrical **54** corresponds to 136 mV (which is not solely ascribed to the interactions between the Fc groups but also because of the different chemical environments). The CV of the appropriate tetraferrocenyl bithiophene **57** is characterized by four reversible one-electron redox events. The $\Delta E^{0'}$ values in this compound ($\Delta E_1^{0'}$ = 212 mV, $\Delta E_2^{0'}$ = 262 mV, and $\Delta E_3^{0'}$ = 243 mV) are larger, when compared with the appropriate diferrocenyl analogs (vide supra). *In situ* UV–vis/NIR experiments indicated that metal–metal interactions are mainly attributed to electrostatic properties [6]. In contrast, compounds Fc(Th)$_{1\text{-}2}$Fc

(Th = thienyl) possess electronic coupling between the iron centers when oxidized, while FcC≡C(ThC≡C)$_{1-2}$Fc indicates the absence of intramolecular electrochemical communication [91]. Also, diverse dihedral-angle-controlled 2,2′-bithiophene derivatives with end-grafted Fc ligands were designed in 2013 [92] and 2015 [93]. The electrochemical properties of these compounds specified that the interaction between the Fc groups is markedly affected by the twist in the π-conjugated systems. While the oxidized diferrocenyl compounds comprise a more flat bithiophene moiety with electronic communication between the Fc/[Fc]$^+$ entities (see above), the more twisted derivatives show no intramolecular metal–metal interaction when oxidized [91–93].

In 2018, electrochemical studies of the multi-step multi-electron redox process of tetraferrocenyl bi- and tri-thiophenes with electron-donor and electron-acceptor abilities were made on the examples of 4,4′,5,5′-tetra-ferrocenyl-2,2′-bithiophene (**58**) and 4,4″,5,5″-tetra-ferrocenyl-2,2′:5′,2″-terthiophene (**59**) [94]. The electron transfer is thereby dominated by the chain length as well as the π-conjugation length of the α-oligo-thiophene bridge (two-step two-electron reversible oxidation process vs. a four-step four-electron oxidation). After the formation of the tetra-ferrocenium cation, a further one-electron oxidation and a one-step reversible reduction occurs at the α-terthiophene bridge, proving that an increase in the number of thiophene units decreases the electronic interaction between Fc/[Fc]$^+$ but increases the electron-donor and electron-acceptor abilities within the oligo-thiophene [6, 94]. Unsymmetrically substituted quarter-thiophenes featuring two terminal Fc groups were also prepared [95–98]. The anisotropy of the chemical structure was introduced by methoxy as well as hexyl groups attached to the thiophene units. The electrochemical behavior of the compounds is reported [94–98]. A delocalization-to-localization charge transition in Fc$_2$-substituted oligo-thienylene-vinylene molecular wire molecules as a function of size by using Raman spectroscopy and electrochemistry shows that the dimer and tetramer members display a full charge delocalized mixed-valence system, while in the octamer, charge resonance disappears and the cation is localized at the connectivity center, and hence, the mixed-valence property vanishes. The hexameric compound is at the delocalized-to-localized turning point. Oxidation potentials (50–800 mV, o-DCB/acetonitrile (ratio 4 : 1, v/v), 0.1 M [Bu$_4$N][ClO$_4$] as the supporting electrolyte) decrease with the chain size, which is understandable by the increasing participation in the bridging unit in the charge resonance. The oxidation potentials become more similar for the longer wire molecules. The shorter species, however, display simultaneous two-electron extractions approving that the divalent systems are more stable than the radical cations [99].

Recently, a series of (multi)ferrocenyl fused thiophene (**60**, **61**, and **64**) [100], benzodithiophene (**62** and **63**) [100–102], and benzotrithiophene (**65**) [101, 102] compounds including their synthesis as well as their chemical and physical properties was reported (Figure 6.11). In **60–65**, the redox-active Fc groups are bridged by fused oligo-thiophene cores and hence were applied as new types of Fc-based conjugated systems for the study of oxidation processes via the mixed-valence state. Compound **65** corresponds to a three-blade propeller-shaped molecule.

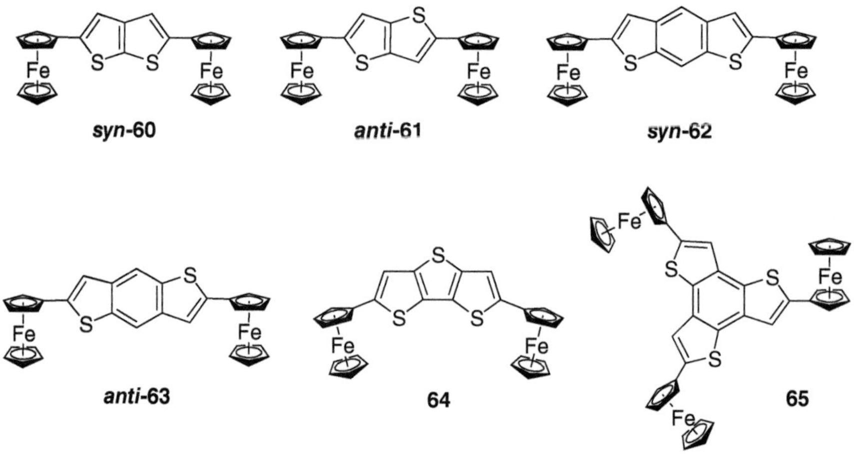

Figure 6.11 Fused Fc-thiophene, Fc-benzodithiophene, and Fc-benzotrithiophene compounds **60–65**. Source: Adapted from [100–102].

The electrochemical properties were examined by CV. For example, compounds **60–64** show reversible Fc-based two-electron oxidation waves with a potential difference between the first and second oxidation of up to 130 mV (**60**: $E_1^{0'} = 260$ mV, $E_2^{0'} = 330$ mV; **61**: $E_1^{0'} = 230$ mV, $E_2^{0'} = 360$ mV; **62** and **63**: $E_1^{0'}/E_2^{0'} = 290$ mV; and **64**: $E_1^{0'} = 250$ mV; $E_2^{0'} = 330$ mV; in CH_2Cl_2, 0.1 mol l^{-1} [Bu$_4$N][PF$_6$] as supporting electrolyte) [100]. The authors report that they are able to tune the oxidation process via a mixed-valence state by the length and resonance contribution of the fused oligo-thiophene bridge [100–102].

The profiles of **63** and **65** could be changed from a (badly) *non*-resolved two-electron oxidation to two processes by changing from [Bu$_4$N][PF$_6$] to [Bu$_4$N][B(C$_6$F$_5$)$_4$] as the supporting electrolyte [101]. In contrast to smaller counterions ([PF$_6$]$^-$), [B(C$_6$F$_5$)$_4$]$^-$ stabilizes (highly) charged species in solution and minimizes ion-pairing effects (vide supra). (The shielding of the electrostatic interactions between the Fc/[Fc]$^+$ groups is realized by ion pairing with the electrolyte's counterion. Hence, minimization of this effect leads to an increase of the observed redox potential splitting.) The NIR spectra of the chemically prepared mono-cations [**63**]$^+$ ($\nu_{max} = 7041$ cm^{-1}, $\varepsilon_{max} = 3701$ mol^{-1} cm^{-1}, $\Delta\nu_{1/2} = 2728$ cm^{-1}; $\Delta\nu_{1/2(theo)} = 4033$ cm^{-1}) and [**65**]$^+$ ($\nu_{max} = 8582$ cm^{-1}, $\varepsilon_{max} = 10951$ mol^{-1} cm^{-1}, $\Delta\nu_{1/2} = 2602$ cm^{-1}; $\Delta\nu_{1/2(theo)} = 4474$ cm^{-1}) showed typical charge transfer bands, which were rationalized in the framework of the Marcus–Hush theory [101]. The IVCT bands displayed large Γ values of 0.32 and 0.42 and solvent dependence, assigning these species to class II mixed-valence systems. It was further found that the sulfur atom strongly increases the metal–metal electronic coupling in the mixed-valence species when compared with the analogous sulfur-free molecules.

Toward the preparation and the electrochemical behavior of thiazoles, thiadiazoles, and thiazolines [6], the synthesis as well as photophysical, electrochemical, and DFT studies of a series of C_2-symmetric donor–acceptor (D–A) Fc-substituted (bis)thiazoles was reported in 2016 and 2017 by the group of Misra [103, 104]. The corresponding molecules (with the Fc groups directly bonded to the (bis)thiazole core or separated from the heterocyclic backbone by vinyl, alkynyl, phenylalkynyl, etc., connectivities) were prepared by starting from the appropriate dibromo derivatives using either the Pd-catalyzed Suzuki–Miyaura, Heck, or Sonogashira *C,C* cross-coupling reaction. In the series of these compounds, the vinyl linkage showed the best electronic communication, which was also confirmed by TD-DFT calculations at the B3LYP level [103].

2,5-Diferrocenyl-1,3,4-thiadiazole, 2,5-Fc$_2$-^{c}C$_2$N$_2$S (**66**), was prepared by the Pd-catalyzed 2-fold Negishi ferrocenylation of 2,5-Br$_2$-^{c}C$_2$N$_2$S with FcZnCl [105]. In addition, spacer units such as C≡C and C$_6$H$_4$C≡C between the heterocycle and the Fc groups could be built in by Sonogashira cross-coupling of 2,5-Br$_2$-^{c}C$_2$N$_2$S or 2,5-(C$_6$H$_4$-4′-I)$_2$-^{c}C$_2$N$_2$S with FcC≡CH (compounds **67** and **68**). In the solid state, none of the Fc ligands are coplanar with the thiadiazole core. Compounds **66**, 2,5-(FcC≡C)$_2$-^{c}C$_2$N$_2$S (**67**), and 2,5-(C$_6$H$_4$-4′-C≡CFc)$_2$-^{c}C$_2$N$_2$S (**68**) were subjected to CV, SWV, LSV, and *in situ* UV–vis/NIR measurements (Figure 6.12) [105]. It was found that for **66**, the Fc ligands showed a reversible electrochemical behavior (in CH$_2$Cl$_2$, 0.2 mol l^{-1} [Bu$_4$N][B(C$_6$F$_5$)$_4$]) at slow scan rates with i_{pc}/i_{pa} approaching unity and ΔE_p of 66 mV, while LSV confirmed two well-separated one-electron transfer processes with $E^{0\prime}$ = 190 mV for the [**66**]/[**66**]$^+$ redox couple and 340 mV for the [**66**]$^+$/[**66**]$^{2+}$ redox couple with $\Delta E^{0\prime}$ = 150 mV and K_c = 350. However, in the presence of [Bu$_4$N][PF$_6$] as supporting electrolyte, compound **66** ($E^{0\prime}$ = 220 mV) shows simultaneous oxidation of the two Fc termini. As expected, Fc oxidations for **67** and **68** occurred basically simultaneously in the presence of [Bu$_4$N][B(C$_6$F$_5$)$_4$], which is due to the greater distance of the conjugated linking units between the Fc and ^{c}C$_2$N$_2$S core.

Spectroelectrochemical studies of **66** in an OTTLE cell in CH$_2$Cl$_2$ solutions (1.0 mmol l^{-1}) and applying [Bu$_4$N][B(C$_6$F$_5$)$_4$] (0.1 mol l^{-1}) as electrolyte in the range of −200 to 1200 mV vs. Ag/AgCl confirmed two wavelength regions of absorbance changes upon stepwise oxidation (0.5 mV s^{-1}) (Figure 6.12). Strong UV–vis (560 and 690 nm) and NIR absorptions (1560 nm) appear/disappear as **66** is oxidized to [**66**]$^{2+}$, which is related to the formation of charge-localized [Fc]$^+$ species [105]. The IVCT band at 6410 cm^{-1} (ε_{max} = 311 l mol^{-1} cm^{-1}) ([**66**]$^+$) approves that **66** can be classified as a class II species according to Robin and Day [9–11].

6.2.5 Five-Membered Heterocyclic Compounds with Transition Metal Elements

Within this section, five-membered metalla-heterocycles will be discussed. The study of the role of the transition metal center with regard to electron transfer between redox-active ferrocenyls at the cycle periphery is outlined, as well as their use, for example, for olefin homo- or copolymerization. Furthermore, metalla-heterocycles pave the field into cluster chemistry. We here complete this family of compounds by adding some further examples by focusing on especially five-membered ferrocenyl-substituted metalla-cycles containing titanocene and zirconocene building blocks within the heterocycle [106, 107]. The synthesis,

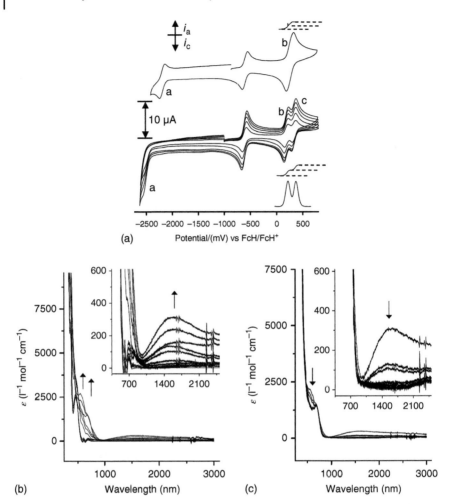

Figure 6.12 (a) Top: LSV and CV of **68** at 100 mV s^{-1}; bottom: CVs at scan rates 100, 200, 300, 400, and 500 mV s^{-1}, LSV at 1 mV s^{-1} and SWV of 0.5 mmol l^{-1} solutions of **66** in CH$_2$Cl$_2$ (0.2 mol l^{-1} [Bu$_4$N][B(C$_6$F$_5$)$_4$]) at a glassy carbon electrode (LSV current values were multiplied by 2 for clarity). (b, c) NIR spectra of **66** at increasing potentials [b: −200–550 mV; c: 550–1200 mV vs. Ag/AgCl, 25 °C, in 1.0 mmol l^{-1} CH$_2$Cl$_2$, supporting electrolyte [Bu$_4$N][B(C$_6$F$_5$)$_4$] (0.1 mol l^{-1})]. Arrows indicate the increase or decrease of absorptions. Source: [105]. Reproduced with permission of Elsevier.

(spectro)electrochemical properties, and structural features of such compounds will be highlighted. For further metalla-heterocycles until 2011, see reference [6] and references cited therein.

Five-membered metalla-heterocycles of titanium (compound **70**) or zirconium (**71**) featuring terminal Fc groups can be prepared by the alkyne exchange reaction of [Ti](η^2-Me$_3$SiC≡CSiMe$_3$) ([Ti] = Ti(η^5-C$_5$H$_5$)$_2$) with FcC≡C—C≡CFc or [Zr](py)(η^2-Me$_3$SiC≡CSiMe$_3$) ([Zr] = Zr(η^5-C$_5$H$_5$)$_2$) with FcC≡CFc as shown in Scheme 6.3 [106, 107]. The compounds are exceptional examples of diferrocenyl-

(**69** and **70**) or tetraferrocenyl small heterocycles (**71**), and the Fc units allowed for the electronic interaction between different metals to be studied. The titana ring of **70** can best be ascribed as a titana-cyclocumulene with three connected CC double bonds, the four titanium–carbon bonds are typical for Ti–C single bonds, which is characteristic for such species, i.e. [Ti](η^4-^{t}BuC$_4^t$Bu) [108]. Supercrowded **71** is set up by a zirconacyclopenta-2,4-diene core with four terminal Fc ligands [107]. However, when [Zr](py)(η^2-Me$_3$SiC≡CSiMe$_3$) was reacted with FcC≡C—C≡CFc in an analogous manner to the respective reaction of [Ti](η^2-Me$_3$SiC≡CSiMe$_3$), then a reaction mixture was obtained from which {[Zr](μ-η^1:η^2-C≡CFc)}$_2$ (**72**) could be isolated [107]. Treatment of [Zr](py)(η^2-Me$_3$SiC≡CSiMe$_3$) with FcC≡CPh produced **73** and **74**, respectively (Scheme 6.3).

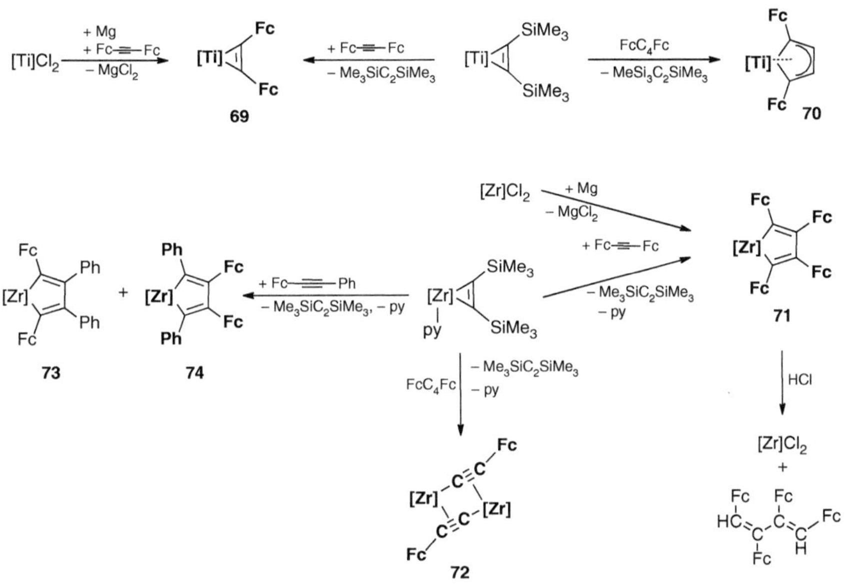

Scheme 6.3 Synthesis of compounds **69–74**. Source: Adapted from [106, 107].

The redox properties of **69–74** were studied by CV (supporting electrolyte 0.1 M [Bu$_4$N][B(C$_6$F$_5$)$_4$] in thf) (Figure 6.13). It was found that for **69** and **70**, three redox events were observed. While the processes for **69** were irreversible indicating that this molecule is not stable under oxidizing conditions, the ones for **70** are reversible and correspond to Fc oxidations as well as the metalla-cycle [106].

The first two oxidations of **70** occur at the Fc termini (Figure 6.13). Computational studies confirm structural changes after the second process. As outlined in Figure 6.13, the TiC$_4$ ring rearranges whereby the C$_4$ unit gets linear. Furthermore, it was found that the geometry optimization of [**70**]$^{2+}$ goes from the cyclic form straight into an open-chain arrangement without any barrier [106]. The third event corresponds to a simultaneous two-step one-electron redox process with spin density at iron. A proposed oxidation mechanism of **70** is depicted in Figure 6.13 [106].

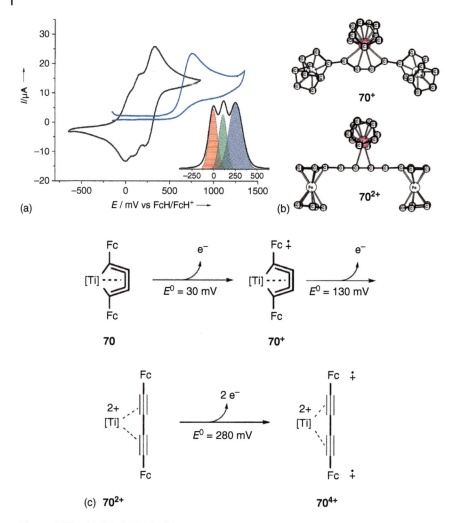

Figure 6.13 (a) CV of **70** [(25 °C, supporting electrolyte 0.1 M [Bu$_4$N][B(C$_6$F$_5$)$_4$] in thf, and scan rate 100 mV) (the inset displays the SWV), the different peak areas show a 1 : 1 : 2 ratio]. (b) BP86/TZVP-optimized structures of [**70**]$^+$ and [**70**]$^{2+}$. (c) Proposed mechanism for the oxidation of **70** to [**70**]$^{2+}$. Source: [106]. Reproduced with permission of Wiley-VCH.

Dicationic [**70**]$^{2+}$ can be considered as a combination of [Ti]$^{2+}$ and FcC≡C—C≡CFc with a reduced C,C triple bond characteristic as proven by *in situ* IR spectroscopy.

The redox properties of **71–74** were studied by CV; however, it was found that upon oxidation, decomposition of the compounds occurred to give FcC≡CH (decomposition of **72**) or butadiene C$_4$H$_2$Fc$_4$ (from **71**) [107]. Moreover, [M*](η^2-Me$_3$SiC≡CSiMe$_3$) (M* = Ti(η^5-C$_5$Me$_5$)$_2$ and Zr(η^5-C$_5$Me$_5$)$_2$) was reacted with the nitriles FcC≡N and FcC≡C—C≡N, giving unusual nitrile–nitrile C,C couplings (FcC=N) by forming 1-metalla-2,5-diaza-cyclopenta-2,4-dienes (M = Ti, **75a**; M = Zr, **75b**), while alkyne–nitrile couplings (FcC≡C—C≡N) produced 1-metalla-2-aza-cyclopentadiens (M = Ti, **76a**; M = Zr, **76b**) [109].

Nevertheless, after oxidation of these compounds, they decompose to different redox-active products.

In addition, a series of Fc-substituted Ru(CO)$_3$(η^4-ruthenoles) [110] and ferracyclopentadienes [111–113] were synthesized but not electrochemically characterized.

6.3 Ferrocenyl-Functionalized Six-Membered Heterocycles

In the following, multi-ferrocenyl-functionalized pyridine, pyrimidine, quinoxaline, and 1,3,5-triazine compounds are envisaged. However, compounds of this type are, to the best of our knowledge, only less studied. Since 2011, the Fc-substituted pyridine derivatives 1-R^1-2-R^2-3-R^3-4-R^4-5-R^5-^{c}C$_6$N (**77a**, R^1 = R^5 = Fc, R^2 – R^4 = H; **77b**, R^2 = R^4 = Fc, R^1 = R^3 = R^5 = H; **77c**, R^2 = R^5 = Fc, R^1 = R^3 = R^4 = H; **77d**, R^1 = R^3 = Fc, R^2 = R^4 = H, R^5 = Me; **77e**, R^1 = R^3 = Fc, R^2 = R^4 = H, R^5 = C$_6$H$_5$; **77f**, R^1 = R^4 = Fc, R^2 = R^3 = H, R^5 = Me; **77g**, R^1 = R^4 = Fc, R^2 = R^3 = H, R^5 = C$_6$H$_5$; **78**, R^1 = R^3 = R^5 = Fc, R^2 = R^4 = H; **79a**, R^1 = R^3 = Fc, R^2 = R^4 = FcC≡C, R^5 = CH$_3$; **79b**, R^1 = R^3 = Fc, R^2 = R^4 = FcC≡C, R^5 = C$_6$H$_5$; **79c**, R^1 = R^4 = Fc, R^2 = R^3 = FcC≡C, R^5 = C$_6$H$_5$; and **77d**, R^2 = R^4 = Fc, R^1 = R^3 = FcC≡C, R^5 = C$_6$H$_5$) were prepared either by Pd-based Negishi C,C cross-coupling protocols [114], the reaction of 3,5-dibromopyridine with FcB(OH)$_2$ by means of the Suzuki–Miyaura approach [115], treatment of FcC≡CH with RC≡N (R = Me and Ph) in refluxing toluene [116], or by the Co(η^5-C$_5$H$_5$)(CO)$_2$-mediated [2+2+2] cycloaddition of FcC≡C—C≡CFc with RC≡N [117]. The electrochemical characterization of these compounds was done by CV, SWV, and (partly) by *in situ* UV–vis/NIR spectroscopy.

Characteristic for the Fc$_2$-substituted compounds **77a–g** is that they show two reversible one-electron redox events (with exception of **77b**) in CV and SWV between $E^{0\prime}$ = 5–840 mV (this broad range is due to different measurement conditions applied) with $\Delta E^{0\prime}$ values between 40 and 180 mV [114–116]. The Fc$_3$-functionalized molecule **78** is marked by three reversible redox waves at $E^{0\prime}$ = 0, 160, and 330 mV ($\Delta E^{0\prime}$ = 160 mV, 170 mV) [114]. *In situ* UV–vis/NIR measurements were carried out on **77a,c** and **78** [114]. The mixed-valence species [**77a,c**]$^+$, [**78**]$^+$, and [**78**]$^{2+}$ showed weak IVCT absorptions at v_{max} = 6710 cm^{-1} (ε_{max} = 50 l mol^{-1} cm^{-1}, $\Delta v_{1/2}$ = 5525 cm^{-1}) for [**77a**]$^+$, 6320 cm^{-1} (ε_{max} = 550 l mol^{-1} cm^{-1}, $\Delta v_{1/2}$ = 4890 cm^{-1}) for [**77c**]$^+$, 6010 cm^{-1} (ε_{max} = 30 l mol^{-1} cm^{-1}, $\Delta v_{1/2}$ = 7515 cm^{-1}) for [**78**]$^+$, and 6290 cm^{-1} (ε_{max} = 65 l mol^{-1} cm^{-1}, $\Delta v_{1/2}$ = 7550 cm^{-1}) for [**78**]$^{2+}$, indicating that these molecules are weak coupled class II systems [114]. Theoretical calculations reveal that the electronic structures of the compounds are comparable near the Fermi level. For compounds **79a–d** featuring two Fc and two FcC≡C units bonded to the pyridine core, only $\Delta E^{0\prime}$ values have been reported (**79a**, 130 mV; **79b**, 115 mV; **79c**, 102 mV; and **79d**, 88 mV) [117].

In contrast, 2-amino-4,6-diferrocenyl pyrimidine (**80**) shows a quasi-reversible redox event at $E^{0\prime}$ = 588 mV (in CH$_2$Cl$_2$, 0.2 mol l^{-1} [Bu$_4$N][ClO$_4$] as supporting electrolyte, 25 °C). The authors point out that the insolubility of the oxidation product prevented the second step of electron transfer [118].

2,3-Diferrocenyl quinoxaline (**81**) was accessible by the reaction of 2,3-diaminobenzene and diferrocenylethane-1,2-dione [119]. This compound was applied as a highly selective probe for calorimetric and redox sensing of Hg(II). The redox properties were examined by CV, LSV, and OSWV in acetonitrile containing 0.15 M [Bu$_4$N][ClO$_4$] as supporting electrolyte between 0 and 900 mV. Two reversible one-electron redox events at $E_1^{0'} = 470$ mV and $E_2^{0'} = 580$ V were observed, demonstrating the existence of a weak metal–metal interaction; however, spectroelectrochemical studies have not been undertaken.

Moreover, the synthesis, characterization, and electrochemical properties of 2,4,6-Fc$_3$-1,3,5-triazine (**82a**) were discussed [114]. It shows three well-defined reversible redox waves at $E_1^{0'} = 115$ mV, $E_2^{0'} = 255$ mV, and $E_3^{0'} = 440$ mV with $\Delta E^{0'} = 140$ and 185 mV, demonstrating that each of the Fc groups can be separately oxidized. *In situ* UV–vis/NIR studies validate that the partly oxidized mixed-valence systems show IVCT absorptions, approving that electrostatic interactions between Fc/[Fc]$^+$ exist [114]. Other 1,3,5-tri-ferrocenyl triazine compounds in which the Fc ligands carry additional Br or PAr$_2$ (Ar = C$_6$H$_5$ and C$_6$F$_5$) coordinating groups were published in a series of papers in 2020 by Hey-Hawkins and coworkers [120–122]. Their complexation behavior toward BH$_3$ and Cu(I), Ag(I) and Au(I) chlorides was studied as well. Because of the presence of three redox-active Fc groups, the electrochemical properties of the compounds were also determined, which is discussed on the examples of 2,4,6-Br$_3$- (**82b**) and 2,4,6-(1-diphenylphosphanyl-1′-ferrocenylene)$_3$-1,3,5-triazine (**82c** and its adducts with BH$_3$ (**83**) and AuCl (**84**)) in the following [120]. The characteristic is that **82b** shows reversible redox activity in weakly coordinating electrolytes including [B(C$_6$F$_5$)$_4$]$^-$ and [B(3,5-(CF$_3$)$_2$-C$_6$H$_3$)$_4$]$^-$. The first iron-centered oxidation potential is observed at $E_1^{0'} = 386$ mV with $\Delta E_p = 127$ mV. Nevertheless, **82c** is irreversibly oxidized in supporting electrolytes such as [Bu$_4$N][BF$_4$] and [Bu$_4$N][B(C$_6$F$_5$)$_4$]. This is in accordance with observations made earlier for Fc-substituted phosphanes and phosphanes in general [123, 124]. In contrast, the BH$_3$ adduct **83** shows three reversible oxidations in the weakly coordinating electrolyte [Bu$_4$N][B(C$_6$F$_5$)$_4$]. The electrochemical feature of the respective AuCl complex **84** as well as the Cu(I) and Ag(I) species display again an irreversible first oxidation between 250 and 350 mV, followed by further oxidations. They show a temperature-dependent oxidation performance linked to an EC mechanism. In addition, adducts **82c**-(BH$_3$)$_3$ and **82c**-(Au(I))$_3$ display three reversible one-electron redox events (first iron-centered oxidation $E_1^{0'} = 421$ mV with $\Delta E_p = 100$ mV for **82c**-(BH$_3$)$_3$ and $E_1^{0'} = 535$ mV with $\Delta E_p = 108$ mV for **82c**-(Au(I))$_3$). The redox-responsive nature of the molecules was exploited in the catalytic ring closing isomerization of *N*-(2-propyn-1-yl) benzamide [120–122].

6.4 Conclusion and Outlook

This chapter refers to the synthesis and detailed electro- (CV, LSV, SWV) and spectroelectrochemical (UV–vis/NIR, IR) features of especially (multi) ferrocenyl-substituted five- and six-membered heterocyclic compounds featuring Group 13–16 main group elements or d-block transition metal fragments such as titanocene or zirconocene. Where appropriate, the application potential of the respective compounds is discussed. The metal–metal interaction between Fc/[Fc]$^+$ (Fc = Fe(η^5-C$_5$H$_4$)(η^5-C$_5$H$_5$)) was proven by UV–vis/NIR measurements, revealing significant differences in the IVCT transitions. More or less, the di-, tri-, and tetraferrocenyl-functionalized compounds, with exception of 2,3,4,5-Fc$_4$-^{c}C$_4$S (resembling to a class I species according to Robin and Day [9–11]), could be classified as weakly to strong coupled class II systems. In addition, these compounds show a linear relationship between the $\Delta E^{0\prime}$ values and the oscillator strength f of the IVCT absorptions as predicted by theoretical hypothesis for a series of compounds, for example, 2,5-Fc$_2$-^{c}C$_4$H$_2$N(C$_6$H$_4$-4-R$^\prime$), 2,5-Fc$_2$-^{c}C$_4$H$_2$E and 2,3,4,5-Fc$_4$-^{c}C$_4$E (E = O, S, NR; R = Me, Ph; and R$^\prime$ = electron-withdrawing, electron-donating groups, H) with similar geometries and consequently similar electronic properties [6, 7, 18, 25, 40, 45]. This relation opens the possibility to estimate r_{ab} (effective electron transfer distance), which is otherwise difficult to be obtained experimentally. In general, more electron-rich molecules exhibit larger $\Delta E^{0\prime}$ values, which can be ascribed to a greater tendency of the Fc iron centers interacting with each other in the mixed-valence oxidation state (Fc/[Fc]$^+$). Furthermore, the electrostatic interactions between the redox-active Fc/[Fc]$^+$ groups in the respective bi- and multi-ferrocenyl-functionalized compounds is discussed in detail, evolving in finding consistent models. Also, inter- and intramolecular impacts on the electrostatic properties and the investigation of different ion-pairing capabilities of several electrolytes including [Cl]$^-$, [PF$_6$]$^-$, [ClO$_4$]$^-$, [B(C$_6$F$_5$)$_4$]$^-$, and [B(3,5-(CF$_3$)$_2$-C$_6$H$_3$] in order to receive information on electrostatic repulsion energies are emphasized [6, 7, 18, 26].

It would be interesting in many ways to see if the uses of the above-discussed bi- and multi-ferrocenyl-functionalized heterocyclic compounds may achieve a comparable state as carbon-only-bridged homo- and hetero-substituted transition metal complexes [8, 28, 42, 125, 126]. It remains to be seen if the herein discussed compounds will play a significant role as, for example, molecular wires and conducting polymers and in a broader sense allow us to design electroactive materials, in general.

Acknowledgment

We gratefully acknowledge Drs. Dominique Gottwald, Alexander Hildebrandt, Marcus Korb, Ulrike Pfaff, Matthäus Speck, Dipl.-Chem. Steve Lehrich, Frank Strehler and M. Sc. Xianming Liu, Julia Mahrholdt, Qing Yuang, and Sebastian Walz for their excellent, creative, outstanding, and very fruitful experimental contributions to this chapter. It was a pleasure with you to work in a team for the past years! We would also like to acknowledge the generous financial support coming from the Deutsche Forschungsgemeinschaft (DFG) and the Fonds der Chemischen Industrie (FCI).

References

1 Döberreiner, J.W. (1832). Über die medicinische und chemische Anwendung und die vorteilhafte Darstellung der Ameinsensäure. *Ann. Pharm.* 3: 141–146.
2 Fischer, E. (1899). Synthese in der Puringruppe. *Ber. Dtsch. Chem. Ges.* 32: 435–504.
3 Runge, F.F. (1834). Über einige Produkte der Steinkohlendestillation. *Ann. Phys.* 107: 65–80.
4 Namsheer, K. and Chandra, S.R. (2021). Conducting polymers: a comprehensive review on recent advances in synthesis, properties and applications. *RSC Adv.* 11: 5659–5697.
5 Santiago, M.G. and Low, P.J. (2016). Molecular electronics: history and fundamentals. *Aust. J. Chem.* 69: 244–253.
6 Hildebrandt, A., Pfaff, U., and Lang, H. (2011). 5-Membered heterocycles with directly-bonded sandwich and half-sandwich termini as multi-redox systems: synthesis, reactivity, electrochemistry, structure and bonding. *Rev. Inorg. Chem.* 31: 111–141.
7 Hildebrandt, A. and Lang, H. (2013). (Multi)ferrocenyl five-membered heterocycles: excellent connecting units for electron transfer studies. *Organometallics* 32: 5640–5653. (review).
8 Milan, D.C., Vezzoli, A., Planje, I.J., and Low, P.J. (2018). Metal bis(acetylide) complex molecular wires: concepts and design strategies. *Dalton Trans.* 47: 14125–14138.
9 Robin, M.B. and Day, P. (1967). Mixed valence chemistry: a survey and classification. *Adv. Inorg. Radiochem.* 10: 247–422.
10 Day, P. (1969). Cooperative effects in the electronic spectra of inorganic solids. *Inorg. Chim. Acta Rev.* 3: 81–97.
11 Day, P., Hush, N.S., and Clark, R.J. (2008). Mixed valence: origins and developments. *Phil. Trans. R. Soc., A* 366: 5–14.
12 Hush, N.S. (1967). Metal-ligands and metal-metal coupling elements. *Prog. Inorg. Chem.* 8: 391–444.
13 Hush, N.S. (1968). Homogeneous and heterogeneous optical and thermal electron transfer. *Electrochim. Acta* 13: 1005–1023.
14 Allen, G.C. and Hush, N.S. (1967). Intervalence charge transfer and electron exchange studies of dinuclear ruthenium complexes. *Prog. Inorg. Chem.* 8: 357–389.
15 Brunschwig, B.S. and Sutin, N. (1999). Energy surfaces, reorganization energies, and coupling elements in electron transfer. *Coord. Chem. Rev.* 187: 233–254.
16 Brunschwig, B., Creutz, S., Sutin, C., and N. (2002). Optical transitions of symmetrical mixed-valence systems in the Class II–III transition regime. *Chem. Soc. Rev.* 31: 168–184.
17 D'Alessandro, D.M. and Keene, F.R. (2006). Current trends and future challenges in the experimental, theoretical and computational analysis of intervalence charge transfer (IVCT) transitions. *Chem. Soc. Rev.* 35: 424–440.

18 Hildebrandt, A., Miesel, D., and Lang, H. (2018). Electrostatic interactions within mixed-valent compounds. *Coord. Chem. Rev.* 371: 56–66.

19 Creutz, C. and Taube, C.H. (1969). Direct approach to measuring the Franck-Condon barrier to electron transfer between metal ions. *J. Am. Chem. Soc.* 91: 3988–3989.

20 Cowan, D.O. and Kaufman, F. (1970). The organic solid state. electron transfer in a mixed valence salt of biferrocene. *J. Am. Chem. Soc.* 92: 219–220.

21 Levanda, C., Bechgaard, K., and Cowan, D.O. (1976). Mixed valence cations. Chemistry of π-bridged analogues of biferrocene and biferrocenylene. *J. Org. Chem.* 41: 2700–2704.

22 LeVanda, C., Bechgaard, K., Cowan, D.O. et al. (1976). Bis(fulvalene)diiron, its mono- and dications. Intramolecular exchange interactions in a rigid system. *J. Am. Chem. Soc.* 98: 3181–3187.

23 Breuer, R. and Schmittel, M. (2013). Unsymmetrically substituted 1,1′-biferrocenylenes maintain class III mixed-valence character. *Organometallics* 32: 5980–5987.

24 Santi, S., Bisello, A., Cardena, R., and Donoli, A. (2015). Key multi(ferrocenyl) complexes in the interplay between electronic coupling and electrostatic interaction. *Dalton Trans.* 44: 5234–5257.

25 Miesel, D., Hildebrandt, A., and Lang, H. (2018). Molecular electrochemistry of multi-redox functionalized 5-membered heterocycles. *Curr. Opin. Electrochem.* 8: 39–44.

26 Hildebrandt, A., Miesel, D., Yuan, Q. et al. (2019). Anion and solvent dependency of the electronic coupling strength in mixed-valent class II systems. *Dalton Trans.* 48: 13162–13168.

27 Astruc, D. (2017). Why is Ferrocene so exceptional? *Eur. J. Inorg. Chem.* 6–29.

28 Bildstein, B. (2000). Cationic and neutral cumulene sp-carbon chains with ferrocene termini. *Coord. Chem. Rev.* 206-207: 369–394.

29 Ohrenberg, C. and Geiger, W.E. (2000). Electrochemistry in benzotrifluoride: redox studies in a "noncoordinating" solvent capable of bridging the organic and fluorous phases. *Inorg. Chem.* 39: 2948–2950.

30 Camire, N., Mueller-Westerhoff, T.U., and Geiger, W.E. (2001). Improved electrochemistry of multi-ferrocenyl compounds: investigation of biferrocene, terferrocene, bis(fulvalene)diiron and diferrocenylethane in dichloromethane using [NBu$_4$][B(C$_6$F$_5$)$_4$] as supporting electrolyte. *J. Organomet. Chem.* 637–639: 823–826.

31 Barrière, F. and Geiger, W.E. (2006). Use of weakly coordinating anions to develop an integrated approach to the tuning of $\Delta E_{1/2}$ values by medium effects. *J. Am. Chem. Soc.* 128: 3980–3989.

32 Geiger, W.E. and Barrière, F. (2010). Organometallic electrochemistry based on electrolytes containing weakly-coordinating fluoroarylborate anions. *Acc. Chem. Res.* 43: 1030–1039.

33 Krejcik, M., Danek, M., and Hartl, F. (1991). Simple construction of an infrared optically transparent thin-layer electrochemical cell: Applications to the redox reactions of ferrocene, Mn$_2$(CO)$_{10}$ and Mn(CO)$_3$(3,5-di-*t*-butylcatecholate). *J. Electroanal. Chem.* 317: 179–187.

34 Gritzner, G. and Kuta, J. (1984). Recommendations on reporting electrode potentials in nonaqueous solvents. *Pure Appl. Chem.* 56: 461–466.

35 Noviandri, I., Brown, K.N., Fleming, D.S. et al. (1999). The decamethylferrocenium/decamethylferrocene redox couple: a superior redox standard to the ferrocenium/ferrocene redox couple for studying solvent effects on the thermodynamics of electron transfer. *J. Phys. Chem., B* 103: 6713–6722.

36 Chen, J., Murillo Parra, D.A., Lalancette, R.A., and Jäkle, F. (2015). Redox-switchable chiral anions and cations based on heteroatom-fused biferrocenes. *Organometallics* 34: 4323–4330.

37 Wrackmeyer, B., Kenner-Hofmann, B.H., Milius, W. et al. (2006). Ferrocenylethynyltin compounds – characterization and reactivity towards triethylborane. *Eur. J. Inorg. Chem.* 101-108.

38 Lehrich, S.W., Hildebrandt, A., Rüffer, T. et al. (2014). Synthesis, characterization, electrochemistry, and computational studies of ferrocenyl-substituted siloles. *Organometallics* 33: 4836–4845.

39 Miesel, D., Hildebrandt, A., Korb, M. et al. (2013). Synthesis and (spectro)electrochemical behavior of 2,5-diferrocenyl-1-phenyl-1*H*-phosphole. *Organometallics* 32: 2993–3002.

40 Hildebrandt, A., Schaarschmidt, D., Claus, R., and Lang, H. (2011). Influence of electron delocalization in heterocyclic core systems on the electrochemical communication in 2,5-di- and 2,3,4,5-tetraferrocenyl thiophenes, furans and pyrroles. *Inorg. Chem.* 50: 10623–10632.

41 Li, Y., Josowicz, M., and Tolbert, L.M. (2010). Diferrocenyl molecular wires. The role of heteroatom linkers. *J. Am. Chem. Soc.* 132: 10374–10382.

42 Ward, M.D. (1995). Metal-metal interactions in binuclear complexes exhibiting mixed valency; molecular wires and switches. *Chem. Soc. Rev.* 24: 121–134.

43 Tour, J.M. (2000). Molecular electronics. Synthesis and testing of components. *Acc. Chem. Res.* 33: 791–804.

44 Chen, F. and Tao, N.J. (2009). Electron transport in single molecules: from benzene to graphene. *Acc. Chem. Res.* 42: 429–438.

45 Hildebrand, A. and Lang, H. (2011). Influencing the electronic interaction in diferrocenyl-1-phenyl-1*H*-pyrroles. *Dalton Trans.* 40: 11831–11837.

46 Hu, Y.Q., Han, L.M., Zhu, N. et al. (2013). Synthesis of 2,5-diferrocenyl five-membered heterocyclic compounds and their electrochemistry. *J. Coord. Chem.* 66: 3481–3497.

47 Pfaff, U., Korb, M., and Lang, H. (2016). Crystal structure of 3-ferrocenyl-*N*-phenylpyrrole [Fe(η^5-C$_5$H$_4{}^c$C$_4$H$_3$NPh)(η^5-C$_5$H$_5$)]. *Acta Crystallogr., Sect. A* E72: 92–95.

48 Lehrich, S.W., Hildebrandt, A., Korb, M., and Lang, H. (2015). Electronic modification of redox active ferrocenyl termini and their influence on the electron transfer properties of 2,5-diferrocenyl-*N*-phenyl-1*H*-pyrroles. *J. Organomet. Chem.* 792: 37–45.

49 Pfaff, U., Hildebrandt, A., Schaarschmidt, D. et al. (2013). Molecular wires using (oligo)pyrroles as connecting units: an electron transfer study. *Organometallics* 32: 6106–6117.

50 Pfaff, U., Hildebrandt, A., Korb, M., and Lang, H. (2015). The influence of an ethynylspacer on the electronic properties in 2,5-ferrocenyl-substituted heterocycles. *Polyhedron* 86: 2–9.

51 Goetsch, W.R., Solntsev, P.V., van Stappen, C. et al. (2014). Electron-transfer processes in 3,4-diferrocenylpyrroles: insight into a missing piece of the polyferrocenyl-containing pyrroles family. *Organometallics* 33: 145–157.

52 Korb, M., Pfaff, U., Hildebrandt, A. et al. (2014). 3,4-Ferrocenyl-functionalized pyrroles: synthesis, structure and (spectro)electrochemical studies. *Eur. J. Inorg. Chem.* 1051-1061.

53 Hildebrandt, A., Schaarschmidt, D., and Lang, H. (2011). Electronically Intercommunicating iron centers in di- and tetraferrocenyl pyrroles. *Organometallics* 30: 556–563.

54 Didukh, N.O., Zatsikha, Y.V., Rohde, G.T. et al. (2016). NIR absorbing diferrocene-containing meso-cyano-BODIPY with a UV-Vis-NIR spectrum remarkably close to that of magnesium tetracyanotetraferrocenyltetraazaporphyrin. *Chem. Commun.* 52: 11563–11566.

55 Zatsikha, Y.V., Holstrom, C.D., Chanawanno, K. et al. (2017). Observation of the strong electronic coupling in near-infrared-absorbing tetraferrocene *aza*-dipyrromethene and *aza*-BODIPY with direct ferrocene–α- and ferrocene–β-pyrrole bonds: toward molecular machinery with four-bit information storage capacity. *Inorg. Chem.* 56: 991–1000.

56 Ziegler, C.J., Chanawanno, K., Hasheminsasab, A. et al. (2014). Synthesis, redox properties, and electronic coupling in the diferrocene aza-dipyrromethene and azaBODIPY donor–acceptor dyad with direct ferrocene–α-pyrrole bond. *Inorg. Chem.* 53: 4751–4755.

57 Vecchi, A., Gallani, P., Floris, B. et al. (2015). Metallocenes meet porphyrinoids: consequences of a "fusion". *Coord. Chem. Rev.* 291: 95–171.

58 Lehrich, S.W., Mahrholdt, J., Korb, M. et al. (2020). Synthesis, characterization and electrochemistry of diferrocenyl-β-diketones, -diketonates and pyrazoles. *Molecules* 25: 4476–4503.

59 Zhuo, J.B., Li, H.D., Lin, C.X. et al. (2014). Ferrocene-based sulfonyl dihydropyrazole derivatives: synthesis, structure, electrochemistry and effect on thermal decomposition of NH_4ClO_4. *J. Mol. Struct.* 1067: 112–119.

60 Li, H.D., Ma, Z.H., Yang, K. et al. (2012). Synthesis, crystal structure and redox properties of dihydropyrazole-bridged ferrocene-based derivatives. *J. Mol. Struct.* 1024: 40–46.

61 Wei, B., Gao, Y., Lin, C.X. et al. (2011). Synthesis, structure and electrochemical behaviour of 2,2-diferrocenylpropane-substituted dihydropyrazole derivatives. *J. Organomet. Chem.* 696: 1574–1578.

62 Otón, F., del Carmen González, M., Espinosa, A. et al. (2012). Synthesis, structural characterization, and sensing properties of clickable unsymmetrical 1,1′-disubstituted ferrocene–triazole derivatives. *Organometallics* 31: 2085–2096.

63 Romero, T., Orenes, R.A., Tárraga, A., and Molina, P. (2013). Preparation, structural characterization, electrochemistry, and sensing properties toward anions and cations of ferrocene-triazole derivatives. *Organometallics* 32: 5740–5753.

64 Ganesh, V., Sudhir, V.S., Kundu, T., and Chandrasekaran, S. (2011). 10 Years of click chemistry: synthesis and applications of ferrocene-derived triazoles. *Chem. Asian J.* 6: 2670–2694.

65 Manck, S., Röger, M., van der Meer, M., and Sarkar, B. (2017). Heterotri- and heteropentanuclear copper(I)-ferrocenyl complexes assembled through a "cick" strategy: a structural, electrochemical, and spectroelectrochemical investigation. *Eur. J. Inorg. Chem.* 477-482.

66 Solntsev, P.V., Dudkin, S.V., Sabin, J.R., and Nemykin, V.N. (2011). Electronic communications in (Z)-bis(ferrocenyl)ethylenes with electron-withdrawing substituents. *Organometallics* 30: 3037–3046.

67 Hildebrandt, A., Lehrich, S.W., Schaarschmidt, D. et al. (2012). Ferrocenyl maleimides - synthesis, (spectro-)Electrochemistry, and solvatochromism. *Eur. J. Inorg. Chem.* 1114-1121.

68 Märkl, G. and Potthast, R. (1967). A simple synthesis of phospholes. *Angew. Chem. Int. Ed.* 6: 86.

69 Miesel, D., Hildebrandt, A., Korb, M. et al. (2015). Influence of P-bonded bulky substituents on electronic interactions in ferrocenyl substituted phospholes. *Chem. A Eur. J.* 21: 11545–11559.

70 Miesel, D., Hildebrandt, A., Korb, M. et al. (2015). Transition-metal carbonyl complexes of 2,5-diferrocenyl-1-phenyl-1H-phosphole. *Organometallics* 34: 4293–4304.

71 Miesel, D., Hildebrandt, A., Korb, M., and Lang, H. (2016). Electronic interactions in gold(I) complexes of 2,5-diferrocenyl-1-phenyl-1H-phosphole. *J. Organomet. Chem.* 803: 104–110.

72 Keglevich, G., Böcskei, Z., Keserü, G.M. et al. (1997). 1-(2,4,6-Tri-tert-butylphenyl)-3-methylphosphole: a phosphole with a significantly flattened phosphorus pyramid having pronounced characteristics of aromaticity. *J. Am. Chem. Soc.* 119: 5095–5099.

73 Bird, C.W. (1985). A new aromaticity index and its application to five-membered ring heterocycle. *Tetrahedron* 41: 1409–1414.

74 Egan, W., Tang, R., Zon, G., and Mislew, K. (1970). The low barrier to pyramidal inversion in phospholes. A measure of aromaticity. *J. Am. Chem. Soc.* 92: 1442–1444.

75 Hildebrandt, A. (2011) (Spektro-)Elektrochemisches Verhalten Metallocenyl-funktionalisierter Fünfring-Heterocyclen. Ph.D. thesis, TU Chemnitz.

76 Molina, P., Tárraga, A., Curiel, D., and Velasco, M.D. (2001). Characterization and electrochemical study of bis(ferrocenes) with a furan spacer and ferrocenophanes prepared from α-bromoacetyl substituted ferrocenes. *J. Organomet. Chem.* 637-639: 258–265.

77 Gidron, O., Diskin-Posner, Y., and Bendikov, M. (2013). High charge delocalization and conjugation in oligofuran molecular wires. *Chem. Eur. J.* 19: 13140–13150.

78 Pfaff, U., Hildebrandt, A., Korb, M., and Lang, H. (2015). The influence of an ethynyl spacer on the electronic properties in 2,5-ferrocenyl-substituted heterocycles. *Polyhedron* 86: 2–9.

79 Tauchman, J., Trnka, J., Císařová, I., and Štěpnička, P. (2010). Synthesis, crystal structures and electrochemistry of ferrocenyl-substituted 1,3,4-oxadiazoles. *Collect. Czech. Chem. Commun.* 75: 1023–1040.

80 Speck, J., Claus, R., Hildebrandt, A. et al. (2012). Electron transfer studies on ferrocenylthiophenes: synthesis, properties, and electrochemistry. *Organometallics* 31: 6373–6380.

81 Speck, J., Korb, M., Rüffer, T. et al. (2014). Substituent influence on charge transfer interactions in α,α′-diferrocenylthiophenes. *Organometallics* 33: 4813–4823.

82 van der Westhuizen, B., Speck, J.M., Korb, M. et al. (2014). (Spectro)electrochemical investigations on (ferrocenyl)thiophenes modified by tungsten Fischer carbenes. *J. Organomet. Chem.* 772-773: 18–26.

83 Speck, J.M., Korb, M., Schade, A. et al. (2015). Ferrocenes bridged by ethylenediamino thiophene: varying charge transfer properties in a series of 3,4-di-*N*-substituted 2,5-diferrocenyl thiophenes. *Organometallics* 34: 3788–3798.

84 Hildebrandt, A., Rüffer, T., Erasmus, E. et al. (2010). A star-shaped supercrowded 2,3,4,5-tetraferrocenylthiophene: synthesis, solid-state structure and electrochemistry. *Organometallics* 29: 4900–4905.

85 Speck, J.M., Korb, M., Hildebrandt, A., and Lang, H. (2018). Ferrocenyl-functionalized η^5-thiophene Cr(CO)$_3$ half-sandwich compounds. *Eur. J. Inorg. Chem.* 4566–4572.

86 Speck, J.M., Korb, M., Hildebrandt, A., and Lang, H. (2019). Synthesis and electrochemical investigations of [Ru(η^5-ferrocenyl-thiophene)(η^5-C$_5$R$_5$)]$^+$ sandwich compounds. *Eur. J. Inorg. Chem.* 2419–2429.

87 Mathur, P., Singh, K., Chatterjee, S. et al. (2010). Metal carbonyl-promoted reactions of ferrocenyl acetylene with sulphur to form thiophene, dithiine, thioketone and vinylthioketone derivatives. *J. Organomet. Chem.* 695: 950–954.

88 Jin, B., Tao, F., and Leo, P. (2008). Rapid-scan time-resolved FT-IR spectroelectrochemistry – study on the electron transfer of ferrocene-substituted thiophenes. *J. Electroanal. Chem.* 624: 179–185.

89 Mathwig, K., Zafarani, H.R., Speck, J.M. et al. (2016). Potential-dependent stochastic amperometry of multiferrocenylthiophenes in an electrochemical nanogap transducer. *J. Phys. Chem. C* 120: 23262–23267.

90 Speck, J.M., Schaarschmidt, D., and Lang, H. (2012). Atropisomeric 3,3′,4,4′,5,5′-hexaferrocenyl-2,2′-bithiophene: synthesis, solid-state structure, and electrochemistry. *Organometallics* 31: 1975–1982.

91 Roy, S.S. and Patra, S.K. (2019). Synthesis and characterization of diferrocenyl conjugates: varying π-conjugated bridging ligands and its consequence on electrochemical communication. *Eur. J. Inorg. Chem.* 2193–2201.

92 Sato, M., Arita, S., Kawajiri, K., and Isayama, A. (2013). Oxidation of dihedral-angle-controlled 2,2′-bithiophenes with terminal ferrocenyl groups. *Chem. Lett.* 42: 1571–1573.

93 Sato, M., Arita, S., and Kawajiri, K. (2015). One-electron-oxidized states of dihedral-angle-controlled 2,2′-bithiophenes with terminal ferrocenyl groups. *Bull. Chem. Soc. Jpn.* 88: 262–270.

94 Muraoka, H., Ozawa, K., and Ogawa, S. (2018). Electrochemical studies of the multi-step multi-electron redox process of tetraferrocenyloligothiophenes with electron donor and acceptor abilities. *Heteroat. Chem* 29: 21455–21462.

95 Sato, M., Fukui, K., Sakamoto, M. et al. (2001). Charge transfer between two ferrocene groups linked by oligothiophene. *Thin Solid Films* 393: 210–216.

96 Sato, M. and Fukui, K. (2007). Oxidized states of methoxy- and hexyl-oligothiophenes with ferrocenyl groups. *Synth. Met.* 157: 619–626.

97 Sato, M. and Kamine, H. (2009). Oxidation of unsymmetrically substituted quaterthiophene with two terminal ferrocenyl groups. *Chem. Lett.* 38: 924–925.

98 Sato, M., Kamine, H., and Kato, T. (2010). Oxidation states of unsymmetrically substituted quaterthiophenes with two terminal ferrocenyl groups. *Bull. Chem. Soc. Jpn.* 83: 1539–1544.

99 González, S.R., Delgado, M.C.R., Caballero, R. et al. (2012). Delocalization-to-localization charge transition in diferrocenyl-oligothienylene-vinylene molecular wires as a function of the size by Raman spectroscopy. *J. Am. Chem. Soc.* 134: 5675–5681.

100 Muraoka, H., Watanabe, Y., Takahashi, A. et al. (2014). Synthesis of new types of ferrocene dimers bridged by a fused oligothiophene spacer and study of their electrochemical oxidation process via a mixed-valence state. *Heteroat. Chem* 25: 473–480.

101 Rossi, S., Bisello, A., Cardena, R., and Santi, S. (2018). Testing the conjugative properties of benzodithiophene and benzotrithiophene in charge transfer multi(ferrocenyl) systems. *Organometallics* 37: 4242–4249.

102 Rossi, S., Bisello, A., Cardena, R. et al. (2017). Benzodithiophene and benzotrithiophene as π cores for two- and three-blade propeller-shaped ferrocenyl-based conjugated systems. *Eur. J. Org. Chem.* 5966–5974.

103 Maragani, R., Gautam, P., Mobin, S.M., and Misra, R. (2016). C_2-Symmetric ferrocenyl bisthiazoles: synthesis, photophysical, electrochemical and DFT studies. *Dalton Trans.* 45: 4802–4809.

104 Maragani, R., Bijesh, S., Sharma, R., and Misra, R. (2017). C_2-Symmetric donor-acceptor bis(thiazole)s: synthesis and photophysical, electrochemical and computational studies. *Asian J. Org. Chem.* 6: 1408–1414.

105 Hildebrandt, A., Schaarschmidt, D., van As, L. et al. (2011). Diferrocenes containing thiadiazole connectivities. *Inorg. Chim. Acta* 374: 112–118.

106 Kaleta, K., Hildebrandt, A., Strehler, F. et al. (2011). Ferrocenyl-substituted metallacycles of titanocenes – a new family of attractive oligocyclopentadienyl complexes with promising properties. *Angew. Chem. Int. Ed.* 50: 11248–11252.

107 Kaleta, K., Strehler, F., Hildebrandt, A. et al. (2012). Synthesis and characterization of multiferrocenyl-substituted group 4 metallocene complexes. *Chem. Eur. J.* 18: 12672–12680.

108 Burlakov, V.V., Ohff, A., Lefeber, C. et al. (1995). The first titanacyclic five-membered cumulene. Synthesis, structure, and reactivity. *Chem. Ber.* 128: 967–971.

109 Becker, L., Strehler, F., Korb, M. et al. (2014). Unusual nitrile-nitrile-alkyne coupling of Fc-C≡N and Fc-C≡C-C≡N. *Chem. Eur. J.* 20: 3061–3068.

110 Mathur, P., Rai, D.K., Joshi, R.K. et al. (2014). Synthesis of novel allene-coordinated, phosphido-bridged Ru_2Pt clusters involving enyne to allene transformation. *Organometallics* 33: 3857–3866.

111 Mathur, P., Tauqeer, M., Ji, R.S. et al. (2014). Some unusual reactions of metal carbonyls with (Z)-1-ferrocenyltelluro-1-ferrocenyl-4-ferrocenyl-1-buten-3-yne. *RSC Adv.* 4: 6878–6885.

112 Mathur, P., Ji, R.S., Boodida, S. et al. (2010). Photochemical reactions of 1-ferrocenyl-4-phenyl-1,3-butadiyne with $Fe(CO)_5$ and CO. *J. Organomet. Chem.* 695: 1986–1992.

113 Xie, R.J., Han, L.M., Zhu, N. et al. (2012). Synthesis of ortho-diferrocenyl-benzene with cobalt clusters as reaction precursors: crystal structure and electrochemical properties. *Polyhedron* 38: 7–14.

114 Pfaff, U., Hildebrandt, A., Schaarschmidt, D. et al. (2012). Di- and triferrocenyl (hetero)aromatics: synthesis, characterization, (spectro-)electrochemistry and calculations. *Organometallics* 31: 6761–6771.

115 Wright, J.R., Shaffer, K.J., McAdam, C.J., and Crowley, J.D. (2012). 3,5-Diferrocenylpyridine: synthesis characterisation, palladium(II) dichloride complex and electrochemistry. *Polyhedron* 36: 73–78.

116 Wang, Y.Q., Han, L.M., Suo, Q.L. et al. (2013). Synthesis, structure and electrochemistry of new diferrocenyl pyridine derivatives. *Polyhedron* 54: 221–227.

117 Wang, Y.Q., Han, L.M., Suo, Q.L. et al. (2013). Synthesis and structural characterization of new multi-ferrocenyl pyridine derivatives. *Polyhedron* 71: 42–46.

118 Zou, Y., Zhang, Q., Hossain, A.M.S. et al. (2012). Synthesis, crystal structures, electrochemistry and nonlinear optical properties of a novel (D–A–D) biferrocenyl derivative: 2-amino-4,6-diferrocenylpyrimidine. *J. Organomet. Chem.* 720: 66–72.

119 Zapata, F., Caballero, A., Molina, P., and Tarraga, A. (2010). A ferrocene-quinoxaline derivative as a highly selective probe for colorimetric and redox sensing of toxic mercury(II) cations. *Sensors* 10: 11311–11321.

120 Straube, A., Coburger, P., Dütsch, L., and Hey-Hawkins, E. (2020). Triple the fun: tri(ferrocenyl)arene-based gold(I) complexes for redox-switchable catalysis. *Chem. Sci.* 11: 10657–10665.

121 Straube, A., Coburger, P., Ringenberg, M.R., and Hey-Hawkins, E. (2020). Tri-coordinated coinage metal complexes with a redox-active tris(ferrocenyl)triazine backbone feature triazine-metal interactions. *Chem. Eur. J.* 26: 5758–5764.

122 Straube, A., Coburger, P., Michak, M. et al. (2020). The core of the matter – arene substitution determines the coordination and catalytic behaviour of tris(1-phosphanyl-1′-ferrocenylene)arene gold(I) complexes. *Dalton Trans.* 49: 16667–16682.

123 Schaarschmidt, D., Hildebrandt, A., Bock, S., and Lang, H. (2014). Planar-chiral phosphino alkenylferrocenes – synthesis, solid-state structure and electrochemistry. *J. Organomet. Chem.* 751: 742–753.

124 Grecchi, S., Arnaboldi, S., Korb, M. et al. (2020). Widening the scope of "inherently chiral" electrodes: enantiodiscrimination of chiral electro-active probes with planar stereogenicity. *ChemElectroChem* 7: 3429–3438.

125 Paul, F. and Lapinte, C. (1998). Organometallic molecular wires and other nanoscale-sized devices. An approach using the organoiron (dppe)Cp*Fe building block. *Coord. Chem. Rev.* 178-180: 431–509.

126 Szafert, S. and Gladysz, J.A. (2003). Carbon in one dimension: structural analysis of the higher conjugated polyynes. *Chem. Rev.* 103: 4175–4205.

7

Electronic Coupling and Electron Transfer in Mixed-Valence Systems with Covalently Bonded Dimetal Units

Chun Y. Liu[1], Nathan J. Patmore[2], and Miao Meng[1]

[1] Jinan University, College of Chemistry and Materials Science, Department of Chemistry, 601 Huang-Pu Avenue West, Guangzhou 510632, China
[2] University of Huddersfield, School of Applied Sciences, Department of Chemical Sciences, Queensgate, Huddersfield HD1 3DH, UK

7.1 Introduction

Coordination chemistry of covalently bonded dimetal centers started with Cotton's discovery of metal–metal covalent bonds in the $[Re_2Cl_8]^{2-}$ ion in 1964 [1], by which the highest bond order for two atoms was increased from 3 for main group elements to 4 for transition metals. A $[M_2]^{2n+}$ unit behaves as a single-metal ion but has an electronic configuration that is completely different from that for the associated mononuclear ion M^{n+}, which also gives rise to a distinct coordination geometry. With the emergence of dimetal complexes, the coordinational topology for mononuclear metal complexes defined by Werner was extended. Formation of the covalent bonds between two metal ions breaks the degeneracy of the d orbitals that is commonly found in complexes with high symmetry. A quadruply bonded M_2 unit formed by d^4 metal ions, such as $[M_2]^{4+}$ (M = Cr, Mo, and W) and $[M_2]^{6+}$ (M = Tc and Re), possesses a unique close shell electronic configuration, namely, $\sigma^2\pi^4\delta^2$, derived from d orbital bonding interactions between the two M atoms [2, 3], as shown in Figure 7.1.

In the ideal eclipsed geometry, while four of the five d orbitals (d_{z^2}, d_{xz}, d_{yz}, and d_{xy}) of each atom are used to form the quadruple bonds, the empty $d_{x^2-y^2}$ orbitals are usually very high in energy to participate in MM bonding but are used for coordination to the various supporting ligands. For a M_2 (M = Cr, Mo, and W) center, there are 10 total coordination positions available, 8 strong equatorial binding positions and 2 axial sites that bind weakly to donors. Typically, three-atom bridging ligands, such as, carboxylate and formamidinate anions, are used, generating $Mo_2(O_2CCH_3)_4$ and $Mo_2(DArF)_4$ (DArF = N, N',-di(p-aryl)formamidinate) with the well-known paddlewheel structure in D_{4h} symmetry, as shown in Scheme 7.1.

The electronic structure not only determines the coordination geometry of a M_2 unit but more importantly dominates the chemical and spectroscopic properties of the complex molecule. For paddlewheel quadruply bonded M_2 complexes,

Mixed-Valence Systems: Fundamentals, Synthesis, Electron Transfer, and Applications, First Edition.
Edited by Yu-Wu Zhong, Chun Y. Liu, and Jeffrey R. Reimers.
© 2023 WILEY-VCH GmbH. Published 2023 by WILEY-VCH GmbH.

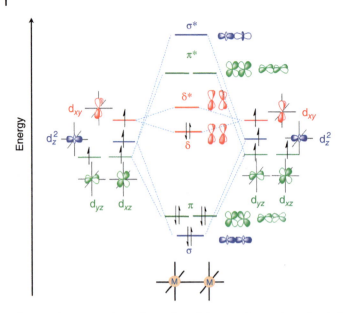

Figure 7.1 Electronic configuration of the metal–metal quadruple bonds.

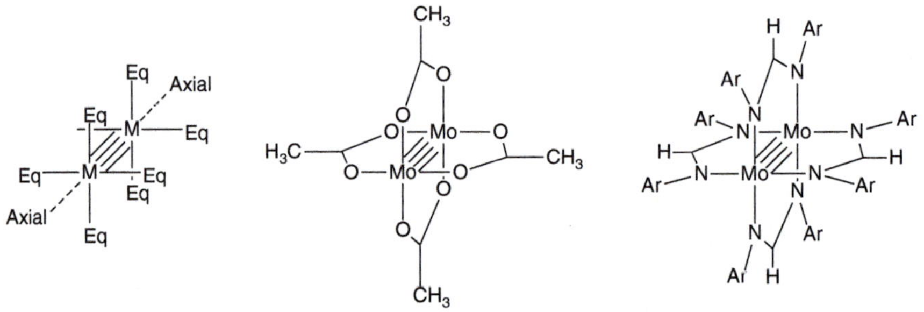

Scheme 7.1 Coordinational topology of a quadruply bonded M_2 unit (left) and the molecular structures of $Mo_2(O_2CCH_3)_4$ (middle) and $Mo_2(DArF)_4$ (right).

the HOMO is δ orbital and the LUMO is δ* orbital. Vertical transition from the HOMO to the LUMO is the so-called δ→δ* transition [1]. The HOMO level and thus the HOMO–LUMO gap are highly sensitive to the auxiliary ligands, which offer the complex characteristic and tunable electrochemical and spectroscopic parameters. The δ electrons are discriminated from the d electrons in other bonding orbitals by energy and symmetry and delocalized within the M_2 unit through the δ bond. This makes the molecule a one-electron active system in chemical transformation between ground and excited states, which behaves as a single atom with two valence electrons. For example, for dimolybdenum carboxylates and formamidinates, the redox potentials for $Mo_2^{4+} \rightarrow Mo_2^{5+}$ fall in the range of 0.2–0.8 V (AgCl/Ag), and the δ→δ* transition occurs at 400–500 nm [2]. In general, the redox potential for $Mo_2^{5+} \rightarrow Mo_2^{6+}$ is very high and not accessible chemically. One approach to achieve double oxidation of a Mo_2 core from Mo_2(II,II) to Mo_2(III,III) is to use strong electron-donating ligands that raise the δ orbital energy [4, 5]. It is worth noting that the oxidation potential of a M_2^{4+} center varies depending on the electronic nature of the ancillary ligands and specifically the chelating atoms because these factors govern the δ orbital energy. For example, a remote electron-donating group on the formamidinate ligand raises the δ electron energy, while an electron-withdrawing group imposes the opposite effect. Substitution of the oxygen chelating atoms with sulfur atoms significantly lowers the redox potential. This turns out to be the important synthetic strategy by which the donor and acceptor potentials and the free energy change (ΔG^0) are controlled for the electron transfer reaction in mixed-valence complexes [6].

Bridged M_2 dimers (M = Mo and W), so-called *dimer of dimers*, were first synthesized by Chisholm in 1989 [7], which initiated study of electronic coupling using M_2–BL–M_2 (BL = Bridging ligand) complex models. These compounds were synthesized by reaction of paddlewheel dimetal tetracarboxylate $M_2(O_2C^tBu)_4$ with a dicarboxylic acid ($HO_2C(X)CO_2H$) (X = conjugated spacer), producing the neutral complexes $[M_2(O_2C^tBu)_3]_2(\mu\text{-}O_2C(X)CO_2)$. The bridged M_2 dimers exhibit two redox waves, corresponding to the single oxidation of each M_2 unit from M_2(II,II) to M_2(II,III). For the oxalate (oxa, X = nothing), the strong electronic coupling was manifested by the large oxidation potential separations, $\Delta E_{1/2}$ = 280 mV for [Mo_2]–oxa–[Mo_2] and 717 mV for [W_2]–oxa–[W_2], indicating stronger electronic coupling for the ditungsten dimer [8]. The thermodynamic stability of the mixed-valence ion in solution is described by the comproportionation reaction (Eq. (7.1)).

$$[M_2] - BL - [M_2] + \{[M_2] - BL - [M_2]\}^{2+} \rightleftharpoons 2\{[M_2] - BL - [M_2]\}^+ \quad (7.1)$$

$$K_C = e^{\Delta E_{1/2}/25.69}$$

The comproportionation constant (K_c) is determined from the separation of the redox potentials ($\Delta E_{1/2}$ in mV) for the two successive one-electron oxidations occurring on each of the two [M_2] units, according to Taube and Richardson [9]. The mixed-valence complex $\{[Mo_2]$–oxa–$[Mo_2]\}^+$ exhibits a charge resonance band at 4000 cm^{-1} (H_{ab} = 2000 cm^{-1}), which shows evidence of a low energy cut-off characteristic of a fully delocalized system. The corresponding charge resonance band for $\{[W_2]$–oxa–$[W_2]\}^+$ at 5960 cm^{-1} (H_{ab} = 2980 cm^{-1}) is more Gaussian in shape but is also relatively narrow and intense ($\Delta v_{1/2}$ = 940 cm^{-1} and ε_{IT} = 5620 M^{-1}cm^{-1}) and also fully delocalized [10]. The increased coupling seen for W vs. Mo is due to the increased energy of the W_2-δ orbitals, which results in greater interaction with the oxalate π^* orbitals [8]. These results established that dimers of metal–metal-bonded M_2 units are suitable electron donors and acceptors for constructing the D–B–A experimental models for study of intramolecular electronic coupling (EC) and electron transfer (ET). Quadruply bonded M_2^{4+} units (M = Mo and W) are preferable over other dimetal building blocks, for example, Re_2^{6+}, because they yield neutral bridged M_2 dimers when dianion bridging ligands are used, which have synthetic advantages. Such a [M_2]–BL–[M_2] system is electronically unique because of the $\sigma^2\pi^4\delta^2$ electron configuration of the dimetal core. The electronic coupling and

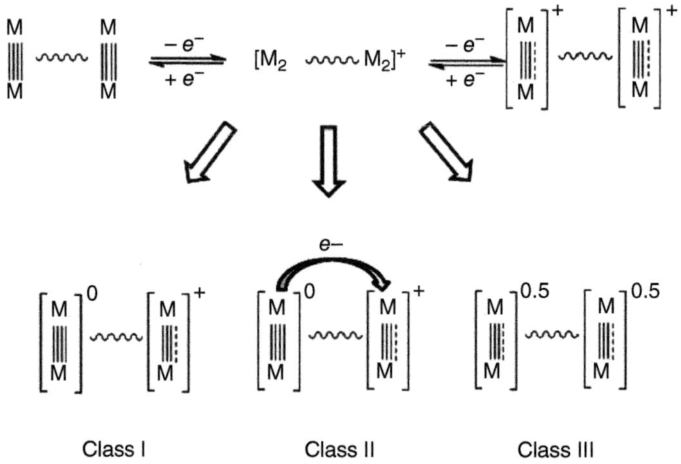

Scheme 7.2 Oxidation of $[M_2]$–BL–$[M_2]$ with quadruply bonded M_2 units (M = Mo and W) and the mixed valency of the cationic dimers in terms of the Robin–Day classification scheme.

electron transfer in the MV system can be described by Scheme 7.2 in terms of the Robin–Day classification [11].

From Scheme 7.2, it can be observed that one-electron oxidation of the neutral compound yields the mixed-valence complex, while Class I and Class II have the two M_2 units with different formal M–M bond orders and thus different oxidation states. According to the Robin–Day classification scheme [11], for Class I, one M_2 unit is quadruply bonded, while the other has a bond order of 3.5; the two dimetal units are electronically uncoupled. On the other extreme, for those in the Class III regime, both of the M_2 units have an averaged bond order of 3.75. For study of intramolecular electron transfer under the contemporary theoretical frameworks, the mixed-valent compounds in Class II are suitable experimental models. Electron transfer from the donor to the acceptor may occur optically through Franck–Condon excitation and/or thermally crossing the transition state via the superexchange mechanism [12], depending on the strength of coupling (H_{ab}).

Dicarboxylates $[O_2C-X-CO_2]^{2-}$ are perfect bridging ligands for assembling electronically coupled *dimers of dimers* and are the most common bridging ligands employed in equatorially bridged $[M_2]$–BL–$[M_2]$ systems. The first structurally characterized dicarboxylate-bridged dimolybdenum dimer, $[Mo_2(DAniF)_3]_2(\mu\text{-}O_2CCO_2)$ (DAniF = N,N'-di(p-anisyl)formamidine), was reported by Cotton and coworkers [13], in which N-donor formamidinate anions were used as the ancillary ligands for the Mo_2 units. Detailed systematic experimental and computational studies of BL effects on electronic coupling were performed by employing various dicarboxylates of different lengths and conjugations [14, 15]. Later, the categories of bridging ligands have been extended to other organic three-atom bridging dianions, as shown in Scheme 7.3. These typically include diamidate, dithiodiamidate, and thiolated dicarboxylates with various conjugated central groups. Extensive studies on these types of compounds have been carried out mainly by the groups

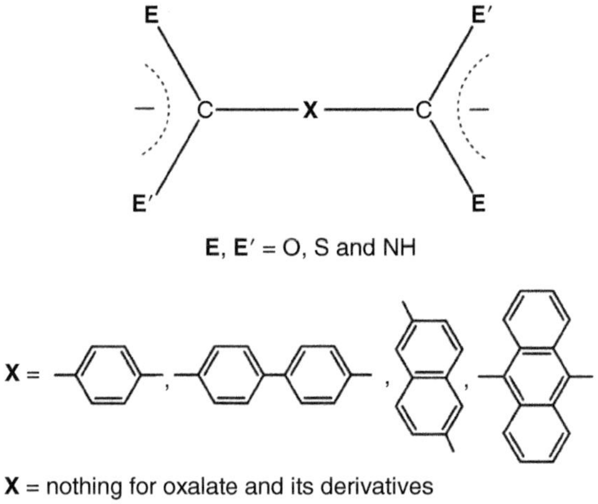

E, E′ = O, S and NH

X =

X = nothing for oxalate and its derivatives

Scheme 7.3 Dicarboxylates and their derivatives used as bridging ligands for the dimetal dimers.

of Chisholm (with carboxylate auxiliary ligands) [16] and Cotton and Liu (with formadinate auxiliary ligands) [17, 18], which showed that the length, donor atom, conjugation, and conformation of the bridging ligand impose a significant impact on the electronic communication between the two [M_2] units.

7.2 Synthesis and Characterization

Several generalized synthetic routes have been used for synthesis of the *dimers of dimers*. For dimetal complexes with carboxylate ancillary ligands, synthesis typically involves metathesis between the dimetal tetracarboxylate and a diprotic bridging ligand such as a dicarboxylic acid (Route A, Scheme 7.4). This route has been extensively employed by the Chisholm group to generate compounds of the form [$Mo_2(O_2C^tBu)_3]_2(\mu-O_2C(X)CO_2$) from the readily available $Mo_2(O_2C^tBu)_4$ building block [8, 19]. An issue with this reaction is that it is divergent because the four carboxylates on the dimetal complex are equal in reactivity and substitutionally labile. Therefore, the reaction can produce undesired oligomers and polymers as impurities affecting the reaction yields [20]. To overcome this, the reactions are generally performed in toluene, from which the desired product precipitates, preventing further reaction to form polymeric species. This strategy can also be used to generate tungsten analogs of the dimetal dimers. For bridging ligands that are not sufficiently basic, [$Mo_2(O_2C^tBu)_3(MeCN)_4][BF_4$] can be used as an alternative starting material (Route B, Scheme 7.4) [8]. Although the equatorially coordinated acetonitrile molecules are good leaving groups, the products can still undergo ligand scrambling reactions, resulting in reaction impurities that need to be separated.

Compared to the labile carboxylate ancillary ligands, formamidinates are more Lewis basic and therefore do not readily undergo ligand exchange reactions. The use of formamidinate ancillary ligands therefore allows convergent

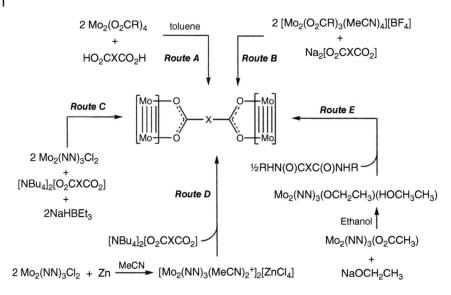

Scheme 7.4 General synthetic routes to dimolybdenum *dimers of dimers*.

synthesis of the target compounds. However, dimetal dimers cannot be readily obtained from dimetal teraformadinates; hence, different synthetic strategies have been developed. In 1998, two compounds, [Mo$_2$(DAniF)$_3$]$_2$(μ-O$_2$CCO$_2$) and [Mo$_2$(DAniF)$_3$]$_2$(μ-O$_2$CC$_6$F$_4$CO$_2$) (DAniF = *N,N*′-di(*p*-anisyl)formamidinate), were prepared by reaction of the oxidized precursor Mo$_2$(DAniF)$_3$Cl$_2$ with a dicarboxylate in the presence of reducing agent (Route C, Scheme 7.4), and their structures were determined by single-crystal X-ray diffraction [13]. This procedure was optimized by using Zn dust as the reducing reagent and acetonitrile as solvent, as Route D in Scheme 7.4 [14, 21, 22].

However, there exist some synthetic disadvantages in employing the oxidized precursor for preparing the *dimers of dimers*. Preparation of the dichloro complex Mo$_2$(DAniF)$_3$Cl$_2$ from Mo$_2$(DAniF)$_4$ takes about one week and a large amount of solvent (THF) and ends with a low yield [21]. More importantly, its application is limited to assembling dimolybdenum units with dicarboxylate bridging ligands. More basic ligands such as diamidates are not suitable for this method because the deprotonated bridging ligand would initiate a nucleophilic attack on the coordinated acetonitrile molecules in the [Mo$_2$(DAniF)$_3$(NCCH$_3$)$_2$]$^+$ intermediate [18]. This recognition was obtained after many efforts were made to synthesize the diamidate-bridged analogs by following a similar approach [23], which led to designed synthesis of the mixed-ligand complex Mo$_2$(DAniF)$_3$(O$_2$CCH$_3$) (see Figure 7.2a) as the starting material. This building block can be obtained from a one-pot reaction of HDAniF and sodium ethoxide with Mo$_2$(O$_2$CCH$_3$)$_4$, as shown in Eq. (7.2), in good yield and purity [24].

$$Mo_2(O_2CCH_3)_4 + 3HDAniF + 3NaOCH_2CH_3 \rightarrow$$
$$Mo_2(DAniF)_3(O_2CCH_3) + 3NaO_2CCH_3 + 3CH_3CH_2OH \quad (7.2)$$

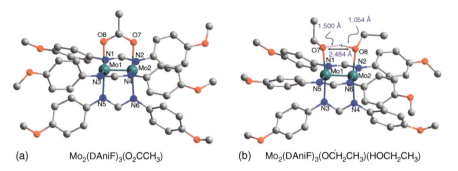

Figure 7.2 X-ray crystal structures for $Mo_2(DAniF)_3(O_2CCH_3)$ (a) and $Mo_2(DAniF)_3$ $(OCH_2CH_3)(HOCH_2CH_3)$ (b). The hydrogen atoms have been omitted, except for the hydroxyl group on the ethanol molecule. Source: Lei et al. [18]/Reproduced with permission of Elsevier.

This Mo_2 building block was used to develop a more generalized method for convergent assembly of two dimolybdenum units with a diverse group of bridging ligands [25]. In preparing the dimolybdenum *dimer of dimers*, $Mo_2(DAniF)_3(O_2CCH_3)$ is treated with a stoichiometric amount of sodium ethoxide in ethanol, generating the highly reactive intermediate, $Mo_2(DAniF)_3(OCH_2CH_3)(HOCH_2CH_3)$, which can then be reacted with a BL, for example, dicarboxylates and dicarboxylamidates (see Route E in Scheme 7.4). The $Mo_2(DAniF)_3(OCH_2CH_3)(HOCH_2CH_3)$ intermediate was isolated and structurally characterized, as shown in Figure 7.2b, with the $CH_3CH_2O^-$ anion and CH_3CH_2OH molecule filling a pair of equatorial coordination sites in the Mo_2 unit, with a hydrogen bond between them making them act in a similar manner to a three-atom bridging ligand [18]. In the subsequent reaction with a bridging ligand, the coordinated ethoxide provides one equivalent of base for deprotonation of the bridging ligand. The ethanol molecules are kinetically labile and can be easily substituted by the incoming bridging ligand. Using this or the slightly modified procedures, a number of dimolybdenum dimers with a broad range of BLs have been synthesized.

7.3 d(δ)(M₂)-p(π)(Ligand) Conjugation

The X-ray crystal structures of $[Mo_2(DAniF)_3]_2(\mu\text{-}O_2CCO_2)$ and $[Mo_2(DAniF)_3]_2$ $(\mu\text{-}O_2CC_6F_4CO_2)$ show that the molecules have a bridge architecture [13], with the four vertical DAniF ligands acting as the bridge piers, and the Mo–Mo units and the bridging ligand are co-planar with the two horizontal DAniF ligands. The Mo—Mo bonds are ~2.09 Å in length, which is typical for a Mo_2 unit supported by formamidinate ligands [3, 26]. These two structures represent the basic molecular geometry for neutral and mixed-valent M_2 dimers through equatorial linkage. This structure visualizes the picture of a charge transfer platform that transports the δ electrons from one Mo_2 unit to the other [13]. Interestingly, this existence of charge transfer platform is evidenced by the ^1H NMR signals of the vertical and horizontal methine protons on the DAniF ligands, namely, $H_\perp$ and $H_\parallel$, respectively. The strongly coupled $[Mo_2(DAniF)_3]_2(\mu\text{-}O_2CCO_2)$ shows chemical shifts $\delta_\perp$ 8.52 ppm and $\delta_\parallel$ 8.47 ppm, but for the more weakly coupled $[Mo_2(DAniF)_3]_2(\mu\text{-}O_2CC_6F_4CO_2)$, $\delta_\perp$ (8.52 ppm) and $\delta_\parallel$ (8.51 ppm) are close. The downfield shift of $H_\parallel$ resonance is indicative of the shielding effect resulting from charge delocalization in the conjugated π system involving δ electrons from the two Mo_2 units (Scheme 7.5). These properties of Mo_2 dimers with formamidinate ligands have been systematically studied via experimental and theoretically approaches in the O- and S-bridged dimers [27, 28]. The results are well interpreted by the electric ring current model, as described in Figure 7.3. It is found that the displacement between $\delta_\perp$ and $\delta_\parallel$, i.e. $\Delta\delta = \delta_\parallel - \delta_\perp$, is correlated with the extent of electron delocalization over the two $[Mo_2]$ sites. The larger the $\Delta\delta$ value, the stronger the coupling. A linear correlation of $\Delta\delta$ with the Hammett constants (σ_X) is observed in a series of Mo_2 dimers having the same terephthalate BL [29]. The variation of $\Delta\delta$ as a function of electronic coupling strength is parallel to those for electrochemical ($\Delta E_{1/2}$) and spectroscopic (λ_{MLCT}) parameters, as shown in Figure 7.4. Therefore, in the M_2 dimeric systems with formamidinate ancillary ligands, the ^1H NMR spectrum of the methine protons serves as additional probe to gauge the extent of electronic coupling of the system.

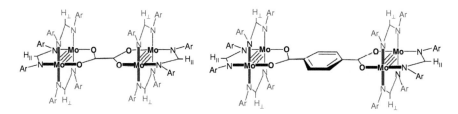

Scheme 7.5 Schematics of diaryl formamidinate (DArF)-supported dimolybdenum dimers, showing the horizontal charge transfer platform (black). The methine protons on the orthogonal DArF ligands are denoted as $H_\parallel$ (black) and $H_\perp$ (gray).

With covalently bonded M_2 units in different oxidation states serving as the electron donors and acceptors in the MV complex, one would expect that the M–M bond distances are the structural indicator of electronic coupling strength. This advantage of M_2 building blocks is demonstrated by a pair of structural isomers derived from the oxamidate bridging ligand. Assembling oxamidate with $[Mo_2(DAniF)_3]^+$ yielded two complexes differing in coordination geometries [24, 30], namely, α and β, as shown in Figure 7.5a. In the α form, the bridging ligand alligates the Mo_2 units with the ONC three-atom chelating groups, and the Mo—Mo bonds are orthogonal to one another because of the steric hindrance of the bulky aryl group on the amidate. Coordination of the four-atom chelating group OCCN to the Mo_2 units gave rise to the β form, which has the central C—C bond parallel and co-planar with the two Mo—Mo bonds. The MV complexes of both isomers were prepared by chemical oxidation and structurally characterized by single-crystal X-ray diffraction. In the α isomer, the Mo—Mo distances are 2.0920(6) and 2.1291(6) Å, corresponding to bond orders of 4.0 and 3.5 [2, 30, 31], respectively. In contrast, in the β

7.3 d(δ)(M₂)-p(π)(Ligand) Conjugation | 237

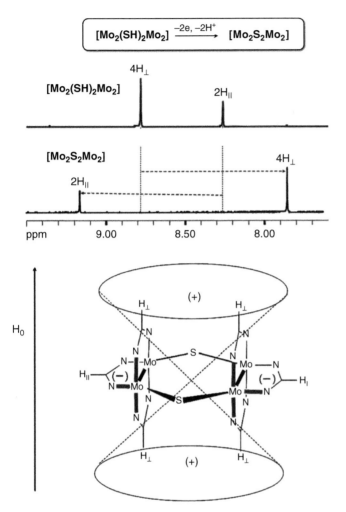

Figure 7.3 ^{1}H NMR spectra (top) and diamagnetic anisotropy (bottom) in the aromatic [Mo$_2$S$_2$Mo$_2$] ring system-induced d(δ)-p(π) conjugation.

Figure 7.4 Plots of $\Delta E_{1/2}$, λ max (MLCT), and $\Delta\delta_{\parallel-\perp}$ against the Hammett constant (σ_x), showing the control of Mo$_2$–Mo$_2$ coupling by the remote substituents.

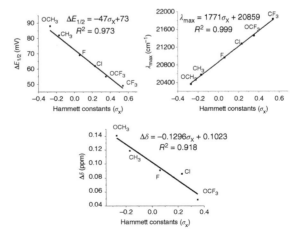

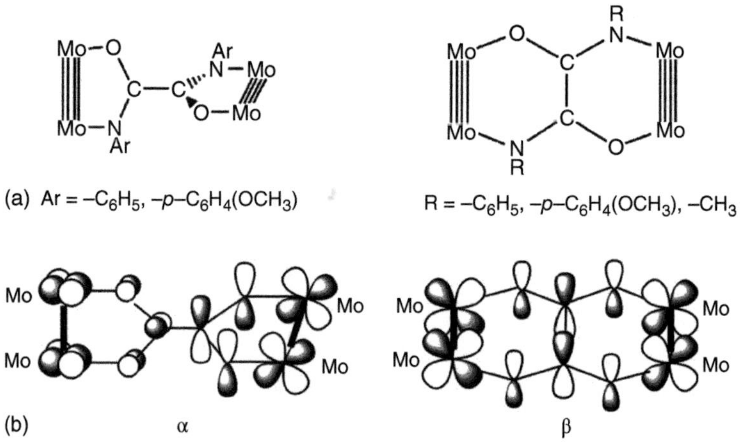

(a) Ar = –C$_6$H$_5$, –p–C$_6$H$_4$(OCH$_3$)

R = –C$_6$H$_5$, –p–C$_6$H$_4$(OCH$_3$), –CH$_3$

(b) α β

Figure 7.5 δ–δ electronic interaction between the two Mo$_2$ units controlled by bridge conjugation. Source: Cotton et al. 2007 [30]/Reproduced with permission of the American Chemical Society.

structure, the two Mo—Mo bonds, 2.1116(7) and 2.1140(6) Å in length, are nearly identical [30]. Therefore, based solely on the molecular structures, the MV complexes in α and β geometries belong to Class I and Class III, respectively, as defined by Robin and Day [11]. In the bridged d$^{5-6}$ [M]–BL–[M] dinuclear MV complex systems, attempts have been made to X-ray crystallography, specifically for the Creutz–Taube ion [32]. However, the results were unsatisfactory because of disorder and the small changes expected in the metal–ligand bond distances involving light atoms such as nitrogen (N). DFT calculations have shown that in the α form, the orbital overlaps between two dimetal units are eliminated because of the orthogonality of the two chelating groups of the bridging ligand, but for the β isomer, delocalized molecular orbitals spread across between the two Mo$_2$ centers with significant involvement of the ligand π* orbitals (Figure 7.5b).

7.4 Electronic and Intervalence Transitions and DFT Calculations

For the quadruply bonded M$_2$ dimers with conjugated bridges, the d(δ)(M$_2$)-p(π)(BL) orbital interactions can be qualitatively described by Figure 7.6. Single-point DFT calculations are usually performed on the simplified models, for which the aryl groups on the formamidinate ligand are replaced with H atoms to reduce the computational expense [6, 15, 31]. This simplification does not introduce significant errors or deviations to the results because the model retains the same symmetry as the molecule, and electron density contributions from the aryl groups are negligible. The simplified frontier MO diagram shown in Figure 7.6 is very helpful in understanding the metal–ligand and the metal–metal interactions. The HOMO results from the out-of-phase (δ − δ) combination of the δ orbitals with a filled π orbital of the

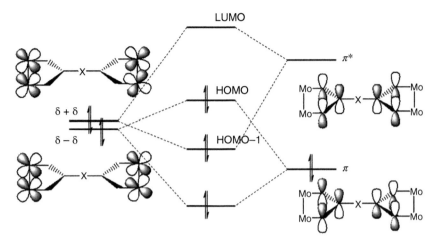

Figure 7.6 A qualitative molecular orbital diagram showing the orbital interactions between the dimetal units (d_δ) and the bridging ligand (p_π).

bridging ligand, while the HOMO-1 is obtained by the in-phase ($\delta + \delta$) combination of the δ orbitals with an empty π^* orbital from the bridging ligand. These two occupied metal-based orbitals are nondegenerate because of the metal–ligand interactions. The LUMO is contributed by the bridging ligand [6, 33]. While the HOMO–LUMO energy gap ($\Delta E_{H\text{-}L}$) predicts the metal to ligand (ML) transition energy, the energy difference between HOMO and HOMO-1, or $\Delta E_{H\text{-}H\text{-}1}$, signals the strength of the electronic interaction between the two [Mo$_2$] units. Small $\Delta E_{H\text{-}L}$ value indicates a low energy for metal-to-ligand charge transfer (MLCT), which enhances the metal(M$_2$)–metal(M$_2$) coupling, in accordance with the superexchange formalism [12]. The stronger the metal–metal interaction becomes, the larger the magnitude of $\Delta E_{H\text{-}H\text{-}1}$ becomes. Clearly, in M$_2$–BL–M$_2$ systems, only the δ electrons from the M$_2$ center are involved in the metal-to-ligand as well as the metal-to-metal interactions, and the other d electrons confined in the M–M σ and π bonds are precluded for symmetry and energy reasons (Figure 7.1). The frontier MO picture (Figure 7.6) highlights the difference in electronic structure between dimetal M$_2$–BL–M$_2$ and monometallic [M]–BL–[M] mixed-valence ions. For d$^{5\text{-}6}$ monometallic MV complexes, the HOMO is based in one of the metal dπ orbitals, which are degenerated in a distorted octahedral ligand field, and can lead to multiple IVCT (Intervalence Charge Transfer) and IC transitions that complicate analysis [34].

The neutral complexes display an intense metal-to-bridging ligand (ML) absorption band in the visible region, which in most cases can be distinguished from the weak $\delta\rightarrow\delta^*$ transition by energy and intensity [15, 35]. While the $\delta\rightarrow\delta^*$ transition occurs at relatively high energy and in a narrow spectral window, for example, ~450 nm for formamidinate supporting ligands [2, 26], the MLCT band vary in a broad region of visible spectrum, depending on the nature of the bridging ligand [15]. The MLCT transition energy is in good agreement with the calculated $\Delta E_{H\text{-}L}$ value [15]. For the series [Mo$_2$(DAniF)$_3$]$_2$(μ-EE$'$CC$_6$H$_4$CEE$'$) (EE$'$ = OO, OS, and SS), the calculated MLCT transition energies are close to the experimentally observed spectral energies ($\pm 2000 \, \text{cm}^{-1}$) [6]. Most importantly, for the three complexes in the series, the trends in the experimental and calculated data match, that is, a decrease in the energy of the MLCT with more S atoms introduced to the charge transfer platform [6, 33]. In these M_2–BL–M_2 systems, the MV complexes generated upon one-electron oxidation usually exhibit MLCT absorptions with the same transition energy, but reduced intensity, by comparison to the spectra of the neutral compounds. For the MV complexes in Class II, a spectrum arising from bridge ligand (π) to metal (δ) charge transfer (LMCT) on the acceptor side is usually observed in the visible region [35]. Both the MLCT and LMCT bands resulting from transitions from/to a discrete δ orbital of the transferring electron are not complicated by other d electron transitions of a similar energy. While the MLCT spectrum represents transferring an electron from Mo_2 to the bridge, the LMCT band accounts for an electron moving from the bridge to Mo_2. The simultaneous presence of these two absorption bands in the spectra of Class II compounds allows for evaluation of both the electron transfer (LUMO-mediated) and hole transfer (HOMO-mediated) superexchange mechanisms for electron transfer [12, 35]. The unique electronic structure for M_2–BL–M_2 mixed-valence systems therefore endows them with a great advantage over M–BL–M MV systems in optical analysis of electronic coupling.

Like other MV systems, the most important optical feature for the M_2–BL–M_2 systems is the intervalence transition from the donor to acceptor, that is, the so-called IVCT band in the near-mid IR region. This IVCT absorption is broad, well separated from other electronic transitions, and therefore easy to be identified, unlike d^{5-6} MV complexes that may have multiple IVCT transitions. These optical properties allow accurate determination of coupling constant (H_{ab}) and study of electron transfer energetics and kinetics based on the established theories [36–38]. The theoretical basis of single-point DFT calculations for the cationic radicals is Koopmans' theorem, interpreting the molecular orbital energies of the HOMO-1 (bonding) orbital and the HOMO (antibonding) orbital [39]. Specifically, the transition energies for the radical cations are calculated using the "neutral in cation geometry" (NCG) method, which has been used in organic MV systems by Nelsen and coworkers [40, 41]. For the strongly coupled Mo_2–Mo_2 MV systems [42, 43], this approach yields theoretical results fully consistent with experimental observations, demonstrating that in the Class III system, the vertical transition energy (IVCT) is determined by $\Delta E_{\text{H-H-1}}$ [39].

7.5 Transition in Mixed Valency Between Robin–Day Classes

The terephthalate-bridged dimolybdenum dimer [$Mo_2(DAniF)_3$](μ-$O_2CC_6H_4CO_2$) is one of the first synthesized molecules of this type [44] and exhibits d(δ)–p(π) conjugation between the dimetal center and BL. A series of analogs, with similar bridging ligands of form [**EE'**–ph–**EE'**] (E, E' = O, S, and NH), as shown in Scheme 7.6, were synthesized by substitution of the O chelating atoms with N and S to systematically examine the EC effects between the two [Mo_2] units [6, 35, 45].

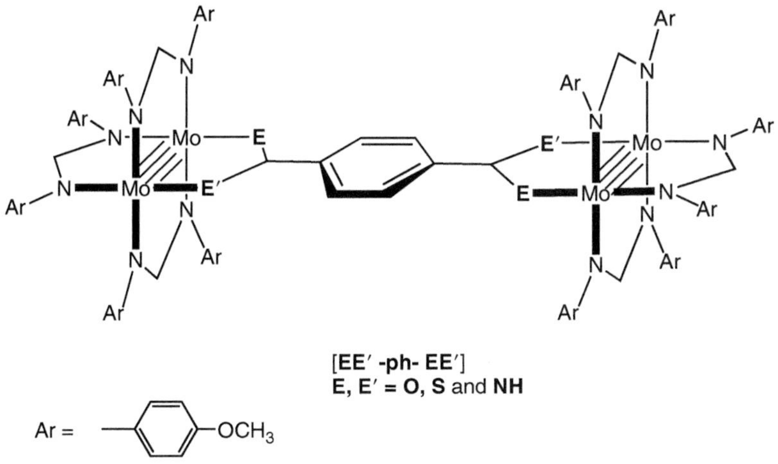

7.5 Transition in Mixed Valency Between Robin–Day Classes

[Molecular structure diagram of Mo₂ dimers showing two Mo₂ units connected by a terephthaloyl bridge, with Ar groups where Ar = p-methoxyphenyl (–C₆H₄–OCH₃)]

[EE′-ph- EE′]
E, E′ = O, S and NH

Ar = –C₆H₄–OCH₃

Scheme 7.6 Molecular skeleton of Mo₂ dimers [EE′–ph–EE′] (E,E′ = O, S, and NH).

Assembling two [Mo$_2$(DAniF)$_3$]$^+$ blocks with terephthaloyldiamidate produced the Mo$_2$ dimers **[ON(R)–ph–ON(R)]**, with R = aryl group or H atom, which exhibit electronic coupling strength similar to that of **[OO–ph–OO]**. For these complexes, the potential separation between the first two oxidations ($\Delta E_{1/2}$) was found to be approximately 100 mV [45]. The shape and intensity of the IVCT transitions in the MV complexes were found to be consistent with the predictions for Class II systems [46]. The IVCT transition for **[O(NH)–ph–O(NH)]**$^+$ was observed at 4651 cm^{-1}, slightly higher in energy but less intense than that for **[OO–ph–OO]**$^+$ (4240 cm^{-1}), indicating that electronic coupling in the diamidate derivative is slightly weak. The decoupling effect of the N(H) group is further evidenced by the terephthalamidinate analog **[(NH)$_2$–ph–(NH)$_2$]**, which has a smaller $\Delta E_{1/2}$ value of 80 mV and E_{IT} 4980 cm^{-1} for the IVCT transition of the analog [45]. Given the well-defined, characteristic IVCT band for each of these complexes, the electronic coupling matrix elements (H_{ab}) were calculated from the Mulliken–Hush expression (Eq. (7.3)) [34, 35, 46, 47]

$$H_{ab} = 2.06 \times 10^{-2} \frac{(E_{IT} \mathcal{E}_{IT} \Delta v_{1/2})^{1/2}}{r_{ab}} \tag{7.3}$$

where E_{IT} and $\Delta v_{1/2}$ are in wavenumber (cm^{-1}) and r_{ab} is the effective electron transfer distance in angstrom (Å). Considering that the δ electrons are fully delocalized within the Mo$_2$ unit and onto the "CEE" group of the ligand, the effective ET distance (r_{ab}) of 5.8 Å is adopted from the length of the "–CC$_6$H$_4$C–" group. Replacing the O chelating atoms in **[OO–ph–OO]**$^+$ with N(H) groups results in a decrease of H_{ab} from 589 to 537 cm^{-1}. These examples show that in the Mo$_2$–BL–Mo$_2$ systems, the electronic coupling between two Mo$_2$ units is highly sensitive to the chelating atoms of the bridging ligands. This property allows modification of the chemical potentials of the donor and acceptor and modulation of electronic coupling without significantly changing the molecular electronic configuration and molecular structure, which retains similar inner nuclear reorganization energy [34].

Substitution of the S atoms for the O atoms on the carboxylate group of the bridging ligand can largely enhance the electronic coupling. Thus, a series of four Mo_2–Mo_2 complexes with terephthalate and its thiolated derivatives as the bridging ligands were synthesized and studied. This series includes **[OO–ph–OO]**, **[OS–ph–OS]**, **[OO–ph–SS]**, and **[SS–ph–SS]** [6, 35]. For the three symmetric complexes, the potential separations of the Mo_2 oxidations increase from 91 (OO) to 116 (OS) to 195 (SS) mV as the S atom is introduced stepwise, indicating an increasing stabilization of the MV state. A large $\Delta E_{1/2}$ (360 mV) is found for the asymmetric **[OO–ph–SS]** because of the intrinsic potential difference (ΔE_{ip}) of the two [Mo_2] units. To determine the ΔE_{ip} value, two reference compounds $Mo_2(DAniF)_3(O_2CC_6H_5)$ and $Mo_2(DAniF)_3(S_2CC_6H_5)$ corresponding to the acceptor and the donor, respectively, were prepared; their potentials for oxidation $Mo_2^{4+} \rightarrow Mo_2^{5+}$ are measured to be 375 and 651 mV, respectively. Therefore, the internal potential difference (ΔE_{ip}) for **[OO–ph–SS]** is estimated to be 276 mV. Subtracting this value from the total potential separation $\Delta E_{1/2}$ (c. 360 mV) gives a "net" potential displacement of 84 mV, which measures the electronic coupling strength of this species in the series [35]. The neutral complexes were singly oxidized using one equivalent of oxidizing reagent to yield the resultant MV complexes for optical analysis of electronic coupling. For all the complexes studied, the EPR spectra displayed a single isotropic peak with $g = 1.95$, indicating an odd electron residing on one of the Mo_2 δ orbitals.

The striking spectral feature for these MV complexes is the IVCT bands in the near-mid-IR regions. The intervalence transition energy (E_{IT}), molar extinction coefficient (ε_{IT}), and half-height bandwidth ($\Delta v_{1/2}$) vary substantially depending on the number of S atoms on the charge transfer platform (Table 7.1). In the MV series, as shown in Figure 7.7, while **[OO–ph–OO]**$^+$ exhibits a weak IVCT band ($E_{IT} = 4240$ cm^{-1} and $\varepsilon_{IT} = 1470$ M^{-1} cm^{-1}) that is nearly Gaussian in shape, while for **[SS–ph–SS]**$^+$, this band features high intensity ($\varepsilon_{IT} = 12660$ M^{-1} cm^{-1}), low energy ($E_{IT} = 2640$ cm^{-1}), and large asymmetry (nearly half cut-off of the absorption) that is typical for compounds on the Class II–III border. [34, 46, 48, 49] Optical analysis based on the Hush formalism (Eq. (7.3)) [35, 47] yielded $H_{ab} = 600$–900 cm^{-1} for the series [35]. The coupling parameters were also calculated using the "CNS" model introduced by Creutz, Newton, and Sutin based on the superexchange formalism [50]. For the three symmetric complexes in the series, the coupling parameters obtained from the two theoretical models are in excellent agreement with the ratio $H_{ab}/H_{CNS} \approx 1$ [35]. Successful application of the CNS approach in the Mo_2–Mo_2 systems verifies the superexchange mechanism for the electronic coupling and electron transfer [12]. With the availability of H_{ab}, electron transfer kinetics and energetics are studied in application of the Marcus–Hush theory [35, 47, 51, 52]. It is found that thermal electron transfer occurs at rate constant $k_{et} = 10^{11}$–10^{12} s^{-1} for the symmetrical complexes. For the asymmetrical **[OO–ph–SS]**$^+$ [35], the downhill electron transfer from the OO site to the SS site proceeds at a rate of $\sim 10^{12}$ s^{-1}, and for the reverse electron transfer, $k_{et} = 4.1 \times 10^7$ s^{-1}. Importantly, these results are in good agreement with the theoretical predictions in mixed-valence chemistry and accounts well for the

Table 7.1 Electronic coupling parameters for dimolybdenum *dimers of dimers* supported by formamidinate ligands.

Compound	$\Delta E_{1/2}$ (mV)	E_{ML} (nm)	E_{LM} (nm)	E_{IT} (cm^{-1})	ε_{IT} (M^{-1}cm^{-1})	$\Delta\nu_{1/2(\text{exp})}$ (cm^{-1})	$\Delta\nu^{\circ}_{1/2}$ (cm^{-1})	H_{ab} (cm^{-1})	H_{CNS} (cm^{-1})	$2H_{ab}/\lambda$
[OO–ph–OO]	100	485	0	4240	1470	4410	3190	589	551	0.28
[OS–ph–OS]	116	623	811	3440	3690	3290	2820	727	764	0.42
[SS–ph–SS]	195	722	941	2640	12660	1770	2470	864	864	0.65
[N(NH)–ph–N(NH)]	80	465	—	4980	520	8840	3392	537	626	0.21
[NO–ph–NO]	96	490	—	4651	1171	5242	3278	600	661	0.26
[NS–ph–NS]	115	560	—	3182	3589	3688	2711	697	704	0.44
[OO–thi–OO]	76	464	566	4150	1232	3998	5404	561	—	0.27
[NS–thi–NS]	118	640	837	2630	7933	2480	3940	892	—	0.68
[OS–thi–OS]	184	729	924	2254	11398	2009	3931	1260	—	1.1
[SS–thi–SS]	348	1016	—	3290	20261	1266	1741	1645	—	3.4

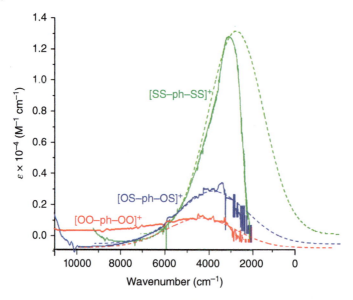

Figure 7.7 Intervalence transition absorptions (IVCT bands) of the MV complexes **[EE′–ph–EE′]**$^+$ (E,E′ = O, S) in the near- to mid-infrared region. The dashed lines simulate the Gaussian-shaped band profiles to show the spectral asymmetry or spectral "cut-off" resulting from strong donor–acceptor electronic coupling. Source: Adapted from Xiao et al. [6].

spectral width of intramolecular electron transfer transitions. As a MV system on the Class II/III borderline, **[SS–ph–SS]**$^+$ has $k_{et} = 3.4 \times 10^{12}\, s^{-1}$, close to the averaged nuclear vibrational frequency ($5 \times 10^{12}\, s^{-1}$), as expected [53].

Employing a thienylene bridging ligand, the series **[E$_1$E$_2$–thi–E$_1$E$_2$]** (E$_1$, E$_2$ = O, N(H), and S, see Scheme 7.7) was developed [54, 55], which allowed observation of

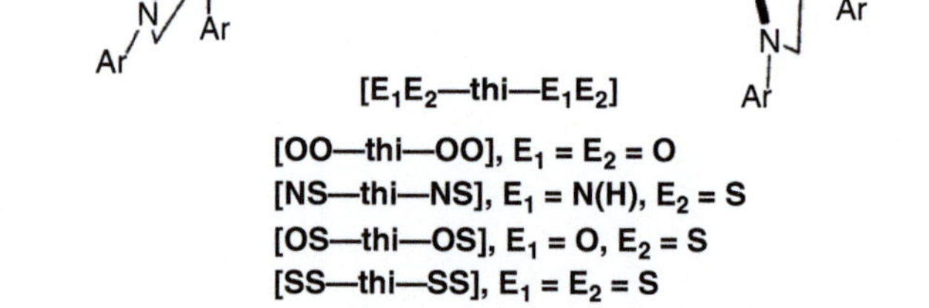

[E$_1$E$_2$—thi—E$_1$E$_2$]

[OO—thi—OO], E$_1$ = E$_2$ = O
[NS—thi—NS], E$_1$ = N(H), E$_2$ = S
[OS—thi—OS], E$_1$ = O, E$_2$ = S
[SS—thi—SS], E$_1$ = E$_2$ = S

Scheme 7.7 Molecular skeleton of Mo$_2$ dimers **[E$_1$E$_2$–thi–E$_1$E$_2$]** (thi = C$_4$H$_2$S and E$_1$, E$_2$ = O, S, and NH). Source: Wu et al. [54]/Royal Society of Chemistry.

the mixed-valency transition from Class II to Class III crossing over the Class II–III borderline [34, 46, 48, 49]. Electrochemical voltammograms (Figure 7.8) show clearly that the chemical potential separations ($\Delta E_{1/2}$) increase progressively in the series, despite the electrostatic interaction remaining nearly constant because of the similar $Mo_2 \cdots Mo_2$ distances. The largest $\Delta E_{1/2}$ (348 mV) is found for **[SS–thi–SS]**, which is significantly larger than that for **[SS–ph–SS]** (195 mV, see Table 7.1). **[OS–thi–OS]** has a $\Delta E_{1/2} = 184$ mV, larger than that of **[NS–thi–NS]** (118 mV), showing stronger resonant interaction. The continuous decrease of IVCT band energies is observed from **[OO–thi–OO]**$^+$ (4150 cm^{-1}) to **[NS–thi–NS]**$^+$ (2630 cm^{-1}) to **[OS–thi–OS]**$^+$ (2254 cm^{-1}) (Figure 7.8a–d) with increase of electronic coupling, in agreement with the electrochemical results. Interestingly, for **[SS–thi–SS]**$^+$, the intervalence transition energy E_{IT} increases to 3290 cm^{-1} as more S atoms are introduced, showing a turning point of E_{IT} variation in the series of continuously increasing the degree of electronic coupling. These results indicate explicitly that **[SS–thi–SS]**$^+$ has crossed the Class II–III borderline and entered the Class III regime. Therefore, the neighboring systems, **[NS–thi–NS]**$^+$ and **[OS–thi–OS]**$^+$, are the intermediates of the transition from Class II to Class III, as indicated by variation of the IVCT transition energies (Table 7.1).

Importantly, this series presents the MV systems in different Robin–Day classes with distinct optical characteristics as predicted by the established theory in mixed-valence chemistry [46]. First, for strongly coupled Class III systems, the IVCT absorption arises from the electronic resonance between delocalized molecular orbitals involving the donor and acceptor, but there is no net electron or charge transfer from donor to acceptor. The high energy is reflective of the strong electronic coupling involved. Secondly, in Class II systems, the transition energies ($E_{IT} = \lambda$) become independent of the coupling (H_{ab}) strength, with the coupling strength serving to provide the observed spectral intensity. These features allow comparison of $2H_{ab}$ with λ that characterize the coupling extent in different Robin–Day's regimes [46, 48, 49]. Calculations from the Hush model found, as listed in Table 7.1, that for **[OO–thi–OO]**$^+$ with $H_{ab} = 561$ cm^{-1}, $2H_{ab}$ (1122 cm^{-1}) $\ll \lambda$ (4150 cm^{-1}) or $2H_{ab}/\lambda = 0.27$ ($\ll 1$), whereas for **[NS–thi–NS]**$^+$, $2H_{ab}$ (1784 cm^{-1}) $< \lambda$ (2630 cm^{-1}) or $2H_{ab}/\lambda = 0.68$. For **[OS–thi–OS]**$^+$, calculation from Eq. (7.3) yields $H_{ab} = 1260$ cm^{-1}, and thus, $2H_{ab}/\lambda = 1.1$ (Table 7.1). Remarkably, this H_{ab} value is slightly larger than 1127 cm^{-1} predicated from the two-state model, that is, $H_{ab} = E_{IT}/2$ for the Class III limit ($2H_{ab}/\lambda = 1$), thus confirming its Class III characteristic. These results provide verification and illustration to the mixed-valence theories, including the Hush model.

It should be noted that in the strong coupled regime, while $2H_{ab} = E_{IT}$ still holds, $E_{IT} \neq \lambda$ because the reorganization energy is largely decreased. Brunschwig and Sutin proposed a combined electrochemical spectroscopic approach to estimate the reorganization energy of a strong coupling system [56]. It is demonstrated that strongly and very strongly coupled systems can be distinguished by comparison of the magnitudes of $\Delta E_{1/2}$ and E_{IT}. For borderline Class III compounds, $\Delta E_{1/2} \approx E_{IT}/2$,

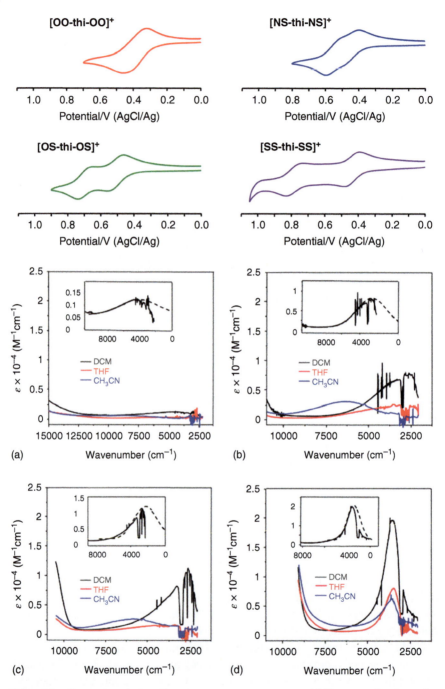

Figure 7.8 Electrochemical cyclic voltammograms and near-mid-IR spectra for the series of **[E₁E₂-thi-E₁E₂]** (E_1, E_2 = O, S, and NH) in CH_2Cl_2, CH_3CN, and THF, which shows the solvent dependence of the IVCT envelopes in different Robin–Day's regimes. (a) **[OO-thi-OO]⁺**. (b) **[NS-thi-NS]⁺**. (c) **[OS-thi-OS]⁺**. (d) **[SS-thi-SS]⁺**. Source: Wu et al. [54]/Reproduced with permission of the Royal Society of Chemistry.

and for fully delocalized systems, $\Delta E_{1/2} \approx E_{IT}$, and reorganization energy can be calculated by

$$\lambda = 2(E_{IT} - \Delta E_{1/2}) \tag{7.4}$$

[OS–thi–OS]$^+$ has a $\Delta E_{1/2}$ of 184 mV (1484 cm^{-1}), larger than $E_{IT}/2$ (1127 cm^{-1}), in agreement with its Class III characteristic ($2H_{ab}/\lambda = 1.1$). For [SS–thi–SS]$^+$, λ is calculated to be 966 cm^{-1}; therefore, $2H_{ab}/\lambda = 3.4$ (Table 7.1). With $2H_{ab}/\lambda \gg 1$, the MV system is brought into a new regime of mixed valency, which we will discuss in the later section. Another transforming characteristic for Class III is that the IVCT band is narrower and less asymmetric in comparison with those in the neighboring regime, Class II/III or borderline Class III systems. In this series, [OS–thi–OS]$^+$ shows a IVCT band nearly half cut-off (49%) at $2H_{ab}$, while a cut-off area of 27% is found in the IVCT spectrum of [SS–thi–SS]$^+$. Moreover, transition of mixed valency in this series is clearly indicated by solvent effects of the complexes (Figure 7.8a–d) [34]. [OS–thi–OS]$^+$ exhibits variable IVCT absorptions in different solvents (DCM, THF, and CH$_3$CN); in contrast, the E_{IT} value for [SS–thi–SS]$^+$ in these solvents remain constant, showing solvent independence of the IVCT transition because of loss of charge transfer [54]. Therefore, this work demonstrates that solvent independence of the IVCT absorption is not a valid experimental criterion for Class II/III.

7.6 Distance Dependence of Electronic Coupling and Electron Transfer

Increasing electron transfer distance in MV compounds is known to reduce the electronic coupling. [M$_2$–BL–M$_2$]$^+$ systems with an oxalate as the bridging ligand, the shortest dicarboxylate bridge, tend to be fully delocalized, Class III in nature. Increasing the length of the dicarboxylate bridge by inserting a phenylene group and using a terephthalate bridge reduces coupling, typically resulting in Class II systems. This can be seen using EPR spectroscopy, which is a convenient technique for studying electron localization in M$_2$–BL–M$_2$$^+$ systems as Mo has two isotopes with $I = 5/2$ (^{95}Mo and ^{97}Mo, 25.4% combined natural abundance), which can be used to determine hyperfine coupling constants (A_0). The EPR spectrum of [Mo$_2$(O$_2$C^{t}Bu)$_3$](μ-O$_2$CCO$_2$)$^+$ in THF displays a six-line pattern of hyperfines at $g = 1.937$ with $A_0 = 14.8$ G [57]. The magnitude of the hyperfine coupling indicates that the single electron is delocalized over all four Mo atoms (Class III) on the EPR timescale. By contrast, the hyperfine coupling for the perfluoroterephthalate-bridged [Mo$_2$(O$_2$C^{t}Bu)$_3$](μ-O$_2$CC$_6$F$_4$CO$_2$)$^+$ ($A_0 = 27.2$ G) is nearly twice that of the oxalate, indicating that the single electron is delocalized over one Mo$_2$ unit and is Class II in nature.

Following the comprehensive investigations on the phenylene-bridged Mo$_2$–Mo$_2$ series discussed in Section 7.5, the biphenylene and triphenylene series have been developed and studied with three members of each [58, 59], as shown in Scheme 7.8. The nine compounds of three series allow a detailed study of distance dependence

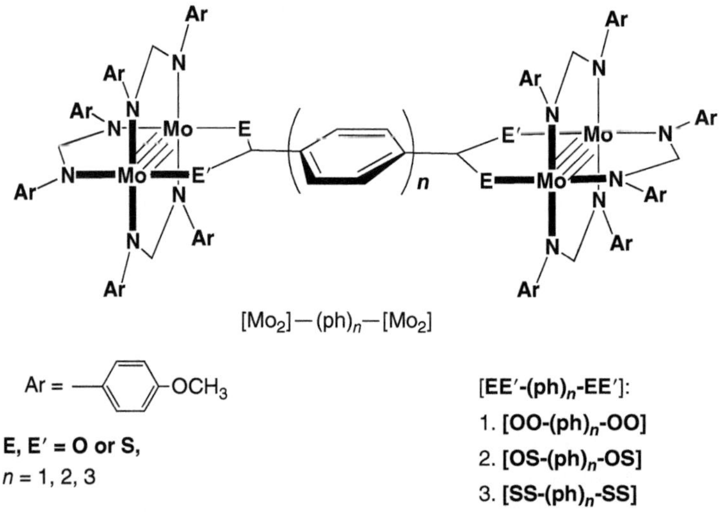

Scheme 7.8 The molecular skeleton of series **[EE′-(ph)$_n$-EE′]** (E,E′ = O, S, and n = 1–3) consisting of nine Mo$_2$ dimers.

of electron transfer. According to the widely accepted decay laws [60], H_{ab} and k_{et} are exponentially related to r_{ab} by an attenuation factor (β) that reflects the intrinsic electronic characteristics of the bridge:

$$H_{ab} = H_0 \exp(-\beta r_{ab}) \tag{7.5}$$

$$k_{et} = k_0 \exp(-\beta r_{ab}) \tag{7.6}$$

where H_0 is the electronic coupling at contact distance and k_0 is a kinetic prefactor. From Eqs. (7.5) and (7.6), a linear relationship in plot of log(H_{ab}) or log(k_{et}) vs. r_{ab} is obtained, and the attenuation factor β can then be determined from the slope. For the three MV series **[OO–(ph)$_n$–OO]$^+$**, **[OS–(ph)$_n$–OS]$^+$**, and **[SS–(ph)$_n$–SS]$^+$** (n = 1, 2, and 3), both H_{ab} and k_{et} decrease exponentially with an increase in the charge transfer distance r_{ab}, as shown in Figure 7.9. The magnitudes of $\beta(H_{ab})$ (0.2–0.3 Å^{-1}) fall in the normal range for unsaturated bridges [61], while the $\beta(k_{et})$ values are larger than those for photoinduced ET crossing a polyphenylene bridge [60, 62]. Nevertheless, these results demonstrate that the optical analysis based on the Marcus–Hush theory is a reliable method for determination of thermal ET kinetics in MV system. The largest attenuation factors are found for the dicarboxylate-bridged series, and the smallest for the fully thiolated series (Figure 7.9), indicating that the S chelating atoms improve substantially the charge transport ability of the bridge.

Remarkably, the complete series spans the whole range of coupling strength in the Class II regime to the Class II–III borderline, in which thermal electron transfer occurs and effective donor–acceptor electronic coupling can be probed optically. Of the nine mixed-valence systems, the strongest (H_{ab} = 856 cm^{-1}) and weakest (H_{ab} = 63 cm^{-1}) coupling belongs to **[SS–ph–SS]$^+$** and **[OO–(ph)$_3$–OO]$^+$**, respectively.

Figure 7.9 Distance dependences of the EC constants and the ET rates, $\beta(H_{ab})$ and $\beta(k_{et})$, respectively, for the three series, $[OO-(ph)_n-OO]^+$, $[OS-(ph)_n-OS]^+$, and $[SS-(ph)_n-SS]^+$ ($n = 1, 2,$ and 3); r_{ab} is the centroid separation between the two Mo_2 units.

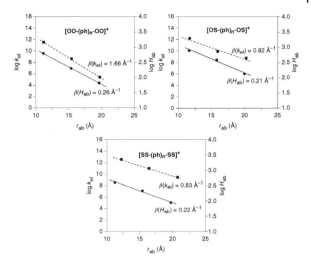

This setup of the molecular systems permits to map the H_{ab} parameters and reorganization energies λ throughout the adiabatic to the nonadiabatic limits, providing a desired test bed of the nonadiabatic transition in the Landau–Zener model. Figure 7.10 shows the variations of the adiabatic potential energy surfaces as a function of the bridging length and with changing the chelating atoms S/O. The nine mixed-valence complexes show, in an order determined by the chelating atoms (S/O) and the bridging length ($n = 1$–3), the smooth systematic transformation from the adiabatic to the nonadiabatic limit. To quantitatively distinguish the nonadiabatic and adiabatic limits, the Landau–Zener models [39] define two parameters: adiabatic parameter γ (Eq. (7.7)) and transition probability P_0 (Eq. (7.8)) with the exponent term being the nonadiabatic transition contribution, which gives the electronic transmission coefficient κ_{el} from Eq. (7.9).

$$\gamma = \frac{H_{ab}^2}{2h\nu_n}\sqrt{\frac{\pi}{\lambda k_B T}} \tag{7.7}$$

$$P_0 = 1 - \exp(-2\pi\gamma) \tag{7.8}$$

$$\kappa_{el} = 2P_0/(1 + P_0) \tag{7.9}$$

When $\gamma \gg 1$, the adiabatic limit is realized, and for thermal ET, $\kappa_{el} \approx 1$, while the nonadiabatic limit prevails with $\gamma \ll 1$. By definition of γ, it is clear that nonadiabatic transition is governed by the electronic and nuclear factors, represented by H_{ab} and ν_n, respectively. In this work [59], the Landau–Zener parameters, γ, P_0, and κ_{el}, are derived by optical analysis of the band shape of the intervalence absorptions.

Figure 7.11 shows that with $\gamma \gg 1$, $P_0 = 1$, and $\kappa_{el} = 1$, **[SS–ph–SS]$^+$** and **[OS–ph–OS]$^+$** are in the adiabatic limit, while for **[OO–(ph)$_3$–OO]$^+$**, $\gamma = 0.013$ ($\ll 1$), $P_0 = 0.076$ ($\ll 1$), and $\kappa_{el} = 0.14$ ($\ll 1$), falling in the nonadiabatic regime. Importantly, two intermediate systems **[SS–(ph)$_3$–SS]$^+$** and **[OS–(ph)$_3$–OS]$^+$** are identified, which show nonadiabatic transitional characteristics with $\kappa_{el} \approx 0.5$ (Figure 7.11 and Table 7.2). This crossover feature is intuitively present for the

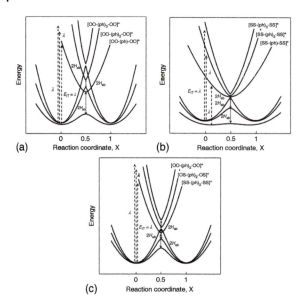

Figure 7.10 Adiabatic potential energy surfaces (PES) calculated from the IVCT data [46, 56] for the mixed-valence complexes **[EE′-(ph)$_n$-EE′]$^+$** (E,E′ = O and S, n = 1–3) [59]. (a) PES diagrams for series **[OO-(ph)$_n$-OO]$^+$** (n = 1–3), showing variation of coupling strength as a function of the bridge length. (b) PES diagrams for series **[SS-(ph)$_n$-SS]$^+$** (n = 1–3), showing variation of coupling strength as a function of the bridge length. (c) PES diagrams for series **[EE′-(ph)$_2$-EE′]$^+$** (EE′ = O and S), showing variation of the coupling strength as the chelating atoms (E) are altered stepwise.

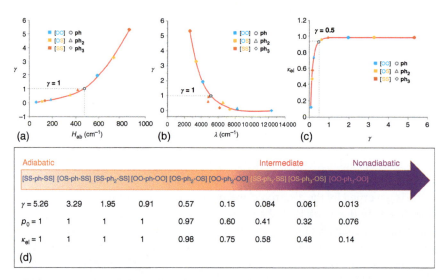

Figure 7.11 Impacts of the electronic (a) and nuclear (b) factors on the Landau–Zener parameters γ. (c) The correlation between κ_{el} and γ in systems **[EE′-(ph)$_n$-EE′]$^+$** (E,E′=O or S, n = 1–3). (d) System transition from adiabatic to nonadiabatic crossing the intermediates in terms of the Landau–Zener parameters.

Table 7.2 Spectroscopic and ET kinetic data[a] and the LZ parameters[b] for mixed-valence complexes [EE'-(ph)$_n$-EE']$^+$ (E E' = O or S and n = 1–3).[c]

	[EE'-ph-EE']$^+$			[EE'-(ph)$_2$-EE']$^+$			[EE'-(ph)$_3$-EE']$^+$		
	[OO]	[OS]	[SS]	[OO]	[OS]	[SS]	[OO][d]	[OS]	[SS]
$r_{c\text{-}c}$[d]	11.24	11.7	12.24	15.44	15.9	16.44	19.74	20.2	20.74
E_{IT} (cm^{-1})	4240	3440	2640	8300	6536	4826	12400	7400	6200
ε_{IT} (M^{-1} cm^{-1})	1470	3690	12660	198	715	1610	50	220	315
$\Delta\nu_{1/2}$ (cm^{-1})	4410	3290	1770	5185	6338	5231	3010	4210	4421
H_{ab} (cm^{-1})[e]	589	726	864	190	354	415	62	120	134
$\lambda/4$ (cm^{-1})	1060	860	660	2075	1634	1206	3100	1850	1550
ΔG^* (cm^{-1})	581	266	79	1889	1299	827	3038	1732	1422
$(\lambda/4 - H_{ab})$	471	134	−204	1885	1288	791	3038	1731	1416
k_{et}(ad) (s^{-1})	3.0×10^{11}	1.4×10^{12}	3.4×10^{12}	4.1×10^{8}	9.3×10^{9}	9.2×10^{10}	3.0×10^{5}	5.6×10^{8}	3.0×10^{9}
k_{et}(nonad) (s^{-1})	7.3×10^{11}	3.2×10^{13}	1.4×10^{14}	4.0×10^{8}	1.3×10^{10}	1.7×10^{11}	2.5×10^{5}	5.0×10^{8}	2.9×10^{9}
ν_{el} (s^{-1})[f]	1.2×10^{14}	2.1×10^{14}	3.3×10^{14}	9.1×10^{12}	3.6×10^{13}	5.7×10^{13}	7.9×10^{11}	3.9×10^{12}	5.2×10^{12}
γ	1.95	3.29	5.32	0.15	0.57	0.91	0.013	0.061	0.084
P_0	1	1	1	0.60	0.97	1	0.076	0.32	0.41
κ_{el}	1	1	1	0.75	0.98	1	0.14	0.48	0.58

a) The spectroscopic and ET kinetic data are cited from Refs. [6, 35, 58, 59].
b) The LZ parameters are calculated from Eq. (7.7)–(7.9).
c) For all the calculations, an average nuclear vibrational frequency, ν_n = 5×10^{12} s^{-1}, is adopted (Ref. [63]).
d) $r_{c\text{-}c}$ refers to the center to center separation between the two Mo$_2$ centers.
e) Electronic coupling constants are determined by optical analysis using the Hush model (Eq. (7.3)).
f) ν_{el} is the electronic transition frequency (Ref. [59]).

intermediate systems but was not pragmatically defined in the theoretical model. Furthermore, this series allows manifestation of the electronic (H_{ab}) and nuclear (λ) factors on the Landau–Zener parameters and the correlation between γ and κ_{el} (Figure 7.11). Experimental results demonstrate that crossover between the adiabatic and nonadiabatic regime occurs at $\gamma \approx 1$ (Figure 7.11a,b), exactly as predicted by the Landau–Zener model, and in the nonadiabatic limit ($\gamma \ll 1$), $\kappa_{el} < 0.5$ (Figure 7.11c). Availability of the energetic parameters attracted from the IVCT bands allowed us to calculate ET rates for systems in different MV regimes. The adiabatic ET rate constants, k_{et}(ad), for the MV systems are calculated from the classical transition-state formalism [64] (Eq. (7.10)) with a pre-exponential factor $\kappa_{el}\nu_n$ and activation energy (ΔG^*) from the Marcus–Hush theory (Eq. (7.11)) [56, 63].

$$k_{et} = \kappa_{el}\nu_n \exp\left(-\frac{\Delta G^*}{k_B T}\right) \tag{7.10}$$

$$\Delta G^* = \frac{(\lambda - 2H_{ab})^2}{4\lambda} \tag{7.11}$$

The nonadiabatic ET rate constants, k_{et}(nonad), are determined from the Levich–Dogonadze–Marcus expression (Eq. (7.11)): [65]

$$k_{et} = \frac{2H_{ab}^2}{h}\sqrt{\frac{\pi^3}{\lambda k_B T}} \exp\left(-\frac{\lambda}{4k_B T}\right) \tag{7.12}$$

In this study, we have seen that for the crossover systems, the adiabatic and nonadiabatic formalisms give the ET kinetic data nearly identical (Table 7.2), but for the adiabatic and nonadiabatic limits, large deviations are found for the results produced from different rate expressions, which validate the contemporary electron transfer theories.

7.7 Conformational Effects of Electronic Coupling and Electron Transfer

The Mo_2–BL–Mo_2 systems have been exploited in study of bridge conformation effect on intramolecular EC and ET. Conformational variation is achieved by steric hindrance between the chelating atoms and the neighboring H atoms or substituent groups on the bridging ligand (Scheme 7.9). The torsion angles from the plane defined that the two Mo_2 units are measured from crystal structures. The conformational effects can be evaluated by the changes of electrochemical and spectral data and the derived coupling and ET kinetic parameters, in comparison with those from the analogs having the same Mo_2 building blocks and bridging group, for example, those with terephthalate or derivatives that share the same phenylene (ph) bridging group that controls equal electrostatic effect. For example, **[O(NH)-1,4-naph-O(NH)]** with torsion angles of 34–50° has $\Delta E_{1/2}$ 61 mV, lower than those of **[OO-ph-OO]** (91 mV) and **[O(NH)-ph-O(NH)]** (80 mV) and does not exhibit an IVCT in the visible region, although these molecules have similar $Mo_2\cdots Mo_2$ separations. Compared to **[SS-ph-SS]** (torsion angles: 24°) with $\Delta E_{1/2} = 195$ mV, the large torsion angle (65°) in **[SS-(2,5-dimethylph-SS)**

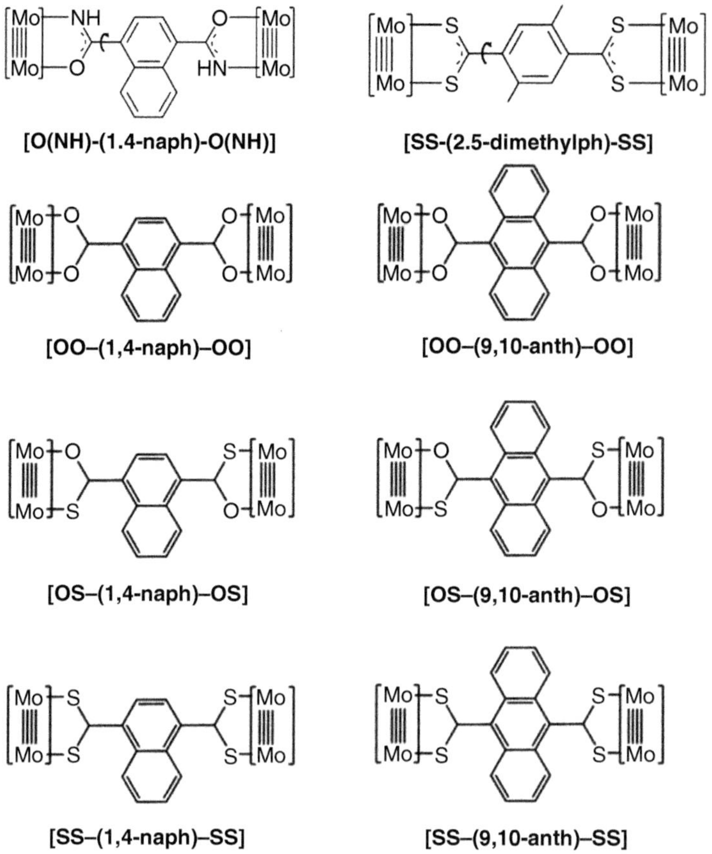

[Structures shown in Scheme 7.9:]

[O(NH)-(1,4-naph)-O(NH)]   [SS-(2,5-dimethylph)-SS]

[OO–(1,4-naph)–OO]   [OO–(9,10-anth)–OO]

[OS–(1,4-naph)–OS]   [OS–(9,10-anth)–OS]

[SS–(1,4-naph)–SS]   [SS–(9,10-anth)–SS]

Scheme 7.9 Mo_2 dimers studied to investigate the conformational dependence of electronic coupling of the bridge. For these complexes, the ancillary ligands for the dimolybdenum units are DAniF.

(Scheme 7.9) lowers the $\Delta E_{1/2}$ value by half [66]. For the MV complexes, the IVCT band is blue shifted from 2640 cm^{-1} for **[SS–ph–SS]**$^+$ to 4100 cm^{-1} for the latter, accompanied by an increase in bandwidth and a decrease in the absorption intensity; accordingly, the H_{ab} value is lowered from 895 cm^{-1} to 240 cm^{-1} [66].

Replacing the ph bridging group in the terephthalate-type series **[EE′–ph–EE′]** (E, E′ = O, S) [6] with a 1,4-naphthalene group produced a series of three compounds **[EE′–(1,4-naph)–EE′]** (E, E′ = O, S) (Scheme 7.9). While the $Mo_2\cdots Mo_2$ separation remains similar, the central plane deviates the charge transfer plane because of the geometric hindrance between the chelating atoms and the nearest H atoms on the 1,4-naphthalene bridge. The steric hindrance increases progressively with stepwise substitution of the O chelating atoms with large S atoms, which enlarges the torsion angles from ~25° to ~58° to ~63° (Figure 7.12) [67]. From DFT calculations, HOMO–HOMO-1 gaps are reduced relative to the ph series, indicating a decoupling effect induced by the conformation variations. By contrast to the ph series [6], **[SS–(1,4-naph)–SS]** has the smallest ΔE_{H-H-1}, 0.07 eV, even smaller than that for **[OO–(1,4-naph)–OO]**. The fully thiolated complex with the largest torsion

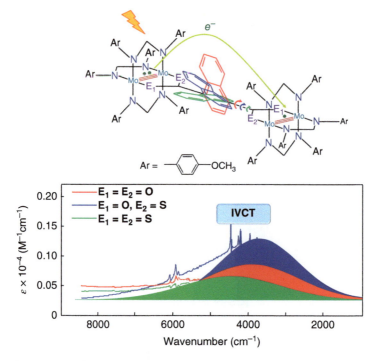

Figure 7.12 Control of electronic coupling between the two bridged Mo$_2$ units by tuning the torsion angle of the bridge with respect to the plane defined by the dimetal units. Source: Adapted from Kang et al. [67].

angle suffers the largest decoupling effect, which lowers the $\Delta E_{1/2}$ value to 107 mV from 195 mV for **[SS–ph–SS]** and increases the IVCT energy to 3680 cm^{-1} from 2640 cm^{-1} for **[SS–ph–SS]**$^+$. The coupling constant and ET rates are determined from the Marcus–Hush theory [35, 51, 52], which shows that the magnitude of ET rates decreases with the increase of the torsion angles [67]. The results show that the decoupling effects induced by the steric hindrance offset the enhancement of S atoms on electronic coupling [6, 59], which obviously diminishes the through-bond orbital interactions. Figure 7.12 illustrates the conformational control of optical electron transfer in the series of **[EE′-(1,4-naph)-EE′]** (E,E′ = O, S).

Assembling two [Mo$_2$(DAniF)$_3$]$^+$ building blocks with 9,10-anthracenedicarboxylate and its thiolated derivatives generated the anthracene-bridged series **[EE′-(9,10-anth)-EE′]**, as shown in Scheme 7.9 [68]. In this series, the three members with successively increased S chelating atoms have the torsion angles 50°, 70°, and 76°, which diminish largely the donor–acceptor coupling. For each of them, electrochemical measurements show a two-electron process in the voltammograms. For the mixed-valence complexes, there is no IVCT absorptions detected in the spectra. Accordingly, the three MV complexes of the series are assigned to Robin–Day's Class I. Therefore, these Mo$_2$–BL–Mo$_2$ systems show a systematic transition in mixed valency from Class I to Class II–III through conformational modification.

Conjugation between M$_2$ units can also be controlled by temperature. The torsional angle between the dimetal units in the oxalate-bridged complex [M$_2$(O$_2$C^{t}Bu)$_3$](μ-O$_2$CCO$_2$) can be changed because of the rotation around the oxalate C—C bond, with calculated rotational barriers of just 3.0 (M = Mo) and 5.8 kcal mol^{-1} (M = W) between the ground-state planar (D_{2h}, $\theta = 0°$) and twisted (D_{2d}, $\theta = 90°$) forms [69]. Variable temperature electron absorption spectra were obtained between 300 and 2 K in 2-MeTHF solutions. The intense MLCT absorption in the visible region of these compounds red shifts and increases significantly in intensity as the temperature is reduced. The reduction in temperature reduces rotation around the oxalate C—C bond, favoring the planar ground state and maximizing the conjugation between the M$_2$-δ and oxalate π* orbitals. A similar effect was seen in the variable temperature spectra of [M$_2$(O$_2$C^{t}Bu)$_3$](μ-O$_2$CC$_6$X$_4$CO$_2$) (X = H and F), in which the planar arrangement of the phenyl ring in the bridging ligand is favored at low temperature, maximizing the conjugation between the two M$_2$ units [70].

Rotation of the central phenyl ring in terephthalate-bridged derivatives can also be controlled using hydrogen bond interactions. This is seen in both [Mo$_2$(O$_2$C^{t}Bu)$_3$](μ-C$_6$H$_2$-2,5-(OH)$_2$-1,4-(CO$_2$)$_2$) and [Mo$_2$(O$_2$C^{t}Bu)$_3$](μ-C$_6$H$_2$-2,5-(NHPh)$_2$-1,4-(CO$_2$)$_2$) in which OH··· or NH···O$_{carboxylate}$ hydrogen bond interactions favor the planar ground state [70, 71]. For [Mo$_2$(O$_2$C^{t}Bu)$_3$](μ-C$_6$H$_2$-2,5-(OH)$_2$-1,4-(CO$_2$)$_2$), deprotonation of the OH groups was achieved by reaction with NBu$_4$OH, resulting in the disappearance of the MLCT transition in the visible region because of the twisting of the phenyl ring to minimize unfavorable O$^-$···O$_{carboxylate}$ interactions. The loss of the MLCT transition indicates the removal of Mo$_2$-δ–bridge π* conjugation, which can be restored by reaction with H$^+$, as shown in Figure 7.13. These results indicate that dimetal complexes can act as molecular rheostats (temperature) and switches (deprotonation/protonation) in molecular devices.

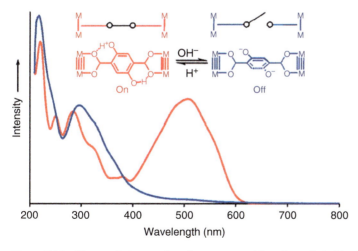

Figure 7.13 Electronic interactions between two M$_2$ units switchable through adjustment of the pH value in solution. Source: Chisholm et al. [70]/Reproduced with permission of the American Chemical Society.

7.8 Class III and Beyond

Extremely strong coupling has been observed in the M_2–BL–M_2 systems with the shortest conjugated bridge, oxalate or its thiolated derivatives. Oxalate-bridged M_2 (M = Mo and W) dimers were the first synthesized in early 1990s [8]. Their MV properties were thoroughly studied and well understood many years later as this field became mature [10, 72]. A very interesting series of [MM′]–ox–[MM′] ([MM′] = {MM′$(O_2C^tBu)_3$}) were reported by Chisholm and coworkers, which includes the two homonuclear (MM = Mo_2 and W_2) and one heteronuclear (MM′ = MoW) dimers of dimers, as shown in Figure 7.14. In the series, the weakest coupling system is **[Mo_2-ox-Mo_2]$^+$** with E_{IT} = 3900 cm^{-1}, while the strongest coupling system is **[W_2-ox-W_2]$^+$** with E_{IT} = 5988 cm^{-1}. The MV complex with mixed-dimetal (MoW) units has an E_{IT} of 5076 cm^{-1}, being the intermediate between the two homonuclear systems.

Liu and coworkers studied another very strongly coupled series of Mo_2 dimers bridged by oxalate and thiolated oxalate, namely, **[EE′-ox-EE′]**, where E, E′ = O or S. In this series, each MV complex exhibits a characteristic IVCT band, with E_{IT} = 4077 cm^{-1} for **[OO-ox-OO]$^+$**, 4149 cm^{-1} for **[OS-ox-OS]$^+$**, and 4926 cm^{-1} for **[SS-ox-SS]$^+$** [42]. Significantly, the two series, developed by variation of the constitution of the dimetal units for one and by alternation of the chelating atoms S/O for the other, are parallel, showing that the increase of the coupling strength shifts the IVCT band toward high energy. This spectral behavior characterizes explicitly the two series belonging to the Robin–Day's Class III and beyond because for those in Class II, increase of the coupling strength results in a red shift of the IVCT absorption. This feature is predicted by theory in mixed-valence chemistry,

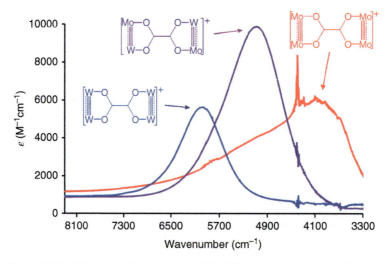

Figure 7.14 The intervalence spectra of the MV complexes in series [M_2-ox-M_2]$^+$ (M_2 = MoMo, MoW, and WW). Source: Chisholm et al. [10]/Reproduced with permission of the American Chemical Society.

showing that the physical origin of IVCT absorption changes from intervalent vibronic to vertical electronic transition. For genuine Class III (fully delocalized) system, it is predicted theoretically that the "IVCT" band actually arises from electronic transition between two delocalized molecular orbitals [34, 46], involving the δ orbitals of the dimetal centers in the current case. Indeed, for **[OS–ox–OS]$^+$** and **[SS–ox–SS]$^+$**, the measured IVCT energies (E_{IT}) equal almost the DFT calculated HOMO–HOMO-1 energy gaps ($\Delta E_{H\text{-}H\text{-}1}$) [42]. Another important optical feature that signals the transition of mixed valency is the shape of IVCT band defined by the half-highest bandwidth $\Delta v_{1/2}$. In the spectra for **[OO–ox–OO]$^+$**, the IVCT band exhibits a large cut-off (34%), while the IVCT bands for the other two analogs are more symmetric, which is also seen in the [MM′]–ox–[MM′] series (Figure 7.14). This occurs as the increased coupling reduces the anharmonicity to produce more regularly shaped spectra. Furthermore, the energy of the IVCT band **[OS–ox–OS]$^+$** and **[SS–ox–SS]$^+$** is solvent independent, as expected as there is no net charge transfer associated with transitions between delocalized orbitals; in contrast, **[OO–ox–OO]+** exhibits a solvent-dependent IVCT absorption associated with the interaction of the transferred charge with the solvent [42]. Assembling two [Mo$_2$(DAniF)$_3$]$^+$ with oxamidate and its dithiolated derivative gave another two strongly coupled Mo$_2$ dimers, **[NO–oxam–NO]** and **[NS–oxam–NS]** [42]. **[NS–oxam–NS]$^+$** exhibits an intense IVCT band at 3823 cm^{-1} (ε_{IT} = 19 480 M^{-1} cm^{-1}), higher in energy and intensity than those for **[NO–oxam–NO]$^+$** (E_{IT} = 2910 cm^{-1} and ε_{IT} = 3500 M^{-1} cm^{-1}), in parallel with the oxalate analogs. Optical analyses of the IVCT bands indicate $2H_{ab}/\lambda \approx 1$ for **[NO–oxam–NO]$^+$**, belonging to the Class II–III or the Class III limit, while **[NS–oxam–NS]$^+$** with $2H_{ab}/\lambda = 3.5 \gg 1$ [42]. In the previous section, we have seen that **[SS–thi–SS]$^+$** ($2H_{ab}/\lambda = 3.4$) presents similar optical properties and parameters. These very strongly coupled MV systems exhibit distinct spectral envelopes arising from the changing electronic structure, in agreement with the basic predictions made using the two-state model of electron transfer.

Note, however, that standard harmonic-based spectral simulation approaches such as the use of Franck–Condon (Huang–Rhys) factors fail as the anharmonicity of the ground-state and excited-state potential energy surfaces, as well as the vibronic coupling that occurs between them, must be taken into account [73–75]. It has been noted that spectra in this domain do not simply fit into expectations based on the Robin–Day scheme [42, 43, 76], demanding its slight adaptation.

7.9 Cross-Conjugation and Quantum Destructive Effect

To examine the destructive quantum interference of the bridge, the Mo$_2$–BL–Mo$_2$ series with 1,3(*meta*)-terephthalate and its thiolated derivatives, namely, **[EE′-(1,3-ph)–EE′]** (E,E = O, S) were developed [77], as shown in Scheme 7.10. The EC and ET were studied, in comparison with the 1,4(*para*)-terephaphthalate-bridged series. It was found that for the cross-conjugated bridge (1,3-ph), the potential separations $\Delta E_{1/2}$ are appreciably reduced, although the Mo$_2$⋯Mo$_2$ spatial

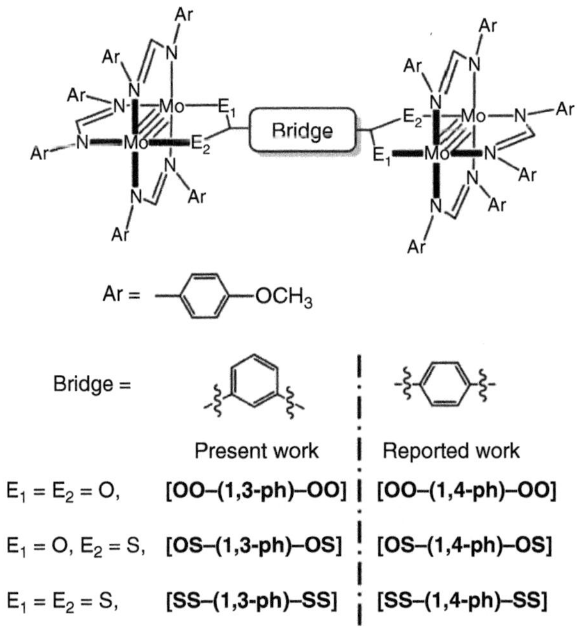

Scheme 7.10 Molecular scaffolds of the Mo$_2$ dimers bridged by a meta-phenylene bridge (left), in comparison with the analogs with a para-phenylene bridge (right). Source: Gao et al. [77]/John Wiley & Sons.

separation is even shorter, which enhances the electrostatic interaction. The IVCT energies are dramatically increased, and the absorptions are extremely weak, for example, 6746 cm^{-1} (ε_{IT}, 343 M^{-1} cm^{-1}) for **[SS–(1,3-ph)–SS]$^+$** vs. 2640 cm^{-1} (ε_{IT}, 1.3 × 10^4 M^{-1} cm^{-1}) for the *para* analog. Compared to the data for the *para* analog, the H_{ab} values are reduced by half and the electron transfer rates (k_{et}) are lowered by 2 orders of magnitude [77]. For the cross-conjugated mixed-valence complexes, nearly identical k_{et} data were obtained from the adiabatic and nonadiabatic treatments, implicating high nonadiabaticity of electron transfer in the system. This study demonstrates that with a *meta* phenylene bridge, the large orbital amplitude of the S-containing systems does not contribute to enhance the donor–acceptor coupling, in agreement with the DFT calculation results that indicate a cancelling of the orbital phases between charge transfer channels because of symmetry violations [77]. This study has also shown that when the π-conjugated charge transfer is interrupted, through-σ-bond and/or through-space pathways are likely to be operative.

7.10 Electronic Coupling and Electron Transfer Across Hydrogen Bonds

Dimers of Mo$_2$ have been exploited to examine the electronic coupling and electron transfer crossing a hydrogen bond (HB) interface. The complexes

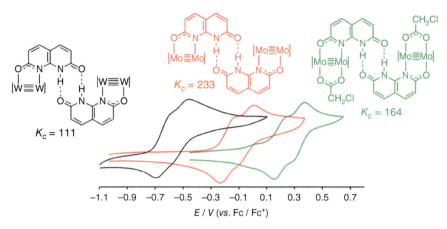

Figure 7.15 Molecular structures and cyclic voltammograms of the hydrogen-bonded dimers $W_2(TiPB)_3(HDON)$, $Mo_2(TiPB)_3(HDON)$, and $Mo_2(TiPB)_2(O_2CCH_2Cl)(HDON)$ in DCM. Source: Wilkinson et al. [78]/Reproduced with permission of the American Chemical Society.

$Mo_2(TiPB)_3(HDON)$ (TiPB = 2,4,6-triisopropylbenzoate and H_2DON = 2,7-dihydroxynaphthyridine) and $Mo_2(TiPB)_3HDOP$ (H_2DOP = 3,6-dihydroxypyridazine) dimerize in DCM [78], producing the corresponding H-bonded Mo_2 dimers, as shown in Figure 7.15 for the HDON derivatives. The cyclic voltammograms of these complexes exhibit two separated one-electron oxidations, demonstrating stabilization of the MV state. Unfortunately, the lack of the IVCT band in their near-IR spectra prevents quantitative evaluation of the degree of electronic coupling.

Liu's group approached this issue through designed synthesis of the Mo_2 acetamide complexes, $Mo_2(DAr(X)F)_3(acetamide)$, where DAr(X)F = diarylformamidinate and X is the *para* substituent playing a role in tuning the electronic properties of the ligand [79]. A series of four $Mo_2(DAr(X)F)_3(acetamide)$ complexes with X = $N(CH_3)_2$, $CH(CH_3)_2$, OCH_3, and CH_3 were synthesized. There are two chelating groups on the acetamide ligand that favor a dimetal coordination, carboxylate and amidate. The great difference in pK_a between them allows the carboxylate selectively binding to the Mo_2 unit, leaving a free amide group for dimerization through amide–amide hydrogen bonding linkage. The four HB-assembled Mo_2 dimers were structurally characterized by X-ray diffraction, showing the co-planar structure of the amide–amide central moiety with the two Mo_2 unit, as shown in Scheme 7.11a. The N–H···O distances are ~2.9 Å, indicating weak hydrogen bonding interactions. The $\Delta E_{1/2}$ of ~100 mV indicates that the two Mo_2 centers bridged by the HB interface are moderately strongly coupled, comparable with the degree of electronic coupling in the covalently bonded terephthalate-bridged analog that has a similar charge transfer distance [35]. By contrast to the HDON/HDOP series discussed above, the mixed-valence state for each of these hydrogen-bonded dimers shows a characteristic IVCT band. This allows optical evaluation of the EC strength and the ET rate based on the Marcus–Hush theory. It was found that $200\,\text{cm}^{-1} > H_{ab} > 100\,\text{cm}^{-1}$, except for the system with strong electronic donating groups (X = $N(CH_3)_2$). The magnitude of k_{et} falls in the range of 10^{10}–$10^{11}\,\text{s}^{-1}$,

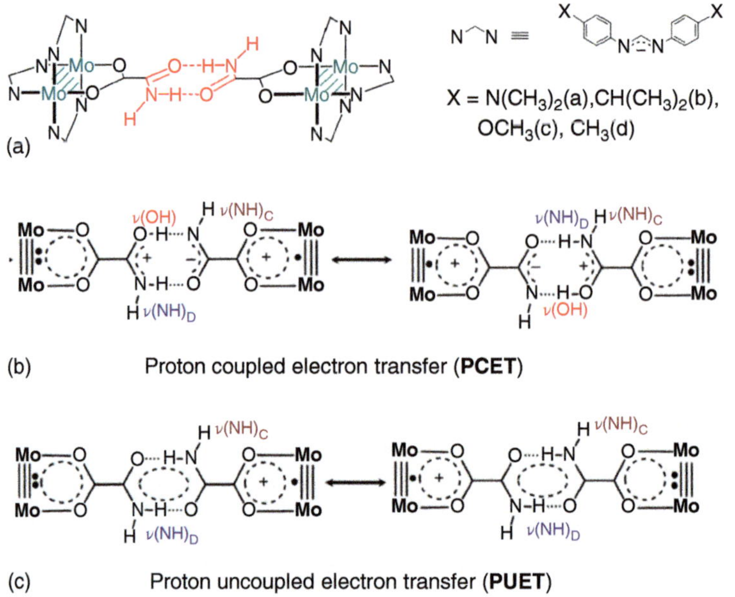

Scheme 7.11 (a) Molecular systems of the Mo$_2$ dimers (X = the *para* substituents on the formamidinate ligands) bridged by amide–amide hydrogen bonds. (b) The proton-coupled (PCET) and proton-uncoupled (PUET) electron transfer mechanisms proposed for the systems.

comparable but reasonably smaller than those for the terephthalate-bridged analogs (~10^{11} s^{-1}). This timescale for ET crossing over the HB is confirmed by simulation of v(NH) vibrational band broadening [79]. Importantly, analysis of the v(NH) and OH vibrational modes of the amide group in the IR spectra reveals that electron transfer across a hydrogen bond may proceed via the known proton-coupled pathway as well as an overlooked proton-uncoupled pathway that does not involve proton transfer (Scheme 7.11b,c). The mechanistic choice is determined by the extent of EC and the hydrogen bonding strength. These results have important implications in the design of self-assembled hydrogen-bonded materials that have interesting charge transport properties.

7.11 Mixed-Valence Diruthenium Dimers

Although the focus of this chapter is on the use of quadruply bonded MM units as the electron donor/acceptor, related dimetal systems have been used in the assembly of [M$_2$–BL–M$_2$] MV complexes. Perhaps, the most notable is the use of [(4,0)-Ru$_2$(ap)$_4$]$^+$ (ap = 2-anilinopyridinate) as buildings blocks, as reported by Ren and coworkers. The diruthenium(II,III) unit has three unpaired electrons and a $\sigma^2\pi^4\delta^2(\pi^*\delta^*)^3$ electronic configuration. In contrast to quadruply bonded complexes, this electron configuration results in strongest metal ligand coupling in the axial positions of the paddlewheel because of the M$_2$(π^*)–ligand(π) interactions. As such,

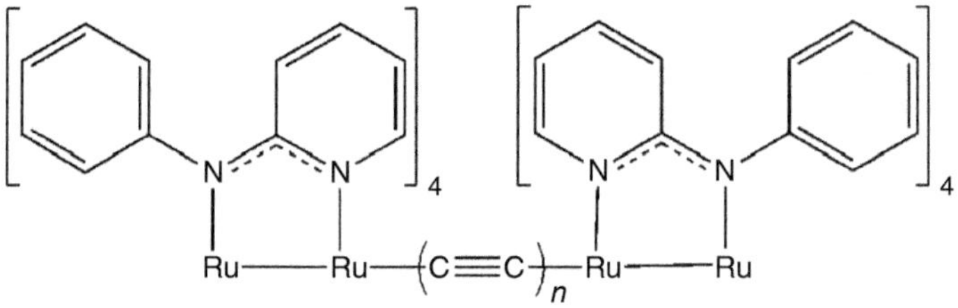

Scheme 7.12 Structure of [Ru$_2$(ap)$_4$]-(C≡C)$_n$-[Ru$_2$(ap)$_4$] (n = 1–6).

polyyn-diyls have found to be efficient bridges for charge transport between the diruthenium units. The crystal structures of the [Ru$_2$(ap)$_4$]-(C≡C)$_n$-[Ru$_2$(ap)$_4$] (n = 1–4 and 6, see Scheme 7.12) series of compounds show a linear rigid rod topology of the polyyn-diyl bridging ligand [80, 81]. These compounds can undergo one-electron oxidation, or reduction, to generate the MV state. When n = 1, cyclic voltammetry shows a 0.36 V (K_c = 5.4 × 10^5) separation between the 0/1+ and 1+/2+ redox couples and a much larger 0.66 V (K_c = 1.4 × 10^{11}) separation between the 0/1− and 1−/2− redox couples. This suggests strong coupling between the Ru$_2$ units, with greater coupling observed upon reduction. This is attributed to the $(\pi^*)^4$ ground state for Ru$_2$(II,II) as opposed to the $(\pi^*)^2$ ground state for Ru$_2$(III,III), which couples more weakly with the polyyn-diyl bridge. As n increases, the Kc values decrease as expected.

Substitution of ap with Xap [Xap = 2-(3,5-dimethoxyanilino)pyridinate or 2-(3-butoxy- anilino)pyridinate] in [Ru$_2$(ap)$_4$]$_2$(μ-C$_6$) increases the solubility of the complexes; voltammetric and spectroelectrochemical studies show significant coupling between the two Ru$_2$ termini in the MV complex monoanion [82]. Reaction with tetracyanoethane (TCNE) yields the cycloaddition/insertion product [Ru$_2$(ap)$_4$]$_2$(μ-C≡CC(C(CN)$_2$)C(C(CN)$_2$)C≡C). Spectroelectrochemical studies reveal the absence of an IVCT transition upon oxidation and reductions that are centered on the bridging ligand. However, reaction of the tetraruthenium C$_6$ complex with Co$_2$(dppm)(CO)$_6$ yields an η^2-Co$_2$ adduct on the middle —C≡C—, which has larger $\Delta E_{1/2}$ values for the redox couples than the parent complex. Surprisingly, despite the larger $\Delta E_{1/2}$ values, the Co$_2$ adduct displays an IVCT transition that is consistent with a weakly coupled Class II system, suggesting that factors other than resonance exchange contribute to K_C.

The ability of polyyn-diyl bridges to electronically couple diruthenium through long distances (>10 Å) was demonstrated in the [Ru$_2$(Xap)$_4$]$_2$(μ-C$_{4k}$) (k = 2–4) series of compounds, whose crystal structures reveal Ru$_2$···Ru$_2$ separations that range from 13 Å (k = 2) to 23 Å (k = 4) [83]. Spectroelectrochemical studies reveal that the MV complexes generated upon one-electron oxidation are Class II, with H_{ab} values that decrease from 380 to 88 cm^{-1} as the length of the bridge increases. Evaluation of electronic coupling in the phenylene-bridged Ru$_2$(II,III) complex [Ru$_2$(ap)$_4$]$_2$(μ-C$_6$H$_4$) was thwarted by its poor solubility and air sensitivity [84]. Reaction with [NBu$_4$][CN] resulted in the formation of the more soluble and stable Ru$_2$(III,III) complex [(CN)Ru$_2$(ap)$_4$]$_2$(μ-C$_6$H$_4$). Cyclic voltammetry showed two successive one-electron oxidations ($\Delta E_{1/2}$ = 291 mV) and reductions

($\Delta E_{1/2}$ = 174 mV). Despite the appreciable separation in redox processes upon reduction, IR and UV/vis/NIR spectroelectrochemical studies on the reduced MV complex indicate a localized (Class I) system on the IR timescale. By contrast, spectroelectrochemical studies on the MV species generated upon one-electron oxidation show that it is strongly coupled (Class III) in nature. DFT calculations showed that the HOMO in the neutral complex is a Ru_2-π^* that mixes strongly with a phenylene π-orbital, whereas the LUMO is a Ru_2-δ^* orbital that does not mix with the phenylene-π, accounting for the difference in coupling between the two MV states.

7.12 Conclusions and Outlook

Since the 1990s when metal–metal quadruple bond chemistry met with mixed-valence chemistry through seminal contributions from Cotton and Chisolm, a diversity of mixed-valence compounds in the form of M_2–BL–M_2 or MM′–BL–MM′ have been prepared, in which covalently bonded dimetal units are used as the electronic donor (D) and acceptor (A) and studied in terms of electronic coupling and electron transfer. In the case where the two metal atoms are quadruply bonded, the donor and acceptor have a formal MM bond order of 4.0 and 3.5 for the MV complexes, respectively, and intramolecular EC and ET involve solely the δ electrons. The unique electronic configuration of a dimetal center (M_2) makes the M_2–BL–M_2 molecule a truly single electron transferring system, in accordance with the semiclassical two-state models. The structural and electronic features render these MV systems most suitable for X-ray diffraction, electrochemical and optical studies to characterize the MV complexes within the Robin–Day classification scheme. Experimental and theoretical studies have shown that in the dimetal MV systems, EC and ET can be tuned in different regimes from Class I to Class III, and beyond, by judicious choice of the MM center (MoMo, MoW, and WW), its supporting ligands, and by the chemical composition of the bridging ligand (the chelating atoms and the central group). Recent investigations have examined various structural factors that affect the EC and ET, including ancillary ligands, chelating atoms, length, and conjugation of the BL. In various $[Mo_2$–BL–$Mo_2]^+$ mixed-valence systems, the EC and ET are quantitatively evaluated by optical analysis based on the Marcus–Hush theory. Although effective ET distances (r_{ab}') can be determined experimentally by measuring dipole moment changes for the IVCT by Stark effect spectroscopy, these measurements are difficult to obtain. The results on $[Mo_2$–BL–$Mo_2]^+$ systems show that estimation of r_{ab}' from the geometrical length of the central moiety of the bridging ligand, as opposed to the $Mo_2\cdots Mo_2$ internuclear separation, gives reasonable and consistent results in calculation of the coupling parameter (H_{ab}) from the Mulliken–Hush expression. The CNS model derived from the superexchange theory has been validated by the dimetal MV systems and becomes an alternative for determination of the exchange integral. The combination of metal–metal bond chemistry with mixed-valence chemistry has contributed to advances in our knowledge in molecular science. Moving forward in this direction, it is expected that the structural, electronic, and optical features of dimetal units and the understanding gained in study of M_2 dimers can be exploited in development and application of innovative optoelectronic devices and materials.

Acknowledgments

We thank the National Science Foundation of China (no. 21971088, 20871093, 90922010, 21371074, and 21301070), the Natural Science Foundation of Guangdong Province (2018A030313894), and the Jinan University and Tongji University for financial support. We are grateful to Prof. Jeffrey Reimers (University of Technology Sydney) for comments and suggestions on the manuscript.

References

1 Cotton, F.A., Curtis, N.F., Harris, C.B. et al. (1964). Mononuclear and polynuclear chemistry of rhenium (III): its pronounced homophilicity. *Science* 145: 1305–1307.
2 Cotton, F.A. and Nocera, D.G. (2000). The whole story of the two-electron bond, with the δ bond as a paradigm. *Acc. Chem. Res.* 33: 483–490.
3 Cotton, F.A., Murillo, C.A., and Walton, R.A. (2005). *Multiple Bonds between Metal Atoms*, 3e. Springer: New York.
4 Cotton, F.A., Daniels, L.M., Liu, C.Y. et al. (2002). How to make a major shift in a redox potential: ligand control of the oxidation state of dimolybdenum units. *Inorg. Chem.* 41: 4232–4238.
5 Cotton, F.A., Daniels, L.M., Murillo, C.A. et al. (2002). The extraordinary ability of guanidinate derivatives to stabilize higher oxidation numbers in dimetal units by modification of redox potentials: structures of Mo_2^{5+} and Mo_2^{6+} compounds. *J. Am. Chem. Soc.* 124: 9249–9256.
6 Xiao, X., Liu, C.Y., He, Q. et al. (2013). Control of the charge distribution and modulation of the class II–III transition in weakly coupled Mo_2-Mo_2 systems. *Inorg. Chem.* 52: 12624–11263.
7 Cayton, R.H. and Chisholm, M.H. (1989). Electronic coupling between covalently linked metal-metal quadruple bonds of molybdenum and tungsten. *J. Am. Chem. Soc.* 111: 8921–8923.
8 Cayton, R.H., Chisholm, M.H., Huffman, J.C., and Lobkovsky, E.B. (1991). Metal-metal multiple bonds in ordered assemblies. 1. Tetranuclear molybdenum and tungsten carboxylates involving covalently linked metal-metal quadruple bonds. Molecular models for subunits of one-dimensional stiff-chain polymers. *J. Am. Chem. Soc.* 113: 8709–8724.
9 Richardson, D.E. and Taube, H. (1981). Determination of $E_2°-E_1°$ in multistep charge transfer by stationary-electrode pulse and cyclic voltammetry: application to binuclear ruthenium ammines. *Inorg. Chem.* 20: 1278–1285.

10 Chisholm, M.H. and Patmore, N.J. (2007). Studies of electronic coupling and mixed valency in metal–metal quadruply bonded complexes linked by dicarboxylate and closely related ligands. *Acc. Chem. Res.* 40: 19–27.

11 Robin, M.B. and Day, P. (1967). Mixed valence chemistry-a survey and classification. *Adv. Inorg. Chem. Radiochem.* 10: 247–422.

12 McConnell, H.M. (1961). Intramolecular charge transfer in aromatic free radicals. *J. Chem. Phys.* 35: 508–515.

13 Cotton, F.A., Lin, C., and Murrillo, C.A. (1998). Coupling Mo_2^{n+} units via dicarboxylate bridges. *J. Chem. Soc. Dalton Trans.* 19: 3151–3153.

14 Cotton, F.A., Donahue, J.P., and Murillo, C.A. (2003). Polyunsaturated dicarboxylate tethers connecting dimolybdenum redox and chromophoric centers: syntheses, structures, and electrochemistry. *J. Am. Chem. Soc.* 125: 5436–5450.

15 Cotton, F.A., Donahue, J.P., Murillo, C.A., and Pérez, L.M. (2003). Polyunsaturated dicarboxylate tethers connecting dimolybdenum redox and chromophoric centers: absorption spectra and electronic structures. *J. Am. Chem. Soc.* 125: 5486–5492.

16 Chisholm, M.H. (2013). Mixed valency and metal–metal quadruple bonds. *Coord. Chem. Rev.* 257: 1576–1583.

17 Cotton, F.A., Li, Z., Liu, C.Y., and Murillo, C.A. (2007). Modulating electronic coupling using O- and S-donor linkers. *Inorg. Chem.* 46: 7840–7847.

18 Lei, H., Xiao, X., Meng, M. et al. (2015). Dimolybdenum dimers spaced by phenylene groups: the experimental models for study of electronic coupling. *Inorg. Chim. Acta* 424: 63–74.

19 Cayton, R.H., Chisholm, M.H., Huffman, J.C., and Lobkovsky, E.B. (1991). Mo-Mo quadruple bonds bridged by 1,8-naphthyridinyl-2,7-dioxide: an insight into the nature of a parallel-linked stiff-chain polymer: $\{M^nM\sim\sim\sim M^nM\}_x$. *Angew. Chem. Int. Ed.* 30: 862–864.

20 Chisholm, M.H., Epstein, A.J., Gallucci, J.C. et al. (2005). Oligothiophenes incorporating metal–metal quadruple bonds. *Angew. Chem. Int. Ed.* 40: 6695–6698.

21 Chisholm, M.H., Cotton, F.A., Daniels, L.M. et al. (1999). Compounds in which the Mo_2^{4+} unit is embraced by one, two or three formamidinate ligands together with acetonitrile ligands. *J. Chem. Soc. Dalton Trans.* 9: 1387–1392.

22 Cotton, F.A., Donahue, J.P., Lin, C., and Murillo, C.A. (2001). The simplest supramolecular complexes containing Pairs of $Mo_2(formamidinate)_3$ units linked with various dicarboxylates: preparative methods, structures, and electrochemistry. *Inorg. Chem.* 40: 1234–1244.

23 Cotton, F.A., Daniels, L.M., Donahue, J.P. et al. (2002). The first designed syntheses of bis-dimetal molecules in which the bridges are diamidate ligands. *Inorg. Chem.* 41: 1354–1356.

24 Cotton, F.A., Liu, C.Y., Murillo, C.A. et al. (2003). Modifying electronic communication in dimolybdenum units by linkage isomers of bridged oxamidate dianions. *J. Am. Chem. Soc.* 125: 13564–13575.

25 Cotton, F.A., Liu, C.Y., Murillo, C.A., and Wang, X. (2003). Trapping tetramethoxyzincate and -cobaltate(II) between Mo_2^{4+} units. *Inorg. Chem.* 42: 4619–4623.
26 Lin, C., Protasiewicz, J.D., Smith, E.T., and Ren, T. (1996). Linear free energy relationships in dinuclear compounds. 2. Inductive redox tuning via remote substituents in quadruply bonded dimolybdenum compounds. *Inorg. Chem.* 35: 6422–6428.
27 Fang, W., He, Q., Tan, Z.F. et al. (2011). Experimental and theoretical evidence of aromatic behavior in heterobenzene-like molecules with metal–metal multiple bonds. *Chem. Eur. J.* 17: 10288–10296.
28 Tan, Z.F., Liu, C.Y., Li, Z. et al. (2012). Abnormally long-range diamagnetic anisotropy induced by cyclic d_δ–p_π π conjugation within a six-membered dimolybdenum/chalcogen ring. *Inorg. Chem.* 51: 2212–2221.
29 Cheng, T., Meng, M., Lei, H., and Liu, C.Y. (2014). Perturbation of the charge density between two bridged Mo_2 centers: the remote substituent effects. *Inorg. Chem.* 53: 9213–9221.
30 Cotton, F.A., Liu, C.Y., Murillo, C.A., and Zhao, Q. (2007). Electronic localization versus delocalization determined by the binding of the linker in an isomer pair. *Inorg. Chem.* 46: 2604–2611.
31 Cotton, F.A., Liu, C.Y., Murillo, C.A. et al. (2004). Strong electronic coupling between dimolybdenum units linked by the N,N'-dimethyloxamidate anion in a molecule having a heteronaphthalene-like structure. *J. Am. Chem. Soc.* 126: 14822–14831.
32 Fürholz, U., Bürgi, H.-B., Wagner, F.E. et al. (1984). The Creutz-Taube complex revisited. *J. Am. Chem. Soc.* 106: 121–123.
33 Zhu, G.Y., Meng, M., Tan, Y.N. et al. (2016). Electronic coupling between two covalently bonded dimolybdenum units bridged by a naphthalene group. *Inorg. Chem.* 55: 6315–6322.
34 Demadis, K.D., Hartshorn, C.M., and Meyer, T.J. (2001). The localized to-delocalized transition in mixed-valence chemistry. *Chem. Rev.* 101: 2655–2686.
35 Liu, C.Y., Xiao, X., Meng, M. et al. (2013). Spectroscopic study of δ electron transfer between two covalently bonded dimolybdenum units via a conjugated bridge: adequate complex models to test the existing theories for electronic coupling. *J. Phys. Chem. C* 117: 19859–19865.
36 Hush, N.S. (1967). Intervalence-transfer absorption. Part 2. Theoretical considerations and spectroscopic data. *Prog. Inorg. Chem.* 8: 391–444.
37 Marcus, R.A. (1956). The theory of oxidation–reduction reactions involving electron transfer. I. *J. Chem. Phys.* 24: 966–978.
38 Marcus, R.A. and Sutin, N. (1985). Kinetics of water transport in sickle cells. *Biochim. Biophys. Acta* 812: 811–815.
39 Newton, M.D. (1991). Quantum chemical probes of electron-transfer kinetics: the nature of donor-acceptor interactions. *Chem. Rev.* 91: 767–792.
40 Nelsen, S.F., Weaver, M.N., Zink, J.I., and Telo, J.P. (2005). Optical spectra of delocalized dinitroaromatic radical anions revisited. *J. Am. Chem. Soc.* 127: 10611–10622.

41 Nelsen, S.F., Konradsson, A.E., and Telo, J.P. (2005). Pseudo-para-dinitro[2.2]paracyclophane radical anion, a mixed-valence system poised on the class II/class III borderline. *J. Am. Chem. Soc.* 127: 920–925.

42 Tan, Y.N., Cheng, T., Meng, M. et al. (2017). Optical behaviors and electronic properties of Mo_2–Mo_2 mixed-valence complexes within or beyond the class III regime: testing the limits of the two-state model. *J. Phys. Chem. C* 121: 27860–27873.

43 Cheng, T., Tan, Y.N., Zhang, Y. et al. (2015). Distinguishing the strength of electronic coupling for Mo_2-containing mixed-valence compounds within the class III regime. *Chem. Eur. J.* 21: 2353–2357.

44 Cotton, F.A., Donahue, J.P., Lin, C. et al. (2002). A molecular pair having two quadruply bonded dimolybdenum units linked by a terephthalate dianion. *Acta Crystallogr.* E58: m298–m300.

45 Shu, Y., Lei, H., Tan, Y.N. et al. (2014). Tuning the electronic coupling in Mo_2–Mo_2 systems by variation of the coordinating atoms of the bridging ligands. *Dalton Trans.* 43: 14756–14765.

46 Brunschwig, B.S., Creutz, C., and Sutin, N. (2002). Optical transitions of symmetrical mixed-valence systems in the Class II–III transition regime. *Chem. Soc. Rev.* 31: 168–184.

47 Hush, N.S. (1968). Homogeneous and heterogeneous optical and thermal electron transfer. *Electrochim. Acta* 13: 1005–1023.

48 Nelsen, S.F. (2000). "Almost delocalized" intervalence compounds. *Chem. Eur. J.* 6: 581–588.

49 D'Alessandro, D.M. and Keene, F.R. (2006). Current trends and future challenges in the experimental, theoretical and computational analysis of intervalence charge transfer (IVCT) transitions. *Chem. Soc. Rev.* 35: 424–440.

50 Creutz, C., Newton, M.D., and Sutin, N. (1994). Metal-ligand and metal-metal coupling elements. *J. Photochem. Photobiol. A* 82: 47–59.

51 Marcus, R.A. (1956). Electrostatic free energy and other properties of states having nonequilibrium polarization. I. *J. Chem. Phys.* 24: 979–989.

52 Marcus, R.A. (1965). On the theory of electron-transfer reactions. VI. Unified treatment for homogeneous and electrode reactions. *J. Chem. Phys.* 43: 679–701.

53 Crutchley, R.J. (1994). Intervalence charge transfer and electron exchange studies of dinuclear ruthenium complexes. *Adv. Inorg. Chem.* 41: 273–325.

54 Wu, Y.Y., Meng, M., Wang, G.Y. et al. (2017). Optically probing the localized to delocalized transition in Mo_2–Mo_2 mixed-valence systems. *Chem. Commun.* 53: 3030–3033.

55 Mallick, S., Cao, L., Chen, X. et al. (2019). Mediation of electron transfer by quadrupolar interactions: the constitutional, electronic, and energetic complementarities in supramolecular Chemistry. *iScience* 22: 269–287.

56 Brunschwig, B.S. and Sutin, N. (1999). Energy surfaces, reorganization energies, and coupling elements in electron transfer. *Coord. Chem. Rev.* 187: 233–254.

57 Chisholm, M.H., Pate, B.D., Wilson, P.J., and Zaleski, J.M. (2002). On the electron delocalization in the radical cations formed by oxidation of MM quadruple bonds linked by oxalate and perfluoroterephthalate bridges. *Chem. Commun.* 10: 1084–1085.

58 Xiao, X., Meng, M., Lei, H., and Liu, C.Y. (2014). Electronic coupling and electron transfer between two dimolybdenum units spaced by a biphenylene group. *J. Phys. Chem. C* 118: 8308–8315.

59 Zhu, G.Y., Qin, Y., Meng, M. et al. (2021). Crossover between the adiabatic and nonadiabatic electron transfer limits in the Landau-Zener model. *Nat. Comm.* 12: 456.

60 Albinsson, B., Eng, M.P., Pettersson, K., and Winters, M.U. (2007). Electron and energy transfer in donor–acceptor systems with conjugated molecular bridges. *Phys. Chem. Chem. Phys.* 9: 5847–5864.

61 Carter, M.T., Rowe, G.K., Richardson, J.N. et al. (1995). Distance dependence of the low–temperature electron transfer kinetics of (ferrocenylcarboxy)-terminated alkanethiol monolayer. *J. Am. Chem. Soc.* 117: 2896–2899.

62 Helms, A., Heiler, D., and McLendon, G. (1992). Electron transfer in bis–porphyrin donor–acceptor compounds with polyphenylene spacers shows a weak distance dependence. *J. Am. Chem. Soc.* 114: 6221–6238.

63 Creutz, C. (1983). Mixed-valence-complexes-of d^5d^6-metal-centers. *Prog. Inorg. Chem.* 30: 1–73.

64 Marcus, R.A. and Sutin, N. (1985). Electron transfers in chemistry and biology. *Biochim. Biophys. Acta* 811: 265–322.

65 Levich, V.G. (1966). Present state of the theory of oxidation-reduction in solution (bulk and electrode reactions). *Adv. Electrochem. Electrochem. Eng.* 4: 249–371.

66 Zhang, H.L., Zhu, G.Y., Wang, G. et al. (2015). Electronic coupling in [Mo_2]–bridge–[Mo_2] systems with twisted bridges. *Inorg. Chem.* 54: 11314–11322.

67 Kang, M.T., Meng, M., Tan, Y.N. et al. (2016). Tuning the electronic coupling and electron transfer in Mo_2 donor–acceptor systems by variation of the bridge conformation. *Chem. Eur. J.* 22: 3115–3126.

68 Chen, H.W., Mallick, S., Zou, S.F. et al. (2018). Mapping bridge conformational effects on electronic coupling in Mo_2–Mo_2 mixed-valence systems. *Inorg. Chem.* 57: 7455–7467.

69 Bursten, B.E., Chisholm, M.H., Clark, R.J. et al. (2002). Oxalate-bridged complexes of dimolybdenum and ditungsten supported by pivalate ligands:$(tBuCO_2)_3M_2(\mu\text{-}O_2CCO_2)M_2(O_2CtBu)_3$. Correlation of the solid-state, molecular, and electronic structures with Raman, resonance Raman, and electronic spectral data. *J. Am. Chem. Soc.* 124: 3050–3063.

70 Chisholm, M.H., Feil, F., Hadad, C.M., and Patmore, N.J. (2005). Electronically coupled MM quadruply-bonded complexes (M = Mo or W) employing functionalized terephthalate bridges: toward molecular rheostats and switches. *J. Am. Chem. Soc.* 127: 18150–18158.

71 Chisholm, M.H. and Patmore, N.J. (2007). 2, 5-Dianilinoterephthalate bridged MM quadruply bonded complexes of molybdenum and tungsten. *Dalton Trans.* 1: 91–96.

72 Chisholm, M.H. and Patmore, N.J. (2005). Electronically-coupled MM quadruply-bonded complexes of molybdenum and tungsten. *Chem. Record* 5: 308–320.

73 Piepho, S.B., Krausz, E.R., and Schatz, P.N. (1978). Vibronic coupling model for calculation of mixed valence absorption profiles. *J. Am. Chem. Soc.* 100: 2996–3005.

74 Reimers, J.R. and Hush, N.S. (1991). Electric field perturbation of electronic (vibronic) absorption envelopes: application to characterization of mixed-valence states. In: *Mixed Valence Systems: Applications in Chemistry, Physics, and Biology* (ed. K. Prassides) Chapter 2, 29–50. Dordrecht: Kluwer Academy Publishers.

75 Reimers, J.R. and Hush, N.S. (1996). The effects of couplings to symmetric and antisymmetric modes and minor asymmetries on the spectral properties of mixed-valence and related charge-transfer systems. *Chem. Phys.* 208: 177–193.

76 Lear, B.J. and Chisholm, M.H. (2009). Oxalate bridged MM (MM = Mo_2, MoW, and W_2) quadruply bonded complexes as test beds for current mixed valence theory: looking beyond the Intervalence charge transfer transition. *Inorg. Chem.* 48: 10954–10971.

77 Gao, H., Mallick, S., Cao, L. et al. (2019). Electronic coupling and electron transfer between two Mo_2 units through *meta*-and *para*-phenylene bridges. *Chem. Eur. J.* 25: 3930–3938.

78 Wilkinson, L.A., McNeill, L., Scattergood, P.A., and Patmore, N.J. (2013). Hydrogen bonding and electron transfer between dimetal paddlewheel compounds containing pendant 2-pyridone functional groups. *Inorg. Chem.* 52: 9683–9691.

79 Cheng, T., Shen, D.X., Meng, M. et al. (2019). Efficient electron transfer across hydrogen bond interfaces by proton-coupled and-uncoupled pathways. *Nat. Commun.* 10: 1531.

80 Ren, T., Zou, G., and Alvarez, J.C. (2000). Facile electronic communication between bimetallic termini bridged by elemental carbon chains. *Chem. Comm.* 13: 1197–1198.

81 Xu, G.L., Zou, G., Ni, Y.H. et al. (2003). Polyyn-diyls capped by diruthenium termini: a new family of carbon-rich organometallic compounds and distance-dependent electronic coupling therein. *J. Am. Chem. Soc.* 125: 10057–10065.

82 Xi, B., Liu, I.P.C., Xu, G.L. et al. (2011). Modulation of electronic couplings within Ru_2-polyyne frameworks. *J. Am. Chem. Soc.* 133: 15094–15104.

83 Cao, Z., Xi, B., Jodoin, D.S. et al. (2014). Diruthenium–polyyn-diyl–diruthenium wires: electronic coupling in the long distance regime. *J. Am. Chem. Soc.* 136: 12174–12183.

84 Miller-Clark, L.A., Raghavan, A., Clendening, R.A., and Ren, T. (2022). Phenylene as an efficient mediator for intermetallic electronic coupling. *Chem. Comm.* 58: 5478–5481.

8

Mixed-Valence Electron Transfer of Cyanide-Bridged Multimetallic Systems

Shao-Dong Su, Xin-Tao Wu, and Tian-Lu Sheng

Chinese Academy of Science, Fujian Institute of Research on the Structure of Matter, State Key Laboratory of Structural Chemistry, West Yangqiao Road 155, Fuzhou 350002, China

8.1 Introduction

Cyanide-bridged mixed-valence (MV) complexes have a long history. As the most representative one, Prussian Blue, $Fe^{III}_4[Fe^{II}(CN)_6]_3 \cdot xH_2O$, which is also the oldest known man-made coordination compound, was found unexpectedly in the early eighteenth century in Berlin [1]. Since then, this deep blue cyanide-bridged complex has attracted the interest of many scientists to get an understanding of its chemical and physical properties [2, 3], and the chemistry of cyanide-bridged complexes has gradually grown to be a major research effort in inorganic and organometallic chemistry. Especially in recent years, the popularity of the chemistry of materials has further motivated the investigation of cyanide-bridged complexes on account of their potential applications in electrical conductors and magnetic materials [4, 5].

Electron transfer widely exists in chemical, physical, and biological systems, which has been the intense interest of scientists from various disciplines to elucidate the mechanism of this ubiquitous phenomenon. MV compounds, usually being described as D–B–A systems where two redox sites act as the electron donor and acceptor linked via a bridge, are considered as the ideal model systems for the investigation because they are featured by intramolecular electron transfer and electronic coupling behaviors and could be used to measure the electron transfer rate constants and activation barriers that are difficult to study by direct measurement in other systems [6]. Accordingly, many efforts have been devoted to the synthesis and characterization of MV compounds [7–9], and some important theories have been developed to interpret the electron transfer processes [10, 11]. Particularly, it shows that the degree of electronic interaction between the electron donor and acceptor in MV compounds is primarily controlled by the electronic characteristics of the bridging ligand, which can mediate the electronic interaction through its π-orbitals [12].

For the cyanide-bridged MV complex, its bridging ligand cyanide ion CN⁻ has some interesting features. Firstly, the bridging CN displays a characteristic of

Mixed-Valence Systems: Fundamentals, Synthesis, Electron Transfer, and Applications, First Edition.
Edited by Yu-Wu Zhong, Chun Y. Liu, and Jeffrey R. Reimers.
© 2023 WILEY-VCH GmbH. Published 2023 by WILEY-VCH GmbH.

strong stretching vibration absorptions in the IR spectrum within the range of c. 1900–2200 cm^{-1}, which generally avoids the interference from other infrared-active group. Besides, because of the π back-bonding effect, the cyanide bridge stretching frequency is very sensitive to the electron density of its binding metal, which can directly reflect the bonding mode and electron distribution of cyanide-bridged MV complexes [13]. These relative advantages with respect to other MV compounds make IR spectroscopy a very useful tool in the determination and identification of cyanide-bridged MV complexes. The second but more important is the very different binding properties of the two ends of cyanide. As the asymmetric electronic distribution and orbital shape, the C-bonded cyanide exhibits the characteristic of a typical strong-field ligand, which possesses good π-acceptor properties and affinity for relatively electron-rich metals, whereas the N-bonded cyanide behaves like a medium-field ligand with predominant σ-donor properties and has an affinity for relatively electron-deficient metals [7]. A classic example of these asymmetric binding properties is the aforementioned complex Prussian Blue. X-ray structural analysis of this complex reveals that it is composed of infinite repeat unit of Fe^{II}–CN–Fe^{III}, and the intense blue color is attributed to photoinduced electron transfer from the t_{2g} orbital of low-spin Fe^{II} ion to t_{2g} orbital of high-spin Fe^{III} ion through the bridging cyanide ligand [2] (Figure 8.1).

Taking advantage of the asymmetric binding properties of bridging unit CN, cyanide-bridged MV complexes possess a high degree of designability. The dinuclear cyanide-bridged MV complex, for instance, should be the simplest one; however, there exist two isomers of cyanide and isocyanide, M_A–CN–M_B/M_A–NC–M_B, which could have very different properties. When trying to construct trinuclear complexes, the situation would be more interesting. For example, as shown in Figure 8.2, the combination of an octahedral metal M_A in the center and two metals M_B in the terminal via cyanide bridges is supposed to generate six theoretical isomers of *trans/cis*-[M_B–NC–M_A–CN–M_B], *trans/cis*-[M_B–CN–M_A–NC–M_B], and *trans/cis*-[M_B–CN–M_A–CN–M_B]. As more metal units are added, the number of isomers of cyanide-bridged complexes will increase exponentially. In general, the design strategy for controlling the linkage of cyanide bridge has provided a very large family from di-, tri-, and higher nuclear cluster complexes to one-, two-, and three-dimensional polymers, thereby making the cyanide-bridged MV complex an ideal model system for the investigation of electron transfer and electron coupling.

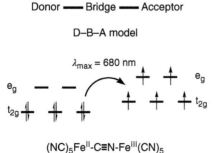

Figure 8.1 D–B–A model and the electron transfer process of Prussian Blue.

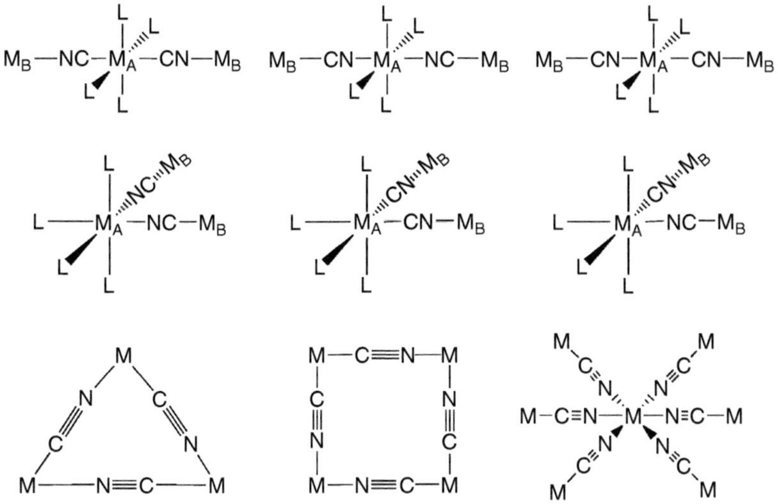

Figure 8.2 Examples of coordination mode of cyanide-bridged complexes.

Over the past decades, a great number of cyanide-bridged MV complexes have been synthesized and investigated by Vogler [14], Haim [15], Scandola [16], Bocarsly [17], Meyer [18], Denning [19], Connelly [20], Endicott [21], Vahrenkamp [22], Baraldo [23] et al., who have made important contributions in this field. The factors influencing electron transfer and electron coupling have been subjected to a systematic study from the nature of the complexes, which involves the nature of the metals and their ligands, the orientation of the cyanide bridges, and the shape and length of the chains, to the environmental effects, such as solvent effects and salt effects. Meanwhile, of particular interest in this field is the synthesis of delocalized cyanide-bridged MV complexes. As is well known, Robin and Day [24] have classified the MV compounds into three categories by the degree of electronic interaction: *Class I*, where the electronic interaction is very weak; *Class II*, where the electronic interaction is moderate but the charge is still localized; and *Class III*, where the electronic interaction is very strong and the charge is delocalized. In addition to the Robin and Day classification, a new intermediate regime category of *Class II–III* was proposed by Meyer [6], where the charge is localized, but transition absorption is solvent independent. Although the delocalized MV compounds are attractive for their unique physical properties [25], it is still a challenge to synthesize and characterize the *Class III* complexes in a cyanide-bridged system and aim at which the groups of Baraldo [26–28] and Sheng [29–31] have reported some significant work in recent years.

The aim of this chapter is to elucidate the factors affecting electron transfer and electron coupling in cyanide-bridged MV complexes and how they work by reviewing studies conducted over the years. The discussion herein is limited to oligonuclear cyanide-bridged complexes, as they are easily modeled by current theories in comparison to complicated one-, two-, and three-dimensional polymers. Meanwhile, some of the latest research will be covered in this chapter.

8.2 Dinuclear Cyanide-Bridged Mixed-Valence Complex

As the most accessible one, dinuclear cyanide-bridged MV complex, which is composed of two metal fragments in different oxidation states connected by cyanide bridge, provides the simple and intelligible model for the investigation of the fundamental properties of electron transfer. Particularly, their characteristics of photoinduced metal–metal charge transfer (MMCT) absorptions are linked to the thermally induced electron transfer by the theories of Marcus and Hush [10, 11], which give a good description of electron transfer behaviors and make it possible to quantitatively determine the extent of electron coupling. As a result, the most concerned metal–metal interaction parameters H_{ab} and α^2 can be easily calculated from the data of spectrographic, electrochemical, and crystallographic measurements employing these theoretical formulations.

The investigation of electron transfer of cyanide-bridged MV complexes was inspired by the discovery of the famous Creutz–Taube ion [{Ru(NH$_3$)$_5$}(pyrazine){Ru(NH$_3$)$_5$}]$^{5+}$ [32], which was considered as the first fully delocalized *Class III* complex. The investigation started with Haim, Vogler, and Scandola et al. by replacing the pyrazine of the Creutz–Taube ion and its analog with cyanide bridge to compare their abilities of intermediating electron interaction. Thus, a series of homobinuclear and heterobinuclear cyanide-bridged MV complexes, which consist of simple inorganic metal units, such as [(CN)$_5$M$_A^{II}$CNM$_B^{III}$(NH$_3$)$_5$]$^-$ (**1**) (M$_A$ = Fe, Ru, Os, and Co; M$_B$ = Ru, Os, Cr, and Co), have been synthesized and characterized [15, 23, 33, 34]. The investigation reveals that their electron spectra are featured by intense MMCT transition absorptions from the divalent metal to the trivalent metal in the visible or near-infrared region, and the electron coupling parameters H_{ab} of them calculated by Hush equation [11] range from about 1000 to 2000 cm^{-1}, which appears larger than that of the corresponding pyrazine-bridged analog. Interestingly, this phenomenon can be usually observed in *Class II* system, in which the cyanide-bridged complexes possess a typical H_{ab} of about 1000–2000 cm^{-1}, while the pyrazine-bridged complexes possess a typical H_{ab} of about 300–500 cm^{-1} [35]. Therefore, although the use of pyrazine bridge gave rise to the first *Class III* complex, cyanide bridges have been proposed to bring about a definitely stronger metal–metal electronic coupling than pyrazine in *Class II* system, resulting from its short bridge length and good orbitals matching the attribute for πd–πd overlap [34].

In addition to the simple inorganic metal units, a wide range of organometallic fragments have been used to construct the dinuclear cyanide-bridged MV complex, and the influences of the nature of metals, ligands, as well as the orientation of cyanide bridge on electronic properties have been further studied [20, 36–42]. For example, Connelly has reported the intermetallic interaction in a dinuclear cyanide-bridged system involving *trans/cis*-[Mn(CO)$_2$(L$_p$)(L$_{pp}$)] (**2**) (L$_p$ = PEt$_3$, P(OPh$_3$), and P(OEt$_3$); L$_{pp}$ = 1,2-bis(diphenylphosphino)ethane (dppe), bis(diphenylphosphino)methane (dppm), and 1,2-bis(dimethylphosphino)ethane (dmpe)) fragments, which indicates that the potential difference ($\Delta E_{1/2}$) between the two metal units dominates the transition energy of MMCT [20, 36]. Interestingly,

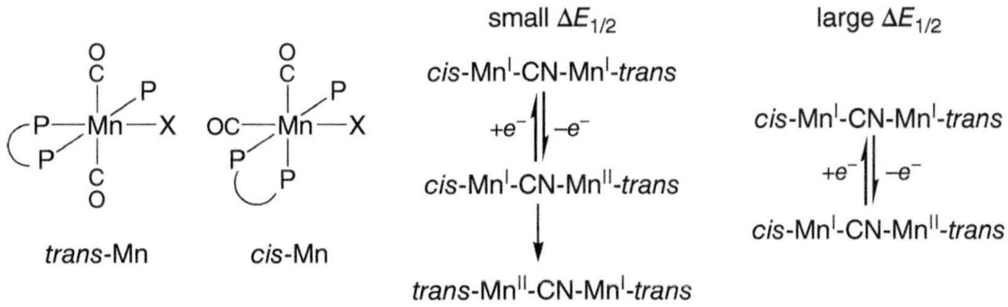

Figure 8.3 The oxidative isomerization of the dimanganese complexes.

the oxidative isomerization of a series of dimanganese complexes has been observed. As shown in Figure 8.3, when $\Delta E_{1/2}$ of the two manganese fragments is small, as in the case of cis, trans-[Mn^I(CO)$_2$(PEt$_3$)(dppe)–CN–Mn^I(CO)$_2$(P(OPh$_3$))(dppm)]$^+$ (**3**) ($\Delta E_{1/2} \approx 0.5$ V), one-electron oxidation of the trans-Mn$^{(I)}$ site would induce intramolecular electron transfer toward cis-Mn$^{(I)} \rightarrow$ trans-Mn$^{(II)}$, leading to the isomerization of cis-Mn$^{(II)}$ fragment to the trans-configuration. Conversely, as in the case of cis, trans-[Mn^I(CO)$_2$(P(OPh$_3$))(dppm)–CN–Mn^I(CO)$_2$(P(OPh$_3$))(dppm)]$^+$ (**4**) where the $\Delta E_{1/2}$ is about 1.0 V, the configuration of cis-Mn$^{(I)}$–CN–trans-Mn$^{(II)}$ is retained upon one-electron oxidation of the same trans-Mn$^{(I)}$ site [20]. The fact that the isomerization of the cis-Mn fragment can be triggered by oxidation of its neighbor trans-Mn$^{(I)}$ when $\Delta E_{1/2}$ is small clearly suggests that the electron interaction is present in these dimanganese complexes and is enhanced as the redox potentials of the two fragments become close.

Also, the groups of Vahrenkamp and Sheng have synthesized numbers of dinuclear cyanide-bridged complexes by using organometallics fragments such as [Cp(dppe)M] and [Cp(CO)$_2$M] [37–41]. Among those complexes, a few pairs of cyanide/isocyanide isomers including the first stable and structurally characterized isomers [(CO)$_5$Cr–CN–Fe(dppe)Cp] (**5a**) and [(CO)$_5$Cr–NC–Fe(dppe)Cp] (**5b**) have been reported [37–39]. Given the thermodynamic isomerizations of M–CN–M′ → M–NC–M′ usually take place in the cyanide-bridged complex, such cyanide–isocyanide isomers provide the chance to investigate the influence of the orientation of cyanide bridge in MV systems. The discussion on this topic shows that the effect of cyanide–isocyanide configuration can work in two ways for different pairs of isomers. On the one hand, for a given pair of isomers such as [Cp(dppe)Fe^{II}–CN–Fe^{II}(dppp)Cp]$^+$ (**6a$^+$**) and [Cp(dppe)Fe^{II}–NC–Fe^{II}(dppp)Cp]$^+$ (**6b$^+$**) [39] where the first oxidation process always occurs at the cyanido-N bound metal fragment, the orientation of cyanide bridge can determine the order of oxidation of the two metal center, and hence, the expected direction of MMCT between these metals would reverse. On the other hand, in the case of the pair of isomers such as trans-[(dppm)(CO)$_2$(P(EtO)$_3$)Mn^I–CN–Mn^I(CNBu^t)(NO)(CpMe)]$^+$ (**7a$^+$**) and trans-[(dppm)(CO)$_2$(P(EtO)$_3$)Mn^I–NC–Mn^I(CNBu^t)(NO)(CpMe)]$^+$ (**7b$^+$**) [36] where a certain metal center is always oxidized firstly, such as this trans-Mn$^{(I)}$ site in **7**, the direction of MMCT in the cyanide or isocyanide configuration is unaltered upon one-electron oxidation, and the orientation of cyanide bridge would affect the redox potential difference ($\Delta E_{1/2}$) between the two metal units, which results in

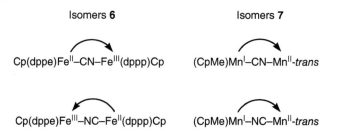

Figure 8.4 The direction of MMCT influenced by the orientation of cyanide bridge in two different situations.

variation of the intermetallic electron interaction, especially the energy of MMCT transition (E_{op}) of such a pair of isomers. Clearly, the effect of cyanide–isocyanide configuration in the two situations originates from the asymmetric electronic distribution and binding properties of the two ends of cyanide bridge, which provides an additional way to modulate electronic interactions in the MV system compared to other symmetric bridge ligands, e.g. pyrazine or alkyne (Figure 8.4).

More recently, Baraldo and coworkers have reported their research of fine-tuning the electronic coupling between metal centers in the dinuclear cyanide-bridged system by varying the ligand of organometallics fragments [27, 28, 42]. To verify the hypothesis that the small energy difference ΔE_0 between the initial and final states for electron transfer (be approximated by the redox potential difference of the two metal centers, $\Delta E_{1/2}$) would be beneficial to increase the electron coupling, four complexes $trans$-[Ru^{II}(L$_T$)(bpy)(μ-NC)Ru^{II}(L)$_4$(CN)]$^{2+}$(**8a^{2+}** and **8b^{2+}**, L = py and L$_T$ = tris(1-pyrazolyl)methane (tpm) and 2,2′;6′,2″-terpyridine (tpy); **8c^{2+}** and **8d^{2+}**, L = MeO-py and L$_T$ = tpm and tpy) and their oxidation products have been synthesized and characterized [42]. The electrochemical and spectroscopy measurements reveal that the $\Delta E_{1/2}$ of Ru(py)$_4$-contained complexes **8a^{2+}** and **8b^{2+}** are smaller than those of the Ru(MeO-py)$_4$-contained complexes **8c^{2+}** and **8d^{2+}**, and one-electron oxidation at the Ru(py)$_4$ fragment of the former results in the almost solvent-independent and stronger MMCT absorptions than those of the latter complexes, suggesting the presence of strong electronic coupling in **8a^{2+}** and **8b^{2+}**. This is further supported by the DFT calculation, which shows that the spin density is localized on one ruthenium ion of **8c^{3+}** and **8d^{3+}** while more delocalized over the two ruthenium ions of **8a^{3+}** and **8b^{3+}**, especially [Ru^{II}(tpm)(bpy)(μ-NC)Ru^{III}(py)$_4$(CN)]$^{3+}$ (**8a^{3+}**), with spin densities of 0.447 and 0.493 at the two ruthenium ions, respectively. Thus, the Ru(py)$_4$-contained **8a^{3+}** and **8b^{3+}** with smaller $\Delta E_{1/2}$ are assigned to strong electronic interaction *Class II–III* species compared to **8c^{3+}** and **8d^{3+}** with weak electronic coupling.

Furthermore, a more systematic study has been carried on with two similar series of complexes [Ru^{II}(tpm)(bpy)(μ-CN)Ru^{II}(py)$_4$L]$^{3/4+}$ (**9**) and [Ru^{II}(tpm)(bpy)(μ-CN)Ru^{II}(bpy)$_2$L]$^{3/4+}$ (**10**) (L = Cl$^-$, NCS$^-$, DMAP, and ACN) as shown in Figure 8.5 [27, 28]. For both the series of complexes, their electrochemistry behaviors are dominated by the ability of electron donating of ligand L, that is, the $\Delta E_{1/2}$ between the ruthenium ions decreases as the ligand L varies from the strongly donating chloride to the weakly donating acetonitrile, such as the $\Delta E_{1/2}$ of [Ru^{II}(tpm)(bpy)(μ-CN)Ru^{II}

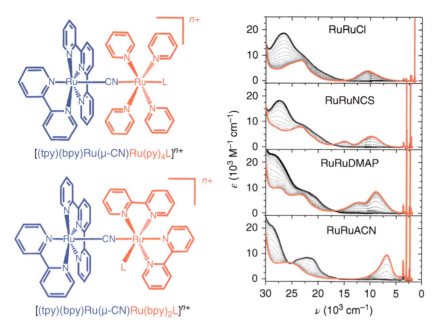

Figure 8.5 Left: The diagram of complexes **9** and **10** (L = Cl$^-$, NCS$^-$, DMAP, and ACN). Right: Vis–NIR spectroelectrochemistry of the series of [Ru^{II}(tpm)(bpy)(μ-CN)Ru^{II}(py)$_4$L]$^{3/4+}$. Source: Domínguez et al. 2020 [28]/Reproduced with permission of the Royal Society of Chemistry.

(bpy)$_2$L]$^{3/4+}$ are 0.83, 0.63, 0.58, and 0.53 V for RuRuCl, RuRuNCS, RuRuDMAP, and RuRuACN, respectively. After one-electron oxidation of the cyanido-N-bound Ru^{II} fragment, the MV species of the two series of complexes exhibit evident MMCT bands in the NIR region, and as the $\Delta E_{1/2}$ decreases, the intensity of these MMCT bands increase while their energies shift to the lower energy region, which indicates that the electronic coupling in the two series is enhanced as the ligand L varies from Cl$^-$ to acetonitrile. Meanwhile, the DFT calculation gives a picture of the localized to delocalized transition in the two series, which is in agreement with the previous result. Moreover, it is observed that complexes [Ru^{II}(tpm)(bpy)(μ-CN)Ru^{III}(py)$_4$L]$^{4/5+}$ with the ligand L lying *trans* to cyanide bridge display higher degree of delocalization than [Ru^{II}(tpm)(bpy)(μ-CN)Ru^{III}(bpy)$_2$L]$^{4/5+}$ with the same L lying *cis* to cyanide bridge. Such phenomenon can be attributed to the overlap of different orbitals of the two series, in which the ligand L lying *trans* to cyanide bridge is easier to participate in the delocalized orbitals composed of the two Ru ions and the cyanide bridge compared to the *cis* one, leading to a more even charge distribution as supported by the DFT calculation. Likewise, the similar trend that electronic coupling varied by different ligand has been observed by the group of Sheng in complexes [Ru^{II}(tpm)(L)(μ-CN)Fe^{II}(dppe)Cp]$^{2+}$ (**11**) [40] and [Ru^{II}(tpm)(L)(μ-CN)Ru^{II}(dppe)Cp*]$^{2+}$ (**12**) [41] where the bipyridine of [Ru^{II}(tpm)(bpy)]$^{2+}$ fragment were replaced by the weak donor ligand phenanthroline (phen) and the strong donor ligand 4,4′-dimethyl-bipyridine (dbpy).

8.3 Trinuclear Cyanide-Bridged Mixed-Valence Complex

Compared with the dinuclear systems, trinuclear cyanide-bridged MV complexes consist of two terminal metals and a centrally located cyanide or isocyanide metal, formulated as M_A–CN/NC–M_B–CN/NC–M_C, and therefore, the electron interactions of these complexes could be more complicated because there are not only the electron transfers between cyanide-bridged adjacent atoms but also the remote or long-range electron transfer between the non-directly bridged terminal metals. It is obvious that the latter one is of much attractiveness for the potential applications of "molecular wires" and other molecular devices, and the investigation could help to understand the long-range electron transfer in chemical, physical, and biological systems.

The research started from Meyer, Bignozzi, Scandola et al. [16, 18, 34] with the synthesis and characterization of a series of Ru(py)$_4$-centered and Ru(bpy)$_2$-centered trimetallic MV complexes, for instance, [(NH$_3$)$_5$Ru^{III}–NC–Ru^{II}(bpy)$_2$–CN–Ru^{II}(NH$_3$)$_5$] (**13**) (the oxidation state is simplified as (III, II, II)) [34]. The investigation reveals that there are two electron transfer processes in this complex. One is the normal electron transfer from the central Ru^{II} to its neighboring Ru^{III} and the other is the long-range electron transfer from the terminal Ru^{II} to the remote Ru^{III} at the other end, confirmed by the comparison with its reduction state (II,II, II) and the two-electron oxidation state (III, II, III). The spectral parameters estimated for the long-range electron transfer, according to Hush model, give an electronic interaction H_{ab} of about 340 cm^{-1}, which suggests that it belong to *Class II* species. Although experimental results were very clear, the origin of such a remarkable electronic coupling is still puzzling at that moment. Hence, Scandola has proposed that the superexchange mechanism mediated by the pyridyl-type dicyano-ruthenium bridge can be responsible for the long-range electron transfer [16], which is further supported by Endicott on the basis of analyzing plenty of Ru-pyridyl-centered trimetallic complexes. In addition, Endicott and coworkers also reported a series of trinuclear complexes containing the tetraazamacrocyclic ligand on the central metal, which are not in favor of the superexchange mechanism, suggesting that the electron coupling mechanism could be variable depending on the nature of the bridging M(L)(CN/NC)$_2$ fragment [21, 43] (Figure 8.6).

After preliminarily answering the "what?" and "why?" questions on long-range electron coupling within cyanide-bridged systems, how to construct and manipulate such interesting electron interaction in the trinuclear cyanide-bridged MV complex

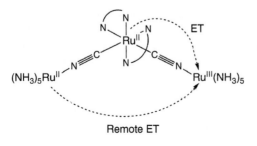

Figure 8.6 The diagram of two metal-to-metal electron transfer processes in complex **13**.

would be next to consider. The influence factors could be more complicated than in dinuclear systems, including the nature of the metals and their ligands, the orientation of the μ-CN bridge (cyanide or isocyanide), and especially the geometry at the central metal unit (e.g. square-planar, tetrahedral, octahedral, and cis/trans-configuration). Hence, many efforts have been devoted to discuss these factors on this subject.

To select appropriate metals and their ligands, a great deal of attempts has been made so far [21, 22, 44–52]. For the central bridging metal unit, almost all of the common transition elements with different oxidation states, including Cr(III), Mn(II)/(III), Fe(II)/(III), Ru(II)/(III), Os(II)/(III), Co(II)/(III), Rh(III), Ni(II), Pt(II)/(IV), Cu(I)/(II), Ag(I), Au(I), Zn(II), Cd(II), Hg(II), and a series of ligands, such as pyridine-type, tetraazamacrocyclic-type, and salen-type, have been used (Figure 8.7). For the terminal unit, it usually involves some redox-active units, such as $[L(NH_3)_4Ru]$ (L = NH_3, py, and Cl), [Cp(dppe)M] (M = Fe and Ru), $[Cp(PPh_3)_2Ru]$, $[Cp(CO)_2M]$ (M = Fe and Mn), $[(phen)(CO)_3Re]$, $[Mn(CO)_2(dppe/dppm)(P(OR)_3)]$, $[M(CO)_5]$ (M = Cr, Mo, and W), $[M(CN)_5]$ (M = Fe, Ru, and Os), etc. The combination of these central and terminal metal units via cyanide bridges has given rise to hundreds of trinuclear complexes with or without the long-range electron interaction. As a consequence, it is not straightforward to define which metal or ligand supports the long-range electron interaction because there often exists the exception; for example, in many cases, the Fe atom and pyridine ligand seem to favor this interaction [45, 53], but in complex trans-$[Cp(PPh_3)_2Ru-CN-Fe(py)_4-NC-Ru(PPh_3)_2Cp](PF_6)_2$ (**14**) [48], no such long-range electron coupling can be observed. However, even so, a general trend might be concluded from the hundreds of trinuclear cyanide-bridged complexes that the electronic-rich

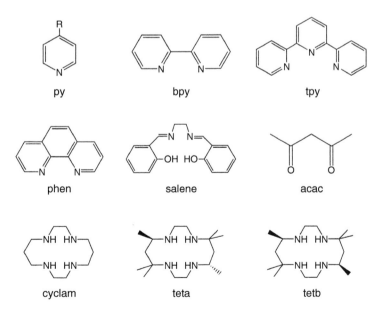

Figure 8.7 The diagram of common ligands used for the central metal unit.

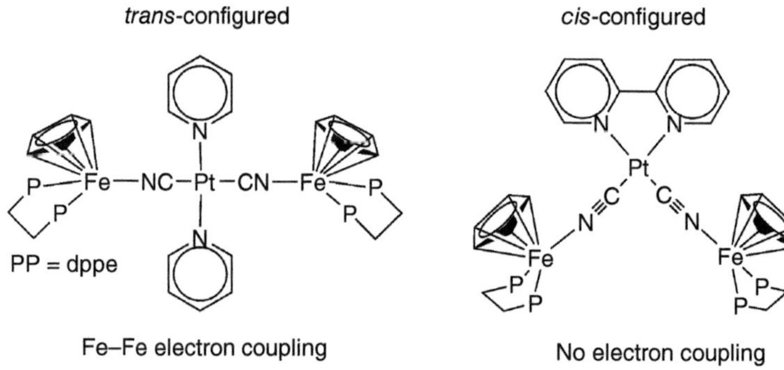

Figure 8.8 The diagram of *trans-* and *cis-*isomeric complexes **15a** and **15b**.

metal unit with good redox activity, such as pyridine-type Ru or [Cp(dppe)Fe/Ru], could be beneficial to the long-range electron interaction within trimetallic system.

The next influencing factor for the long-range electron interaction is concerned with the geometry of the central metal unit. The ideal object for such investigation should be the *cis-* and *trans-*isomeric complexes that share the identical terminal [ML$_n$] units and the central bridging metal M′ as well as its ligands. The first example aiming on this issue is two pairs of isomeric square-planar platinum(II) complexes: *trans-*[ML–NC–Pt^{II}(py)$_2$–CN–ML] and *cis-*[ML–NC–Pt^{II}(bpy)–CN–ML] (**15a** and **15b**, ML = Cp(dppe)Fe), *trans-*[ML–CN–Pt^{II}(py)$_2$–NC–ML], and *cis-*[ML–CN–Pt^{II}(bpy)–NC–ML] (**16a** and **16b**, ML = Cp(dppe)Fe), which are very approximate to the ideal one (Figure 8.8) [50]. The cyclic voltammograms of both the *trans-*configured **15a** and **16a** display the potential split between two terminal Cp(dppe)Fe units, indicating the presence of long-range electron coupling. The clean electron transfer absorptions from the terminal Fe^{II} to Fe^{III} of their one-electron oxidation products were observed in NIR region, without the interference of imaginary Pt^{II}→Fe^{III} transition because of the weak π-donor property of platinum(II). In contrast, there is no potential split between two terminal Fe units in the *cis-*configured **15b** and **16b**, which means the *cis-*configuration at the central metal is failed to mediate the long-range electron interaction. Therefore, the comparison of these pairs of *cis-* and *trans-*isomeric complexes indicates that only *trans-*configuration of the square-planar platinum(II) complex is in support of the long-range electron interaction other than the *cis-*configuration.

Further discussion on this subject could be carried on with a pair of real *cis/trans-*isomeric octahedral ruthenium(II)-centered complexes, *cis/trans-*[Cp(dppe)Fe^{II}–NC–Ru^{II}(phen)$_2$–CN–Fe^{II}(dppe)Cp][PF$_6$]$_2$ (**17a**$^{2+}$ for *trans-* and **17b**$^{2+}$ for *cis-*), and their one- and two-electron oxidation products (**17a**$^{3/4+}$ and **17b**$^{3/4+}$) [54]. Differentiated from the square-planar platinum(II) complexes, these *cis/trans-*isomeric complexes exhibit very similar properties in the electrochemistry, IR, EPR, and electron absorption spectra. Especially, both of the *cis/trans-*isomeric **17a**$^{2+}$ and **17b**$^{2+}$ show the potential split of about 110 mV for the symmetric terminal Cp(dppe)Fe^{II} units, which demonstrates the presence of the long-range electron coupling between

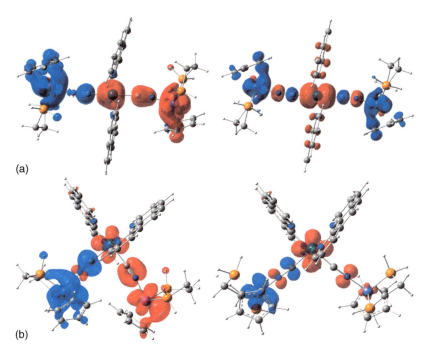

Figure 8.9 (a) Redistribution of electron densities of **17a**$^{3+}$ (left) and **17a**$^{4+}$ (right) in the calculated MMCT band. (b) Redistribution of electron densities of **17b**$^{3+}$ (left) and **17b**$^{4+}$ (right) in the calculated MMCT band. (The blue and red areas represent gain and losses of density). Source: Ma et al. 2014 [54]/Reproduced with permission of John Wiley & Sons.

them. For the one- and two-electron oxidation products, *cis/trans*-Fe^{III}–Ru^{II}–Fe^{II} and *cis/trans*-Fe^{III}–Ru^{II}–Fe^{III}, the MMCT transition from terminal Fe^{III} to Fe^{II} mediated by central Ru^{II} and the transition of centered Ru^{II}→Fe^{III} were observed respectively, which is supported by the TDDFT calculations and well visualized by the plot of redistribution of electron densities (Figure 8.9). The only main difference between the *cis*- and *trans*-complexes is that the transition energies of *trans* complexes **17a**$^{3/4+}$ are lower than those of the corresponding *cis* complexes **17b**$^{3/4+}$, respectively, because of the better symmetry of relevant molecular orbitals of the *trans*-configuration. Hence, although some experimental phenomena of other trinuclear MV complexes support an idea that *trans*-configuration might be more suitable to mediate the long-range electron interaction, given this example that long-range electron interactions take place in both *cis/trans*-isomeric complexes, more studies would have to be carried out in order to establish the generality of this idea.

When the influence of the orientation of cyanide bridge on electronic interactions is discussed within trimetallic systems, it seems a challenge to obtain the applicable pairs of cyanide/isocyanide isomers because the isomerization of the target product toward the thermally stable complex often happens, such as [(OC)$_5$Cr–CN–Hg–NC–Cr(CO)$_5$] (**18a**) ↔ [(OC)$_5$Cr–NC–Hg–CN–Cr(CO)$_5$] (**18b**) and [(NC)(cyclam)Cr–CN–Ru(bpy)$_2$–NC–Cr(cyslam)(CN)] (**19a**) → [(NC)(cyclam)

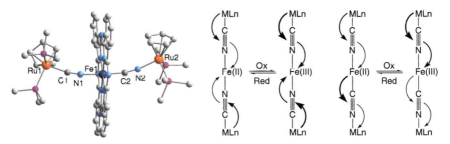

Figure 8.10 The structure of [Cp(PPh$_3$)$_2$Ru–CN–Fe(Pc)–CN–Ru(PPh$_3$)$_2$Cp](SbF$_6$) (left, Ph at PPh$_3$ is omitted for clarity). Schematic diagram of donator–acceptor interactions in **20** and **22** (right). Source: Adapted from [55].

Cr–NC–Ru(bpy)$_2$–CN–Cr(cyslam)(CN)] (**19b**) [51, 52]. However, fortunately, a series of Fe(Pc) (Pc = phthalocyaninato^{2-})-centered trinuclear complexes with all three possible cyanide orientations, ML–CN–Fe(Pc)–NC–ML (**20**), ML–NC–Fe(Pc)–CN–ML (**21**), and the unusual ML–CN–Fe(Pc)–CN–ML (**22**), have been synthesized and characterized by Vahrenkamp and coworkers [55, 56]. The investigation reveals that the long-range electron coupling between two terminal units exists in all the complexes **20–22**. More importantly, the comparison of different orientations of cyanide bridge in **20–22** on the basis of analyzing their electrochemistry and IR gives a picture of changes in cyanide orientation, influencing the donor–acceptor properties of the bridging metal units and altering the flow direction of electron density along the M–(μ-CN)–Fe(Pc)–(μ-CN)–M chains (Figure 8.10). That is, because of the σ-donation from the cyanide N termini and the π-acceptance at the cyanide C termini, the motion direction of electron density could be from the C-bonded metal to the N-bonded metal, and the degree of redistribution of electron densities would be enhanced by oxidation of the N-bonded metal and weakened by oxidation of the C-bonded metal, which is supported by the MMCT absorption spectra of **20–22**. Therefore, it appears to be a feasible approach to manipulate the property of electron interaction within trimetallic systems by inverting the orientation of the cyanide bridge, which is able to alter the redox potential and donor–acceptor characteristic of the bridging metal units, resulting in the relative energy changed of them. This idea has also been supported by the group of Sheng with the comparison of a pair of cyanide/isocyanide isomeric trinuclear MV complexes, which will be discussed below [29].

Until now, it has been investigated a lot about tuning of the electronic coupling and electron transfer in trinuclear cyanide-bridged MV complexes by altering metals and ligands, *cis/trans*-configuration, as well as the orientation of cyanide bridge [53, 57–61]. As previously mentioned, the cyanide CN has been proven to be an ideal bridge ligand, which provides a substantial amount of metal–metal electronic coupling in MV complexes. However, detailed electrochemical, spectroscopic, and photophysical studies of complexes usually give a valence-localized picture; that is, most of the cyanide-bridged MV complexes belong to *Class II* species according to Robin and Day. The exploration of fully valence-delocalized

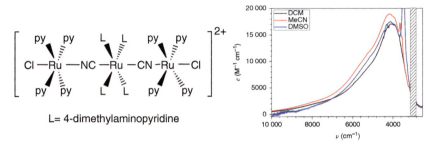

Figure 8.11 The structure diagram of **23²⁺** (left). Electronic absorption spectra of **23³⁺** in different solvents (right). Source: [26]/Reproduced with permission of John Wiley & Sons.

Class III cyanide-bridged complexes was successful until recently, when the group of Baraldo and Sheng has found the missing piece of the puzzle in the trinuclear cyanide-bridged system [26, 29–31].

In 2014, Baraldo and coworkers have reported the first *Class III* cyanide-bridged MV complex, *trans*-[Ru(dmap)$_4${(μ-CN)Ru(py)$_4$Cl}$_2$]³⁺ (**23³⁺**, dmap = 4-dimethylaminopyridine) (Figure 8.11) [26]. The cyclic voltammogram of this trinuclear complex reveals three well-separated one-electron redox processes, which indicates the existence of strong electronic coupling. Only one broad cyanide stretching absorption was observed at 1980 cm⁻¹, implying an ultrafast charge resonance beyond the IR timescale (10⁻¹¹ seconds), as expected for a completely delocalized symmetrical ground state. The most powerful evidence comes from its NIR spectrum; that is, **23³⁺** displays a narrow, intense, and solvent-insensitive electron absorption band at 4150 cm⁻¹ (ε_{max} = 19500 M⁻¹ cm⁻¹) with a "cutoff" shape at lower energies, which correspond to the behaviors of a typical *Class III* MV complex. Moreover, this work established the method to construct a delocalized complex by tuning the energy of the central bridging fragment in the trinuclear complex. However, it would somehow be a pity for this delocalized complex as the lack of crystallographic characterization.

As the exploration goes on, Sheng and coworkers have reported two cyanide/isocyanide isomeric trinuclear MV complexes *trans*-[Cp*(dppe)Ru^{III}(μ-NC)Ru^{II}(dmap)$_4$(μ-CN)Ru^{II}(dppe)Cp*]³⁺ (**24³⁺**) and *trans*-[Cp*(dppe)Ru^{II}(μ-CN)Ru^{III}(dmap)$_4$(μ-NC)Ru^{II}(dppe)Cp*]³⁺ (**25³⁺**) (Figure 8.12) [29], and the effect of the different orientations of cyanide bridges on electronic interaction has been studied. Interestingly, the investigation reveals that complex **24³⁺** with the configuration of Ru^{III}–NC–Ru^{II}–CN–Ru^{II} is fully valence delocalized, which shows (i) centrosymmetric crystallography structure with two strictly equivalent terminal Ru atoms, (ii) a single broad cyanide stretching absorption, (iii) a narrow, intense, and solvent-insensitive electron absorption, (iv) the characteristics of isotropic free electron and temperature-independent behavior in the EPR spectra, being in accordance with the behaviors of other *Class III* systems. Furthermore, this is also the first example of the crystallographically characterized *Class III* cyanide-bridged MV complex. By contrast, for the isocyanide complex **25³⁺**, the temperature-dependent EPR behavior and X-ray crystallography demonstrate that it belongs to *Class II–III*

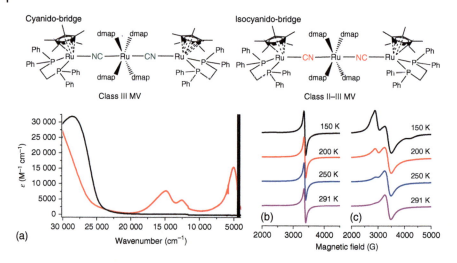

Figure 8.12 The structure diagram of **24³⁺** and **25³⁺** (top). The electronic spectra of **24³⁺** (red) and **24²⁺** (black) (a). The variable temperature EPR of **24³⁺** (b) and **25³⁺** (c). Source: Yang et al. 2018 [29]/Reproduced with permission of John Wiley & Sons.

MV complex, suggesting that the orientation of the cyanide bridge plays an important role in the electron interaction.

To undertake a more systematic research on the electron transfer process in MV complexes with strong electron interactions, Sheng and coworkers have further synthesized and characterized a series of cyanide-bridged complexes [CpMe$_n$Fe(dppe)CNRu(dmap)$_4$NC(dppe)FeCpMe$_n$][PF$_6$]$_2$ (**26a²⁺–26e²⁺**, n = 0, 1, 3, 4, and 5, Figure 8.13) with different numbers of methyl substitutions on the terminal Cp ligand and their one-electron and two-electron oxidation products [30]. The spectral and theoretical analyses suggest that the series of one-electron oxidation products **26a³⁺–26e³⁺** belong to *Class II–III* species, while the degree of delocalization of the two-electron oxidation products **26a⁴⁺–26e⁴⁺** increases as the donor effect of the terminal CpMe$_n$ becomes stronger, leading to the transition of their properties from the valence-localized *Class II* species to the completely delocalized *Class III* species.

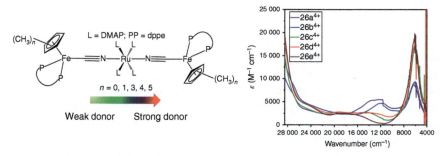

Figure 8.13 The structure diagram of the Fe–CN–Ru–NC–Fe array (left) and the electronic spectra of **26a⁴⁺–26e⁴⁺** (right). Source: Yang et al. 2020 [30]/Reproduced with permission of John Wiley & Sons.

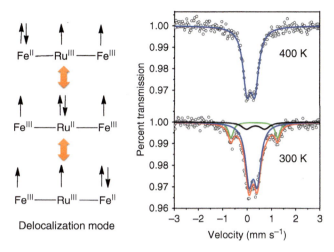

Figure 8.14 Description of delocalization of the excess electron over three metal centers of **27^{4+}** (left), and the variable temperature Mössbauer spectra of **27^{4+}** (right). Source: Ma et al. 2017 [31]/Reproduced with permission of John Wiley & Sons.

Thus, it could be concluded that for such MV system with low-lying energy bridge, the energy difference between the bridge state and MV states could be effectively tuned by the fine modification of donor substitution.

Another interesting example of the *Class III* species within the cyanide-bridged system has been observed in the trinuclear complexes *trans*-[Cp(dppe)Fe(μ-NC)Ru(o-bpy)(μ-CN)Fe(dppe)Cp][PF$_6$]$_n$ (**27^{n+}**, $n = 2$, 3, and 4, Figure 8.14) [31]. For the one-electron oxidation product **27^{3+}**, it behaves as the typical *Class II* MV complexes, which displays a moderate MMCT absorption originated from the Fe(II)→Fe(III) transition, but the electron is still localized on the timescale of IR and EPR spectra. In the case of the two-electron oxidation product **27^{4+}**, it starts out like the normal valence-localized complex; however, when temperature increases from 300 to 400 K, a noticeable change of its IR and EPR behavior can be observed unexpectedly, implying the change in internal electronic structure of this complex. Further investigation reveals that **27^{4+}** possesses an unusually stable delocalized MV state of [Fe^{III}–NC–Ru^{III}–CN–Fe^{II}] after a thermal treatment at 400 K, and the thermally induced electron transfer from the central Ru^{II} to the terminal Fe^{III} should be responsible for that, which is confirmed by magnetic data and Mössbauer spectra. This finding of the first thermally induced fully delocalized cyanide-bridged MV complex encourages the further exploration of *Class III* species.

In addition to the above studies of electron interactions, other studies related to electron transfer process within trinuclear cyanide-bridged MV complexes have also been reported. For example, the group of Bocarsly has synthesized and characterized a series of complexes [L(NC)$_4$M^{II}–CN–Pt^{IV}(NH$_3$)$_4$–NC–M^{II}(CN)$_4$L] (**28**) (M = Fe, Ru, and Os and L = CN$^-$ or other σ-donor ligands) [17]. These complexes can undergo a photoinduced one-electron transfer (M^{II}→Pt^{IV}) to generate the [M^{III}–CN–Pt^{III}(NH$_3$)$_4$–NC–M^{II}] intermediate; however, because of the instability of Pt^{III} species, the [M^{III}–CN–Pt^{III}(NH$_3$)$_4$–NC–M^{II}] intermediate would further proceed with the thermodynamics-induced one-electron transfer ($M^{II} \rightarrow Pt^{III}$), resulting in the [$M^{III}$–CN–$Pt^{II}$ (NH_3)$_4$–NC–M^{III}], which is called the photoinduced multielectron charge transfer process. Moreover, the energy transfer accompanied by MMCT has also been discussed in the cyanide-bridged MV system by Scandola, Endicott, Baraldo, et al. [62–64].

8.4 Tetranuclear and Higher Nuclear Cyanide-Bridged Mixed-Valence Complex

As the complexity of cyanide-bridged system increases, the synthesis of tetranuclear and higher nuclear cyanide-bridged MV complexes becomes challengeable and needs more experimental skills. However, the attempt on such subject often results in interesting complexes with chemical aesthetics and unexpected physical and chemical properties, which has attracted great attention from chemists. The most obvious way of constructing a high nuclear cyanide-bridged MV complex is connecting metal ions through several cyanide bridges, and this approach has been proved to be effective, resulting in some star-like, square, cubic, or linear cyanide-bridged MV complexes.

The star-like cyanide-bridged complex usually consists of a central metal ion and several cyanometal or isocyanometal fragments as [M-(CN/NC-M′L)$_n$], and the coordination number n can vary from 3 to 6 depending on the property of the central metal. The electronic coupling and MMCT of these complexes have been investigated extensively [65–68]; for example, a series of multinuclear complexes [M-{NC-M′(CO)$_5$}$_n$] (**29**, M = Si^{IV}, Ge^{IV}, Sn^{IV}, Cr^{III}, Mn^{II}, Fe^{II}, Co^{II}, Ni^{II}, Cu^I, Zn^{II}, and M′ = Cr^0, Mo^0, and W^0, n = 3, 5, and 6) were investigated by Fehlhammer (Figure 8.15) [65]. Surprisingly, the expected M′→M electron transition is absent in these complexes, suggesting typical *Class I* species of these complexes. In contrast, Vahrenkamp and coworkers reported a series of cyanide-bridged complexes [Pt-(CN-M′L)$_n$] (M′L = Cp(dppe)Fe for **30a**, Cp(PPh$_3$)$_2$Ru and Cp(dppe)Fe for **30b**, and (tpa)Cu for **30c**), among which complex [Pt(CN){CN-Fe(dppe)Cp}$_2$ {CN-Ru(PPh$_3$)$_2$Cp}](SbF$_6$) (**30b**) has a configuration of the central Pt(CN)$_4$ fragment attached by two Cp(dppe)Fe fragments in the *trans*-position and one Cp(PPh$_3$)$_2$Ru in the *cis*-position (Figure 8.15) [66]. The cyclic voltammograms of **30b** suggest the electronic coupling between the two *trans*-Cp(dppe)Fe fragments, and the MMCT of Fe(II)→Fe(III) was observed in electronic spectra after one-electron oxidation. Interestingly, no MMCT absorption of *cis*-Ru(II)→*trans*-Fe(III) was observed, which indicates that the *cis*-configuration is not in favor of electron transfer in this case, and the same phenomenon has also been found in **30c** [67].

Cyanide-bridged tetrametallic square complexes, such as [Ru^{II}_4], [$Ru^{II}_2Rh^{III}_2$], and [$Fe^{II}_2Ru^{II}_2$] (**31–33**), have been synthesized and characterized [69, 70]. They share similar structure that four metal ions are alternately bridged at the vertices of the square by cyanide groups (Figure 8.16). In terms of electronic coupling, complex [Ru^{II}_4(μ-CN)$_4$(bpy)$_8$](PF$_6$)$_4$ (**31**) gives an appropriate example that its cyclic voltammogram displays four well-separated redox waves,

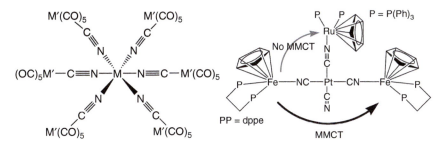

Figure 8.15 The structural diagram of **29** and **30b**.

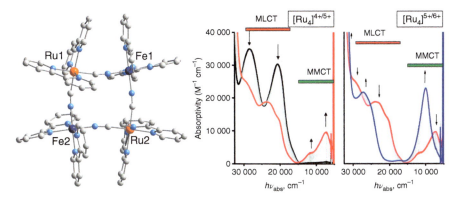

Figure 8.16 The structure of complex **33** and the electronic spectra of one/two-electron oxidation states of **31**. Source: Lin et al. 2011 [69]/Reproduced with permission of the American Chemical Society.

indicating the presence of electronic coupling between the two diagonal Ru atoms. The MV states [$Ru^{II}{}_3Ru^{III}$] and [$Ru^{II}{}_2Ru^{III}{}_2$], generated by the oxidation of Ce^{4+}, exhibit strong MMCT absorptions of Ru(II)→Ru(III), respectively (Figure 8.16) [69]. Similarly, complex [$Ru^{II}{}_2Fe^{II}{}_2(\mu\text{-}CN)_4(bpy)_8$]($PF_6$)$_4$ (**33**) displays four redox waves that correspond to four one-electron processes: [$Ru^{II}{}_2Fe^{II}{}_2$]$^{4+}$ ↔ [$Ru^{II}{}_2Fe^{II}Fe^{III}$]$^{5+}$ ↔ [$Ru^{II}{}_2Fe^{III}{}_2$]$^{6+}$ ↔ [$Ru^{II}Ru^{III}Fe^{III}{}_2$]$^{7+}$ ↔ [$Ru^{III}{}_2Fe^{III}{}_2$]$^{8+}$. Two new MMCT absorption bands appear at 2350 and 1380 nm in spectroelectrochemical measurements, being attributed to electron transitions of Fe(II)→Fe(III) for the [$Ru^{II}{}_2Fe^{II}Fe^{III}$]$^{5+}$ state and Ru(II)→Fe(III) for the [$Ru^{II}{}_2Fe^{III}{}_2$]$^{6+}$ state. Furthermore, the H_{ab} and α^2 of the two MMCT transitions were calculated as 1090, 0.065, and 1990 cm^{-1}, 0.096, respectively, indicating the *Class II* behavior [70].

The cyanide-bridged cages with a cubic structure have long been known but more often as the smallest building units of Prussian Blue analogs. The synthesis of isolated octanuclear cubic complex is a very challenging task that requires a lot of skill and luck. Consequently, only in 2008 did Oshio and coworkers report the first cyanide-bridged cube [$Fe^{II}{}_4Fe^{III}{}_4(\mu\text{-}CN)_{12}(tp)_8$] (**34**, tp = hydrotris(pyrazolyl)borate, Figure 8.17) [71]. This octanuclear MV complex is composed of four low-spin Fe^{II} and four high-spin Fe^{III} locating alternately at the vertices of a cube. The cyclic voltammogram of this complex displays impressive potential split of four

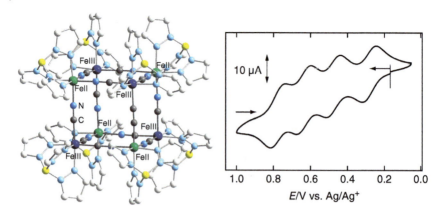

Figure 8.17 The structure and cyclic voltammogram of complex **34**. Source: Nihei et al. 2008 [71]/Reproduced with permission of the American Chemical Society.

quasi-reversible redox waves at 0.28, 0.48, 0.64, and 0.78 V corresponding to the step-by-step oxidation of the four identical low-spin Fe^{II}. Meanwhile, two MMCT bands at 816 and 1000 nm were observed by the spectroelectrochemical method, which are assigned to the adjacent MMCT transitions of [Fe^{II}–CN–Fe^{III}] units and the remote MMCT transitions of [Fe^{II}–CN–Fe^{III}–NC–Fe^{III}] units, respectively.

One example of linear cyanide-bridged MV species is a series of heptanuclear complexes with the formula trans-[L_4Ru^{II}\{(μ-NC)Fe^{III}(NC)$_4$(μ-CN)$Ru^{II}L'_4$(μ-NC)Fe^{III}(CN)$_5$\}$_2$]$^{6-}$ (**35**, L or L′ is substituted pyridine) [72]. Their chemical compositions and the linear shape were confirmed by electrospray ionization mass spectrometry, elemental analysis, and scanning tunneling microscopy. The electrochemical measurements give the evidence for existence of electronic coupling along the cyanide-bridged backbone of these complexes. Especially for the complex with identical {Ru(MeO-py)$_4$} fragments, the terminal Fe fragments of the complex display a potential splitting of about 60 mV, probably suggesting an electronic communication at a distance of more than 3.0 nm. The electronic spectra of these complexes exhibit several overlapping MMCT transitions of $d\pi(Ru^{II})\rightarrow d\pi(Fe^{III})$ in the NIR region, and a qualitative analysis indicates that improving the energy match of the three $d\pi(Ru^{II})$ orbitals in the heptanuclear complex could increase the electronic interaction between the metal centers.

Another approach to construct the cyanide-bridged MV complexes with high nuclearity is the attachment of cyanometal units [LM-(CN/NC)] to the existing clusters, such as $Ru_3(CO)_{10}$, Fe_4S_4, Mo_6Cl_8, Ta_6Cl_{12}, [$Co_3(dpa)_4$]$^{2+}$, and [$Ni_5(tpda)_4$]$^{2+}$ [73–78]. Taking this approach, Vahrenkamp and coworkers reported a series of cyanide-bridged complexes. For example, as shown in Figure 8.18, the attachment of Cp(dppe)Fe^{II}CN fragments to the clusters [Fe_2(salene)$_2$] and [Mn_2(Pc)$_2$O] resulted in two tetranuclear cyanide-bridged complexes, [{Cp(dppe)FeCN}$_2$-Fe_2(salene)$_2$](BPh$_4$)$_2$ (**36**) [76] and [{Cp(dppe)FeCN}$_2$-Mn_2(Pc)$_2$O] (**37**) [77]. The former displays the MMCT transition from Fe^{II} to the central Fe^{III}(salene), and the latter shows the potential split of the two terminal Cp(dppe)Fe^{II} fragments, indicating the long-range electron coupling of them with a distance of 13.2 Å.

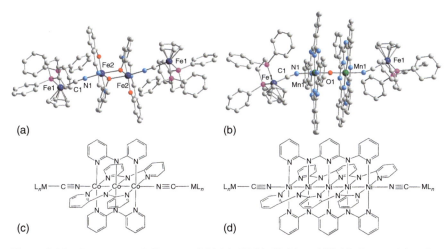

Figure 8.18 The structural diagram of **36** (a), **37** (b), **38** (c), and **39** (d). Source: (a) and (b) Adapted from Geiß et al. [76, 77] and (c) and (d) Sheng et al. [78].

In addition, [Co$_3$(dpa)$_4$]$^{2+}$ and [Ni$_5$(tpda)$_4$]$^{2+}$ were also used for the central unit; however, the obtained penta- and heptanuclear complexes **38** and **39** (Figure 8.18) do not exhibit the long-range electron coupling between the terminal [LM-CN] fragments because of the "electron-sink" properties of these cluster [78].

Further efforts were taken on the famous Fe$_4$S$_4$ cluster, which was attached by many different kinds of [LM-CN] fragments, such as [Cp(CO)$_2$Mn-CN], [Cp(CO)$_2$Fe–CN], [(CO)$_5$W-CN], [Cp(dppe)Fe–CN], [Cp(PPh$_3$)$_2$Ru–CN], and [(PPh$_3$)$_2$Cu–CN], forming a series of octanuclear cyanide-bridged MV complexes **40a–40f** (Figure 8.19) [73, 74]. Although the cyclic voltammogram measurements do not support the long-range electron coupling of the terminal [LM-CN] fragments in these complexes, the MMCT transitions from the cyanometal units to the Fe$_4$S$_4$ core were observed in most of them and resulted in the unusual magnetic properties.

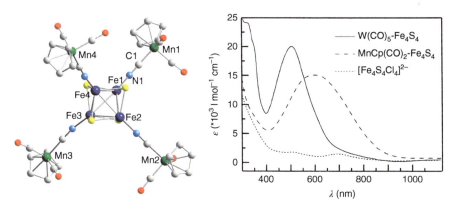

Figure 8.19 The structure of complex **40a** (left) and the electronic spectra of **40a** and **40c** (right). Source: Zhu et al. 1998 [74]/Reproduced with permission of Elsevier.

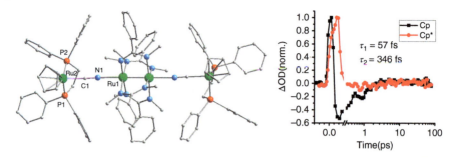

Figure 8.20 The structure of complex **41a** (left) and the kinetic traces of **41a** and **41b** (right). Source: Su et al. 2019 [79]/Reproduced with permission of the Royal Society of Chemistry.

Recently, the group of Sheng has reported the synthesis and characterization of a series of tetranuclear MV complexes by introducing the diruthenium cluster to cyanide-bridged systems. The complexes Ru^{II}–CN–$Ru_2^{III,III}$–NC–Ru^{II}, composed of one central $Ru_2^{III,III}(DMBA)_4$ attached by two ML fragments, were the first to be investigated (DMBA = N,N'-dimethylbenzamidinate, ML = Cp(dppe)Ru^{II}CN for **41a**, Cp*(dppe)Ru^{II}CN for **41b**, Figure 8.20) [79]. In spectroscopy measurements, the Cp*-contained complex **41b** exhibits a stronger electron interaction and lower MMCT transition energy of $Ru^{II} \rightarrow Ru_2^{III,III}$ than those of the Cp-contained complex **41a**, which implies the difference of the lifetime of their MMCT excited states. Therefore, the MMCT dynamics investigation of these complexes was carried on by the ultrafast TA spectral experiments, which reveals that the MMCT rate of **41b** ($0.29*10^{13}$ s^{-1}, $\tau = 3.46*10^{-13}$ s) is slower than that of **41a** ($0.18*10^{14}$ s^{-1}, $\tau = 5.7*10^{-14}$ s). This investigation shows that the MMCT rate or the lifetime of the MMCT excited state could be controlled by slight change of the substituent group on the metal center.

To acquire the knowledge of the influence of substitution effect on the MMCT process, a more systematic study has been conducted on nine tetranuclear cyanide-bridged MV complexes [CpMe$_n$(dppe)Fe^{II}CNRu$_2^{III,III}$(X-DMBA)$_4$NCFe^{II}(dppe)CpMe$_n$][PF$_6$]$_2$ (**42a–42i**, X = H and MeO, CF$_3$, n = 0, 4, and 5, Figure 8.21) with different substituents on the ligand of the bridging Ru_2 or the terminal Fe^{II} unit, respectively [80]. The crystallographic data, IR spectra, and UV/vis/NIR spectra of these complexes vary with different ligands regularly. In particular, for the complexes with the same [Ru$_2$(X-DMBA)$_4$] fragments, their MMCT bands exhibit a larger red shift as the donor Fe^{II} fragment becomes stronger, and for the complexes with the same donor Fe^{II} fragments, their MMCT bands move to lower energy in the order of [Ru$_2$(MeO-DMBA)$_4$]$^{2+}$ > [Ru$_2$(DMBA)$_4$]$^{2+}$ > [Ru$_2$(CF$_3$-DMBA)$_4$]$^{2+}$. This regular variation demonstrates that the strong electron-donating ligand at the donor Fe^{II} fragment and the strong electron-withdrawal ligand at the acceptor Ru$_2^{III,III}$ fragment contribute to decrease the energy barrier of MMCT process. Furthermore, the electrochemical measurement might suggest that the long-range electronic coupling between the two terminal Fe^{II} exists in this Fe^{II}–CN–Ru$_2^{III,III}$–NC–Fe^{II} array.

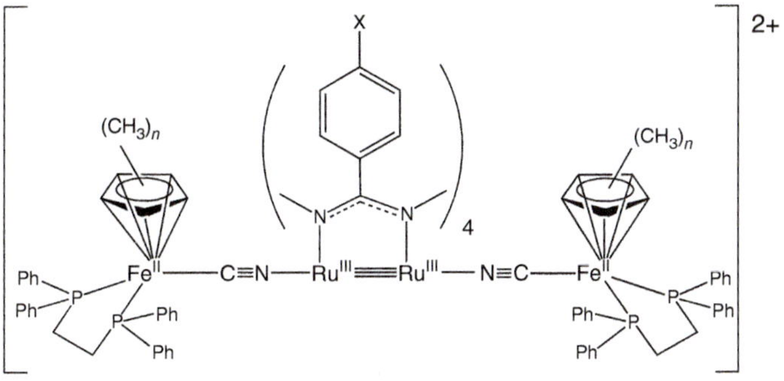

8.4 Tetranuclear and Higher Nuclear Cyanide-Bridged Mixed-Valence Complex

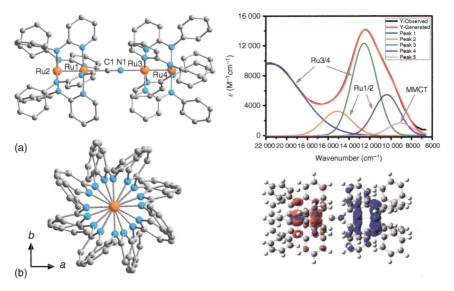

Figure 8.21 The structural diagram of **42**. Source: Su et al. 2021 [80]/Reproduced with permission of John Wiley & Sons.

Figure 8.22 The structure of **43** and its electronic spectra (a) as well as the redistribution of electron densities in the calculated MMCT transition band (b). Source: Su et al. 2019 [81]/Reproduced with permission of John Wiley & Sons.

Moreover, the group of Sheng has reported an interesting tetranuclear linear cyanide-bridged complex [Ru$_2$(μ-ap)$_4$–CN–Ru$_2$(μ-ap)$_4$](BPh$_4$) (**43**, ap = 2-anilinopyridinate, Figure 8.22), which consists of two identical [Ru$_2$(μ-ap)$_4$] clusters linked via cyanide bridge [81]. The crystallographic data, magnetic measurement, IR, EPR, and theoretical calculation results demonstrate that complex **43** possesses the uncommon electronic configuration of Ru$_2$(π^{*3}, $S = 1/2$)–CN–Ru$_2$($\pi^{*2}\delta^{*1}$, $S = 3/2$) because of the asymmetrical π back-bonding effect of cyanide bridge, and this is the first time [Ru$_2$(μ-ap)$_4$] cluster is displaying the uncommon low-spin electronic configuration of $(\pi^*)^3(S = 1/2)$. Meanwhile, the analysis of UV/vis/NIR spectra of this mixed-spin complex reveals a new absorption band at 8597 cm^{-1}, which could be assigned to the MMCT transitions from the cyanido-N-bound high-spin [Ru$_2$(μ-ap)$_4$](S = 3/2) cluster to the cyanido-C-bound low-spin [Ru$_2$(μ-ap)$_4$](S = 1/2) cluster, supported by the TDDFT calculations. As is well known, the intermetallic electron delocalization or MMCT usually occurs between the metal centers with different oxidation states in MV systems. Therefore, complex **43** gives the first example of MMCT between two same metal cluster centers with different spin states but in same oxidation states.

8.5 Conclusion and Outlook

In the past decades, as an important branch of MV chemistry, the experimental and theoretical studies of cyanide-bridged MV complexes have been well developed. It has been proved that the cyanide-bridged MV system is an ideal model for the investigation of electron transfer and electron coupling, not only because of the good ability of cyanide bridges to mediate electron interaction but also because of their high degree of designability. This chapter has briefly reviewed the studies on electron transfer and electron coupling in cyanide-bridged oligonuclear complexes. From the simplest asymmetric binuclear complexes, to trinuclear complexes with long-range electron interactions, and to higher nuclearity complexes with a range of geometries, the discussion of these complexes has demonstrated how their electron interactions are influenced by the nature of the metals and ligands, the orientation of the cyanide bridges, the geometrical construction of the chains, and other factors. Accordingly, the electron interactions of cyanide-bridged MV complexes have been tune successfully by controlling these influencing factors, resulting in the transition of complexes from localized species to fully delocalized *Class III* species. Moreover, some interesting phenomena involving electron transfer have also been introduced briefly, such as photoinduced multielectron charge transfer and a new type of MMCT between two same metal cluster centers with different spin states but in same oxidation states, which may provide new perspectives on the investigation of electron transfer process.

The future challenges in this field may be concentrated on construction of more complicated cyanide-bridged MV complexes to explore the novelty electron transfer behavior and synthesis of more fully delocalized *Class III* species to explore their physical properties for the potential applications in magnetic, electroconductive, and non-linear optical materials. Such research will not only deepen the understanding of the ubiquitous electron transfer phenomena in nature but also help to bring cyanide-bridged complexes out of the laboratory for real applications; therefore, it deserves to devote great efforts to this field in future.

Acknowledgment

Financial support from the National Natural Science Foundation of China (grants 21773243 and 21973095) and the Strategic Priority Research Program of Chinese Academy of Sciences (XDB20010200).

References

1. Powell, H.M. (1959). The beginnings of co-ordination chemistry. *Proc. Chem. Soc.* 3: 73–75.
2. Robin, M.B. (1962). The color and electronic configurations of Prussian Blue. *Inorg. Chem.* 1: 337–342.
3. Ito, A., Suenaga, M., and Ono, K. (1968). Mössbauer study of soluble Prussian Blue, Insoluble Prussian Blue, and Turnbull's Blue. *J. Chem. Phys.* 48: 3597–3599.
4. Ohkoshi, S., Namai, A., and Tokoro, H. (2019). Humidity sensitivity, organic molecule sensitivity, and superionic conductivity on porous magnets based on cyano-bridged bimetal assemblies. *Coord. Chem. Rev.* 380: 572–583.
5. Coronado, E. (2020). Molecular magnetism: from chemical design to spin control in molecules, materials and devices. *Nat. Rev. Mater.* 5: 87–104.
6. Demadis, K.D., Hartshorn, C.M., and Meyer, T.J. (2001). The localized-to-delocalized transition in mixed-valence chemistry. *Chem. Rev.* 101: 2655–2685.
7. D' Alessandro, D.M. and Keene, F.R. (2006). Intervalence charge transfer (IVCT) in trinuclear and tetranuclear complexes of iron, ruthenium, and osmium. *Chem. Rev.* 106: 2270–2298.
8. Hankache, J. and Wenger, O.S. (2011). Organic mixed valence. *Chem. Rev.* 111: 5138–5178.
9. Halet, J.-F. and Lapinte, C. (2013). Charge delocalization vs localization in carbon-rich iron mixed-valence complexes: a subtle interplay between the carbon spacer and the (dppe)Cp*Fe organometallic electrophore. *Coord. Chem. Rev.* 257: 1584–1613.
10. Marcus, R.A. and Sutin, N. (1985). Electron transfers in chemistry and biology. *Biochim. Biophys. Acta* 811: 265–322.
11. Hush, N.S. (1967). Intervalence-transfer absorption. Part 2. Theoretical considerations and spectroscopic data. *Prog. Inorg. Chem.* 8: 391–444.
12. McCleverty, J.A. and Ward, M.D. (1998). The role of bridging ligands in controlling electronic and magnetic properties in polynuclear complexes. *Acc. Chem. Res.* 31: 842–851.
13. Bignozzi, C.A., Argazzi, R., Schoonover, J.R. et al. (1992). Electronic coupling in cyano-bridged ruthenium polypyridine complexes and role of electronic effects on cyanide stretching frequencies. *Inorg. Chem.* 31: 5260–5267.
14. Vogler, A., Osman, A.H., and Kunkely, H. (1985). Photoredox reactions of mixed-valence compounds induced by intervalence excitation. *Coord. Chem. Rev.* 64: 159–173.
15. Burewicz, A. and Haim, A. (1988). Formation and properties of the binuclear complex $(NH_3)_5Ru^{III}NCFe^{II}(CN)_5^-$. *Inorg. Chem.* 27: 1611–1614.
16. Scandola, F., Argazzi, R., Bignozzi, C.A. et al. (1993). Electronic coupling between remote metal centers in cyanobridged polynuclear complexes. *Coord. Chem. Rev.* 125: 283–292.
17. Wu, Y., Pfennig, B.W., Sharp, S.L. et al. (1997). Light-induced multielectron charge transfer processes occurring in a series of Group-8-platinum cyanobridged complexes. *Coord. Chem. Rev.* 159: 245–255.

18 Coe, B.J., Meyer, T.J., and White, P.S. (1995). Cyano-bridged complexes of *trans*-tetrakis(pyridine)ruthenium(II). *Inorg. Chem.* 34: 3600–3609.

19 Laidlaw, W.M., Denning, R.G., Murphy, D.M., and Green, J.C. (2008). Solvent dependence of the g-anisotropy in the ESR of cyanide-bridged mixed-valence complexes. *Dalton Trans.* 6257–6264.

20 Carriedo, G.A., Connelly, N.G., Crespo, M.C. et al. (1991). Oxidation of cyanide-bridged dimanganese(I)complexes: redox-induced isomerisation as a probe of intermetallic interaction. *J. Chem. Soc., Dalton Trans.* 2: 315–323.

21 Endicott, J.F. and Chen, Y.-J. (2013). Electronic coupling between metal ions in cyanide-bridged ground state and excited state mixed valence complexes. *Coord. Chem. Rev.* 257: 1676–1698.

22 Vahrenkamp, H., Geiß, A., and Richardson, G.N. (1997). Cyanide-bridged oligonuclear complexes: features and attractions. *J. Chem. Soc., Dalton Trans.* 3643–3651.

23 Baraldo, L.M., Forlano, P., Parise, A.R. et al. (2001). Advances in the coordination chemistry of $[M(CN)_5L]^{n-}$ ions (M=Fe, Ru, Os). *Coord. Chem. Rev.* 219–221: 881–921.

24 Robin, M.B. and Day, P. (1967). Mixed valence chemistry-a survey and classification. *Adv. Inorg. Chem. Radiochem.* 9: 247–422.

25 Bechlars, B., D'Alessandro, D.M., Jenkins, D.M. et al. (2010). High-spin ground states via electron delocalization in mixed-valence imidazolate-bridged divanadium complexes. *Nat. Chem.* 2: 362–368.

26 Pieslinger, G.E., Alborés, P., Slep, L.D., and Baraldo, L.M. (2014). Class III delocalization in a cyanide-bridged trimetallic mixed valence complex. *Angew. Chem. Int. Ed.* 53: 1293–1296.

27 Oviedo, P.S., Pieslinger, G.E., Cadranel, A., and Baraldo, L.M. (2017). Exploring the localized to delocalized transition in non-symmetric bimetallic ruthenium polypyridines. *Dalton Trans.* 46: 15757–15768.

28 Domínguez, S.E., Pieslinger, G.E., Sanchez-Merlinsky, L., and Baraldo, L.M. (2020). Does geometry matter? Effect of the ligand position in bimetallic ruthenium polypyridine siblings. *Dalton Trans.* 49: 4125–4135.

29 Yang, Y.Y., Zhu, X.Q., Hu, S.M. et al. (2018). Different degrees of electron delocalization in mixed valence Ru-Ru-Ru compounds by cyanido-/isocyanido-bridge isomerism. *Angew. Chem. Int. Ed.* 57: 14046–14050.

30 Yang, Y.Y., Zhu, X.Q., Launay, J.P. et al. (2021). The electron transfer process in mixed valence compounds with a low-lying energy bridge in different oxidation states. *Angew. Chem. Int. Ed.* 60: 4804–4814.

31 Ma, X., Lin, C.S., Zhu, X.Q. et al. (2017). An unusually delocalized mixed-valence state of a cyanidometal-bridged compound induced by thermal electron transfer. *Angew. Chem. Int. Ed.* 56: 1605–1609.

32 Creutz, C. and Taube, H. (1969). A direct approach to measuring the Franck-Condon barrier to electron transfer between metal ions. *J. Am. Chem. Soc.* 91: 3988–3989.

33 Vogler, A., Osman, A.H., and Kunkely, H. (1987). Heterobinuclear transition-metal complexes. Synthesis and optical metal to metal electron transfer. *Inorg. Chem.* 26: 2337–2340.

34 Bignozzi, C.A., Roffia, S., and Scandola, F. (1985). Intervalence transfer in cyano-bridged bi- and trinuclear ruthenium complexes. *J. Am. Chem. Soc.* 107: 1644–1654.

35 Creutz, C. (1983). Mixed valence complexes of d^5-d^6 metal centers. *Prog. Inorg. Chem.* 30: 1–73.

36 Adams, C.J., Anderson, K.M., Connelly, N.G. et al. (2007). The effect of μ-CN linkage isomerism and ancillary ligand set on directional metal-metal charge-transfer in cyanide-bridged dimanganese complexes. *Dalton Trans.* 33: 3609–3622.

37 Zhu, N.Y. and Vahrenkamp, H. (1994). Cyanide-isocyanide isomerism in CN-bridged organometallic complexes. *Angew. Chem. Int. Ed.* 33: 2090–2091.

38 Zhu, N.Y. and Vahrenkamp, H. (1997). Synthesis, redox chemistry, and mixed-valence phenomena of cyanide-bridged dinuclear organometallic complexes. *Chem. Ber.* 130: 1241–1252.

39 Ma, X., Lin, C.S., Zhang, H. et al. (2013). Synthesis, spectral and redox switchable cubic NLO properties of chiral dinuclear iron cyanide/isocyanide-bridged complexes. *Dalton Trans.* 42: 12452–12459.

40 Zhang, L.T., Zhu, X.Q., Su, S.D. et al. (2018). Influence of the substitution of the ligand on MM′CT properties of mixed valence heterometallic cyanido-bridged Ru-Fe complexes. *Cryst. Growth Des.* 18: 3674–3682.

41 Zhang, L.T., Zhu, X.Q., Hu, S.M. et al. (2019). Influence of ligand substitution at the donor and acceptor center on MMCT in a cyanide-bridged mixed-valence system. *Dalton Trans.* 48: 7809–7816.

42 Pieslinger, G.E., Aramburu-Trošelj, B.M., Cadranel, A., and Baraldo, L.M. (2014). Influence of the electronic configuration in the properties of d^6-d^5 mixed-valence complexes. *Inorg. Chem.* 53: 8221–8229.

43 Macatangay, A.V., Song, X., and Endicott, J.F. (1998). What does "through-bond coupling" mean? Observations on simple donor-acceptor systems. *J. Phys. Chem. A* 102: 7537–7540.

44 Brown, N.C., Carpente, G.B., Connelly, N.G. et al. (1996). Intramolecular electron transfer in linear trinuclear complexes of copper(I), silver(I) and gold(I) bound to redox-active cyanomanganese ligands. *J. Chem. Soc., Dalton Trans.* 3977–3984.

45 Zhu, N.Y. and Vahrenkamp, H. (1999). Cyanide bridged trinuclear complexes with bent M-NC-Fe-CN-M and M-CN-Fe-NC-M arrangements-synthesis and physical properties. *J. Organomet. Chem.* 573: 67–72.

46 Chen, Z.N., Appelt, R., and Vahrenkamp, H. (2000). Cyanide-bridged trinuclear complexes based on central bis(hexafluoroacetylacetonate)metal units. *Inorg. Chim. Acta* 309: 65–71.

47 Appelt, R. and Vahrenkamp, H. (2003). Cyanide bridged di- and tri-nuclear complexes with central Cr(III)-, Mn(III)- and Co(III)-salen units. *Inorg. Chim. Acta* 350: 387–398.

48 Sheng, T.L. and Vahrenkamp, H. (2004). Cyanide bridged trinuclear complexes with central metals bearing pyridine ring ligands. *Inorg. Chim. Acta* 357: 1739–1747.

49 Sheng, T.L. and Vahrenkamp, H. (2004). Long range metal-metal interactions along Fe-NC-Ru-CN-Fe chains. *Eur. J. Inorg. Chem.* 1198–1203.

50 Richardson, G.N., Brand, U., and Vahrenkamp, H. (1999). Linear and bent M(μ-CN)Pt(μ-CN)M chains: probes for remote metal-metal interactions. *Inorg. Chem.* 38: 3070–3079.

51 Höfler, M., Jung, W., Scobel, M., and Rademacher, J. (1986). Darstellung und Reaktionen Hg(CN)-verbrückter Metallcarbonyle. *Chem. Ber.* 119: 3268–3275.

52 Bignozzi, C.A., Chiorboli, C., Indelli, M.T. et al. (1994). Molecular structure and linkage isomerism of *cis*-[Ru(bipy)$_2$-{*trans*-Cr(cyclam)(CN)$_2$}$_2$]$^{4+}$ (bipy=2,2'-bipyridine, cyclam =1,4,8,11-tetraazacyclotetradecane). *J. Chem. Soc., Dalton Trans.* 2391–2395.

53 Wang, Y., Lin, C., Ma, X. et al. (2015). Influence of the central diamagnetic cyanidometal on the distant magnetic interaction in cyanidebridged Fe(III)-M(II)-Fe(III) complexes. *Dalton Trans.* 44: 7437–7448.

54 Ma, X., Lin, C.S., Hu, S.M. et al. (2014). Influence of central metalloligand geometry on electronic communication between metals: syntheses, crystal structures, MMCT properties of isomeric cyanido-bridged Fe$_2$Ru complexes, and TDDFT calculations. *Chem. Eur. J.* 20: 7025–7036.

55 Geiss, A. and Vahrenkamp, H. (2000). M(μ-CN)Fe(μ-CN)M′Chains with phthalocyanine iron centers: preparation, structures, and isomerization. *Inorg. Chem.* 39: 4029–4036.

56 Geiss, A., Kolm, M.J., Janiak, C., and Vahrenkamp, H. (2000). M(μ-CN)Fe(μ-CN)M′Chains with phthalocyanine iron centers: redox, spin-state, and mixed-valence properties. *Inorg. Chem.* 39: 4037–4043.

57 Alborés, P., Slep, L.D., Weyhermuller, T., and Baraldo, L.M. (2004). Fine tuning of the electronic coupling between metal centers in cyano-bridged mixed-valent trinuclear complexes. *Inorg. Chem.* 43: 6762–6773.

58 Alborés, P., Rossi, M.B., Baraldo, L.M., and Slep, L.D. (2006). Donor–acceptor interactions and electron transfer in cyano-bridged trinuclear compounds. *Inorg. Chem.* 45: 10595–10604.

59 Pieslinger, G.E., Alborés, P., Slep, L.D. et al. (2013). Communication between remote moieties in linear Ru–Ru–Ru trimetallic cyanide-bridged complexes. *Inorg. Chem.* 52: 2906–2917.

60 Xu, Q.D., Zeng, C., Su, S.D. et al. (2021). Tuning metal to metal charge transfer properties in cyanidometal-bridged complexes by changing the auxiliary ligand on the bridge. *Dalton Trans.* 50: 6161–6169.

61 Xu, Q.D., Zhang, L.T., Zeng, C. et al. (2021). Influence of fine ligand substitution modification of the isocyanidometal bridge on metal-to-metal charge transfer properties in class II–III mixed valence complexes. *Chem. Eur. J.* 27: 11183–11194.

62 Scandola, F., Bignozzi, C.A., Chiorboli, C. et al. (1990). Intramolecular energy transfer in Ru(II)-Ru(II) and Ru(II)-Cr(III) polynuclear complexes. *Coord. Chem. Rev.* 97: 299–312.

63 Endicott, J.F. and Chen, Y.-J. (2007). Charge transfer-excited state emission spectra of mono- and bi-metallic coordination complexes: Band shapes, reorganizational energies and lifetimes. *Coord. Chem. Rev.* 251: 328–350.

64 Oviedo, P.S., Baraldo, L.M., and Cadranel, A. (2021). Bifurcation of excited state trajectories toward energy transfer or electron transfer directed by wave function symmetry. *PNAS* 118 (4), e2018521118: 1–7.

65 Fritz, M., Rieger, D., Bär, E. et al. (1992). Tetra- to heptanuclear carbonyl(cyano) chromato, -molybdato and -tungstato complexes of 3d metals. *Inorg. Chim. Acta* 198-200: 513–526.

66 Richardson, G.N. and Vahrenkamp, H. (2000). Cyanide-bridged tetranuclear complexes with T-shaped $Pt(CN-M)_3$ arrangements. *J. Organomet. Chem.* 597: 38–41.

67 Flay, M.L. and Vahrenkamp, H. (2003). Cyanide-bridged oligonuclear complexes containing Ni-CN-Cu and Pt-CN-Cu linkages. *Eur. J. Inorg. Chem.* 1719–1726.

68 Laidlaw, W.M. and Denning, R.G. (1996). Heterobi-, tri- and tetra-metallic cyanide-bridged complexes based on pentacarbonylcyanochromate(0) and penta-, cis-tetra- and fac-triammineruthenium(III) groups: optical metal-metal charge transfer and electrochemical characteristics. *Inorg. Chim. Acta* 248: 51–58.

69 Lin, J.L., Tsai, C.N., Huang, S.Y. et al. (2011). Nearest- and next-nearest-neighbor Ru(II)/Ru(III) electronic coupling in cyanide-bridged tetra-ruthenium square complexes. *Inorg. Chem.* 50: 8274–8280.

70 Oshio, H., Onodera, H., and Ito, T. (2003). Spectroelectrochemical studies on mixed-valence states in a cyanide-bridged molecular square, $[Ru^{II}_2Fe^{II}_2(\mu\text{-CN})_4(bpy)_8](PF_6)_4 \cdot CHCl_3 \cdot H_2O$. *Chem. Eur. J.* 9: 3946–3950.

71 Nihei, M., Ui, M., Hoshino, N., and Oshio, H. (2008). Cyanide-bridged iron(II,III) cube with multistepped redox behavior. *Inorg. Chem.* 47: 6106–6108.

72 Alborés, P., Slep, L.D., Eberlin, L.S. et al. (2009). From monomers to geometry-constrained molecules: one step further toward cyanide bridged wires. *Inorg. Chem.* 48: 11226–11235.

73 Zhu, N., Pebler, J., and Vahrenkamp, H. (1996). Combination of the Fe_4S_4 and M-CN-Fe redox functions. *Angew. Chem. Int. Ed.* 35: 894–895.

74 Zhu, N., Appelt, R., and Vahrenkamp, H. (1998). Attachment of cyanometal units to the Fe_4S_4 cluster. *J. Organomet. Chem.* 565: 187–192.

75 Kennedy, V.O., Stern, C.L., and Shriver, D.F. (1994). Synthesis, structure, and reactivity of the substitution-labile metal cluster $[Bu_4N]_2[Ta_6Cl_{12}(OSO_2CF_3)_6]$. *Inorg. Chem.* 33: 5967–5969.

76 Geiß, A. and Vahrenkamp, H. (1999). Cyanide-bridged arrays of 2, 3 and 4 metal atoms based on salene-iron complexes-syntheses, structures and metal–metal interactions. *Eur. J. Inorg. Chem.* 1793–1803.

77 Geiß, A., Keller, M., and Vahrenkamp, H. (1997). A highly redox-active tetranuclear complex with a nearly linear Fe-CN-Mn-O-Mn-NC-Fe backbone. *J. Organomet. Chem.* 541: 441–443.

78 Sheng, T.L., Appelt, R., Comte, V., and Vahrenkamp, H. (2003). Chain-like tetra-, penta- and heptanuclear cyanide-bridged complexes by attachment of organometallic cyanides to M2, M3 and M5 units. *Eur. J. Inorg. Chem.* 3731–3737.

79 Su, S.D., Zhu, X.Q., Zhang, L.T. et al. (2019). The MMCT excited state of a localized mixed valence cyanido-bridged Ru^{II}–$Ru_2^{III,III}$–Ru^{II} complex. *Dalton Trans.* 48: 9303–9309.

80 Su, S.D., Zhu, X.Q., Wen, Y.H. et al. (2021). Influence of substitution effect on MMCT in mixed-valence cyanido-bridged Fe^{II}-CN-Ru_2^{III}, ^{III}-NC-Fe^{II} System. *Eur. J. Inorg. Chem.* 3474–3480.

81 Su, S.D., Zhu, X.Q., Wen, Y.H. et al. (2019). A diruthenium-based mixed spin complex Ru_2^{5+}(S=1/2)-CN-Ru_2^{5+}(S =3/2). *Angew. Chem. Int. Ed.* 58: 15344–15348.

9

Organic Mixed-Valence Systems: Toward Fundamental Understanding of Charge/Spin Transfer Materials

Akihiro Ito

Mie University, Graduate School of Engineering, Department of Chemistry for Materials, 1577 Kurimamachiya-cho, Tsu, Mie 514-8507, Japan

9.1 A Brief Sketch of the History of Organic Mixed-Valence Systems

The notion of mixed valence (MV) was originally nurtured by inorganic chemists through grasping coordination compounds containing two or more metal atoms in different oxidation states. In particular, the so-called Creutz–Taube complexes have been widely accepted as a hallmark of MV compounds [1–3]. In those inorganic complexes, the metal atoms with various oxidation states are linked by multidentate ligands as the bridging units. However, generally, the redox-active centers and/or molecular units are not necessarily connected by the bridging units: they can be directly linked by covalent bonds, and more broadly, the mixed valency between adjacent redox-active molecules with different oxidation states in a non-covalently assembled solid can also be discussed in conjunction with the design of organic metals.

According to the case of inorganic MV compounds, the purely organic MV compounds are simply formed by replacing the metal redox centers with organic redox-active centers. However, it becomes often difficult to distinguish what is attributable to "redox-active centers" and what to "bridging units" in organic MV systems. The chemistry of organic MV systems, hence, has been progressing for a long time without being recognized as MV systems, despite the fact that an archetype of an organic MV system, Wurster's blue, was already discovered in 1879 [4]. In 1976, Mazur and coworkers initiated a new approach to investigate the electronic states of stable organic radical ions from the viewpoint of "intramolecular" mixed valency [5], and they used the term "organic mixed-valence species" in their paper [6]. On the other hand, tetracyanoquinodimethane (TCNQ) in 1960 [7] and tetrathiafulvalene (TTF) in 1970 [8] were synthesized as representative organic electron acceptor and donor, respectively, and moreover, in 1973,

Mixed-Valence Systems: Fundamentals, Synthesis, Electron Transfer, and Applications, First Edition.
Edited by Yu-Wu Zhong, Chun Y. Liu, and Jeffrey R. Reimers.
© 2023 WILEY-VCH GmbH. Published 2023 by WILEY-VCH GmbH.

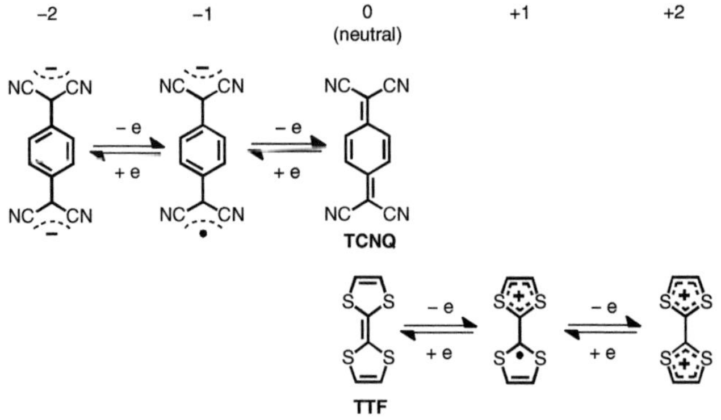

Figure 9.1 TCNQ as two-stage Wurster-type redox system and TTF as two-stage Weitz-type redox system.

their charge transfer complex (TTF-TCNQ) as well as non-stoichiometric TCNQ radical anion and TTF radical cation salts was reported to exhibit a high electrical conductivity [7–9]. Since then, "intermolecular" mixed valency has been started to discuss in those ion radical salts as "synthetic metals" [10, 11], and in this context, the term "organic mixed-valence compounds" first appeared in the title of Matsunaga's paper [12]. For those organic redox systems such as TCNQ and TTF, Hünig categorized their structural characteristics caused by the multistage redox reactions: two-stage Wurster-type redox system and two-stage Weitz-type redox system [13]. As shown in Figure 9.1, TCNQ is classified as two-stage Wurster-type electron acceptor, where the end groups (dicyanomethane groups) are part of a cyclic π-system that exhibits aromatic character in the reduced form, while TTF is classified as two-stage Weitz-type electron donor, where the end groups (1,3-dithiol groups) include a cyclic π-system that exhibits aromatic character in the oxidized form. Apparently, one-electron-reduced or one-electron-oxidized semi-quinone (TCNQ⁻ or TTF⁺) form can be considered a (intramolecular) mixed-valence compound, and they were presently known as charge-delocalized MV compounds.

In the 1990s, Nelsen's and Launay's groups independently opened a new trend to analyze intervalence charge transfer (IVCT) optical absorption bands of purely organic MV compounds on the basis of Marcus–Hush theory, thereby enabling to gain information about electronic couplings [14–16]. Following these early works, various researchers, particularly Lambert and coworkers, have thoroughly investigated the MV states between redox-active centers separated by various bridging units over the past two decades, and the organic MV compounds have recently gained traction in a broad range of chemical science [17, 18].

The aim of this chapter is to highlight several topics on purely organic MV systems, although the basic concepts that are essential to understand the details of organic MV chemistry are also briefly explained, simply because comprehensive overview of this field is already covered much more adequately in other review papers [17, 18]. Charge localization and delocalization on π-conjugated organic molecular systems are discussed in terms of key parameters: electronic coupling, reorganization energy, and activation barrier.

9.2 A Glossary for This Chapter

Before we take a closer look at the details of organic MV systems, in this section, a few basic concepts should be concisely given for understanding of organic MV systems [17–19]. For the sake of simplicity, we will here confine ourselves to the "degenerate" MV systems with only two redox-active centers that are identical except for their degrees of oxidation, connected by a symmetric bridging unit like bis(diarylamino)-based radical cations as a typical organic MV system, as shown below.

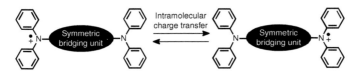

The intramolecular charge transfer (CT) reaction can be described by the motion of the system on a potential energy surface from the left (right) to the right (left) through the transition state. The distortion of the process is modeled by harmonic potentials with equal force constants (2λ) along a dimensionless reaction coordinate from 0 (left) to 1 (right). When there is no electronic communication between the two redox-active centers, the diabatic free energy surfaces (FESs) are represented by the two harmonic potentials with a CT barrier $\lambda/4$ at the crossing point, as shown in the dashed curves in Figure 9.2. In contrast, there exists electronic communication between them, the two diabatic harmonic potentials are coupled to form adiabatic FESs through the electronic interaction (H_{ab}) between the diabatic states, $|1\rangle$ and $|2\rangle$, as shown in the broken curves in Figure 9.2. These two important quantities, λ and H_{ab}, with a dimension of energy are called reorganization energy and electronic coupling, respectively. The reorganization energy stands for the whole energy change involved in the structural change of both the MV system itself (internal) and the surroundings such as solvent molecules (external) through the CT process.

As is apparent from Figure 9.2, when the value of H_{ab} increases at a constant value of λ, the separation $2H_{ab}$ between the two adiabatic potentials at the crossing point of the two diabatic FESs is gradually widening up. Hence, the barrier (ΔG^*) between the two minima decreases with increasing H_{ab}, and finally, the two minima merge into a single minimum for $2H_{ab} \geq \lambda$. At this point, the MV state can be classified by the magnitude of the electronic coupling H_{ab}, as done by Robin and Day [20]:

(Class I)	The electronic coupling H_{ab} is negligibly small (the adiabatic FESs resemble the diabatic ones).
	The charge/spin is statically localized in one redox center.
(Class II)	The electronic coupling H_{ab} has an intermediate value ($0 < 2H_{ab} < \lambda$).
	The charge/spin is dynamically delocalized between two redox centers (valence localization).
(Class III)	The electronic coupling H_{ab} is larger than $\lambda/2$.
	The charge/spin is statically delocalized between two redox centers (valence delocalization).

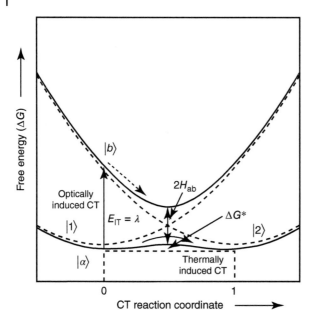

Figure 9.2 Schematic representation of diabatic (broken curves) and adiabatic (solid curves) FESs for a degenerate two-state MV system.

As shown in Figure 9.2, the CT process in MV systems is induced by optical excitation and/or thermal fluctuation. The optically induced CT is detected as the IVCT band corresponding to λ in absorption spectrum. In particular, for bis(triarylamine) radical cations, the IVCT band often appears in the near-infrared (NIR) region, and more importantly, is not concealed by other intense bands, as shown in Figure 9.3. From the IVCT band analysis, we can evaluate both λ and H_{ab}. First of all, the reorganization energy λ is simply determined by the absorption maximum $\tilde{v}_{max}$ ($\equiv E_{IT}$) (in cm^{-1}). Next, the evaluation method of electronic coupling H_{ab} depends on the MV behavior of the studied compounds. In the case of delocalized Class III system where $2H_{ab}$ becomes larger than λ, the absorption maximum $\tilde{v}_{max}$ equals with $2H_{ab}$, enabling a direct evaluation of H_{ab} from the IVCT band maximum.

In contrast, for the Class II case, the determination of electronic coupling H_{ab} (in cm^{-1}) is conducted on the basis of several theoretical treatments. It should be noted here that the IVCT band shape changes from a nearly Gaussian shape in the weak electronic coupling case to an asymmetric band shape where the lower energy side appeared to be cut off at $\tilde{v} = 2H_{ab}$, simply due to no electronic transitions exceeding the photon energy $2H_{ab}$, in the strong electronic coupling or "almost delocalized" MV case [21].

9.2.1 Hush Analysis [22]

In the special case where the IVCT band is genuinely Gaussian shaped, that is, the diabatic FESs are strictly harmonic potentials.

$$H_{ab} = \frac{0.0206(\varepsilon_{max} E_{IT} \Delta\tilde{v}_{1/2})^{1/2}}{R} \tag{9.1}$$

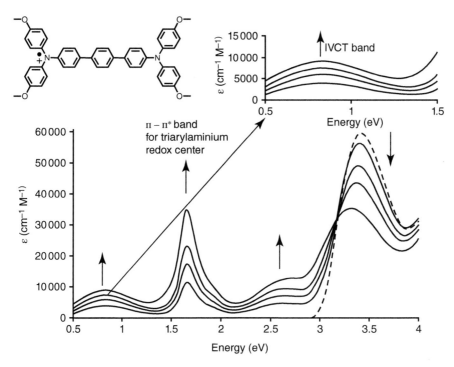

Figure 9.3 Archetypal absorption spectroscopic change upon one-electron oxidation of a degenerate two-state MV system **1** in CH_2Cl_2 at room temperature.

where ε_{max} is the maximum molar extinction coefficient (in M^{-1} cm^{-1}), $\Delta\tilde{\nu}_{1/2}$ is the bandwidth at half-height (in cm^{-1}), and R is the CT distance (in Å), which is often substituted by the distance between redox centers.

9.2.2 Mulliken–Hush Two-State Analysis [19, 23]

More versatile treatment than Eq. (9.1)

$$H_{ab} = \frac{|\mu_{ab}|}{|\Delta\mu_{12}|} E_{IT} \tag{9.2}$$

where μ_{ab} is the transition dipole moment between the two adiabatic states, $|a\rangle$ and $|b\rangle$, as shown in Figure 9.2, and the value of $|\mu_{ab}|^2$ is obtained by the integration of the observed IVCT band. $\Delta\mu_{12}$ is the diabatic dipole moment difference $\mu_2 - \mu_1$ and is given by $\sqrt{(\mu_b - \mu_a)^2 + 4\mu_{ab}^2}$ using the adiabatic dipole moment difference $\mu_b - \mu_a$ [23], which is often estimated by quantum chemical calculations.

9.2.3 Mulliken–Hush Two-Mode Analysis [24]

The above treatment is based on the CT process along one dimension connecting the two redox centers assisted by an averaged asymmetric vibrational mode. However, an average totally symmetric vibrational mode does exist in the studied systems. In the two-mode analysis, the diabatic potentials are approximated up to quadratic and linear terms for diagonal and off-diagonal elements, respectively, within two-dimensional reaction coordinate system, and the parameters including H_{ab} are fitted to the observed IVCT band [25–27].

9.2.4 Generalized Mulliken–Hush Three-State Analysis [28]

In the case of two-state models, only superexchange (or coherent) mechanism mediated by the bridging unit is taken into consideration for the CT process. However, when the frontier MO energy of the bridging unit approaches that of the redox centers, the frontier MOs of the whole MV systems extend over the bridging unit, thus indicating the importance of hopping (or incoherent) mechanism through the "bridge-centered" state, in addition to the above superexchange mechanism in the CT process [29]. By fitting to the observed IVCT band between the two redox units as well as so-called "bridge band" corresponding to the transition from the ground state to the "bridge-centered" state on the basis of this generalized Mulliken–Hush three-state (+ two-mode model), we can extract the CT parameters such as H_{ab} and λ more accurately [30].

Another important quantity ΔG^* (in cm^{-1}) is closely related to the thermally induced CT process as shown in Figure 9.2 and can be determined directly by the variable-temperature electron spin resonance (VT-ESR) measurements or indirectly by using the separately determined values of both H_{ab} and λ, following Eq. (9.3), assuming the harmonic approximation for the diabatic FESs.

$$\Delta G^* = \frac{\lambda}{4} - H_{ab} + \frac{H_{ab}^2}{\lambda} \tag{9.3}$$

It should be noted that the ESR spectrum observed for a spin-localized Class II MV compound shows the same hyperfine structure as that for a spin-delocalized Class III MV compound, when the intramolecular spin transfer process becomes fast in comparison with the ESR time scale in the measured temperature range, resulting that we cannot discern whether the MV system is of Class II or Class III.

As an example for the IVCT band, the UV–vis–NIR absorption spectrum observed for the radical cation of 4,4″-terphenylene-bridged bis(dianisylamine) **1** is shown in Figure 9.3 [31]. The bis(diarylamine)s or bis(triarylamine)s as redox-active centers are marked by (i) the appearance of the IVCT band in the NIR region, mainly due to low reorganization energy [32–34], and (ii) the separation of the IVCT band from the other low-energy bands (Figure 9.3), thus facilitating a detailed analysis of the MV state. In this case, both electronic coupling (H_{ab}^{MH}) and activation barrier (ΔG^*_{ESR}) in association with the thermal intramolecular CT were determined by the Mulliken–Hush two-state analysis of the IVCT band and the VT-ESR measurements, respectively, as shown in Figure 9.4a, and radical cation **1**$^{·+}$ is regarded as a valence-localized Class II MV system.

It may be worth pointing out, in passing, that the use of B3LYP hybrid functional in the quantum chemical calculations of the Class II MV compounds often leads to erroneous results, in association with the problem of self-interaction error [35].

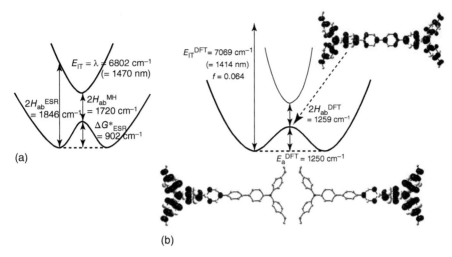

Figure 9.4 Schematic energy diagrams and key parameters for the intramolecular CT of **1**$^{·+}$ determined (a) experimentally from the IVCT band analysis and the variable-temperature ESR spectroscopy and (b) theoretically from density functional theory calculations [(TD-)UBLYP35/SVP+CPCM(CH$_2$Cl$_2$)].

Instead, the BLYP35 functional with 35% exact exchange term, which was proposed by Kaupp [36, 37], affords the reasonable results by including solvent effects such as conductor-like polarizable continuum model (CPCM), if the experiment is conducted in solution phase. In Figure 9.4b, the important parameters computed at (TD-)UBLYP35/SVP//UBLYP35/SVP+CPCM(CH$_2$Cl$_2$) for the Class II MV radical cation **1**$^{·+}$ are summarized schematically. The computed values for the electronic coupling, activation energy, and the IVCT band maximum (H_{ab}^{DFT}, E_a^{DFT}, and E_{IT}^{DFT}) show a good agreement with experimentally determined ones (Figure 9.4a) [31].

Lastly, there are a few comments to make concerning the MV system with two redox centers:

1. The IVCT band disappears when the transition dipole moment, μ_{ab}, is strictly or nearly zero, apparently from Eq. (9.2).
2. The IVCT band maximum shifts in accordance with the reorganization energy λ; in the MV compounds with high rigidity, the internal contribution in λ becomes very small, thereby resulting that the IVCT band maximum shifts out of the NIR region to lower energy region. In Class II MV systems, the IVCT band maximum in solution changes according to the solvent polarity, simply due to the external contribution in λ.
3. The higher the reorganization energy λ, the higher the thermally assisted CT barrier ΔG^*, thus leading to the Class I MV system. Conversely, the small reorganization energy leads to lowering the activation barrier. As mentioned above, the reorganization energy comprised internal and external contributions, and therefore, the solvent polarity impacts the CT rates of the Class II MV systems, when they are measured in solution phase.

9.3 Relationship Between Bridging Units and Electronic Coupling

The strength of the electronic coupling H_{ab} is closely related to the orbital interaction between the fragment molecular orbitals (FMO) of the redox centers and the bridging unit, which is mainly determined by the energy level matching between them, as shown schematically in the case of radical cation MV system in Figure 9.5 [38]. When the energy level of the FMO of the bridging unit is far below that of the redox centers, the direct orbital interaction between the FMOs of the redox centers results in the small electronic coupling H_{ab}, thus expecting the Class I MV state. However, as the bridge FMO is raised in energy, one of the FMOs of the redox centers strongly interacts with the bridge MO, leading to an electronic coupling large enough to realize the Class III MV state. Finally, if the FMO level of the bridging unit exceeds that of the redox centers, the bridging unit switches the role to the redox-active unit, thus resulting in a bridge-localized radical cation that is hardly analyzed in the context of MV framework.

For instance, in cyclopenta-ring-fused rylene π-bridged bis(dianisylamine) radical cations, the perylene-bridged $2^{•+}$ was found to be a delocalized Class III compound with $H_{ab} = 3464$ cm^{-1}, in line with case (c) in Figure 9.5. However, in π-extended quarter-rylene- and hexarylene-bridged $3^{•+}$ and $4^{•+}$, it was confirmed that the positive charge is exclusively localized in the central rylene moieties, due to the high lying FMO of the bridging unit [39].

As has already been mentioned, whether the intramolecular CT mechanism in organic MV systems is superexchange, hopping, or both is determined by a subtle balance between the FMO energy levels of the redox centers and the bridging unit. However, as far as the optically induced CT process is concerned, the superexchange mechanism is considered to be dominant. When the distance between two redox

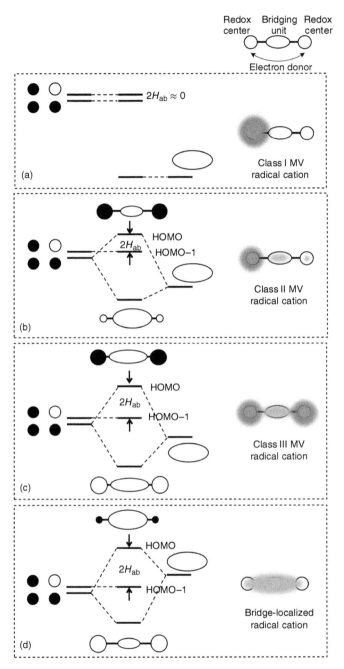

Figure 9.5 Schematic drawings of orbital interaction in terms of fragment MOs of redox centers and bridging unit (left) and the instantaneous spin distribution on the radical cation of organic MV systems with two redox centers (right). Pictures from (a) to (d) indicate snapshots corresponding to Class I to Class III MV and bridge-localized radical cations, respectively. The resulting energy gap between HOMO and HOMO-1 is approximately equal to the value of $2H_{ab}$.

centers is lengthened by a linear π-bridge with identical repeating bridging units, the electronic coupling H_{ab} decays exponentially with distance in accordance with Eq. (9.4) [29]:

$$H_{ab} = H_{ab}^0 \exp\left(-\frac{\beta}{2}Na\right) \qquad (9.4)$$

where N is the number of bridging units with unit length a and β is an attenuation factor (or decay constant).

The bridge length dependence of electronic coupling H_{ab} in organic MV systems was firstly examined by several six bis(triarylamine) radical cations with different numbers of bonds between the nitrogen redox centers though a true homologous series of bridging unit was not utilized [40, 41]. The attenuation factor β for radical cations of compounds **5–8** shown below was estimated to be a low value of 0.16 Å^{-1}, indicating quite longer π-bridges can mediate the intramolecular CT through sizable electronic coupling [41].

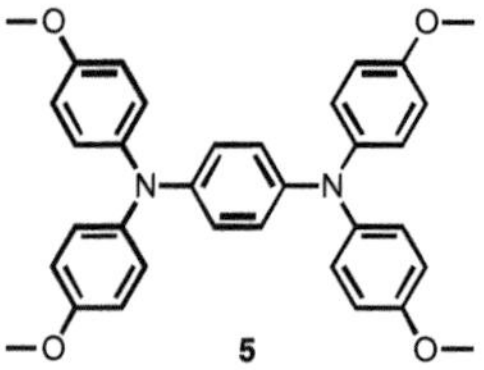

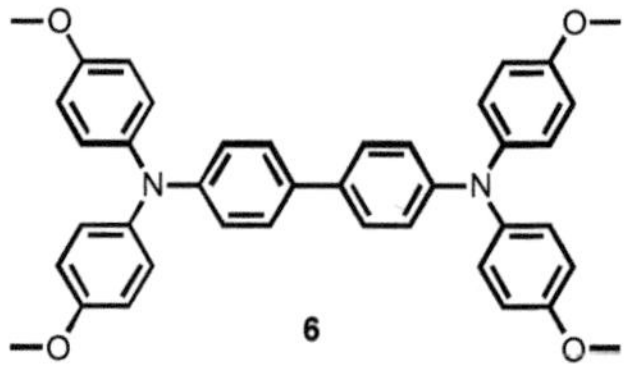

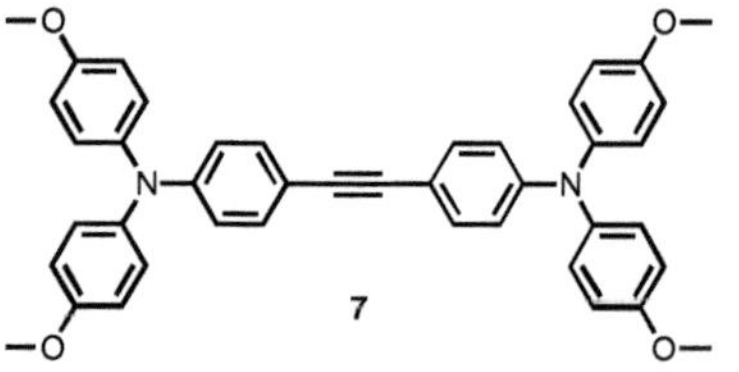

An investigation of the bridge length dependence of H_{ab} by using a homologous series of π-bridges has been initiated by Barlow and coworkers for radical cations of oligo(p-phenylenevinylene)-bridged (OPV-bridged) bis(dianisylamine) system **9–12** [42]. The low attenuation factor (0.01 Å^{-1}) has previously been reported for photo-induced CT between donor–acceptor (DA) systems through the OPV wires [43]. Unfortunately, the IVCT band analysis was possible only for **9**$^{·+}$ (2890 cm^{-1}) and **10**$^{·+}$ (1510 cm^{-1}) on the basis of Mulliken–Hush two-state analysis [42].

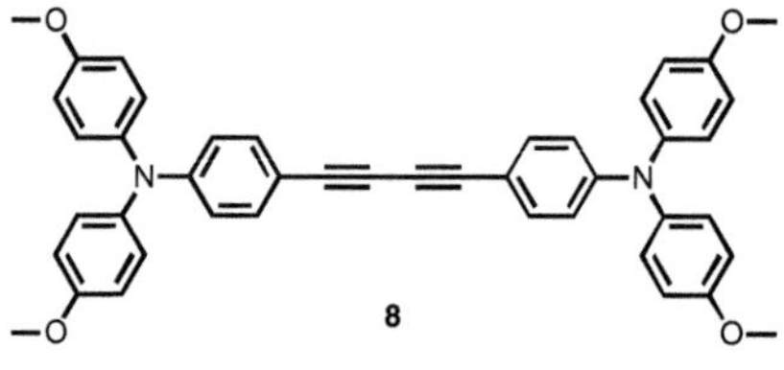

9.3 Relationship Between Bridging Units and Electronic Coupling | 307

In 2017, the exponential dependence of electronic coupling has been reinvestigated with a homologous series of radical cations of bis(diarylamine) system **13–16**, which are bridged by carbon-bridged planar OPV, and the attenuation factor was successfully determined to be 0.092 Å^{-1} [44].

Oligo(p-phenylene) (OPP) has also been recognized as an important class of π-bridging unit for the understanding of the CT process of donor–bridge–acceptor (DBA) systems, and the bridge length dependence of the photo-induced CT rates has been thoroughly examined [45]. The effective conjugation length for the OPP wire is confirmed to be around 6 repeating *para*-phenylene units by means of optical absorption spectroscopy [46]. Kochi and coworkers were the first to tackle the elucidation of OPP-bridged organic MV systems [47, 48], and a series of OPP derivatives **17–21** have been studied utilizing the 2,5-dimethoxy-4-methylphenyl group as a redox center. While the electronic coupling for the radical cations **17**$^{·+}$–**20**$^{·+}$ were evaluated to be 2330, 760, 430, and 217 cm^{-1}, respectively, on the basis of Hush analysis, the IVCT band was no longer visible for the *p*-quaterphenylene-bridged **21**$^{·+}$, approaching the Class I/II boundary [48]. Judging from the obtained H_{ab}, the value of β is estimated to be 0.32 Å^{-1}.

Following this early seminal paper, both optically induced and thermally induced CT processes have been re-examined through radical cations of OPP-bridged bis(dianisylamine)s **1**, **22**, and **23** in addition to the hexabenzocorone-bridged **24** [47]. The electronic couplings of **1**$^{·+}$ and **22**$^{·+}$–**24**$^{·+}$ were estimated to be 860, 530, 240, and 980 cm^{-1}, respectively, within the frame of the Mulliken–Hush two-state model, and the determined attenuation factor of 0.35 Å^{-1} coincides with the value by Kochi group. Moreover, it was pointed out that *p*-quinquephenyl-bridged **23**$^{·+}$ was on the Class I/II border.

Oligoacene-bridged MV systems are of interest from the view point of bridge length dependency of electronic couplings due to π-extended planar bridging units. For a series of oligoacene-bridged bis(dianisylamine)s **25–27**, the optical absorption spectroscopic analysis has been demonstrated for their radical cations [48]. The MV radical cations **25**·⁺–**27**·⁺ showed the electronic coupling of 1009, 796, and 561 cm⁻¹, respectively, and the linear regression fit to the data vs. separation length between the two nitrogen redox centers gave the attenuation factor 0.28 Å⁻¹, indicating that the electronic communication through the oligoacene bridges is not so strong as compared with the OPP-bridged case.

As described in Section 9.2, the MV state can also be tuned by using the redox centers with small or large internal reorganization energy. Recently, the organic MV system with two planar-constrained triarylamines (*N*-heterotriangulenes) **28–30** has been investigated to probe the effect of low reorganization energy [49]. The values of H_{ab} are not worthy of a special mention [**28**·⁺ (825 cm⁻¹), **29**·⁺ (566 cm⁻¹), and **30**·⁺ (261 cm⁻¹)]. However, the activation barrier ΔG^* of 4.5 kJ mol⁻¹ in **27**·⁺ was found to be almost half as small as that of the reference compound **7**·⁺ (8.9 kJ mol⁻¹ [50]), anticipating the faster thermally assisted CT rate.

Conversely, what will happen in case of MV systems having redox centers with considerably large reorganization energy? Generally, fully charge-localized Class I MV system is hard to prepare simply because the electronic communication can be induced by even a little electronic interaction between the redox centers, and hence, the genuine Class I MV compound is still rare. The only exception is the *para*-phenylene-bridged bis(hydrazine) radical cation **31**·⁺ studied by Nelsen and coworkers [51]. It is conceivable that such charge localization on one redox center in **31**·⁺ is caused by much larger internal reorganization energy of the hydrazine moiety instead of the ion pairing effect.

9.4 Where to Attach Redox Centers

In conjunction with the developing field of single-molecule electronics, it is often said that *para*- and *meta*-phenylene bridges give rise to constructive and destructive quantum interference effects on electrical conductance through a single molecule, respectively [52]. Such a concept (*meta-para*-paradigm) is, in organic MV chemistry, interpreted as the fact that *meta*-phenylene-bridged MV systems are of localized Class II, whereas *para*-phenylene-bridged MV systems are of delocalized Class III, as exemplified by **5**$^{•+}$ and **32**$^{•+}$. However, it turns out that things are not so easy. Grampp, Lambert, and coworkers demonstrated that *meta*-phenylene-connected double-tolan-bridged bis(dianisylamine) MV radical cations undergo a change in electronic coupling comparable to that for the *para*-phenylene-connected double-tolan-bridged ones by the introduction of electron-donating substituents at the right positions of the central phenylene moiety. For instance, the methoxy-substituted *meta*- and *para*-conjugated MV radical cations, **33**$^{•+}$ and **34**$^{•+}$, exhibited comparable electronic couplings, 320 and 550 cm^{-1}, respectively, in nitrobenzene solution, thereby overcoming the *meta–para* paradigm. This observation can be explainable by simple MO consideration. The *meta*-phenylene moiety has two non-bonding MO characteristics, which is included in the β-HOMO and β-LUMO of **33**$^{•+}$, as shown in Figure 9.6, and the electronic

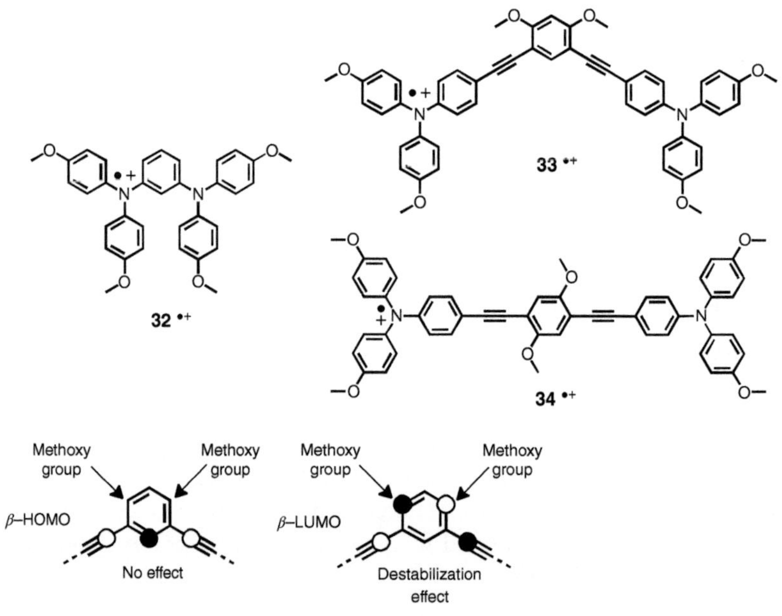

Figure 9.6 Schematic representation of orbital coefficients on the central *meta*-phenylene moiety in **33**$^{•+}$. When the electron-donating methoxy groups are attached at the 4- and 6-positions of *meta*-phenylene, β-LUMO (right) is destabilized as compared with β-HOMO (right), resulting in the large electronic coupling.

transition from β-HOMO and β-LUMO at the symmetrical structure (at the position of 0.5 in the CT reaction coordinate in Figure 9.2) corresponds to $2H_{ab}$. Hence, the electronic coupling H_{ab} increases with destabilizing β-LUMO level. When the electron-donating substituents such as methoxy groups are attached to positions where there are significant orbital coefficients at *meta*-phenylene bridge, the resulting MOs are effectively destabilized. Apparently, in the substitution pattern of **33·+**, only β-LUMO are being selectively destabilized, thus leading to the large electronic coupling. In this way, the electronic coupling in organic MV systems can be tuned by substitution of the right substituents at the right positions of the bridging unit on the basis of MO consideration.

Furthermore, it was revealed that *meta*-bridged **32·+** has an unexpectedly large electronic coupling of 1310 cm^{-1}, which is equal to the tolan-bridged **7·+** (1200 cm^{-1} [40]) [31]. These findings expand the molecular design possibility of organic MV systems.

9.5 Through-Bond or Through-Space?

In the preceding few sections, we have focused on the organic MV compounds exhibiting the intramolecular CT mainly by the through-bond pathway. However, as noted in Section 9.2, the elucidation of CT process proceeding by the through-space pathway, which is presumed to be dominant in intermolecular organic MV systems, is also important for application to the electronic materials. Although the separation between the through-bond and through-space contributions is technically difficult to carry out, some recent progress toward understanding of the through-space CT occurring between spatially adjacent redox centers is described in this section. One of the intriguing examples to mention first is the radical cation of hexaamino-substituted hexaarylbenzene **35** by Kochi and coworkers [53, 54]. The observed IVCT band ranged broadly from 900 to 5000 nm, and the detailed band shape analysis resulted in the large electronic coupling and the weak reorganization energy, indicating the thermally assisted CT barrier between six redox-active dialkylaniline groups is estimated to be quite low. Judging from the geometrical situation of the adjacent redox centers that partially overlap their π-orbitals of the *ipso* carbon atoms (the separation of $C_{ipso}-C_{ipso}$: 2.9 Å), Kochi and coworkers believed that toroidal delocalization through six arene units in a ring-like arrangement takes place in **35·+**. A little earlier than Kochi group's work, Lambert et al. demonstrated that the oxidized species of hexa(dianisylamine)-substituted hexaarylbenzene **36** exhibit the IVCT band mainly due to the tricationic species [55]. Then, the follow-up investigation on **37**, which enables to generate selectively the tricationic species, clarified that the trication has a localized Class II MV system with a significant barrier for the superexchange CT between adjacent triarylamine redox centers (the separation of N–N: 7.1 Å) [56]. These findings tell us that, though it may be obvious, the centers of charge on adjacent redox units should be placed as close as possible for strong electronic coupling enough to realize through-space π-delocalization.

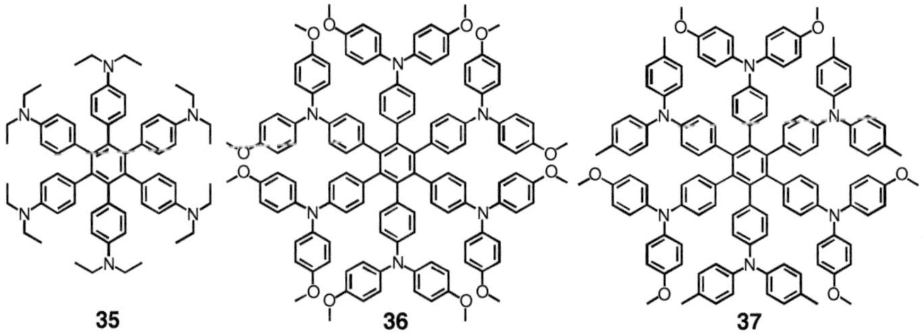

35 **36** **37**

In conjunction with the above-mentioned MV systems, bis(trianisylamine)s **38** and **39**, in which the redox centers are connected by *ortho*-phenylene and *ortho*-carborane clusters, respectively, have been examined to elucidate the difference in the intramolecular CT pathway due to the different bridging units with a similar geometrical arrangement between the two redox centers, and **38**$^{\cdot+}$ and **39**$^{\cdot+}$, were found to be classified into Class II and Class I MV systems, respectively [57]. The additional SQUID measurements for dications of **38** and **39** gave opposite results that no spin–exchange interaction is observed for **39**$^{2+}$, whereas antiferromagnetic interaction for **38**$^{2+}$, thus indicating that through-space interaction is negligibly small in **39**$^{\cdot+}$, while through-bond interaction exists in **38**$^{\cdot+}$, hence leading to a conclusion that through-space CT pathway is unacceptable for **38**$^{\cdot+}$ and the oxidized species of **36**.

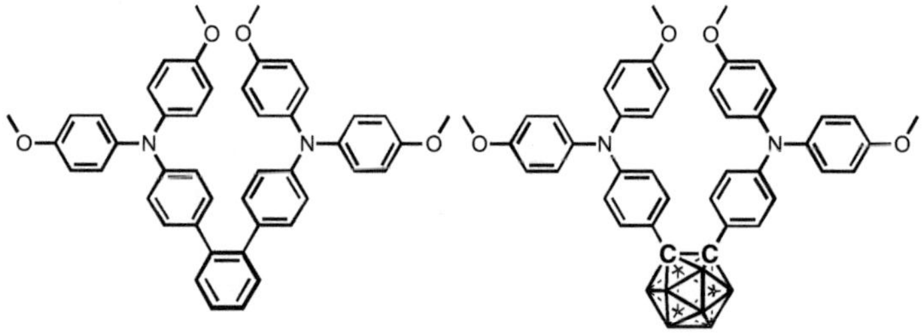

38 **39**

Wenger and coworkers have pursued intramolecular CT pathways utilizing a unique series of organic MV systems. They focused on naphthalene as an underexplored type of π-bridging unit. It was found that the radical cations of 1,8- and 1,5-isomers **40** and **41** of naphthalene-bridged bis(dianisylamine)s showed a similar electronic communication, and the simplistic through-bond CT pathway model fails to explain the similarity in electronic communication for **40**$^{\cdot+}$ and **41**$^{\cdot+}$, indicating that a through-space CT pathway can provide an adequate qualitative explanation for **40**$^{\cdot+}$ [58]. More recently, they hypothesized that rigid symmetrical compounds comprised a central arene unit with two planar redox centers at relatively short distance above and below the arene unit render CT pathways directly across the stacked π-structure where three π-systems form a triple decker [59]. Such compounds with two phenothiazines as redox centers, **42** and **43**, were successfully prepared and the corresponding radical cations were regarded as Class

II compounds with only weak electronic coupling of c. 110 cm^{-1}, and this results suggests that the overall CT pathways include a mixture of through-bond and through-space interactions.

One more interesting example in consideration of coexistence of through-bond and through-space CT pathways in organic MV systems has been reported [60]. The smallest macrocyclic oligoarylamine, tetraaza[1$_4$]o,p,o,p-cyclophane **44**, has two *para*-phenylenediamine (PD) redox-active units that are rigidly fixed in a close proximity. The interannular distance between two PD units for a chair conformation in the crystalline state of **44** was found to be around 3.2 Å (less than twice the van der Waals radius of carbon atom [3.4 Å]), thus enabling the through-space intramolecular CT. Moreover, the radical cation of *ortho*-phenylene-bridged bis(dianisylamine) **45**, as a reference compound of **44**, has been pointed out to be an intriguing MV molecule having the possibility that both through-bond and through-space interactions contribute to the CT process [61]. In the radical cation of **44**, the most stable conformation was estimated to be a boat conformation with charge-localized distribution on one PD unit on the basis of DFT computations. The observed IVCT band was also assigned to the charge-delocalized Class III MV within one charged PD unit. However, the IVCT transition between the eclipsed neutral and charged PD units was predicted to be virtually forbidden. On the other hand, the VT-ESR measurements for **44**$^{•+}$ clearly supported a charge delocalization over the macrocyclic backbone. Thus, the clarification of CT pathways in MV systems is still elusive, in particular for rigid and overcrowded organic molecular systems.

9.6 Control of Spin States Through Mixed-Valence States

In organic molecular systems with several redox-active centers, partial oxidation or reduction has the possibility to produce the multispin MV systems. For those compounds, not only electronic coupling but also spin–exchange interaction should be taken into consideration. Such molecular systems have already been discussed for multinuclear inorganic complexes [62], but the organic analogs are still rare. For instance, in the two-electron-oxidized MV species of three-center cyclopeantadienyliron complex, the singlet and triplet states are possible due to the sign and/or strength of exchange interaction between the two unpaired electrons, as illustrated in Figure 9.7, and two kinds of IVCT bands were observed probably due to energetically distinct transitions corresponding to their spin multiplicity [63]. However, the interplay between the intramolecular CT caused by electronic coupling among the redox centers and magnetic interaction between two spins on the oxidized two redox centers has not been clarified yet. Although similar MV state has been investigated for one-electron reduced MV anion of organic triradical **46** [64], only one IVCT band was observed, suggesting that, in **46⁻**, the high-spin triplet ground state affects the MV state or vice versa. Di(radical cation) of tetrakis(dianisylamine)-substituted tetraazacyclophane **47²⁺** was reported to be in spin-triplet ground state. In the macrocycle **47**, PD moiety is regarded as redox-active centers. Interestingly, an IVCT band via *meta*-phenylene bridges along the macrocyclic backbone was seen at the NIR region in this di(radical cation), strongly suggesting the charge/spin transfer between oxidized and neutral PD redox centers [65].

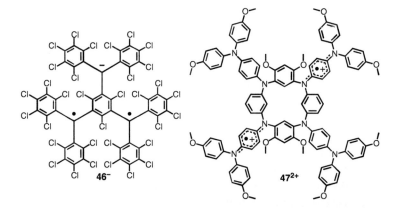

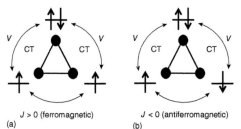

Figure 9.7 Schematic drawings of spin alignment in a three-redox-centered MV system with two unpaired electrons: (a) triplet case (ferromagnetic exchange interaction) and (b) singlet case (antiferromagnetic exchange interaction).

The attempt to control spin preference in the whole molecular system actively utilizing the MV state is currently even more rare. However, we can elaborately make the multispin molecular systems where localized and delocalized spins coexist. In this line, PD derivatives with nitroxide or nitronylnitroxide radicals as localized spin centers have been investigated [66–68] based on the concept of the so-called double-exchange mechanism by Anderson and Hasegawa [69]. In the neutral states of **48** and **49**, localized spins are almost magnetically uncoupled. However, once one-electron oxidation takes place from the central PD moiety, the delocalized MV state generated on the central PD moiety mediates ferromagnetic coupling between the localized spins to align all the spins parallelly in **48⁺** and **49⁺**.

9.7 Future Prospects

In this chapter, we have taken a quick overview of organic MV systems. Charge and/or spin localization and delocalization on organic π-conjugated molecular systems are regarded as a fundamentally important concept. In the organic MV chemistry, specific model systems are devised to probe the actual intra- or intermolecular CT phenomena in organic materials. To get an insight into charge (or polaron) delocalization on polyaniline as one of the conductive polymers [70], various model systems such as linear hexamer and cyclic hexamer models are selectable. However, the results obtained are totally different depending on the conformation, symmetry, type of substituents, and so forth; the radical cation of linear hexamer model was in a charge-localized state [71], while the radical cation of cyclic hexamer model in a charge-delocalized state [72]. So far, a significant number of findings have been accumulated about organic MV model systems, and currently, we know a lot about simple MV compounds with two redox centers. Although there are several future directions of this field, the mixed valency in organic solid-state conductive materials is closely related to the intramolecular MV systems, as described in Section 9.1, and some progress is being obtained [73].

The deeper understanding of multidimensional organic MV systems is also of great importance in association with the application to electronic materials, in particular, for quantum dot cellular automata devices [74, 75]. In this spirit, several kinds of pioneering work have already been published [15, 55, 56, 76–81]. From theoretical point of view, multielectron transfer processes in multidimensional MV systems have been examined for the specialized case of MV systems with square topology that can be extended into two-dimensional molecular grid architectures for information storage, and the results tell us the possibility to control a concerted two-electron transfer or consecutive single-electron transfer by tuning the key parameters shown in Section 9.2 [82]. Recently, dicationic species of a genuine squared organic π-conjugated compound **50** have been isolated as a single crystal salt, and **50**$^{2+}$ was reported to be a charge-delocalized MV system with closed-shell structure probably due to a large electronic coupling [83].

50^{2+}

The late Rathore and coworkers have energetically tackled with the problem on the rational design of novel CT organic materials, in particular, linear oligomeric molecular wires, in which the redox-active units are directly connected to each other, and they have repeatedly emphasized the importance of molecular design on the basis of a simple visual inspection of frontier orbitals [84]. Although much research effort still needs to be made in the field of organic MV chemistry, both rational molecular designs with viewpoint of frontier orbital theory and synthesis of innovative novel molecular systems will lead to steadily advancing this field.

Acknowledgment

I would like to thank my collaborators who have contributed in our research on intriguing organic MV systems and whose names are listed in the references.

References

1 Creutz, C. and Taube, H. (1969). Direct approach to measuring the Franck-Condon barrier to electron transfer between metal ions. *J. Am. Chem. Soc.* 91: 3988–3989.
2 Creutz, C. and Taube, H. (1973). Binuclear complexes of ruthenium ammines. *J. Am. Chem. Soc.* 95: 1086–1094.
3 Taube, H. (1984). Electron transfer between metal complexes – a retrospective view (Nobel lecture). *Angew. Chem. Int. Ed. Engl.* 23: 329–339.
4 Wurster, C. and Sendtner, R. (1879). Über die einwirkung oxydirender agentien auf tetramethylparaphenylenediamin. *Ber. Dtsch. Chem. Ges.* 12: 1803–1807.

5 Mazur, S., Sreekumar, C., and Schroeder, A.H. (1976). Symmetry vs. fluxionality, a radical anion with borderline properties. *J. Am. Chem. Soc.* 98: 6713–6714.
6 Schroeder, A.H. and Mazur, S. (1978). A vibronic model for infrared absorption by a mixed-valence anion radical. *J. Am. Chem. Soc.* 100: 7339–7346.
7 Acker, D.S., Harder, R.J., Hertler, W.R. et al. (1960). 7,7,8,8-tetracyanoquinodimethane and its electrically conducting anion-radical derivatives. *J. Am. Chem. Soc.* 82: 6408–6409.
8 Wudl, F., Smith, G.M., and Hufnagel, E.J. (1970). Bis-1,3-dithiolium chloride: an unusually stable organic radical cation. *Chem. Commun.* 1453–1454.
9 Coleman, L.B., Cohen, M.J., Sandman, D.J. et al. (1973). Superconducting fluctuations and the Peierls instability in an organic solid. *Solid State Commun.* 12: 1125–1132.
10 Mayerle, J.J. (1980). Mixed-valence in the organic solid state. In: *Mixed-valence compounds* (ed. D.B. Brown), 451–473. Dordrecht: Reidel.
11 In 1979, a new international journal, *Synthetic Metals*, has been launched by Elsevier.
12 Matsunaga, Y. and Takayanagi, K. (1980). Electrical and optical properties of the 5,6:11,12-bis(epidithio)naphtalene-bromine system (TTT-Br_n). Formation of organic mixed-valence compounds (TTT)(TTT^+)Br^- and (TTT^+)(TTT^{2+})(Br^-)$_3$. *Bull. Chem. Soc. Jpn.* 53: 2769–2799.
13 Deuchert, K. and Hünig, S. (1978). Multistage Organic Redox Systems – A general structural principle. *Angew. Chem. Int. Ed. Engl.* 17: 875–886.
14 Nelsen, S.F., Chang, H., Wolff, J.J., and Adamus, J. (1993). Polycyclic bis(hydrazine) and bis(hydrazyl) radical cations: high and low inner-sphere reorganization energy organic interval compounds. *J. Am. Chem. Soc.* 115: 12276–12289.
15 Bonvoisin, J., Launay, J.-P., Van der Auweraer, M., and De Schryver, F.C. (1994). Organic mixed-valence systems: intervalence transition in partly oxidized aromatic polyamines. Electrochemical and optical studies. *J. Phys. Chem.* 98: 5052–5057, see also correction (1996) *J. Phys. Chem.*, 100, 18006.
16 Bonvoisin, J., Launay, J.-P., Verbouwe, W. et al. (1996). Organic mixed-valence systems. II. Two-centers and three-centers compounds with meta connections around a central phenylene ring. *J. Phys. Chem.* 100: 17079–17082.
17 Hankache, J. and Wenger, O.S. (2011). Organic mixed valence. *Chem. Rev.* 111: 5138–5178.
18 Heckmann, A. and Lambert, C. (2012). Organic mixed-valence compounds: a playground for electrons and holes. *Angew. Chem. Int. Ed.* 51: 326–392.
19 Brunschwig, B.S., Creutz, C., and Sutin, N. (2002). Optical transitions of symmetrical mixed-valence systems in the class II - III transition regime. *Chem. Soc. Rev.* 31: 168–184.
20 Robin, M.B. and Day, P. (1968). Mixed valence chemistry – A survey and classification. *Adv. Inorg. Chem. Radiochem.* 10: 247–422.
21 Nelsen, S.F. (2000). "Almost delocalized" intervalence compounds. *Chem. Eur. J.* 6: 581–588.

22 Hush, N.S. (1985). Distance dependence of electron transfer rates. *Coord. Chem. Rev.* 64: 135–157.

23 Cave, R.J. and Newton, M.D. (1996). Generalization of the Mulliken-Hush treatment for the calculation of electron transfer matrix elements. *Chem. Phys. Lett.* 249: 15–19.

24 Wong, K.Y. and Schatz, P.N. (1981). A dymanic model for mixed-valence compounds. *Prog. Inorg. Chem.* 28: 369–449.

25 Low, P.J., Paterson, M.A.J., Puschmann, H. et al. (2004). Crystal, molecular and electronic structure of N,N'-diphenyl-N,N'-bis(2,4-dimethylphenyl)-(1,1'-biphenyl)-4,4'-diamine and the corresponding radical cation. *Chem. Eur. J.* 10: 83–91.

26 Szeghalmi, A.V., Erdmann, M., Engel, V. et al. (2004). How delocalized is N,N,N',N'-tetraphenylphenylenediamine radical cation? An experimental and theoretical study on the electronic and molecular structure. *J. Am. Chem. Soc.* 126: 7834–7845.

27 Lambert, C., Risko, C., Coropceanu, V. et al. (2005). Electronic coupling in tetraanisylarylenediamine mixed-valence systems: the interplay between bridge energy and geometric factors. *J. Am. Chem. Soc.* 127: 8508–8516.

28 Cave, R.J. and Newton, M.D. (1997). Calculation of electronic coupling matrix elements for ground and excited state electron transfer reactions: Comparison of the generalized Mulliken-Hush and block diagonalization methods. *J. Chem. Phys.* 106: 9213–9226.

29 Newton, M.D. (1991). Quantum chemical probes of electron-transfer kinetics: the nature of donor-sсceptor interactions. *Chem. Rev.* 91: 767–792.

30 Lambert, C., Nöll, G., and Schelter, J. (2002). Bridge-mediated hopping or superexchange electron-transfer processes in bis(triarylamine) systems. *Nat. Mater.* 1: 69–73.

31 Uebe, M. and Ito, A. (2019). Intramolecular charge transfer in Kekulé- and non-Kekulé-bridged bis(triarylamine) radical cations: missing key compounds in organic mixed-valence systems. *Chem. Asian J.* 14: 1692–1696.

32 Lin, B.C., Cheng, C.P., and Lao, Z.P.M. (2003). Reorganization energies in the transports of hole and electrons in organic amines in organic electroluminescence studied by density functional theory. *J. Phys. Chem. A* 107: 5241–5251.

33 Cias, P., Slugovc, C., and Gescheidt, G. (2011). Hole transport in triphenylamine based OLED devices: from theoretical modeling to properties prediction. *J. Phys. Chem. A* 115: 14519–14525.

34 Comparative study by single crystal X-ray structure analysis on tris(4-bromophenyl)amine and its radical cation, so-called "Magic Blue" revealed that both molecular structures are identical within experimental uncertainty, which may also support the low internal reorganization energy of triarylamine-based redox centersQuiroz-Guzman, M. and Brown, S.N. (2010). Tris(4-bromophenyl)aminium hexachloroantimonate ('Magic Blue'): a strong oxidant with low inner-sphere reorganization. *Acta Cryst. C* 66: m171–m173.

35 Cohen, A.J., Mori-Sánchez, P., and Yang, W. (2012). Challenges for density functional theory. *Chem. Rev.* 112: 289–320.

36 Renz, M., Theilacker, K., Lambert, C., and Kaupp, M. (2009). A reliable quantum-chemical protocol for the characterization of organic mixed-valence compounds. *J. Am. Chem. Soc.* 131: 16292–16302.

37 Kaupp, M., Renz, M., Parthey, M. et al. (2011). Computational and spectroscopic studies of organic mixed-valence compounds: where is the charge? *Phys. Chem. Chem. Phys.* 13: 16973–16986.

38 Barlow, S., Risko, C., Odom, S.A. et al. (2012). Tuning delocalizeation in the radical cations of 1,4-bis[4-(diarylamino)styryl]benzenes, 2,5-bis-[4-(diarylamino)styryl]thiophenes, and 2,5-bis[4-(diarylamino)styryl]pyrroles through substituent effects. *J. Am. Chem. Soc.* 134: 10146–10155.

39 Burrezo, P.M., Zeng, W., Moos, M. et al. (2019). Perylene π-bridges equally delocalize anions and cations: proportioned quinoidal and aromatic content. *Angew. Chem. Int. Ed.* 41: 14467–14471.

40 Lambert, C. and Nöll, G. (1999). The class II/III transition in triarylamine redox systems. *J. Am. Chem. Soc.* 121: 8434–8442.

41 Amthor, S. and Lambert, C. (2006). [2.2]Paracyclophane-bridged mixed-valence compounds: application of a generalized Mulliken-Hush three-level model . *J. Phys. Chem. A* 110: 1177–1189.

42 Barlow, S., Risko, C., Chung, S.-J. et al. (2005). Intervalence transitions in the mixed-valence monocations of bis(triarylamines) linked with vinylene and phenylene-vinylene bridges. *J. Am. Chem. Soc.* 127: 16900–16911.

43 Giacalone, F., Segura, J.L., Martín, N., and Guldi, D.M. (2004). *J. Am. Chem. Soc.* 126: 5340–5341.

44 Burrezo, P.M., Lin, N.-T., Nakabayashi, K. et al. (2017). Bis(aminoaryl) carbon -bridged oligo(phenylenevinylene)s expand the limits of electron couplings. *Angew. Chem. Int. Ed.* 56: 2898–2902.

45 Gilbert, M. and Albinsson, B. (2015). Photoinduced charge and energy transfer in molecular wires. *Chem. Soc. Rev.* 44: 845–862.

46 Meier, H., Stalmach, U., and Kolshorn, H. (1997). Effective conjugation length and UV/vis spectra of oligomers. *Acta Polym.* 48: 379–384.

47 Uebe, M., Kaneda, K., Fukuzaki, S., and Ito, A. (2019). Bridge-length-dependent intramolecular charge transfer in bis(dianisylamino)-terminated oligo(p-phenylene)s. *Chem. Eur. J.* 25: 15455–15462.

48 Zhang, J., Chen, Z., Yang, L. et al. (2016). Elaborately tuning intramolecular electron transfer through varing oligoacene linkers in the bis(diarylamino) systems. *Sci. Rep.* 6: 36310.

49 Krug, M., Fröhlich, N., Fehn, D. et al. (2021). Pre-planarized triphenylamine-based linear mixed-valence charge-transfer systems. *Angew. Chem. Int. Ed.* 60: 6771–6777.

50 Holzapfel, M., Lambert, C., Selinka, C., and Stalke, D. (2002). Organic mixed valence compounds with N,N-dihydrodimethylphenazine redox centres. *J. Chem. Soc., Perkin Trans.* 2: 1553–1561.

51 Nelsen, S.F., Ismagilov, R.F., and Powell, D.R. (1996). Charge localization in a dihydrazine analogue of tetramethyl-p-phenylenediamine radical cation. *J. Am. Chem. Soc.* 118: 6313–6314.

52 Schäfer, J., Holzapfel, M., Mladenova, B. et al. (2017). Hole transfer processes in meta- and para-conjugated mixed valence compounds: unforeseen effects of bridge substituents and solvent dynamics. *J. Am. Chem. Soc.* 139: 6200–6209.

53 Sun, D., Rosokha, S.V., and Kochi, J.K. (2005). Through-space (cofacial) π-delocalization among multiple aromatic centers: toroidal conjugation in hexaphenylbenzene-like radical cations. *Angew. Chem. Int. Ed.* 44: 5133–5136.

54 Rosokha, S.V., Neretin, I.S., Sun, D., and Kochi, J.K. (2006). Very fast electron migrations within p-doped aromatic cofacial arrays leading to three-dimensional (toroidal) π-delocalization. *J. Am. Chem. Soc.* 128: 9394–9407.

55 Lambert, C. and Nöll, G. (1998). One- and two-dimensional electron transfer processes in triarylamines with multiple redox centers. *Angew. Chem. Int. Ed.* 37: 2107–2110.

56 Lambert, C. and Nöll, G. (2002). Optically and thermally induced electron transfer pathways in hexakis[4-(N,N-diarylamino)phenyl]benzene derivatives. *Chem. Eur. J.* 8: 3467–3477.

57 Uebe, M., Kazama, T., Kurata, R. et al. (2017). Recognizing through-bond and through-space self-exchange charge/spin transfer pathways in bis(triarylamine) radical cations with similar geometrical arrangements. *Angew. Chem. Int. Ed.* 56: 15712–15717.

58 Schmidt, H.C., Spulber, M., Neuburger, M. et al. (2016). Charge transfer pathways in three isomers of naphthalene –bridged organic mixed valence compounds. *J. Org. Chem.* 81: 595–602.

59 Schmidt, H.C., Guo, X., Richard, P.U. et al. (2018). Mixed-valent molecular triple deckers. *Angew. Chem. Int. Ed.* 57: 11688–11691.

60 Sakamaki, D., Ito, A., Matsumoto, T., and Tanaka, K. (2014). Electronic structure of tetraaza[1.1.1.1]o,p,o,p-cyclophane and its oxidized states. *RSC Adv.* 4: 39476–39483.

61 Nöll, G. and Avola, M. (2006). Optically induced electron transfer in an N,N,N',N'-tetraanisyl-o-phenylenediamine radical cation. *J. Phys. Org. Chem.* 19: 238–241.

62 D'Alessandro, D.M. and Keene, F.R. (2006). Intervalence charge transfer (IVCT) in trinuclear and tetranuclear complexes of iron, ruthenium, and osmium. *Chem. Rev.* 106: 2270–2298.

63 Weyland, T., Costuas, K., Toupet, L. et al. (2000). Organometallic mixed-valence systems. Two-center and three-center compounds with meta connections around a central phenylene ring. *Organometallics* 19: 4228–4239.

64 Sedó, J., Ruiz, D., Vidal-Gancedo, J. et al. (1996). Intramolecular electron transfer phenomena in purely organic mixed-valence high-spin ions: a triplet anion case. *Adv. Mater.* 8: 748–752.

65 Ito, A., Inoue, S., Hirao, Y. et al. (2008). An N-substituted aza[1_4]metacyclophane tetracation: a spin-quintet tetraradical with four *para*-phenylenediamine-based semi-quinone moieties. *Chem. Commun.* 3242–3244.

66 Ito, A., Nakano, Y., Urabe, M. et al. (2006). Triradical cation of p-phenylenediamine having two nitroxide radical groups: spin alighnment mediated by delocalized spin. *J. Am. Chem. Soc.* 128: 2948–2953.

67 Ito, A., Kurata, R., Sakamaki, D. et al. (2013). Redox modulation of *para*-phenylenediamine by substituted nitronyl nitroxide groups and their spin states. *J. Phys. Chem. A* 117: 12858–12867.

68 Ito, A., Kurata, R., Noma, Y. et al. (2016). Radical cation of an oligoarylamine having a nitroxide radical substitutenet: a coexistent molecular system of localized and delocalized spins. *J. Org. Chem.* 81: 11416–11420.

69 Anderson, P.W. and Hasegawa, H. (1955). Considerations on double exchange. *Phys. Rev.* 100: 675–681.

70 Geniès, E.M., Boyle, A., Lapkowski, M., and Tsintavis, C. (1990). Polyaniline: a historical survey. *Synth. Met.* 36: 139–182.

71 Grossmann, B., Heinze, J., Moll, T. et al. (2004). Electron delocalization in one-electron oxidized aniline oligomers, paradigms for polyaniline. A study by paramagnetic resonance in fluid solution. *J. Phys. Chem. B* 108: 4669–4672.

72 Ito, A., Yokoyama, Y., Aihara, R. et al. (2010). Preparation and characterization of *N*-anisyl-substituted hexaaza[1$_6$]paracyclophane. *Angew. Chem. Int. Ed.* 49: 8205–8208.

73 Talipov, M.R., Navale, T.S., and Rathore, R. (2015). The HOMO nodal arrangement in polychromophoric molecules and assemblies controls the interchromophoric electronic coupling. *Angew. Chem. Int. Ed.* 54: 14468–14472.

74 Tokunaga, K. (2009). Signal transmission through molecular quantum-dot cellular automata: a theoretical study on Creutz-Taube complexes for molecular computing. *Phys. Chem. Chem. Phys.* 11: 1474–1483.

75 Tokunaga, K. (2010). Metal dependence of signal transmission through molecular quantum-dot cellular automata (QCA): a theoretical study on Fe, Ru, and Os mixed-valence complexes. *Materials* 3: 4277–4290.

76 Lambert, C., Nöll, G., and Hampel, F. (2001). Multidimensional electron transfer pathways in a tetrahedral tetrakis{4-[*N,N*-di(4-methoxyphenyl)amino]phenyl}phosphonium salt: one-step vs two-step mechanism. *J. Phys. Chem. A* 105: 7751–7758.

77 Rathore, R., Burns, C.L., and Abdelwashed, S.A. (2004). Hopping of a single hole in hexakis[4-(1,1,2-triphenyl-ethenyl)phenyl]benzene cation radical through the hexaphenylbenzene propeller. *Org. Lett.* 6: 1689–1692.

78 Yan, X.Z., Pawlas, J., Goodson, T. III, and Hartwig, J.F. (2005). Polaron delocalization in ladder macromolecular systems. *J. Am. Chem. Soc.* 127: 9105–9116.

79 Hirao, Y., Ito, A., and Tanaka, K. (2007). Intramolecular charge transfer in a star-shaped oligoarylamine. *J. Phys. Chem. A* 111: 2951–2956.

80 Ito, A., Sakamaki, D., Ichikawa, Y., and Tanaka, K. (2011). Spin-delocalization in charged states of *para*-phenylene-linked dendritic oligoarylamines. *Chem. Mater.* 23: 841–850.

81 Skamaki, D., Ito, A., Furukawa, K. et al. (2012). A polymacrocyclic oligoarylamine with a pseudobeltane motif: towards a cylindrical multispin system. *Angew. Chem. Int. Ed.* 51: 12776–12781.

82 Lambert, C. (2003). Concerted two-electron-transfer processes in mixed-valence species with square topology. *ChemPhysChem* 4: 877–880.

83 Sakamaki, D., Ito, A., Tsutsui, Y., and Seki, S. (2017). Tetraaza[14]- and octaaza[18]paracyclophane: synthesis and characterization of their neutral and cationic states. *J. Org. Chem.* 82: 13348–13358.

84 Ivanov, M.V., Reid, S.A., and Rathore, R. (2018). Game of frontier orbitals: a view on the rational design of novel charge-transfer materials. *J. Phys. Chem. Lett.* 9: 3978–3986.

10

Mixed-Valence Complexes in Biological and Bio-mimic Systems

Xiangmei Kong[1,2], Yixin Guo[1,2], Zijie Zhou[1,2], and Tianfei Liu[1,2,3]

[1]*Nankai University, College of Chemistry, State Key Laboratory of Elemento-organic Chemistry, Weijin Road No. 94, Tianjin 300071, China*
[2]*Haihe Laboratory of Sustainable Chemical Transformations, Tianjin 300 192, China*
[3]*Chinese Academy of Sciences, Shanghai Institute of Organic Chemistry, Key Laboratory of Organofluorine Chemistry, Lingling Road No. 345, Shanghai 200032, China*

10.1 Introduction

Mixed-valence metal complexes or clusters are crucial building blocks present in biological systems (Figure 10.1). These complexes typically contain multinuclear metal centers with different oxidation states, and they serve a variety of functions in biological processes [1, 2]. For instance, they play critical roles as electron transporters and enzyme biocatalysts, through which electrons can be efficiently transferred via the valence changes in metal centers [3–5].

The focus of this chapter is on mixed-valence complexes in biological and biomimetic systems. We start our discussion on the mixed-valence iron–sulfur clusters in Section 10.2.1. These clusters mainly are electron transporters and fundamental building blocks in natural proteins and their mimics, and most recently, their roles in the *magnetic protein biocompass* have attracted a lot of attention. From Section 10.2.2 to Section 10.2.4, the structures and functions of the natural cofactors of *[FeFe]-hydrogenase*, *nitrogenases*, and *carbon monoxide dehydrogenase* (CODH) and their artificial mimics, which are derivatives of iron–sulfur clusters, are summarized and discussed. In Section 10.3, the di-heme and multiheme mixed-valence systems are presented with their mimics. Section 10.4 focuses on the natural mixed-valence multicopper cofactors and their artificial mimics. In addition, the mixed-valence multimanganese cofactors are discussed in Section 10.5, including the natural mixed-valence multimanganese cofactors and artificial mimics of oxygen-evolving complex (OEC) in photosystem II and the Mn/Mn cofactor in *Class I ribonucleotide reductases (RNRs)*.

Mixed-Valence Systems: Fundamentals, Synthesis, Electron Transfer, and Applications, First Edition.
Edited by Yu-Wu Zhong, Chun Y. Liu, and Jeffrey R. Reimers.
© 2023 WILEY-VCH GmbH. Published 2023 by WILEY-VCH GmbH.

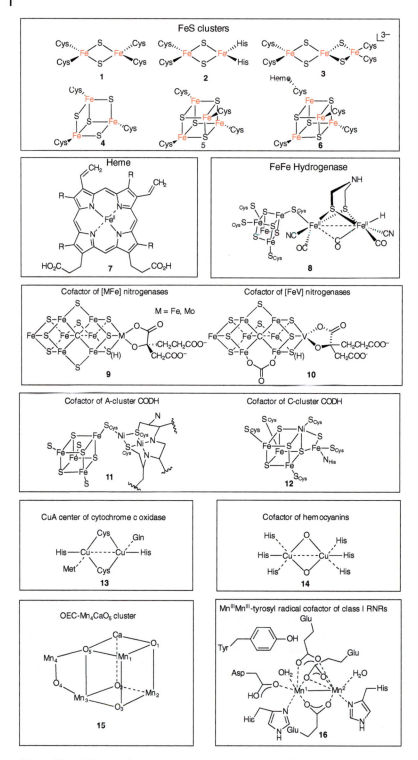

Figure 10.1 Mixed-valence complexes and clusters in biological systems. Source: Adapted from [1, 2].

10.2 Mixed-Valence Iron–Sulfur Clusters in Biological and Bio-mimic Systems

10.2.1 Basic FeS Clusters

Structures **1–6** in Figure 10.1 summarize the structure of three types of basic FeS clusters, including *[2Fe-2S]* clusters, *[3Fe-4S]* clusters, and *[4Fe-4S]* clusters [6, 7]. The *[2Fe-2S]* clusters are the simplest polymetallic systems in this family, which are generally connected by two iron ions bridged by two sulfide ions to four cysteine ligands (cluster **1** in Figure 10.1). There is also a class of heteroligated binuclear clusters in *Rieske proteins,* in which the two imodazole ligands from the histidine residues bind to the same iron atom (cluster **2** in Figure 10.1) [8, 9]. Through the redox cycle between the core oxidation states of 2+ and 1+, *[2Fe-2S] clusters* mediate the electron transfer in a variety of ferredoxins and complex metalloenzymes, and are also inherent components in the photosynthetic and respiratory electron transport chains. The redox potential of the $[2Fe-2S]^{2+/+}$ pair ranges from +380 mV to −460 mV, with Rieske-type centers typically having higher potentials (+380 to −150 mV) than all-cysteinyl-ligated centers (+100 to −460 mV) because of the presence of the more positively charged histidine ligands at the reducible Fe site. The oxidized state $[2Fe-2S]^{2+}$ contains two Fe^{III} ions with an S of 5/2, while the reduced state $[2Fe-2S]^{+}$ contains mixed-valence iron–sulfur clusters, containing one Fe^{II} ($S = 2$) and one Fe^{III} ($S = 5/2$) ion. These iron–sulfur clusters play important roles as reductive bioelectronic carriers because of the presence of iron atoms with variable oxidation states.

There are two kinds of *[3Fe-4S]* clusters in biological systems: linear and cubane types. The linear clusters (cluster **3** in Figure 10.1) contain a $[Fe_3S_4]^+$ core, which is constructed by two Fe_2S_2 rhombs sharing a common vertex [7]. The functions of proteins containing linear clusters *in vivo* need to be further studied. The cubane-type clusters (cluster **4** in Figure 10.1) play a role in mediating electron transfer in organisms, and the redox potential of the $[3Fe-4S]^{+/0}$ couple ranges from +90 to −460 mV. In cubane-type $[3Fe-4S]^{+}$ clusters, the three Fe^{III} ($S = 5/2$) ions are antiferromagnetically coupled to give a ground state with a total S of 1/2. In the $[3Fe-4S]^0$ cluster, because of the antiferromagnetic interaction between a valence-delocalized Fe^{II}/Fe^{III} pair ($S = 9/2$) and a valence-trapped Fe^{III} site ($S = 5/2$), the ground state has a total S of 2.

The cubane-type *[4Fe-4S]* clusters with a typical structure, such as clusters **5** and **6** in Figure 10.1, are one of the most common electron-transfer centers in biology. Biological *[4Fe-4S]* clusters have four oxidation states: 3+, 2+, 1+, and 0. The nitrogenase Fe protein contains a subunit with a bridging [4Fe-4S] cluster that can undergo three stable oxidation states (2+, 1+, and 0) and thus has the ability to act as two-electron donors. Except for the nitrogenase Fe protein, under physiological conditions, all the electron transfers of [4Fe-4S] clusters only carry out a single-electron redox cycle between either the $[4Fe-4S]^{3+/2+}$ or $[4Fe-4S]^{2+/+}$ couples. The magnetic and electrical properties of the [4Fe-4S] center are complex. The $[4Fe-4S]^{3+}$ state with a total S of 1/2 results from Fe^{III}/Fe^{III} ($S = 5$) coupled to

Fe^{II}/Fe^{III} ($S = 9/2$). The $[4Fe-4S]^{2+}$ state with a total S of 0 results from Fe^{II}/Fe^{III} ($S = 9/2$) coupled to Fe^{II}/Fe^{III} ($S = 9/2$). In the case of $[4Fe-4S]^+$ ($S = 1/2$), it can be viewed as the coupling of Fe^{II}/Fe^{II} ($S = 4$) with Fe^{II}/Fe^{III} ($S = 9/2$).

On the basis of the most recent studies on the origin of life, the Fe–S clusters in living organisms may be the ancient remains of Fe-S pores in alkaline hydrothermal vents [10, 11]. The development of synthetic analogs mimicking the active metal center in iron–sulfur proteins can be traced back to the early 1970s. Holm and his colleagues have made significant contributions with pioneering works in this area [7, 12, 13], including the biologically inspired rational ligand design and the synthesis of biomimetic clusters.

Mixed-valence Fe–S clusters can also possess strong magnetic properties, which have unique biophysical characteristics in *magnetic protein biocompass* and important biological significance in animal navigation systems. The topic of *magnetic protein biocompass* has become popular since 2016 [14, 15]. Xie and his colleagues reported a *putative magnetic receptor* (Drosophila CG8198, here named *MagR*) and a multimeric magnetosensing rod-like protein complex. Most recently [16], two forms of iron–sulfur clusters have been identified in pigeon *MagR* and show different magnetic properties. The *[3Fe-4S]-MagR* cluster appears to be superparamagnetic and has saturation magnetization at 5 K, but *[2Fe-2S]-MagR* is paramagnetic; at 300 K, *[3Fe-4S]-MagR* is paramagnetic, but *[2Fe-2S]-MagR* is diamagnetic. This magnetic protein may form the basis of animal magnetoreception and lead to applications across multiple fields.

10.2.2 [FeFe]-Hydrogenase

Hydrogenases include three types of hydrogenase enzymes, which can oxidize hydrogen to protons or reduce protons to hydrogen [17, 18]. The *[FeFe]-hydrogenase* is one of these essential types, which uses a hexanuclear-iron complex, the H-cluster, to catalyse these reactions [18]. A *[FeFe]-hydrogenase* contains a di-iron [2Fe] center with a bridging dithiolate cofactor and a covalently linked proton at the second coordinate sphere of this cofactor (cluster **8** in Figure 10.1); meanwhile, there is a *[4Fe-4S]* cluster playing an essential role in electron transportation. During the catalytic cycle, the two subclusters change oxidation states between $[4Fe-4S]^{2+}{}_H \Leftrightarrow [4Fe-4S]^+{}_H$ and $[Fe(I)Fe(II)]_H \Leftrightarrow [Fe(I)-Fe(I)]_H$, thereby enabling the storage of two electrons to catalyze the reduction of protons to hydrogen [19]. According to the literature report, a *[FeFe]-hydrogenase* can show dramatic activity in the production of molecular hydrogen with a turnover frequency (TOF) in the order of $10\,000\,s^{-1}$ [20], which is higher than the hydrogen production effect of the commonly used platinum catalyst. This high catalytic activity has attracted the attention of scientists around the world.

Meanwhile, it is hoped that the study of model compounds can provide more clues to reveal the reaction mechanism of *[FeFe]-hydrogenase*. Rauchfuss, Song, Wang, Wu, Pickett, Berggren, and Hammarstrom et al. have made significant contributions in this area (Figure 10.2) [21–43]. In 2006, Song et al. synthesized a novel light-driven-type model complex (cluster **21**), which contains a

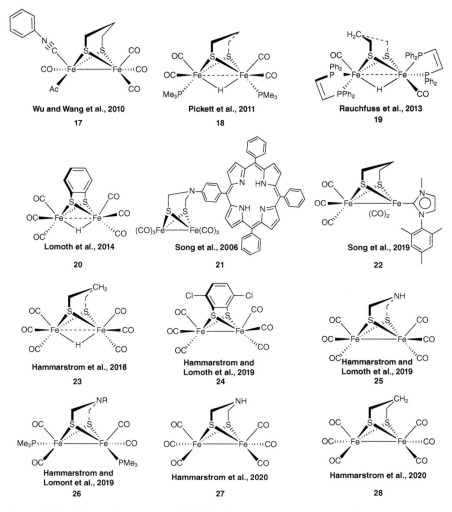

Figure 10.2 Selected examples of *[FeFe]-hydrogenase* model compounds.

tetraphenylporphyrin (TPP) photosensitizer covalently bonded to the N atom of a diiron moiety [42]. In 2011, Pickett et al. [23] reported the spectral characterization of the discrete mixed-valence di-iron μ-hydride species for the first time. Using anthracene anion radical as the reducing agent, the diamagnetic [Fe$^{(II)}$(μ-H)Fe$^{(II)}$] hydride can be reduced to [Fe$^{(1.5)}$(μ-H)Fe$^{(1.5)}$] mixed-valent hydride (**18**). In 2013, Rauchfuss and Stein et al. [22] reported that diferrous hydride was reduced to mixed-valent diiron hydrido complex [(μ-H)Fe$_2$(pdt)(CO)$_2$(dppv)$_2$] (pdt = 1,3-propane-dithiolate and dppv = *cis*-1,2-C$_2$H$_2$[PPh$_2$]$_2$) (**19**) using decamethylcobaltocene as reducing agent. X-ray crystallographic analysis showed that the mixed-valent complex had an asymmetric [Fe(II)(μ-H)Fe(I)] core. In 2019, Lomoth and Hammarstrom et al. [44] reported for the first time the structural and kinetic characterization of the reductive protonation intermediate of complex [FeFe(Cl$_2$-bdt)(CO)$_6$] (Cl$_2$-bdt = 3,6-dichloroben-zene-1,2-dithiolate) (**24**), and the direct spectroscopic detection of the key intermediate for hydrogen production and its conversion process. In the same year, they studied the electron-transfer and proton-transfer reactions of the *[FeFe]-hydrogenase* model complexes [Fe$_2$adt(CO)$_6$] (cluster **25**, adt = azadithiolate) and [Fe$_2$adt(CO)$_4$(PMe$_3$)$_2$] (cluster **26**) using real-time infrared and UV–vis spectroscopic observation to clarify the role of adt-N as a potential proton shuttle in catalytic H$_2$ formation. Later, Hammarstrom et al. [45] reported the reactivity of two hydrogen evolution diiron complexes [Fe$_2$(μ-SCH$_2$NHCH$_2$S)(CO)$_6$] (**27**) and [Fe$_2$(μ-SCH$_2$CH$_2$CH$_2$S)(CO)$_6$] (**28**) with molecular oxygen and reactive oxygen species inspired by the active center of *[FeFe]-hydrogenase*. The reactivity of the cluster **28** without the nitrogen bridge head decreases in the presence of a reductant, which highlights the importance of the second coordination sphere in regulating the oxidation process. These groups investigated not only how the ligands on the first coordinate sphere influenced the bi-iron model complexes but also revealed the applications of these complexes in catalytic reactions. The proton-coupled electron transfer (PCET) [46–49] in the metal hydride intermediates (FeH) is a critical process in the catalytic cycle of the [FeFe] cofactor and catalysts. Some studies have revealed the role of proton relays on the second coordinate sphere in a series of simplified tungsten hydride model complexes [50, 51].

10.2.3 Nitrogenases

Nitrogenases are enzymes that convert nitrogen from the air into nitrogen-containing compounds. These enzymes can be divided into three types, including *molybdenum (Mo) nitrogenase, vanadium (V) nitrogenase,* and *iron (Fe) nitrogenase*, according to the heterometal content of cofactor sites (clusters **9** and **10** in Figure 10.1) [52]. Each *nitrogenase* is composed of two metalloproteins, one of which is iron (Fe) protein and the other is molybdenum–iron (MoFe), vanadium–iron (VFe), or iron–iron (FeFe) protein. The reduction catalyzed by *nitrogenase* involves three basic electron-transfer reactions (Figure 10.3). First, iron (Fe) protein is reduced by electron carriers such as ferridoxin and flavanthoxins *in vivo* or disulfites *in vitro*; then in the presence of MgATP, Fe protein acts as a specific and essential electron carrier to transfer electrons to larger metalloproteins, namely MoFe, VFe, or FeFe protein. Finally, within one of these large dinuclear proteins, electrons are further transferred to the enzyme activity center to reduce the substrate molecule. Thus, the mixed-valence species of the involved cofactors are critical in the catalytic processes.

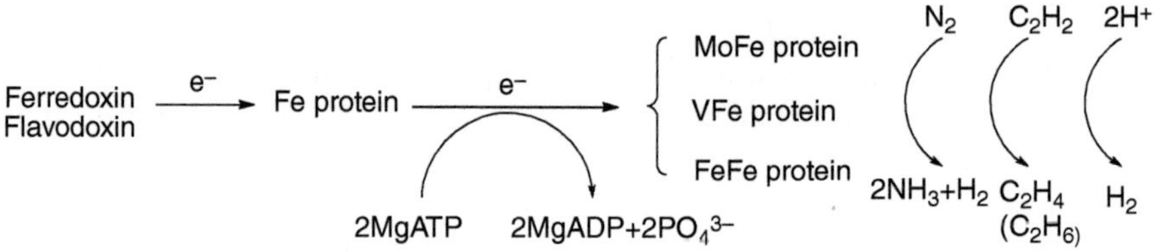

Figure 10.3 A general electron-transfer pathway through the *nitrogenases*. Source: Adapted from [51].

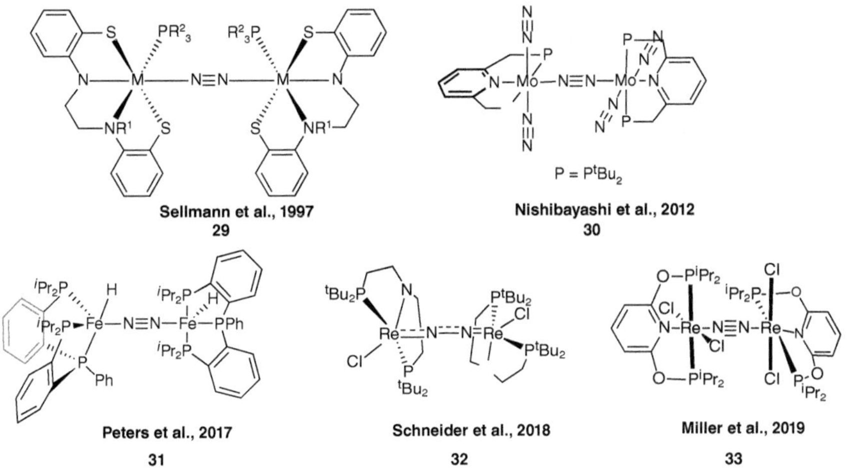

Figure 10.4 Nitrogenase model compounds.

At present, synchrotron radiation and other technologies enable people to have a certain understanding of the CO ligand dissociation and other reactivity of the FeMo-cofactor in *nitrogenase* [54], but the specific electron-transfer and proton-transfer mechanisms and the fate of mixed-valence species in the process of nitrogen activation still need to be further revealed with the help of artificial models [55]. These understandings are also helpful in developing artificial catalysts with excellent catalytic activity. A lot of model compounds (Figure 10.4) were synthesized to mimic and understand the reaction mechanism at the cofactors of *nitrogenases*. Since Schlock published his pioneering modeling complexes with a single Mo center in in 2001 [56], many research groups have been involved in research in this field, including both experimental and theoretical studies of the reaction mechanisms of model compounds. In recent years, Nishibayashi [57, 58], Schneider [59, 60], Miller [61, 62], Peters [63], and other groups have developed a series of nitrogen-activated complexes containing pincer-type ligands. These ligands can stabilize a series of transition metal complexes (such as Mo, Cr, Fe, Co, and Re), and some of these complexes assist in the process of dinitrogen fixation under phototriggered or electrocatalytic conditions. Among them, the mixed-valence species play important roles in these processes.

10.2.4 Carbon Monoxide Dehydrogenase

Carbon monoxide dehydrogenase (referred to as *CODH* for short) can catalyze the two-electron redox reaction of CO [64]. There are two types of CODH, including those with only the CO dehydrogenase subunit and those containing both CO dehydrogenase and acetyl-CoA synthase subunits. The *acetyl-CoA synthase* activity comes from the *A-cluster* (cluster **11** in Figure 10.1), where a Fe_4S_4 cluster is bridged to a proximal nickel (Ni_p) center. This cluster is further bridged with a thiolate ligand to a distal nickel (Ni_d) ion, which is coordinated by two cysteine and two backbone amide ligands. The crystal structure of *Ni-CODH* clearly reveals the existence of five metal clusters for each homodimer enzyme, including two Ni–Fe–S clusters, called *C-clusters* (cluster **12** in Figure 10.1), one Fe_4S_4 *D-cluster*, and two Fe_4S_4 *B-clusters*. The CO/CO_2 interconversion reaction occurs at the *C-cluster*.

As we previously discussed, Holm and his colleagues contributed a lot to the biomimetic synthesis of both the *A-cluster* and the *C-cluster* [6]. After completing the synthesis of the binuclear complex $[Ni^{II}(\mu2\text{-}SR)_2M^{I,II}]$ related to *A-cluster* [65], Holm et al. [66] further explored the method of obtaining the structure of $[e\text{-}(\mu2\text{-}SR)\text{-}Ni^{II}]$. Cluster **34** (Figure 10.5) contains two mutually supportive $[Fe\text{-}(\mu2\text{-}SR)\text{-}Ni^{II}]$ bridges, while cluster **35** possesses one strong and one weaker bridge. For the bionic simulation of *CODH C-Cluster*, Holm et al. synthesized $[MFe_3S_4]$ clusters with a

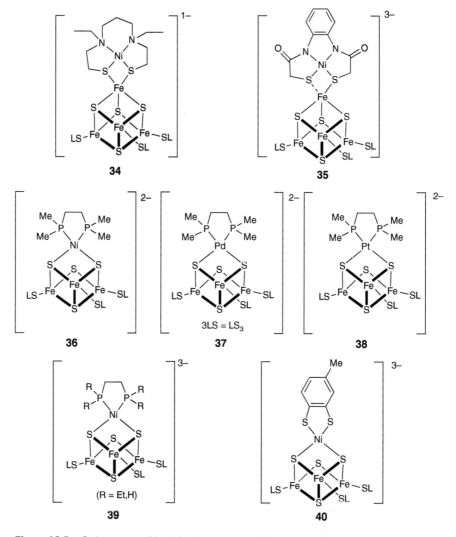

Figure 10.5 Carbon monoxide dehydrogenase model compounds made by Holm's group.

planar M^{II} site (M = Ni, Pd, and Pt; clusters **36**, **37**, and **38**) [67]. Subsequently, Holm et al. [68] synthesized some *CODH C-clusters* through the substitution reaction of the nickel site in a cubane-type cluster. Cluster **39** contains tetrahedral [Ni(μ3-S)$_3$(SR)] sites, whereas cluster **40** contains a cubanoid [NiFe$_3$(μ2-S)(μ3-S)$_3$] core and an approximately planar [Ni(tdt)(μ3-S)$_2$] unit (Figure 10.5).

10.3 Mixed-Valence Systems in Multiheme and Other Multiiron-Contained Biological Systems and Their Mimics

Peroxidase and *catalase* exist widely in animals and plants, and *peroxidase* also exists in many bacteria [69]. *Peroxidase* catalyzes the reduction of peroxide ROOH to ROH and water, while *catalase* converts hydrogen peroxide into H$_2$O and oxygen molecules through catalytic disproportionation reactions. Both of them can convert harmful peroxides and hydrogen peroxide species into harmless species and protect organisms. The common *peroxidases* are *cytochrome c peroxidase*, *horseradish peroxidase*, *chloroperoxidase*, and so on, in which *cytochrome c peroxidase* can catalyze the oxidation of the reduced *cytochrome c* with hydrogen peroxide. There are many similarities in the catalytic active-center structure and catalytic mechanism between *catalase* and various *peroxidases*. Their structures generally contain heme-b auxiliary groups. Other side chains of heme are open, and iron is in a state of trivalent high-spin [70–72].

Ellfolk's group [73] confirmed that the *cytochrome c peroxidases* in *Pseudomonas aeruginosa* contain two heme cofactors with different oxidation states at a long distance inside the protein. The corresponding high-potential electron-transfer (E) heme and the low-potential peroxidatic (P) heme (shown in Scheme 10.1)

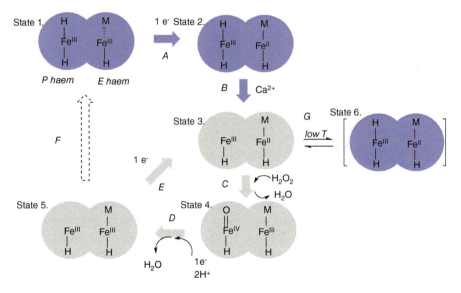

Scheme 10.1 Model of the mechanism of bacterial *cytochrome c peroxidases*.

construct a special long-distance mixed-valence diferrous system. Scheme 10.1 [74] shows a model of the mechanism of bacterial *cytochrome c peroxidases*. In the inactive/resting state (State 1) of *cytochrome c peroxidases*, both hemes are hexa-coordinate, and the central iron atom is trivalent. E heme takes an electron and is reduced to a divalent state (step A), producing the mixed-valence ($Fe^{III}Fe^{II}$) enzyme. A calcium-dependent configuration change of P heme (step B) then occurs to replace the distal histidine with a water molecule (State 3). Histidine can be recoordinated with P heme at low temperatures (State 6). Possible enzyme cycles are indicated in gray. The activated enzyme transfers one electron from P heme and one electron from E heme to hydrogen peroxide, reducing it to water (step C). The enzyme catalytic cycle is then completed through two single-electron-transfer steps (steps D and E). State 5 can slowly transform back to State 1 (step F).

10.4 Mixed-Valence Multicopper Cofactors in Biological and Mimicking Systems

Respiratory complex IV (*cytochrome c oxidase*, *CytcO*) transfers four electrons from four *cytochrome c* to an oxygen molecule and at the same time combines four protons in the internal aqueous phase to produce two water molecules, which is the last oxidation–reduction reaction in the respiratory chain of aerobic organisms. *Complex IV* is a membrane protein complex composed of multiple subunits, thirteen polypeptide chains, two *hemes* (*heme a* and *heme* a_3), one copper ion (Cu_B), and one binuclear copper ion (Cu_A) center (cluster **13** in Figure 10.1), which constructs a mixed-valence multicopper system [75]. Core subunits I, II, and III are encoded by mitochondrial DNA. Among them, subunit I contains *heme a*, *heme* a_3, and Cu_B, and *heme* a_3 and Cu_B form a binuclear Fe–Cu center (a_3-Cu_B), which is similar in structure to [2Fe-2S]; subunit II has a binuclear copper ion (Cu_A) center; and subunit III has no electron transporter and only interacts with subunits I and II to maintain the stability of the electron-transfer center. In the turnover process of *CytcO*, electron transfer from *cytochrome c* to the Cu_A site is followed in time by electron transfer to *heme a* and the *heme* a_3-Cu_B catalytic site (Figure 10.6). Two *cytochrome c* molecules transfer two electrons to the binuclear Fe(III)–Cu(II) (*State O*) via the Cu_A center and *heme a* to obtain Fe(II)–Cu(I) (*State R*). The two-electron-reduced catalytic site binds with O_2 (*State A*). The O–O bond is broken by electron transfer from *heme* a_3 and Cu_B as well as hydrogen transfer from Tyr245, which forms a radical (*State P*). The *State P* uptakes a proton and an electron from the third molecule of cytochrome c, thereby transforming into *State F*. The electrons from the fourth molecule of *cytochrome c* reduce the $Fe^{4+} = O$ to Fe^{3+}, and at the same time, by uptaking a proton, a hydroxide anion is generated (*State O*), returning to the initial cycle. The Cu_A center is a fully delocalized mixed-valence bis(μ-thiolato) dinuclear copper complex that exists in *cytochrome c oxidase* and *nitrous oxidase reductase*. The distance between the two Cu centers in the copper A center is about 2.4–2.5 Å, which is bridged by two thiolate sulfur atoms of two cysteine residues to form an almost planar Cu_2S_2 diamond structure [76].

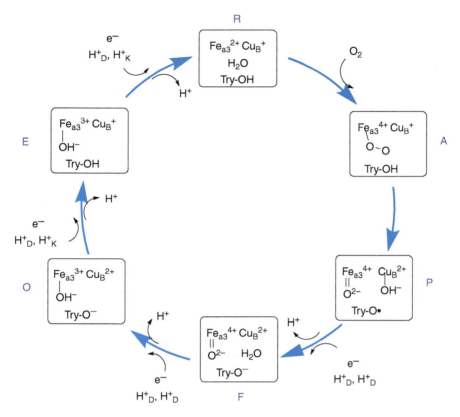

Figure 10.6 Reduction of O_2 at the catalytic site of *cytochrome c oxidase*. Source: [75]/American Chemical Society/Licensed under CC BY 4.0.

Hemocyanins (cofactor **14** in Figure 10.1) are another kind of metalloprotein containing mixed-valence copper cofactors, which can reversibly bind to a single oxygen molecule (O_2). *Hemocyanins* are taking over the job of transporting oxygen in some invertebrates. Oxidation of Cu(I) to Cu(II) results in a color change to blue from colorless. The active site of *hemocyanin* consists of a pair of copper (I) cations that coordinate directly with the protein through the imidazole ring of six histidine residues. *Hemocyanin* consists of many separate subunit proteins. Each of them contains two copper atoms that bind to an oxygen molecule (O_2).

Nitrous oxide reductase is also a copper-containing metallo-enzyme that catalyzes the double-electron reduction of NO_2 to nitrogen. So far, all the reported N_2O reductases are dimers with a molecular weight of 66 kDa, and each monomer contains an average of four copper ions. In the past decade, the mainstream view has been that the four copper ions in N_2O reductase form two binuclear centers. The first center is the Cu_A site (Figure 10.7a), which is similar to *cytochrome c oxidase*. The second one is called Cu_Z (Figure 10.7b) and is considered to be the catalytic site [77].

Multicopper oxidase (MCO) is an enzyme that couples the four-electron reduction of molecular oxygen with oxidation of the substrate. MCO contains at least four Cu sites, which are divided into type 1 (T1), type 2 (T2), and binuclear type 3 (T3)

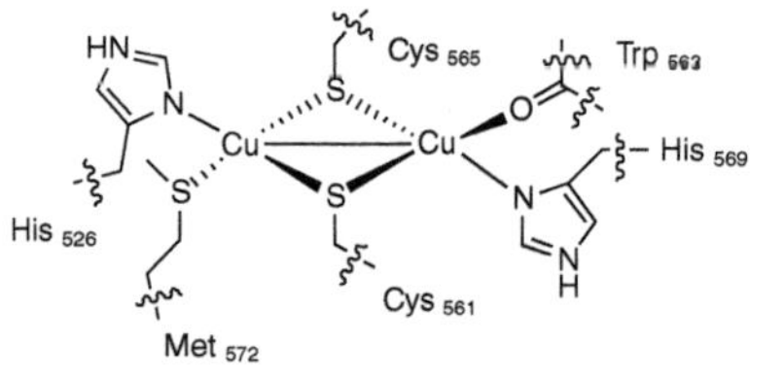

Figure 10.7 (a) Left: Cu_A site in *P. nautica* N_2O reductase. (b) Right: Cu_Z site in *P. nautica* N_2O reductase.

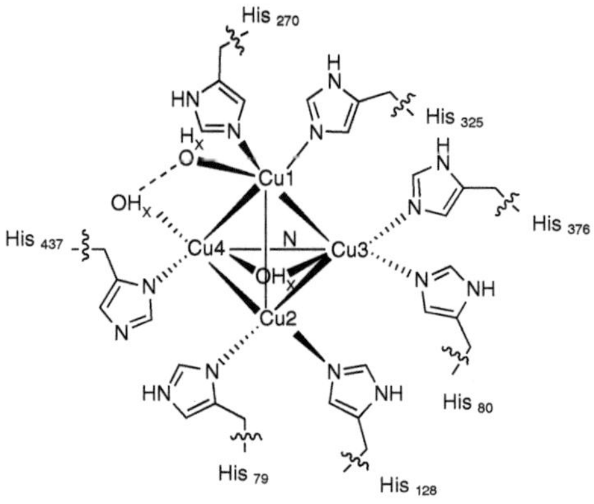

Cu sites. These sites are distinguished based on unique spectral characteristics. Substrate oxidation occurs near T1, and electrons are transferred to the T2/T3 trinuclear copper cluster (TNC) through protein via the Cys–His pathway, where oxygen reduction occurs. Solomon et al. [78] proposed the mechanism of oxygen reduction by *multicopper oxidase* in laccase that can be divided into two bielectronic steps: (i) O_2 reacts rapidly with the fully reduced enzyme ($Cu^I{}_4$) to form PI ($Cu^{(I)}{}_2Cu^{(II)}{}_2(O_2{}^{[2-]})$) and (ii) PI accepts two internal electrons and is reduced to NI (Scheme 10.2).

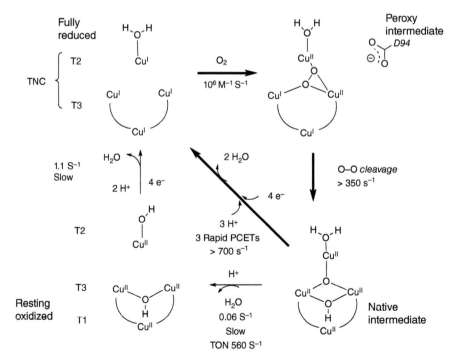

Scheme 10.2 Mechanism of oxygen reduction by *multicopper oxidase* in laccase.

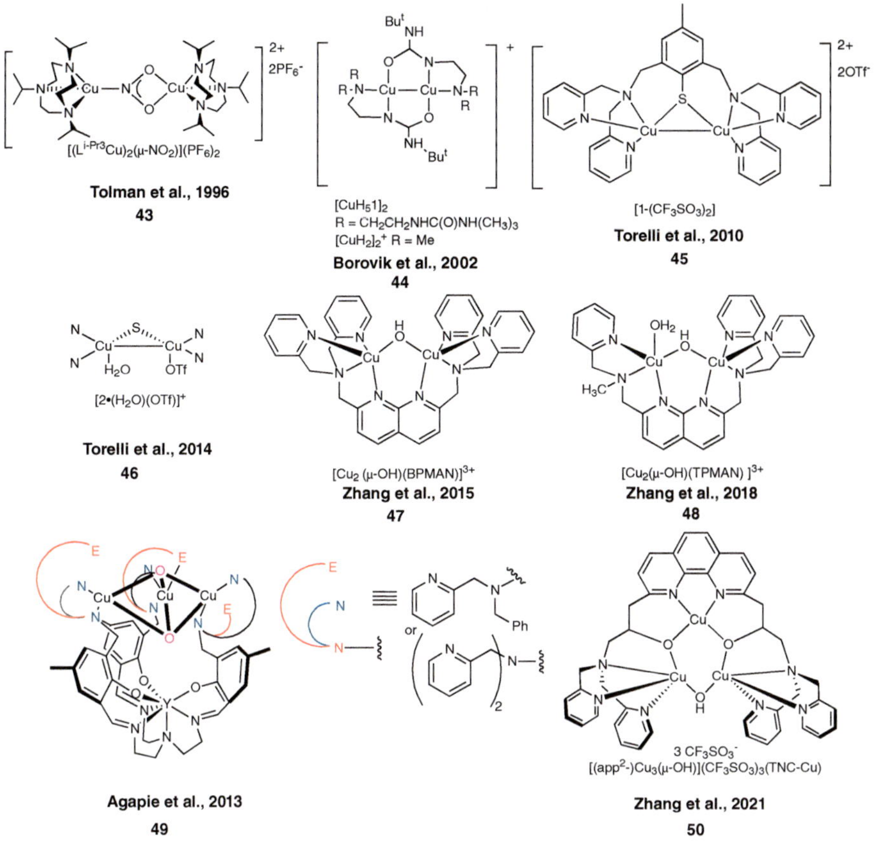

Figure 10.8 The model compounds for mimicking the mixed-valence multicopper clusters in nature.

The artificial molecular models for mimicking the previously mentioned mixed-valence multicopper cofactors have attracted the attention of chemists for decades. In 1996, Tolman et al. synthesized mixed-valent copper complexes [(L^{i-Pr3}Cu)$_2$(μ-NO$_2$)](PF$_6$)$_2$ (cluster **43** in Figure 10.8) in order to study the nitrite reduction mediated by copper protein [79]. In 2002, Borovik et al. [80] synthesized new mixed-valent copper complexes [Cu(H$_5$1)]$_2^+$ and [Cu(H$_2$)]$_2$ (cluster **44**). These complexes have completely delocalized dicopper cores, which are supported by structural and spectral results. The copper–copper distance of both complexes is about 2.39 Å, which is shorter than that of most other binuclear copper complexes. [Cu(H$_5$1)]$_2^+$ has a Cu$^{(1.5)}$Cu$^{(1.5)}$ core, and each copper ion has a square plane coordination geometry. The structural properties of Cu$^{(1.5)}$Cu$^{(1.5)}$ core in [Cu(H$_2$)]$_2^+$ are almost the same as those in [Cu(H$_5$1)]$_2^+$. In 2010, Inspired by the CuZ center, Torelli (cluster **45**) et al. reported the first mixed-valent copper complex with a (Cu$_2$S)$^{2+}$ motif, which can partially represent the Cu_Z mixed-valence resting state. This complex contains copper–copper bonds and shows a high degree of delocalization [81]. In 2014, they reported a new asymmetric mixed-valence dicopper(II,I)[2·(H$_2$O)(OTf)]$^+$ (cluster **46** in Figure 10.8), which contains a (Cu$_2$S)$^{2+}$ core with unstable triflate and water molecules at the copper centers [82]. In 2015, Zhang et al. reported an oxidation-stable robust binuclear Cu complex [Cu$_2$(BPMAN) (μ-OH)]$^{3+}$ (cluster **47**), which is the first copper-based molecular water-oxidation catalyst used in a neutral aqueous solution and does not decompose during long-term electrolysis [83]. Subsequently, in 2018, they introduced an open site on the copper atom using an asymmetric ligand with 3- and 4-coordinate sites on both hands. The synthesized unsymmetrical di-copper complex [Cu$_2$(TPMAN)(μ-OH)]$^{3+}$ (cluster **48**) can be used for electrocatalytic water oxidation under neutral conditions [84]. In 2021, the same group [85] reported a trinuclear copper cluster (cluster **50**), which showed excellent activity for catalytic oxidation of water inspired by *MCOs*. It provided an effective catalyst for the four-electron reduction of O$_2$ to water. Studies have shown that cooperation between multiple metals is an effective strategy to regulate the formation of O—O bonds in water-oxidation catalysis. To simulate *MCO*, Agapie et al. [86] assembled a trinucleating skeleton (cluster **49**) by templating a heptadentate ligand around yttrium and lanthanides, and then synthesized two tricopper complexes. These Cu$_3^I$Y complexes react with oxygen at low temperatures to form Cu$_2^{(II)}$Cu$^{(III)}$(μ$_3$-O)$_2$ motifs. Similar to *MCO*, the close arrangement of three metal centers leads to the synergistic activation of O$_2$. In 2018, Wu et al. [87] reported that two tetranuclear copper catalysts [(L$_{Gly}$-Cu)$_4$] and [(L$_{Glu}$-Cu)$_4$] could be used as an effective catalyst for electrocatalytic water oxidation in aqueous solution (pH = 12). The unique Cu$_4$O$_4$ cubane cores are similar in structure to the oxygen-evolving center Mn$_4$CaO$_5$ cluster in **PSII** and show efficient electrocatalytic activity for water oxidation with a TOF of 267 s^{-1} for [(L$_{Gly}$-Cu)$_4$] at 1.70 V and 105 s^{-1} for [(L$_{Glu}$-Cu)$_4$] at 1.56 V. This high performance may be attributed to the synergism of tetranuclear copper cubane centers and the successive two-electron-transfer process.

10.5 OEC and Other Mixed-Valence Multimanganese Cofactors

The *oxygen evolution complex (OEC)* is a metallo-oxo cluster composed of four manganese ions and a divalent calcium ion, which is an amazing and challenging structure, attracting the attention of chemists and biologists for decades [88–90]. The intermediate structures and the reaction mechanism of *OEC* have been revealed by synchrotron radiation and free-electron lasers [91–96]. The catalytic center of the *OEC* is an asymmetric Mn$_4$CaO$_5$ cluster. When it oxidizes water to produce oxygen and protons, it in turn transfers four electrons from H$_2$O molecules to the side chain of tyrosine, which is then passed on to P$_{680}$ itself. In 1970, Kok et al. [97] proposed a five-state kinetic model for photosynthetic oxygen evolution, known as the Kok's S-state clock or cycle (Figure 10.9). The model comprises four (meta)stable intermediates (S_0, S_1, S_2, and S_3) and one transient S_4 state. During the cycling process, along with the transfer and release of electrons and protons, the spatial structure of the *OEC* itself in the valence state of the four Mn ions will change. Among them, the S_0 state is the initial state, and the valence states of the

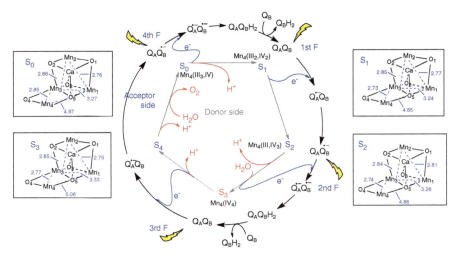

Figure 10.9 The Kok's cycle and all (meta)stable states of OEC in photosystem II. Source: [98]/Springer Nature.

four Mn ions are +3, +3, +3, and +4. The S_1 state is the dark stable state, and the valence states of the four Mn ions are +3, +3, +4, and +4. The S_2 and S_3 states are high oxidation states, and the valence states are +3, +4, +4, +4 and +4, +4, +4, +4, respectively. The S_4 state is a transient state that can bind to water molecules and quickly release oxygen to return to the S_0 state. This cyclic process is coupled with a two-step reduction and protonation of mobile plastoquinone Q_B located on the acceptor side of PS II. In 2018, Messinger et al. [98] visualized all (meta)stable states of Kok's cycle as high-resolution structures (2.04–2.08 Å). They [99] later studied the water and proton channels that connected OEC to the lumenal bulk water.

The artificially synthesized model compounds offered a lot of useful information on the reaction mechanism of OEC (Figure 10.10). Multinuclear manganese clusters have been studied as early as 1983, and some progress has been made since then [100–107]. Some progress has also been made in clusters containing calcium ions in recent years [108–111]. In 2011, Agapie and his colleagues reported the synthesis of $[Mn_3CaO_4]^{6+}$ cubane (cluster **53** in Figure 10.10), which structurally simulated the trimanganese–calcium–cubane subunit of OEC and revealed the potential role of calcium in promoting the high oxidation state of manganese and the assembly of biological clusters [112]. In 2012, Christou et al. reported the synthesis of an asymmetric $[Mn_3CaO_4]$ cubane with additional metal in a protein-free environment (cluster **55**), which can be used as an OEC model compound. All the peripheral ligands are carboxylic acids or carboxylic acid groups [113]. In 2013, a series of highly oxidized tetranuclear dioxin clusters consisting of three manganese centers and one redox inactive metal (M) was reported by Agapie's group (cluster **54**). Crystallographic studies show that the Mn_3M (μ_4-O) (μ_2-O) core remains intact when changing the M or manganese oxidation state [114]. Then, using a similar method, a series of complex clusters of different metal ions were synthesized [115]. Subsequently, they synthesized low-symmetry $Mn^{IV}_3GdO_4$

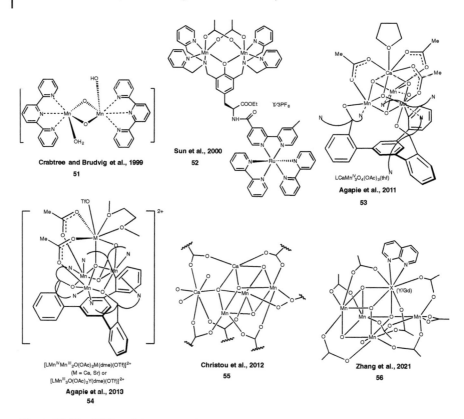

Figure 10.10 OEC model compounds.

and $Mn^{IV}_3CaO_4$ cubanes by means of ligand-substituted desymmetrization, which resulted in significant cubane distortions [116]. In 2015, Zhang et al. synthesized a biomimetic Mn_4CaO_4 cluster $Mn_4CaO_4(Bu^tCO_2)_8$ $(Bu^tCO_2H)_2(Py)$, which is similar to biological *OEC* in structure and properties [117]. The bond-valence sum analysis of the cluster shows that the oxidation states of $Mn1^{3+}$, $Mn2^{4+}$, $Mn3^{3+}$, and $Mn4^{3+}$ are completely similar to the *OEC* in its S_1 state. The $Mn_{3/4}$-O_5 bond lengths (1.85/1.85 Å) of the synthetic cluster are shorter than those in the native *OEC* (2,2/2.3 Å). The Mn1-O5 distance in the native *OEC* (2.7 Å) is much longer than that of the synthetic cluster (2.28 Å). In this fashion, Mn1 becomes effectively five-coordinated, with an open coordination site that may be essential for the coordination of a "substrate" H_2O molecule upon oxidation of Mn4 from the +3 to the +4 state. Consistent with native OEC, the synthetic cluster can undergo four redox transitions and display two magnetic resonance signals attributable to redox and structural isomerism. Then, biomimetic *OEC* clusters Mn_4CaO_4-$(Bu^tCO_2)_8(Py)(Bu^tCO_2H)(CH_3CN)$ and Mn_4CaO_4 $(Bu^tCO_2)_8(Py)(DMF)_2$ were further prepared by peripheral ligand substitution in 2019 [118]. In 2021, the same research group overcame the difficult problem in the synthesis of biomimetic *OEC* containing rare earth ions, and prepared stable Mn_4YO_4 clusters

[Mn$_4$YO$_4$(^{t}BuCO$_2$)$_9$(Napy)] and Mn$_4$GdO$_4$ clusters [Mn$_4$GdO$_4$-(^{t}BuCO$_2$)$_9$(Napy)] (Napy = 1,8-naphthyridine) (cluster **56** in Figure 10.10) containing rare earth ions for the first time [119]. The synthesized biomimetic Mn$_4$XO$_4$ clusters containing rare earth ions simulate not only the geometric and electronic structure of biological *OEC* but also the redox characteristics of biological *OEC*.

Mixed-valence multimanganese cofactors can also be seen in *Class I RNRs*. These enzymes can catalyze the conversion of nucleotides to deoxynucleotides. They can be divided into three categories according to cofactors: Fe$^{(III)}$Fe$^{(III)}$- tyrosyl radical (Y˙) cofactors (class Ia), Mn$^{(III)}$Mn$^{(III)}$-Y˙cofactor (class Ib, cluster 16 in Figure 1), and Mn$^{(IV)}$Fe$^{(III)}$ cofactors (class Ic). The class Ia, Ib, and Ic RNRs are structurally homologous and contain almost the same metal coordination sites. In 1988 [120], Will, Follmann, and Auling purified the first class Ib enzyme, from the RNR of *Brevibacterium ammoniagenes*, and proposed that manganese was needed to exert its activity. At present, there is evidence that Mn$^{(III)}$Mn$^{(III)}$-Y• cofactors are formed in the class Ib RNR in different organisms (*Escherichia coli, C. ammoniagenes*, and *Bacillus subtilis*). Class Ib RNR [121] contains a di-manganese center (Mn$^{(III)}$-O-Mn$^{(III)}$) that produces tyrosyl radical *in vivo*, while the corresponding mixed-valence species may be involved in this reaction process. The detailed mechanism study of these cofactors and their artificial model compounds might become one of the hotspots of research in the future [122].

10.6 Summary

An overview of the rich development during the last 40 years of mixed-valence cofactors in biological systems is provided in this chapter, including iron–sulfur clusters and cofactors, multiple heme systems, multicopper cofactors, and multimanganese cofactors in natural proteins and enzymes. In addition, this chapter also highlights the recent advances in the artificial and biomimetic model complexes of the corresponding mixed-valence cofactors. The research on these mixed-valence complexes in both biological and bio-mimic systems not only provides us with insights into the structures and mechanisms of these natural mixed-valence cofactors but also stimulates us to develop more efficient artificial catalysts for their applications in synthetic chemistry, energy chemistry, and other areas.

Acknowledgement

Funding is provided by the National Key Research and Development Program of China (T.L., grant no. 2022YFB4101800) and the National Natural Science Foundation of China (T. L., grants no. 22171141, 22193010, and 22193014). Supports from State Key Laboratory of Elementoorganic Chemistry, College of Chemistry at Nankai University, the Haihe Laboratory of Sustainable Chemical Transformations, and Key Laboratory of Organofluorine Chemistry at Shanghai Institute of Organic Chemistry, Chinese Academy of Sciences (T. L.) are also acknowledged.

References

1 Liu, J., Chakraborty, S., Hosseinzadeh, P. et al. (2014). Metalloproteins containing cytochrome, iron–sulfur, or copper redox centers. *Chem. Rev.* 114: 4366–4469.
2 Yu, F., Cangelosi, V.M., Zastrow, M.L. et al. (2014). Protein design: toward functional metalloenzymes. *Chem. Rev.* 114: 3495–3578.
3 Solomon, E.I., Xie, X., and Dey, A. (2008). Mixed valent sites in biological electron transfer. *Chem. Soc. Rev.* 37: 623–638.
4 Meyer, T.J. (1980). Electron transfer in mixed-valence compounds. In: *Mixed-Valence Compounds*, NATO Advanced Study Institutes Series, vol. 58 (ed. D.B. Brown), 75–114. Dordrecht: Springer.
5 Demadis, K.D., Hartshorn, C.M., and Meyer, T.J. (2001). The localized-to-delocalized transition in mixed-valence chemistry. *Chem. Rev.* 101: 2655–2686.
6 Lee, S.C., Lo, W., and Holm, R.H. (2014). Developments in the biomimetic chemistry of cubane-type and higher nuclearity iron–sulfur clusters. *Chem. Rev.* 114: 3579–3600.
7 Venkateswara Rao, P. and Holm, R.H. (2004). Synthetic analogues of the active sites of iron–sulfur proteins. *Chem. Rev.* 104: 527–560.
8 Johnson, M.K. and Smith, A.D. (2011). Iron–sulfur proteins. In: *Encyclopedia of Inorganic and Bioinorganic Chemistry*. ISBN: 9781119951438.
9 Zhou, X., Gao, Y., Wang, W. et al. (2021). Architecture of the mycobacterial succinate dehydrogenase with a membrane-embedded Rieske FeS cluster. *Proc. Natl. Acad. Sci. U.S.A.* 118: e2022308118.
10 Hudson, R., de Graaf, R., Strandoo Rodin, M. et al. (2020). CO_2 reduction driven by a pH gradient. *Proc. Natl. Acad. Sci.* 117: 22873–22879.
11 White, L.M., Bhartia, R., Stucky, G.D. et al. (2015). Mackinawite and greigite in ancient alkaline hydrothermal chimneys: identifying potential key catalysts for emergent life. *Earth Planet. Sci. Lett.* 430: 105–114.
12 Stack, T.D.P. and Holm, R.H. (1988). Subsite-differentiated analogs of biological [4Fe–4S]$^{2+}$ clusters: synthesis, solution and solid-state structures, and subsite-specific reactions. *J. Am. Chem. Soc.* 110: 2484–2494.
13 Holm, R.H. and Lo, W. (2016). Structural conversions of synthetic and protein-bound iron–sulfur clusters. *Chem. Rev.* 116: 13685–13713.
14 Cyranoski, D. (2015). Discovery of long-sought biological compass claimed. *Nature* 527: 283–284.
15 Qin, S., Yin, H., Yang, C. et al. (2016). A magnetic protein biocompass. *Nat. Mater.* 15: 217–226.
16 Guo, Z., Xu, S., Chen, X. et al. (2021). Modulation of MagR magnetic properties via iron–sulfur cluster binding. *Sci. Rep.* 11: 23941–23952.
17 Rauchfuss, T.B. (2015). Diiron azadithiolates as models for the [FeFe]-hydrogenase active site and paradigm for the role of the second coordination sphere. *Acc. Chem. Res.* 48: 2107–2116.
18 Schilter, D., Camara, J.M., Huynh, M.T. et al. (2016). Hydrogenase enzymes and their synthetic models: the role of metal hydrides. *Chem. Rev.* 116: 8693–8749.

19 Sommer, C., Adamska-Venkatesh, A., Pawlak, K. et al. (2017). Proton coupled electronic rearrangement within the H-cluster as an essential step in the catalytic cycle of [FeFe] hydrogenases. *J. Am. Chem. Soc.* 139: 1440–1443.

20 Madden, C., Vaughn, M.D., Díez-Pérez, I. et al. (2012). Catalytic turnover of [FeFe]-hydrogenase based on single-molecule imaging. *J. Am. Chem. Soc.* 134: 1577–1582.

21 Wang, S., Aster, A., Mirmohades, M. et al. (2018). Structural and kinetic studies of intermediates of a biomimetic diiron proton-reduction catalyst. *Inorg. Chem.* 57: 768–776.

22 Wang, W., Nilges, M.J., Rauchfuss, T.B., and Stein, M. (2013). Isolation of a mixed valence diiron hydride: evidence for a spectator hydride in hydrogen evolution catalysis. *J. Am. Chem. Soc.* 135: 3633–3639.

23 Jablonskytė, A., Wright, J.A., Fairhurst, S.A. et al. (2011). Paramagnetic bridging hydrides of relevance to catalytic hydrogen evolution at metallosulfur centers. *J. Am. Chem. Soc.* 133: 18606–18609.

24 Mirmohades, M., Pullen, S., Stein, M. et al. (2014). Direct observation of key catalytic intermediates in a photoinduced proton reduction cycle with a diiron carbonyl complex. *J. Am. Chem. Soc.* 136: 17366–17369.

25 Song, L.-C., Chen, J.-S., Jia, G.-J. et al. (2019). Synthetic and structural studies of [FeFe]-hydrogenase models containing a butterfly Fe/E (E = S, Se, or Te) cluster core. electrocatalytic H_2 evolution catalyzed by $[(\mu\text{-SeCH}_2)(\mu\text{-CH}_2\text{NCH}_2\text{Ph})]\text{Fe}_2(\text{CO})_6$. *Organometallics* 38: 1567–1580.

26 Song, L.-C., Hong, D.-J., Guo, Y.-Q., and Wang, X.-Y. (2018). Dinuclear $Fe^{II}Fe^{II}$ biomimetics for the oxidized state active site of [FeFe]-hydrogenases: synthesis, characterization, and electrocatalytic H_2 production. *Organometallics* 37: 4744–4752.

27 Song, L.C., Wang, Y.X., Xing, X.K. et al. (2016). Hydrophilic quaternary ammonium-group-containing [FeFe]-hydrogenase models: synthesis, structures, and electrocatalytic hydrogen production. *Chem. Eur. J.* 22: 16304–16314.

28 Song, L.-C., Luo, F.-X., Liu, B.-B. et al. (2016). Novel ruthenium phthalocyanine-containing model complex for the active site of [FeFe]-hydrogenases: synthesis, structural characterization, and catalytic H_2 evolution. *Organometallics* 35: 1399–1408.

29 Song, L.C., Zhu, A.G., and Guo, Y.Q. (2016). Synthesis, characterization, and H/D exchange of mu-hydride-containing [FeFe]-hydrogenase subsite models formed by protonation reactions of $(\mu\text{-TDT})\text{Fe}_2(\text{CO})_4(\text{PMe}_3)_2$ (TDT = SCH_2SCH_2S) with protic acids. *Dalton Trans.* 45: 5021–5029.

30 Song, L.-C., Gu, Z.-C., Zhang, W.-W. et al. (2015). Synthesis, structure, and electrocatalysis of butterfly $[Fe_2SP]$ cluster complexes relevant to [FeFe]-hydrogenases. *Organometallics* 34: 4147–4157.

31 Song, L.-C., Gai, B., Feng, Z.-H. et al. (2013). Synthesis, structures, and some properties of diiron oxadiselenolate (ODSe) and thiodiselenolate (TDSe) complexes as models for the active site of [FeFe]-hydrogenases. *Organometallics* 32: 3673–3684.

32 Song, L.C., Li, Q.L., Feng, Z.H. et al. (2013). Synthesis, characterization, and electrochemical properties of diiron propaneditellurolate (PDTe) complexes as active site models of [FeFe]-hydrogenases. *Dalton Trans.* 42: 1612–1626.

33 Song, L.-C., Wang, L.-X., Jia, G.-J. et al. (2012). Synthesis, structural characterization, and properties of some functionalized phosphine-containing diiron complexes as models for the active site of [FeFe]-hydrogenases. *Organometallics* 31: 5081–5088.

34 Zhang, Y., Tao, L., Woods, T.J. et al. (2022). Organometallic $Fe_2(\mu\text{-SH})_2(CO)_4(CN)_2$ cluster allows the biosynthesis of the [FeFe]-hydrogenase with only the HydF maturase. *J. Am. Chem. Soc.* 144: 1534–1538.

35 Huynh, M.T., Wang, W., Rauchfuss, T.B., and Hammes-Schiffer, S. (2014). Computational investigation of [FeFe]-hydrogenase models: characterization of singly and doubly protonated intermediates and mechanistic insights. *Inorg. Chem.* 53: 10301–10311.

36 Wang, F., Wang, W.-G., Wang, H.-Y. et al. (2012). Artificial photosynthetic systems based on [FeFe]-hydrogenase mimics: the road to high efficiency for light-driven hydrogen evolution. *ACS Catal.* 2: 407–416.

37 Wang, W.G., Wang, F., Wang, H.Y. et al. (2012). Electron transfer and hydrogen generation from a molecular dyad: platinum(II) alkynyl complex anchored to [FeFe] hydrogenase subsite mimic. *Dalton Trans.* 41: 2420–2426.

38 Wang, W.G., Wang, F., Wang, H.Y. et al. (2010). Photocatalytic hydrogen evolution by [FeFe] hydrogenase mimics in homogeneous solution. *Chem. Asian J.* 5: 1796–1803.

39 Wang, H.Y., Wang, W.G., Si, G. et al. (2010). Photocatalytic hydrogen evolution from rhenium(I) complexes to [FeFe] hydrogenase mimics in aqueous SDS micellar systems: a biomimetic pathway. *Langmuir* 26: 9766–9771.

40 Wang, W.G., Wang, H.Y., Si, G. et al. (2009). Fluorophenyl-substituted Fe-only hydrogenases active site ADT models: different electrocatalytic process for proton reduction in HOAc and HBF_4/Et_2O. *Dalton Trans.* 15: 2712–2720.

41 Rohac, R., Martin, L., Liu, L. et al. (2021). Crystal structure of the [FeFe]-hydrogenase maturase hyde bound to complex-B. *J. Am. Chem. Soc.* 143: 8499–8508.

42 Song, L.C., Tang, M.Y., Su, F.H., and Hu, Q.M. (2006). A biomimetic model for the active site of iron-only hydrogenases covalently bonded to a porphyrin photosensitizer. *Angew. Chem. Int. Ed* 45: 1130–1133.

43 Esmieu, C., Raleiras, P., and Berggren, G. (2018). From protein engineering to artificial enzymes – biological and biomimetic approaches towards sustainable hydrogen production. *Sustain. Energy Fuels* 2: 724–750.

44 Wang, S., Pullen, S., Weippert, V. et al. (2019). Direct spectroscopic detection of key intermediates and the turnover process in catalytic H_2 formation by a biomimetic diiron catalyst. *Chem. Eur. J.* 25: 11135–11140.

45 Wang, V.C., Esmieu, C., Redman, H.J. et al. (2020). The reactivity of molecular oxygen and reactive oxygen species with [FeFe] hydrogenase biomimetics: reversibility and the role of the second coordination sphere. *Dalton Trans.* 49: 858–865.

46 Binstead, R.A., Moyer, B.A., Samuels, G.J., and Meyer, T.J. (1981). Proton-coupled electron transfer between $[Ru(bpy)_2(py)OH_2]^{2+}$ and $[Ru(bpy)_2(py)O]^{2+}$.

A solvent isotope effect (kH$_2$O/kD$_2$O) of 16.1. *J. Am. Chem. Soc.* 103: 2897–2899.

47 Weinberg, D.R., Gagliardi, C.J., Hull, J.F. et al. (2012). Proton-coupled electron transfer. *Chem. Rev.* 112: 4016–4093.

48 Tyburski, R., Liu, T., Glover, S.D., and Hammarström, L. (2021). Proton-coupled electron transfer guidelines, fair and square. *J. Am. Chem. Soc.* 143: 560–576.

49 Zhou, Z., Kong, X., and Liu, T. (2021). Applications of proton-coupled electron transfer in organic synthesis. *Chin. J. Org. Chem.* 41: 3844–3879.

50 Liu, T., Guo, M., Orthaber, A. et al. (2018). Accelerating proton-coupled electron transfer of metal hydrides in catalyst model reactions. *Nat. Chem.* 10: 881–887.

51 Liu, T., Tyburski, R., Wang, S. et al. (2019). Elucidating proton-coupled electron transfer mechanisms of metal hydrides with free energy- and pressure-dependent kinetics. *J. Am. Chem. Soc.* 141: 17245–17259.

52 Van Stappen, C., Decamps, L., Cutsail, G.E. et al. (2020). The spectroscopy of nitrogenases. *Chem. Rev.* 120: 5005–5081.

53 Harry, B.G., Edward, I.S., and Joan, S.V. (2007). *Biological Inorganic Chemistry: Structure and Reactivity*, 1e, 739. Sausalito: University Science Books.

54 Spatzal, T., Perez Kathryn, A., Einsle, O. et al. (2014). Ligand binding to the FeMo-cofactor: structures of CO-bound and reactivated nitrogenase. *Science* 345: 1620–1623.

55 Lindley, B.M., Appel, A.M., Krogh-Jespersen, K. et al. (2016). Evaluating the thermodynamics of electrocatalytic N$_2$ reduction in acetonitrile. *ACS Energy Lett.* 1: 698–704.

56 Greco, G.E. and Schrock, R.R. (2001). Synthesis of triamidoamine ligands of the type (ArylNHCH$_2$CH$_2$)$_3$N and molybdenum and tungsten complexes that contain an [(ArylNCH$_2$CH$_2$)$_3$N]$^{3-}$ ligand. *Inorg. Chem.* 40: 3850–3860.

57 Nishibayashi, Y. (2012). Molybdenum-catalyzed reduction of molecular dinitrogen under mild reaction conditions. *Dalton Trans.* 41: 7447–7453.

58 Ashida, Y., Arashiba, K., Nakajima, K., and Nishibayashi, Y. (2019). Molybdenum-catalysed ammonia production with samarium diiodide and alcohols or water. *Nature* 568: 536–540.

59 Klopsch, I., Finger, M., Würtele, C. et al. (2014). Dinitrogen splitting and functionalization in the coordination sphere of rhenium. *J. Am. Chem. Soc.* 136: 6881–6883.

60 Schendzielorz, F., Finger, M., Abbenseth, J. et al. (2019). Metal-ligand cooperative synthesis of benzonitrile by electrochemical reduction and photolytic splitting of dinitrogen. *Angew. Chem. Int. Ed* 58: 830–834.

61 Bruch, Q.J., Connor, G.P., Chen, C.-H. et al. (2019). Dinitrogen reduction to ammonium at rhenium utilizing light and proton-coupled electron transfer. *J. Am. Chem. Soc.* 141: 20198–20208.

62 Bruch, Q.J., Malakar, S., Goldman, A.S., and Miller, A.J.M. (2022). Mechanisms of electrochemical n2 splitting by a molybdenum pincer complex. *Inorg. Chem.* 61: 2307–2318.

63 Buscagan, T.M., Oyala, P.H., and Peters, J.C. (2017). N$_2$-to-NH$_3$ conversion by a triphos–iron catalyst and enhanced turnover under photolysis. *Angew. Chem. Int. Ed.* 56: 6921–6926.

64 Adam Panagiotis, S., Borrel, G., and Gribaldo, S. (2018). Evolutionary history of carbon monoxide dehydrogenase/acetyl-CoA synthase, one of the oldest enzymatic complexes. *Proc. Natl. Acad. Sci. U.S.A.* 115: 1166–1173.

65 Rao, P.V., Bhaduri, S., Jiang, J., and Holm, R.H. (2004). Sulfur bridging interactions of cis-planar Ni^{II}–S_2N_2 coordination units with nickel(II), copper(I,II), zinc(II), and mercury(II): a library of bridging modes, including niii(μ_2-SR)$_2$MI,II rhombs. *Inorg. Chem.* 43: 5833–5849.

66 Rao, P.V., Bhaduri, S., Jiang, J. et al. (2005). On $[Fe_4S_4]^{2+}$–$(\mu_2$-SR$)$–MII bridge formation in the synthesis of an A-cluster analogue of carbon monoxide dehydrogenase/acetylcoenzyme a synthase. *J. Am. Chem. Soc.* 127: 1933–1945.

67 Panda, R., Berlinguette, C.P., Zhang, Y., and Holm, R.H. (2005). Synthesis of MFe_3S_4 clusters containing a planar M^{II} site (M = Ni, Pd, Pt), a structural element in the C-cluster of carbon monoxide dehydrogenase. *J. Am. Chem. Soc.* 127: 11092–11101.

68 Sun, J., Tessier, C., and Holm, R.H. (2007). Sulfur ligand substitution at the nickel(II) sites of cubane-type and cubanoid $NiFe_3S_4$ clusters relevant to the C-clusters of carbon monoxide dehydrogenase. *Inorg. Chem.* 46: 2691–2699.

69 Yuan, F., Yin, S., Xu, Y. et al. (2021). The richness and diversity of catalases in bacteria. *Front. Microbiol.* 12: 1–11.

70 De Smet, L., Savvides, S.N., Van Horen, E. et al. (2006). Structural and mutagenesis studies on the cytochrome c peroxidase from rhodobacter capsulatus provide new insights into structure-function relationships of bacterial di-heme peroxidases. *J. Biol. Chem.* 281: 4371–4379.

71 Volkov, A.N., Nicholls, P., and Worrall, J.A.R. (2011). The complex of cytochrome c and cytochrome c peroxidase: the end of the road? *Biochim. Biophys. Acta, Bioenerg.* 1807: 1482–1503.

72 Dias, J.M., Alves, T., Bonifácio, C.I. et al. (2004). Structural basis for the mechanism of Ca^{2+} activation of the Di-heme cytochrome c peroxidase from pseudomonas nautica 617. *Structure* 12: 961–973.

73 Ellfolk, N., Rönnberg, M., Aasa, R. et al. (1983). Properties and function of the two hemes in pseudomonas cytochrome c peroxidase. *Biochim. Biophys. Acta, Protein Struct. Mol. Enzymol.* 743: 23–30.

74 Pettigrew, G.W., Echalier, A., and Pauleta, S.R. (2006). Structure and mechanism in the bacterial dihaem cytochrome c peroxidases. *J. Inorg. Biochem.* 100: 551–567.

75 Brzezinski, P., Moe, A., and Ädelroth, P. (2021). Structure and mechanism of respiratory III–IV supercomplexes in bioenergetic membranes. *Chem. Rev.* 121: 9644–9673.

76 Chacón, K.N. and Blackburn, N.J. (2012). Stable Cu(II) and Cu(I) mononuclear intermediates in the assembly of the CuA center of thermus thermophilus cytochrome oxidase. *J. Am. Chem. Soc.* 134: 16401–16412.

77 Rosenzweig, A.C. (2000). Nitrous oxide reductase from CuA to CuZ. *Nat. Struct. Biol.* 7: 169–171.

78 Jones, S.M. and Solomon, E.I. (2015). Electron transfer and reaction mechanism of laccases. *Cell. Mol. Life Sci.* 72: 869–883.

79 Halfen, J.A., Mahapatra, S., Wilkinson, E.C. et al. (1996). Synthetic modeling of nitrite binding and activation by reduced copper proteins. characterization of copper(I)–nitrite complexes that evolve nitric oxide. *J. Am. Chem. Soc.* 118: 763–776.

80 Gupta, R., Zhang, Z.H., Powell, D. et al. (2002). Synthesis and characterization of completely delocalized mixed-valent dicopper complexes. *Inorg. Chem.* 41: 5100–5106.

81 Torelli, S., Orio, M., Pécaut, J. et al. (2010). A $\{Cu_2S\}^{2+}$ mixed-valent core featuring a Cu-Cu bond. *Angew. Chem. Int. Ed.* 49: 8249–8252.

82 Esmieu, C., Orio, M., Torelli, S. et al. (2014). N_2O reduction at a dissymmetric $\{Cu_2S\}$-containing mixed-valent center. *Chem. Sci.* 5: 4774–4784.

83 Su, X.J., Gao, M., Jiao, L. et al. (2015). Electrocatalytic water oxidation by a dinuclear copper complex in a neutral aqueous solution. *Angew. Chem. Int. Ed.* 54: 4909–4914.

84 Hu, Q.-Q., Su, X.-J., and Zhang, M.-T. (2018). Electrocatalytic water oxidation by an unsymmetrical di-copper complex. *Inorg. Chem.* 57: 10481–10484.

85 Chen, Q.-F., Cheng, Z.-Y., Liao, R.-Z., and Zhang, M.-T. (2021). Bioinspired trinuclear copper catalyst for water oxidation with a turnover frequency up to 20000 s^{-1}. *J. Am. Chem. Soc.* 143: 19761–19768.

86 Lionetti, D., Day, M.W., and Agapie, T. (2013). Metal-templated ligand architectures for trinuclear chemistry: tricopper complexes and their O_2 reactivity. *Chem. Sci.* 4: 785–790.

87 Jiang, X., Li, J., Yang, B. et al. (2018). A bio-inspired Cu_4O_4 cubane: effective molecular catalysts for electrocatalytic water oxidation in aqueous solution. *Angew. Chem. Int. Ed.* 57: 7850–7854.

88 Nield, J., Kruse, O., Ruprecht, J. et al. (2000). Three-dimensional Structure of *Chlamydomonas reinhardtii* and *Synechococcus elongatus* photosystem ii complexes allows for comparison of their oxygen-evolving complex organization. *J. Biol. Chem.* 275: 27940–27946.

89 Rhee, K.-H., Morris, E.P., Barber, J., and Kühlbrandt, W. (1998). Three-dimensional structure of the plant photosystem II reaction centre at 8 Å resolution. *Nature* 396: 283–286.

90 Blomberg, M.R.A., Siegbahn, P.E.M., Styring, S. et al. (1997). A quantum chemical study of hydrogen abstraction from manganese-coordinated water by a tyrosyl radical: a model for water oxidation in photosystem II. *J. Am. Chem. Soc.* 119: 8285–8292.

91 Zouni, A., Witt, H.-T., Kern, J. et al. (2001). Crystal structure of photosystem II from *Synechococcus elongatus* at 3.8 Å resolution. *Nature* 409: 739–743.

92 Kamiya, N. and Shen, J.-R. (2003). Crystal structure of oxygen-evolving photosystem II from *Thermosynechococcus vulcanus* at 3.7-Å; resolution. *Proc. Natl. Acad. Sci. U.S.A.* 100: 98–103.

93 Umena, Y., Kawakami, K., Shen, J.-R., and Kamiya, N. (2011). Crystal structure of oxygen-evolving photosystem II at a resolution of 1.9 Å. *Nature* 473: 55–60.

94 Suga, M., Akita, F., Hirata, K. et al. (2015). Native structure of photosystem II at 1.95 Å resolution viewed by femtosecond X-ray pulses. *Nature* 517: 99–103.

95 Young, I.D., Ibrahim, M., Chatterjee, R. et al. (2016). Structure of photosystem II and substrate binding at room temperature. *Nature* 540: 453–457.

96 Suga, M., Akita, F., Sugahara, M. et al. (2017). Light-induced structural changes and the site of O=O bond formation in PSII caught by XFEL. *Nature* 543: 131–135.

97 Kok, B., Forbush, B., and McGloin, M. (1970). Cooperation of charges in photosynthetic O_2 evolution–I. A linear four step mechanism. *Photochem. Photobiol.* 11: 457–475.

98 Kern, J., Chatterjee, R., Young, I.D. et al. (2018). Structures of the intermediates of Kok's photosynthetic water oxidation clock. *Nature* 563: 421–425.

99 Hussein, R., Ibrahim, M., Bhowmick, A. et al. (2021). Structural dynamics in the water and proton channels of photosystem II during the S_2 to S_3 transition. *Nat. Commun.* 12: 6531–6546.

100 Wieghardt, K., Bossek, U., and Gebert, W. (1983). Synthesis of a tetranuclear manganese(IV) cluster with adamantane skeleton: $[(C_6H_{15}N_3)_4Mn_4O_6]^{4+}$. *Angew. Chem. Int. Ed.* 22: 328–329.

101 Wu, J.-Z., De Angelis, F., Carrell, T.G. et al. (2006). Tuning the photoinduced O_2-evolving reactivity of $Mn_4O_4^{7+}$, $Mn_4O_4^{6+}$, and $Mn_4O_3(OH)^{6+}$ manganese–oxo cubane complexes. *Inorg. Chem.* 45: 189–195.

102 Ruettinger, W., Yagi, M., Wolf, K. et al. (2000). O_2 Evolution from the manganese–oxo cubane core $Mn_4O_4^{6+}$: a molecular mimic of the photosynthetic water oxidation enzyme? *J. Am. Chem. Soc.* 122: 10353–10357.

103 Carrell, T., Tyryshkin, A., and Dismukes, G. (2002). An evaluation of structural models for the photosynthetic water-oxidizing complex derived from spectroscopic and X-ray diffraction signatures. *J. Biol. Inorg. Chem.* 7: 2–22.

104 Limburg, J., Vrettos John, S., Liable-Sands Louise, M. et al. (1999). A functional model for O–O bond formation by the O_2-evolving complex in photosystem II. *Science* 283: 1524–1527.

105 Limburg, J., Brudvig, G.W., and Crabtree, R.H. (1997). O_2 evolution and permanganate formation from high-valent manganese complexes. *J. Am. Chem. Soc.* 119: 2761–2762.

106 Sun, L., Raymond, M.K., Magnuson, A. et al. (2000). Towards an artificial model for Photosystem II: a manganese(II,II) dimer covalently linked to ruthenium(II) tris-bipyridine via a tyrosine derivative1Preliminary accounts of this work have been presented as invited lectures at: EUCHEM Conference, Artificial Photosynthesis, May 1998, Sigtuna, Sweden; Fourth Nordic Congress on Photosynthesis, Nov. 1998, Naantali, Finland; EBEC, July 1998, Göteborg, Sweden.1. *J. Inorg. Biochem.* 78: 15–22.

107 Huang, P., Magnuson, A., Lomoth, R. et al. (2002). Photo-induced oxidation of a dinuclear Mn2II,II complex to the Mn2III,IV state by inter- and intramolecular electron transfer to RuIIItris-bipyridine. *J. Inorg. Biochem.* 91: 159–172.

108 Nayak, S., Nayek, H.P., Dehnen, S. et al. (2011). Trigonal propeller-shaped $[Mn^{III}_3M^{II}Na]$ complexes (M = Mn, Ca): structural and functional models for the dioxygen evolving centre of PSII. *Dalton Trans.* 40: 2699–2702.

109 Mishra, A., Wernsdorfer, W., Abboud, K.A., and Christou, G. (2005). The first high oxidation state manganese–calcium cluster: relevance to the water oxidizing complex of photosynthesis. *Chem. Commun.* 1: 54–56.

110 Kotzabasaki, V., Siczek, M., Lis, T., and Milios, C.J. (2011). The first heterometallic Mn–Ca cluster containing exclusively Mn(III) centers. *Inorg. Chem. Commun.* 14: 213–216.

111 Hewitt, I.J., Tang, J.-K., Madhu, N.T. et al. (2006). A series of new structural models for the OEC in photosystem II. *Chem. Commun.* 25: 2650–2652.

112 Kanady Jacob, S., Tsui Emily, Y., Day Michael, W., and Agapie, T. (2011). A synthetic model of the Mn_3Ca subsite of the oxygen-evolving complex in photosystem II. *Science* 333: 733–736.

113 Mukherjee, S., Stull Jamie, A., Yano, J. et al. (2012). Synthetic model of the asymmetric $[Mn_3CaO_4]$ cubane core of the oxygen-evolving complex of photosystem II. *Proc. Natl. Acad. Sci. U.S.A.* 109: 2257–2262.

114 Tsui, E.Y., Tran, R., Yano, J., and Agapie, T. (2013). Redox-inactive metals modulate the reduction potential in heterometallic manganese–oxido clusters. *Nat. Chem.* 5: 293–299.

115 Tsui Emily, Y. and Agapie, T. (2013). Reduction potentials of heterometallic manganese–oxido cubane complexes modulated by redox-inactive metals. *Proc. Natl. Acad. Sci. U.S.A.* 110: 10084–10088.

116 Kanady, J.S., Lin, P.-H., Carsch, K.M. et al. (2014). Toward models for the full oxygen-evolving complex of photosystem II by ligand coordination to lower the symmetry of the Mn_3CaO_4 cubane: demonstration that electronic effects facilitate binding of a fifth metal. *J. Am. Chem. Soc.* 136: 14373–14376.

117 Zhang, C., Chen, C., Dong, H. et al. (2015). A synthetic Mn4Ca-cluster mimicking the oxygen-evolving center of photosynthesis. *Science* 348: 690–693.

118 Chen, C., Chen, Y., Yao, R. et al. (2019). Artificial Mn_4Ca clusters with exchangeable solvent molecules mimicking the oxygen-evolving center in photosynthesis. *Angew. Chem. Int. Ed.* 58: 3939–3942.

119 Yao, R., Li, Y., Chen, Y. et al. (2021). Rare-earth elements can structurally and energetically replace the calcium in a synthetic Mn_4CaO^{4-} cluster mimicking the oxygen-evolving center in photosynthesis. *J. Am. Chem. Soc.* 143: 17360–17365.

120 Willing, A., Follmann, H., and Auling, G. (1988). Ribonucleotide reductase of *Brevibacterium ammoniagenes* is a manganese enzyme. *Eur. J. Biochem.* 170: 603–611.

121 Torrents, E. (2014). Ribonucleotide reductases: essential enzymes for bacterial life. *Front. Cell. Infect. Microbiol.* 4: 1–9.

122 Lee, J.L., Biswas, S., Sun, C. et al. (2022). Bioinspired Di-Fe complexes: correlating structure and proton transfer over four oxidation states. *J. Am. Chem. Soc.* 144: 4559–4571.

11

Control of Electron Coupling and Electron Transfer Through Non-covalent Interactions in Mixed-Valence Systems

Zijie Zhou[1,2], Yixin Guo[1,2], Xiangmei Kong[1,2], Ying Wang[3], and Tianfei Liu[1,2,4]

[1]*Nankai University, College of Chemistry, State Key Laboratory of Elemento-organic Chemistry, Weijin Road No. 94, Tianjin 300071, China*
[2]*Haihe Laboratory of Sustainable Chemical Transformations, Tianjin 300192, China*
[3]*Chinese University of Hong Kong, Department of Chemistry, Ma Lin Building, Shatin, N.T., 999077 Hong Kong SAR, China*
[4]*Chinese Academy of Sciences, Shanghai Institute of Organic Chemistry, Key Laboratory of Organofluorine Chemistry, Shanghai, 200032, China*

11.1 Introduction

Mixed-valence (MV) systems with transition metal centers or organic fragments as the redox-active sites [1] have been widely used to study the fundamental intramolecular electron transfer (ET) process. The extent of electronic coupling of MV systems strongly depends on the electronic property and molecular structure of the bridge [2], such as conjugation, length, and conformation. In Chapter 10, we have discussed that MV structural units play an important role in biological systems. For example, the ferredoxin [2Fe-2S] cofactors act as the key electron transporters in biological transformations. To get a deep insight into these systems, artificial MV compounds are attractive for their abilities to simulate the long-distance electron transfer in metalloenzymes and other MV systems. Among them, those with the degree of electronic coupling being controllable by non-covalent interactions are of great importance because of their relevance to biological systems.

    A variety of MV systems with tunable electron transfer and electronic coupling by non-covalent interactions have been designed and synthesized. Research in this area is aimed at better understanding of the electronic coupling interaction and charge transport across weakly bonded interfaces. In this chapter, the control of electron transfer or electronic coupling by non-covalent interactions of both transition metals and organic systems is discussed, including those mediated by hydrogen bonds, host–guest interactions, and through-space interactions [3–5].

Mixed-Valence Systems: Fundamentals, Synthesis, Electron Transfer, and Applications, First Edition.
Edited by Yu-Wu Zhong, Chun Y. Liu, and Jeffrey R. Reimers.
© 2023 WILEY-VCH GmbH. Published 2023 by WILEY-VCH GmbH.

11.2 Electronic Coupling Through Hydrogen Bonds

11.2.1 Electronic Coupling Between Transition Metal Centers Through Hydrogen Bonds

In hydrogen bond (HB)-bridged MV systems, the electronic coupling and electron transfer between redox sites occur across the HB interface, namely, X–H···X'. To have sufficiently strong HB bridge for the MV systems, systems that are able to form multiple hydrogen bonds are often used. Unlike covalent bond bridges, the HB bridges are structurally more flexible, and the bridging H atom(s) are dynamic, which endow the system with different MV properties and electron transfer dynamics. Patmore and coworkers reported the quadruply bonded dimolybdenum compounds with a pendant lactam functional group, which are dimerized as the "dimers of dimers" (**1**) by self-complementary double hydrogen bonds in the solid state and dichloromethane (DCM) solutions (Figure 11.1) [6]. The cyclic voltammograms (CVs) of **1** in DCM show two consecutive one-electron redox processes corresponding to one-electron oxidation of the Mo_2^{4+} cores, with a comproportionation constant K_c of 233. The addition of dimethyl sulfoxide (DMSO) disrupts the hydrogen bonding, resulting in the formation of corresponding monomers and the appearance of only one two-electron redox couple (Figure 11.1b). Spectroelectrochemical studies do not present intervalence charge transfer (IVCT) bands for the MV radical cations, as a result of the proton-coupled mixed valency. According to the authors, the MV system is stabilized by electronic coupling through the hydrogen bonding motif, or by proton-coupled mixed valency, or both of these mechanisms work in concert.

Kubiak and coworkers developed a carboxylic acid HB MV system **2** with a triruthenium cluster $[Ru_3(\mu_3O)(OAc)_6(CO)(L)(nic)]$ (L = 4-dimethylaminopyridine (dmap) and nic = isonicotinic acid) as the redox site (Figure 11.2). The MV complex results from single-electron reduction of the "dimer of trimers." [7] Electrochemical studies show two apparent overlapping single-electron reductions. For the HB MV complex, an IVCT band is observed in the near-infrared region, which is attributed to the electron transfer between the two Ru_3 cluster centers across the

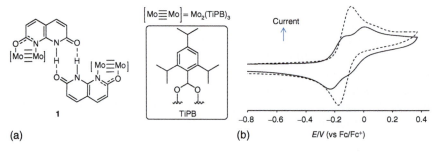

Figure 11.1 (a) Chemical structure of **1**. (b) CVs of **1** in 0.1 M NBu_4PF_6/CH_2Cl_2 before (solid curve) and after (dashed curve) the addition of 100 μl of DMSO. Source: Reproduced with permission [6]. Copyright 2013 American Chemical Society.

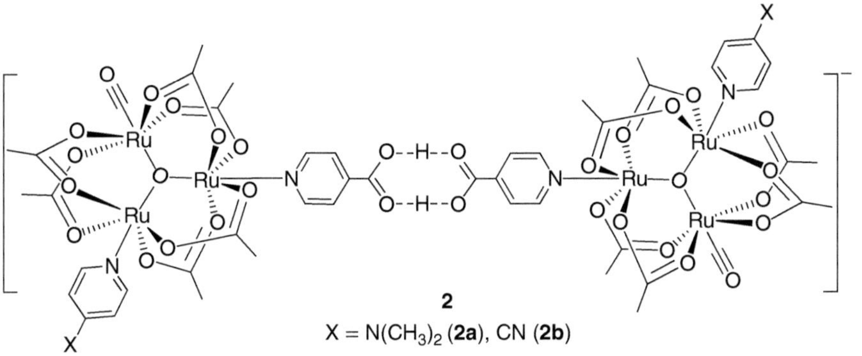

Figure 11.2 Hydrogen bond-bridged dimer of trimer **2**.

Figure 11.3 Hydrogen bond-bridged ferrocene dimer **3**.

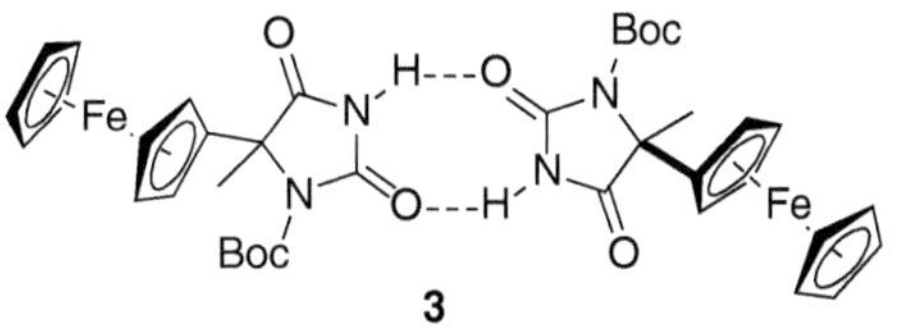

HB; accordingly, the MV system is assigned to the Robin–Day Class II regime. An electron transfer rate of less than 10^{10} s^{-1} is estimated from IR band broadening of the CO groups.

Santi and coworkers reported the synthesis and characterization of ferrocenyl methylhydantoin 5-ferrocenyl-5-methylimidazolidine-2,4-dione and its derivatives with tert-butoxycarbonyl (Boc) protecting groups (Figure 11.3) [8]. The existence of intermolecularly hydrogen-bonded dimers **3** through the double C=O···H—N linkage was confirmed in solid state via X-ray diffraction and electron spray ionization (ESI)-mass spectrometry analyses of the ferrocenyl methylhydantoin as well as in solution by ^{1}H NMR (nuclear magnetic resonance), FT-IR, and cyclic voltammetry experiments. The appearance of overlapping redox peaks was observed for **3** at a high concentration of 10 mM during the electrochemical studies, suggestive of potential electronic coupling between terminal ferrocene groups.

Tadokoro and Isogai studied the hydrogen-bonded metal complex dimers including [Re^{III}Cl$_2$(PnPr$_3$)$_2$(Hbim)]$_2$ (**4**) and [Os^{III}Cl$_2$(PnPr$_3$)$_2$(Hbim)]$_2$ (**5**) in DCM solution (Figure 11.4). CVs show a four-step reversible one-electron transfer process involving two MV states Re^{II}Re^{III} (**4**$^-$) and Re^{III}Re^{IV} (**4**$^+$) for the HB dirhenium and Os^{II}Os^{III} (**5**$^-$) and Os^{III}Os^{VI} (**5**$^+$) for the HB diosmium, respectively, suggesting the presence of efficient electronic coupling through the dual hydrogen bonding. The comproportionation constants K_cs are determined to be 1.34×10^4 and 4.10×10^3 for **4** and 9.04×10^3 and 4.10×10^3 for **5**, respectively. It is believed that the hydrogen bond bridge with a very low energy barrier for proton transfer contributes to stabilize the mixed-valence state by the proton-coupled electron transfer (PCET) [9–11]. However, no IVCT data have been disclosed for these complexes.

In Liu's group, amide–amide dual hydrogen bonds were exploited to construct the HB MV systems **6** with quadruply bonded Mo$_2$ units as the redox-active components

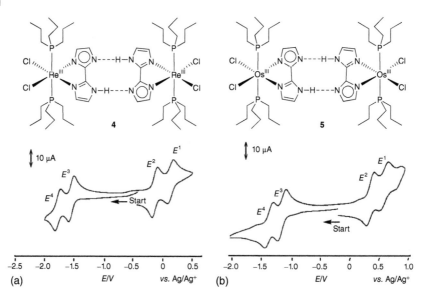

Figure 11.4 Hydrogen bond-bridged dimers (a) **4** and (b) **5** and the corresponding CV in DCM. Source: Reproduced with permission [9]. Copyright 2019 Royal Society of Chemistry.

(Figure 11.5). The neutral HB dimers were structurally characterized in solid state by X-ray diffraction, and the MV states of the HB Mo$_2$ dimers were confirmed by electrochemical and spectroscopic measurements in DCM. It is worthy to note that all reported four HB dimers exhibit a characteristic IVCT band in the near-IR region (Figure 11.5b), while this important optical feature is frequently absent in most reported HB MV systems. With the availability of the IVCT data, the electronic coupling matrix elements, H_{ab} = 130–460 cm^{-1}, were determined from the Mulliken–Hush expression [12]. The rate constants for electron transfer over a distance of 12.5 Å across the HB interface are calculated on the order of ~10^{10} s^{-1} based on Marcus–Hush theory, which are confirmed by simulation of the ν(NH) vibrational band broadening. As is well known, electron transfer in HB systems is usually accompanied by a proton transfer process, namely, PCET, which is the dominant mechanism for ET crossing HB [13]. Interestingly, in this study on a series of HB complexes, while the PCET mechanism is confirmed, ET process without involvement of HB breakage and formation, termed as proton-uncoupled electron transfer (PUET), is evidenced in the IR spectra. The PUET mechanism has been largely overlooked in the past but has been observed in the sequential ET steps along the amino acid residues in photoactivation of the DNA photolyase [14]. According to the study, the mechanistic choice between PCET and PUET is determined by the strengths of electronic coupling and hydrogen bonding. Strong coupling and HB lead to the PUET pathway.

Recently, Winter and coworkers presented the divinylphenylene diruthenium complex **7** and the corresponding tetraruthenium metallacycle **8** containing hydroxyl groups between two ruthenium components in proximity (Figure 11.6) [15]. Two and three reversible redox waves arising from the stepwise oxidations of the

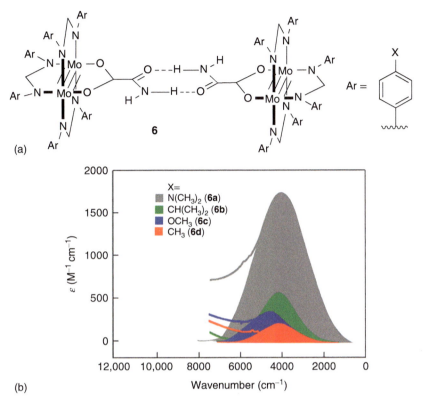

Figure 11.5 (a) Chemical structures of dimer **6** of di-Mo complexes and (b) IVCTs of **6**$^+$ with different terminal substituents indicated.

ruthenium components were observed, as a result of the stabilization of the one- and three-electron oxidized MV states by the intracyclic OH···O hydrogen bonds. In addition, these MV cations display distinct infrared band shifts of their Ru(CO) tags, but no IVCT band could be discerned. This fact was ascribed a redox-induced shuffling of the OH···O hydrogen bonds to the remaining, more electron-rich, reduced redox site (Figure 11.6b).

11.2.2 Electronic Coupling Between Organic Fragments Through Hydrogen Bonds

Electronic coupling and electron transfer between MV organic fragments through non-covalent interactions are another important and intriguing topic. Hydrogen bonds are an important pattern for the non-covalent linking bonds [16]. In 2014, Kikuchi and coworkers synthesized three triarylamine derivatives, which formed as dimers **9** in both solid states and solution through quadruple H-bonds (Figure 11.7) [17]. The formation of the HB MV species is suggested by electrochemical and mechanism investigations. The comproportionation constants for the MV state are calculated to be 1.75×10^{11}, 2.85×10^7, and 211 for those with the terminal R group of H (**9a**), OCH$_3$ (**9b**), and CN (**9c**), respectively.

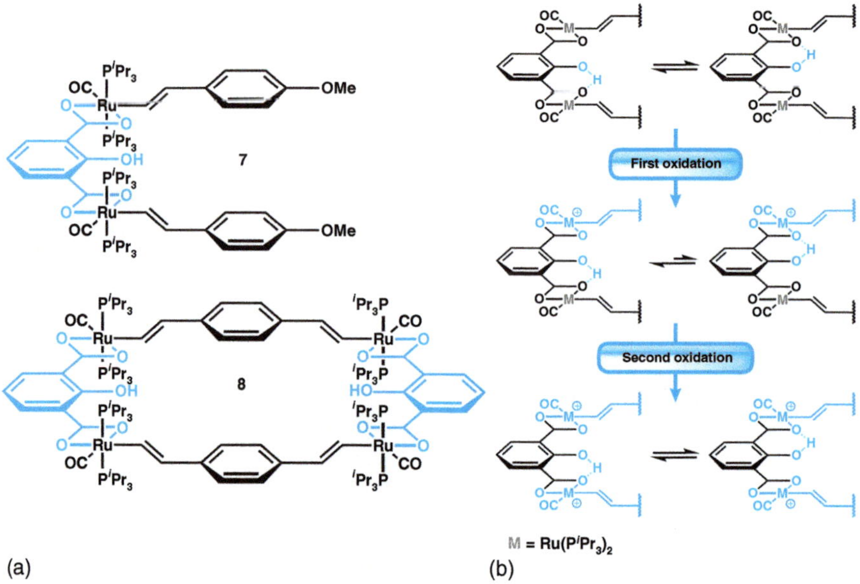

Figure 11.6 **(a)** Chemical structures of **7** and **8**. (b) Schematic representation of the oxidation-induced changes of the OH···O hydrogen bond patterns in **7** and **8**. Source: Reproduced with permission [15]. Copyright 2020 American Chemical Society.

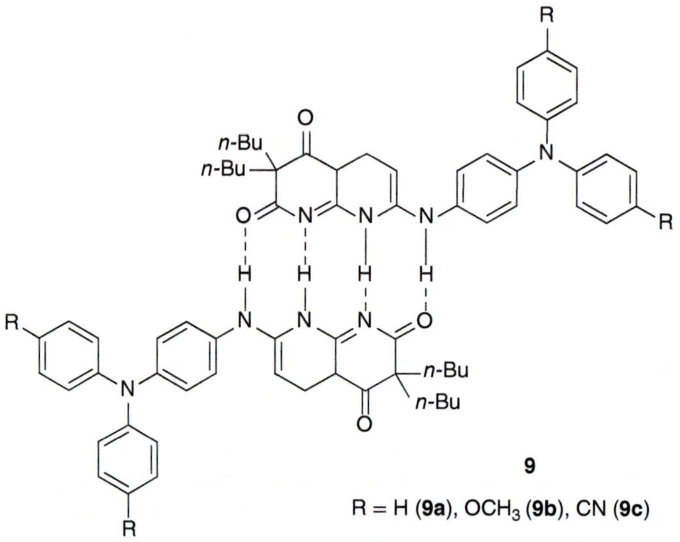

9

R = H (**9a**), OCH_3 (**9b**), CN (**9c**)

Figure 11.7 Quadruple H-bond-bridged dimer **9**.

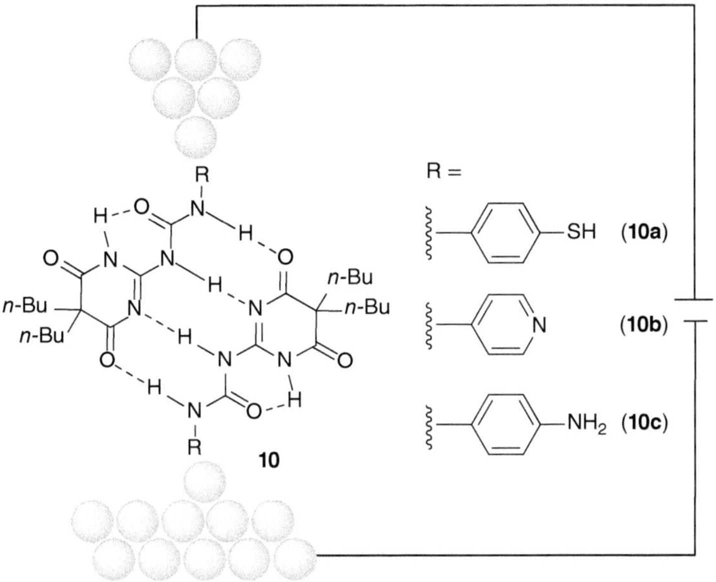

Figure 11.8 The self-complementary UPy derivatives **10** anchored on the probes of the scanning tunneling microscope.

In 2016, a series of similar self-complementary ureido pyrimidinedione (UPy) derivatives **10** modified with different aurophilic anchoring groups have been reported (Figure 11.8) [18]. The formation of quadruple hydrogen bonds was evidenced by ^{1}H-NMR spectrum and controlled experiments, which can be effectively modulated by the polarity of the solvent environment. Most importantly, the electron transport properties through the quadruple hydrogen bonds were studied employing the scanning tunneling microscopy break junction (STMBJ) technique. The molecule terminated with a thiol group (**10a**) shows the optical electron transport properties, with a statistical molecular conductance value approaching 10^{-3} G_0 (G_0 refers to the quantum conductance). This again suggests the effectiveness of hydrogen bonds in mediating electron transfer.

In addition to MV systems, the electron transfer studies between electron donor and acceptor through hydrogen bonds have received much interest [19–27]. These studies mimic the electron transfer via H-bonds that is often seen in biological systems. In such systems, proton transfer or proton delocalization, accompanied by electron transfer, is a fundamental issue that can be modeled by synthetic molecules. For instance, a series of electroactive, urea-containing double hydrogen-bonded dimers **11** have been investigated (Figure 11.9), in which two amide groups provide two H-donors that can bond with two H-acceptors at appropriate space via hydrogen bonds [19, 20]. It is demonstrated that proton transfer may be avoided in electrochemically controlled H-bonding [21, 22]. However, in their study of the phenylenediamine ureas, known as Zimmerman's system (**12**), the result shows that the irreversibility may be evitable when proton transfer accompanies electron transfer in a hydrogen-bonding system, and the proton transfer may also not be

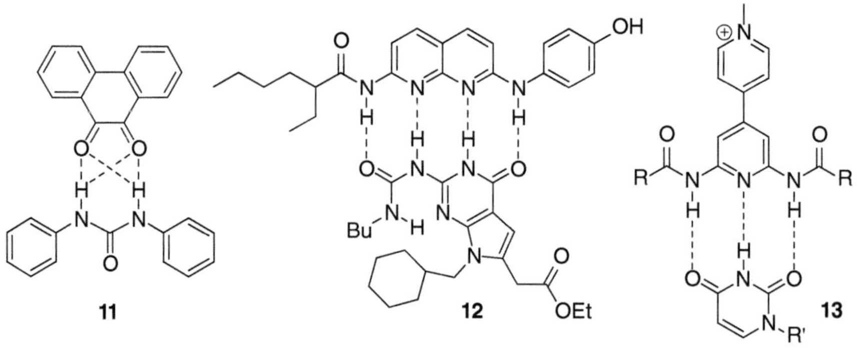

Figure 11.9 Hydrogen bond-linked electron donor–acceptor pair.

avoidable [23]. In a triple hydrogen bond-linked donor–acceptor pair **13**, Smith and coworkers used the ionic component to create favorable H-bonds and strengthen the overall binding significantly [26].

11.3 Modulation of Electronic Coupling via Host–Guest or Through-Space Interaction

The host–guest interactions have been the research focus for decades in supramolecular chemistry. When electro-active units are introduced into either host or guest components, electronic coupling may occur in host–guest systems via through-bond or through-space interactions. It is a highly interesting topic to examine the effect of guest on the electronic coupling of the resulting host–guest systems [28–31]. In 1994, Saalfrank and coworkers reported the Fe(II,III) MV tetranuclear cage complexes **14a** and **14b** with the inclusion of a guest NH_4^+ (Figure 11.10) [28]. These complexes were synthesized by spontaneous self-assembly from the reaction of dialkyl malonates with methyllithium/iron(II) chloride and oxalyl chloride, followed by workup with aqueous ammonium chloride. The host–guest structures were confirmed by single-crystal X-ray analysis. Their MV characteristic was clearly confirmed from their Mössbauer spectra. According to the electron paramagnetic resonance (EPR) spectroscopic results, a weak coupling is observed between the iron metal centers. These complexes show four stepwise one-electron redox waves, probably because of the short Fe–Fe distance in these systems. In contrast, another similar cage complex with a shorter ligand length and without the inclusion of the NH_4^+ guest only display one four-electron redox couple, suggesting the interesting electrochemical feature of the host–guest system.

In 2009, Yamashita and coworkers reported a host–guest complex **17**, which was prepared from the macrocyclic tetranuclear platinum(II) complex **15** and the platinum(IV) complex **16** (Figure 11.11) [32] In complex **17**, the Pt(IV) components act both as the guest molecules of the platinum(II) macrocycle and as the pillar linkers to form quasi-1-D halogen-bridged Pt(II)/Pt(IV) chains. Electronic spectra and polarized reflectance measurements suggest that **17** exhibits Pt(II)/Pt(IV) IVCT bands

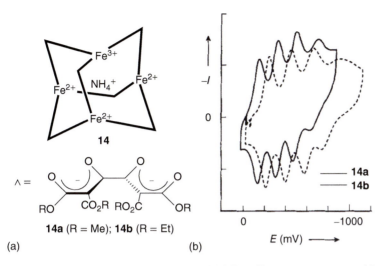

Figure 11.10 (a) Chemical structures and (b) CVs of host–guest systems **14a** and **14b**. Source: Reproduced with permission [28]. Copyright 1994 Wiley VCH.

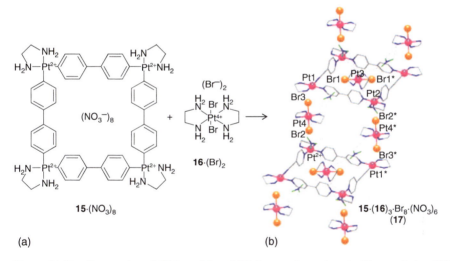

Figure 11.11 Preparation of **17** from **15** and **16**. Source: Reproduced with permission [32]. Copyright 2009 Wiley VCH.

from the isolated [$Pt^{II}\cdots X\text{-}Pt^{IV}\text{-}X\cdots Pt^{II}$] moiety. Although the IVCT transitions of **17** come from the through-bond interactions, this work demonstrates that supramolecular host–guest systems offer an effective means to construct MV materials [33].

Félix, Thomas, and coworkers reported the modulation of the electron transfer properties of a MV system via host–guest chemistry (Figure 11.12) [34]. Studies on the binding and redox properties of the trinuclear Ru(II) macrocycle **18**$^{3+}$ show that specific MV states of the systems are accessible and controllable by host–guest mediation. Electronic interaction in the macrocycles in different MV states is different because of the distinctive molecular architecture. Electron hopping is found in

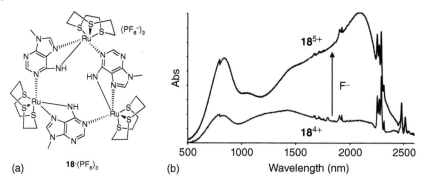

Figure 11.12 (a) Tri-Ru(II) macrocycle host **18**(PF$_6$)$_3$. (b) Absorption spectral changes of **18**$^{4+}$ upon the addition of 1 equiv of fluoride anion. Source: [34]/Royal Society of Chemistry/Licensed under CC BY 3.0.

[Ru^{II}$_2$Ru^{III}]$^{4+}$, while the [Ru^{II}Ru^{III}$_2$]$^{5+}$ state is electronically delocalized. When the counteranion of the system was changed from PF$_6$$^-$ to F$^-$, Cl$^-$, Br$^-$, and I$^-$, the three Ru(II/III) redox waves show potential shifts up to 120 mV because of the formation of different host–guest complexes. As a result, the comproportionation constants for the MV **18**$^{4+}$ and **18**$^{5+}$ states change distinctly. Interestingly, when 1 equiv of F$^-$ was added to the electrochemically generated MV **18**$^{4+}$ state, absorption spectral changes suggested that **18**$^{4+}$ was transformed into the delocalized MV **18**$^{5+}$ state (Figure 11.12b). This means the MV states and the extent of electron delocalization can be switched by an ion without the need to change the electrochemical potential.

In Liu's group, a quadrupolar host–guest interaction was employed for the mediation of electronic coupling and electron transfer in MV Mo$_2$ dimers [35]. In this MV system, covalently bonded dimolybdenum complex **19** linked by a thienylene group was examined (Figure 11.13). It is interesting that with this molecular skeleton, there is a cleft between the two Mo$_2$ units, which is capable of trapping an aromatic solvent molecule through quadrupolar interactions with the thienylene bridge. Therefore, in aromatic solvents such as C$_6$H$_6$, C$_6$H$_5$CH$_3$, and C$_6$F$_6$, the IVCT bands are blue-shifted and decrease in intensity, showing electronic decoupling effect. Optical analyses based on the Marcus–Hush theories indicated that the host–guest interaction gates the intramolecular electron transfer, lowering the electron transfer rate (k_{et}) by about 2 orders of magnitude, for example, from k_{et} = 3.6 × 10^{12} s^{-1} in CH$_2$Cl$_2$ to 6.0 × 10^{10} s^{-1} in C$_6$F$_6$ for one of the compounds in the series. Intermolecular interactions are conventionally considered to be electrostatic in nature. More recently, through a variety of supramolecular MV systems with a C$_6$H$_6$ molecule encapsulated in a series of Mo$_2$ dimers bridged by oxalate or thiolated derivatives, these authors characterized the covalency of intermolecular interactions in the van der Waals distance [36]. It reveals that single solvating benzene molecule decouples dramatically the electronic interactions between MV Mo$_2$ units. Ab initio combined with density functional theory (DFT) calculations suggest that the electronic states of the MV compound can be altered by the intermolecular orbital overlaps between the bridge and C$_6$H$_6$ through intermolecular nuclear dynamics.

11.3 Modulation of Electronic Coupling via Host–Guest or Through-Space Interaction | 359

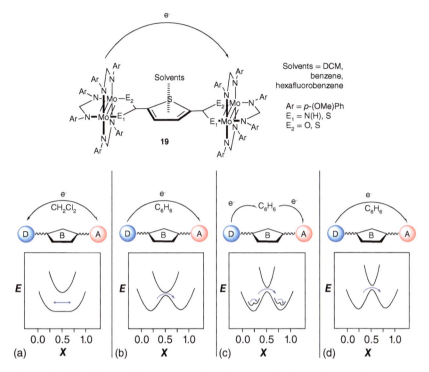

Figure 11.13 Schematic illustration of quadrupole effects on thermal electron transfer in the MV Mo$_2$ dimers. Source: Adapted from ref. [35].

The above results demonstrate the modulation of electronic coupling in inorganic systems via host–guest or through-space interactions. The uses of these interactions to stabilize the MV states or mediate the electronic coupling of organic redox-active systems have received attention as well. For instance, the reaction of the flexible viologen cyclophane host **20**$^{4+}$ and the guest molecule *N,N'*-dimethyl-4,4'-bipyridinium (**21**$^{2+}$), in the presence of Zn, gave a host–guest complex **22** with a formula of [(**20**)(**21**)$_2$]$^{3(+\cdot)(2+)}$ (Figure 11.14a) [37]. In this case, three MV radical cations are shared by four bipyridine components as confirmed by single-crystal X-ray analysis. In another example, the treatment of the viologen cyclophane host **23**$^{4+}$ and **21**$^{2+}$, with the aid of cobaltocene (CoCp$_2$), provided the MV superstructure **24** with a formula of [**21**⊂(**23**)$_2$]$_3$$^{2(+\cdot)}$ (Figure 11.14b) [38]. In this system, two MV radical cation charges are distributed statistically among a total of 15 bipyridine units assembled into two **21**$^{\cdot+}$⊂(**23**^{0})$_2$ and one neutral **21**^{0}⊂(**23**^{0})$_2$. These studies demonstrate that long-range electron delocalization in solid-state structures could be achieved by a supramolecular strategy through precise tuning of the redox states in host–guest systems.

Although the through-space interactions have been used to modulate the electronic coupling or enhance the stability of host–guest supramolecular frameworks, the reports on the direct measurement of the through-space coupling are limited. Two examples are shown in Figure 11.15 (see related discussions in Chapter 9 of this

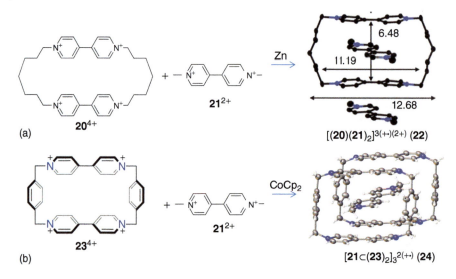

Figure 11.14 (a, b) The preparation of host–guest systems (a) **22** and (b) **24**. Counter anions are PF_6^- for all. Source: Reproduced with permission for (a) [37]. Copyright 2015 Royal Society of Chemistry. Reproduced with permission for (b) [38]. Copyright 2018 American Chemical Society.

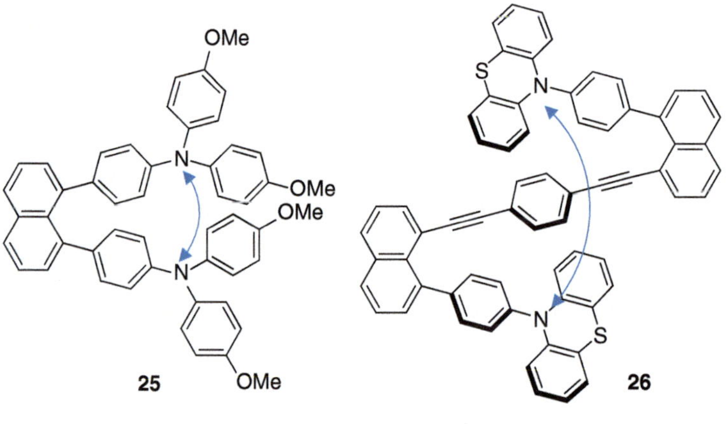

Figure 11.15 Diamine compounds **25** and **26** showing through-space coupling.

book). Compound **25** is a diamine molecule with two identical triarylamine moieties substituted on the 1,8-positions of naphthalene [39]. It shows two well-separated redox waves at +0.18 and +0.35 V vs. $Fc^{0/+}$ (Fc = ferrocene) assigned to the stepwise oxidations of two triarylamine units. In addition, broad and weak IVCT transition is observed in the wavelength region between 1300 and 2200 nm for the MV state **25$^{+\cdot}$**, suggesting that through-space charge transfer pathway plays an important role. Compound **26** contains two phenothiazine (PTZ) moieties connected via 1,8-naphthalene spacers to a central phenyl group to form a PTZ-phenyl-PTZ triple-decker configuration [40]. Although the redox potentials of two PTZ moieties are not well separated, the mono-oxidized **26**$^{+\cdot}$ displays distinct IVCT band in the near-infrared region. The electronic coupling is estimated to be around $110\,\text{cm}^{-1}$, as a result of the through-space coupling directly across the stacked structure. These results illustrate the potential of through-space interaction for long-range electron transfer and future applications in molecular electronic devices.

11.4 Conclusion

In this chapter, mixed-valence systems in which the electronic coupling and electron transfer are mediated by non-covalent interaction such as hydrogen bonding, host–guest interaction, and through-space interaction are surveyed. Through the well-defined molecular systems, hydrogen bonds can efficiently transport electrons between MV redox sites. The host–guest interactions can be used to tune the intramolecular electronic interactions. We have also presented a few of examples showing the through-space electron transfer pathway. However, different from covalently bonded MV systems, non-covalent interactions are intrinsically dynamic and flexible, which raises profound issues in mediating the intra- and intermolecular coupling. The study in this area of mixed-valence chemistry has advanced our knowledge in molecular science, which is helpful in elucidation of redox and charge transfer processes in biological systems.

Acknowledgment

Funding for this study was provided by the National Natural Science Foundation of China (T. L., grant no. 22171141, 22193010, and 22193014). Support from the State Key Laboratory of Elemento-organic Chemistry, College of Chemistry at Nankai University, the Haihe Laboratory of Sustainable Chemical Transformations, and Key Laboratory of Organofluorine Chemistry at Shanghai Institute of Organic Chemistry, Chinese Academy of Sciences (T.L.), and CUHK startup funding (Y.W.) is also acknowledged.

References

1 Day., J. o. t. A. C. S.(1991). *Mixed Valency Systems: Applications in Chemistry, Physics and Biology*. Springer Netherlands.
2 Figueira-Duarte, T.M., Lloveras, V., Vidal-Gancedo, J. et al. (2007). *Chem. Commun.* 4345–4347.
3 Richardson, D.E. and Taube, H. (1984). *Coord. Chem. Rev.* 60: 107–129.
4 Hegner, F.S., Galan-Mascaros, J.R., and Lopez, N. (2016). *Inorg. Chem.* 55: 12851–12862.
5 Krajewska, C.J., Kavanagh, S.R., Zhang, L. et al. (2021). *Chem. Sci.* 12: 14686–14699.

6 Wilkinson, L.A., McNeill, L., Meijer, A.J., and Patmore, N.J. (2013). *J. Am. Chem. Soc.* 135: 1723–1726.
7 Canzi, G., Goeltz, J.C., Henderson, J.S. et al. (2014). *J. Am. Chem. Soc.* 136: 1710–1713.
8 Bisello, A., Cardena, R., Rossi, S. et al. (2017). *Organometallics* 36: 2190–2197.
9 Tadokoro, M., Isogai, K., Harada, S. et al. (2019). *Dalton Trans.* 48: 535–546.
10 Tyburski, R., Liu, T., Glover, S.D., and Hammarström, L. (2021). *J. Am. Chem. Soc.* 143: 560–576.
11 Liu, T., Kong, X., and Zhou, Z. (2021). *Chin. J. Org. Chem.* 41: 3844–3879.
12 Hush, N.S. (1968). *Electrochim. Acta* 13: 1005–1023.
13 Cheng, T., Shen, D.X., Meng, M. et al. (2019). *Nat. Commun.* 10: 1531.
14 Aubert, C., Vos, M.H., Mathis, P. et al. (2000). *Nature* 405: 586–590.
15 Fink, D., Staiger, A., Orth, N. et al. (2020). *Inorg. Chem.* 59: 16703–16715.
16 Heckmann, A. and Lambert, C. (2012). *Angew. Chem. Int. Ed.* 51: 326–392.
17 Tahara, K., Nakakita, T., Katao, S., and Kikuchi, J. (2014). *Chem. Commun.* 50: 15071–15074.
18 Wang, L., Gong, Z.-L., Li, S.-Y. et al. (2016). *Angew. Chem. Int. Ed.* 55: 12393–12397.
19 Smith, D.K. (2017). *Curr. Opin. Electrochem.* 2: 76–81.
20 Ge, Y., Lilienthal, R.R., and Smith, D.K. (1996). *J. Am. Chem. Soc.* 118: 3976–3977.
21 Li, Y., Park, T., Quansah, J.K., and Zimmerman, S.C. (2011). *J. Am. Chem. Soc.* 133: 17118–17121.
22 Ji, X., Jie, K., Zimmerman, S.C., and Huang, F. (2015). *Polym. Chem.* 6: 1912–1917.
23 Clare, L.A. and Smith, D.K. (2016). *Chem. Commun.* 52: 7253–7256.
24 Beijer, F.H., Sijbesma, R.P., Kooijman, H. et al. (1998). *J. Am. Chem. Soc.* 120: 6761–6769.
25 Cedano, M.R. and Smith, D.K. (2018). *J. Org. Chem.* 83: 11595–11603.
26 Choi, H., Baek, K., Toenjes, S.T. et al. (2020). *J. Am. Chem. Soc.* 142: 17271–17276.
27 Lin, C.Y. and Boxer, S.G. (2020). *J. Phys. Chem. B.* 124: 9513–9525.
28 Saalfrank, R.W., Burak, R., Breit, A. et al. (1994). *Angew. Chem. Int. Ed.* 33: 1621–1623.
29 Ibrahim, A.M.A. (1999). *Polyhedron* 18: 2711–2721.
30 Nijhuis, C.A., Dolatowska, K.A., Ravoo, B.J. et al. (2007). *Chem. Eur. J.* 13: 69–80.
31 Zhang, Q., Ma, J.-P., Wang, P. et al. (2008). *Cryst. Growth Des.* 8: 2581–2587.
32 Hirahara, E., Takaishi, S., and Yamashita, M. (2009). *Chem. Asian J.* 4: 1442–1450.
33 Kobayashi, Y., Jacobs, B., Allendorf, M.D., and Long, J.R. (2010). *Chem. Mater.* 22: 4120–4122.
34 Zubi, A., Wragg, A., Turega, S. et al. (2015). *Chem. Sci.* 6: 1334–1340.

35 Mallick, S., Cao, L., Chen, X. et al. (2019). *iScience* 22: 269–287.
36 Mallick, S., Zhou, Y., Chen, X. et al. (2022). *iScience* 25: 104365.
37 Berville, M., Karmazin, L., Wytko, J.A., and Weiss, J. (2015). *Chem. Commum.* 51: 15772–15775.
38 Liu, Z., Frasconi, M., Liu, W.G. et al. (2018). *J. Am. Chem. Soc.* 140: 9387–9391.
39 Schmidt, H.C., Spulber, M., Neuburger, M. et al. (2016). *J. Org. Chem.* 81: 595–602.
40 Schmidt, H.C., Guo, X., Richard, P.U. et al. (2018). *Angew. Chem. Int. Ed.* 57: 11688–11691.

12

Stimulus-Responsive Mixed-Valence and Related Donor–Acceptor Systems

Jiang-Yang Shao[1] and Yu-Wu Zhong[1,2]

[1] Beijing National Laboratory for Molecular Sciences, CAS Key Laboratory of Photochemistry, Institute of Chemistry, Chinese Academy of Sciences, 2 Bei Yi Jie, Zhong Guan Cun, Haidian District, Beijing 100190, China
[2] School of Chemical Sciences, University of Chinese Academy of Sciences, No.19(A) Yuquan Road, Shijingshan District, Beijing 100049, China

12.1 Introduction

Electron transfer is ubiquitous in nature and electronic devices. Since the invention of the Creutz–Taube ion [1], studies of mixed-valence (MV) compounds have attracted intense interest in how charge distribution can change via intramolecular electron transfer [2, 3]. MV compounds, in essence, are electrochemically driven switchable systems. The electron transfer only occurs in the MV state but not in the lower or higher homovalent state. We discuss in this chapter the switchable electron transfer of MV compounds and related donor–acceptor systems apart from the electrochemical stimulus. This is realized by introducing switchable groups into MV frameworks that can respond to photons, chemical species, protons, etc. [4–6]. In some cases, the switchable MV electron transfer is accompanied by distinct changes in the degree of electronic coupling and the intervalence charge transfer (IVCT) absorptions in the visible to infrared region, making them potentially useful in intelligent electronic and sensing devices.

12.2 Photoswitchable Compounds

Among the potential systems that can be regulated by external stimuli, photo-driven switches with a reversible transformation between two different isomers receive increasing attention. Simple MV compounds consist of two redox components connected by a π-bridge. Photoswitchable MV compounds usually use a photoisomerizable unit as the π-bridge [4]. Photo-induced structural isomerization in the bridge changes the degree of π-conjugation and thus reversibly modulates the electronic communication between redox termini [7]. Various photoisomerizable units have been reported to date as the π-bridge for switchable systems, including

Mixed-Valence Systems: Fundamentals, Synthesis, Electron Transfer, and Applications, First Edition.
Edited by Yu-Wu Zhong, Chun Y. Liu, and Jeffrey R. Reimers.
© 2023 WILEY-VCH GmbH. Published 2023 by WILEY-VCH GmbH.

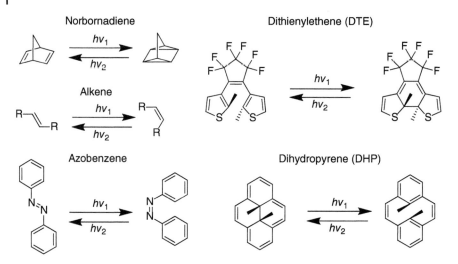

Figure 12.1 Photoisomerizable units investigated in the context of photoswitchable MV compounds.

norbornadiene [8, 9], dithienylethene (DTE) [10, 11], azobenzene [12, 13], alkene [14, 15] and dimethyldihydropyrene [16, 17] (Figure 12.1).

An early example of a photoswitchable MV compound was reported by Launay and coworkers in 1996 [18]. They prepared a bis(pentaammineruthenium) complex **1** containing a dicyanonorbornadiene moiety as the bridge unit. Irradiation of complex **1** with ultraviolet (UV) or visible light induced the photoisomerization of the bridging unit from its initial norbornadiene form to the quadricyclane form to give **1'** (Figure 12.2a). After being transformed into **1'**, the initial Ru(II)/Ru(III) wave of **1** at +0.53 V vs. saturated calomel electrode (SCE) shifted to +0.40 V because of a decreased back-bonding interaction between the metal unit and the coordinated dicyanoquadricyclane ligand. At the MV Ru(II,III) state, the dicyanonorbornadiene complex **1** exhibited an IVCT band at 1400 nm with an estimated electronic coupling (H_{ab}) of 185 cm^{-1} by Hush analysis. In contrast, no IVCT band was detected for the quadricyclane-bridged diruthenium complex **1'** due to the insignificant metal–metal electronic coupling.

Cyclometalated MV systems are known to display stronger metal–metal electronic coupling with respect to their noncyclometalated analogs [19]. Launay and coworkers further connected two cyclometalated ruthenium units to norbornadiene via ethynyl spacers to provide complex **2** (Figure 12.2b) [20]. Complex **2** exhibited a single reversible redox wave at +0.50 V because of the long metal–metal separation, corresponding to the oxidations of two ruthenium sites. This, however, does not preclude the generation of an appreciable concentration of the MV form upon partial oxidation. The comproportionation constant (K_C) for the equilibrium [Ru(II)-Ru(II)]$^{2+}$ + [Ru(III)-Ru(III)]$^{4+}$ ↔ 2 [Ru(II)-Ru(III)]$^{3+}$ was determined to be 13, corresponding to ca. 65% of the MV form at the half-oxidation state. This was accompanied by the appearance of an IVCT band in the wavelength range of 1300–2200 nm. The estimated H_{ab} of complex **2** (0.068 eV) is three times larger

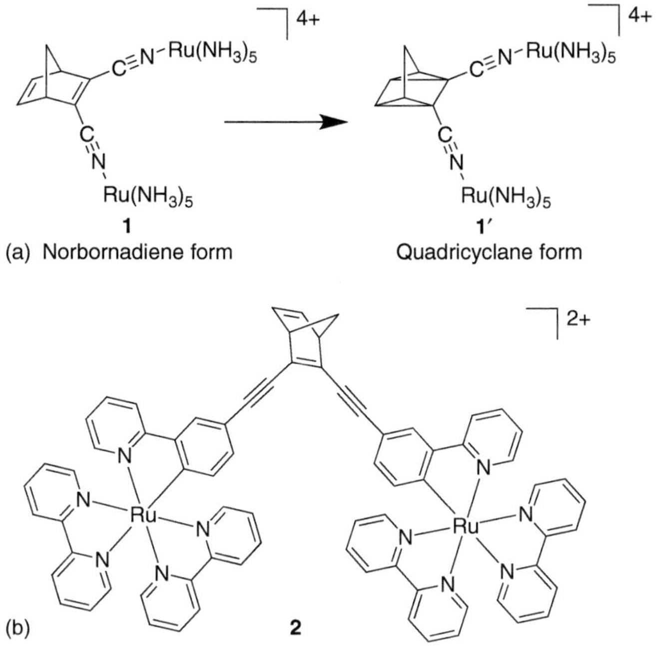

Figure 12.2 (a) Photoswitchable compound **1** based on the isomerization of the norbornadiene and quadricyclane forms. (b) Cyclometalated diruthenium complex **2** bridged by norbornadiene.

than that of complex **1** (0.023 eV; 185 cm^{-1}), which evidences the advantage of the cyclometalated unit in enhancing the electronic coupling. Unfortunately, complex **2** could not be isomerized into the quadricyclane form when it was irradiated at 254 or 365 nm either in methanol or acetonitrile. This was probably caused by the presence of the conjugated phenylethynyl-bridging parts, which quenched the excited state of the norbornadiene unit.

DTEs have received particular attention because of their appealing switching properties, such as long-term chemical stability and pronounced photochromic behavior [21–23]. The DTE unit can reversibly photoisomerize between a weakly π-conjugated open form and a strongly conjugated closed form upon UV or visible light irradiation, respectively. The closed isomer is more effective in mediating electron transfer than the open form. A large number of MV molecular systems with two redox-active centers bridged with the DTE moiety have been reported. It is demonstrated that the redox potential splitting becomes much more pronounced in the ring-closed form compared with that in the open form. Therefore, metal–metal electronic communication can be successfully modulated through alternative UV and visible light irradiation on the photoswitchable DTE moiety.

Similar to complex **2**, an ethynyl-linked cyclometalated ruthenium complex **3** bridged with a DTE unit was reported [20] (Figure 12.3a). Complex **3** can be photoisomerized from the open form to the closed form by UV irradiation with a photocyclization efficiency of around 75%. No detectable metal–metal electronic

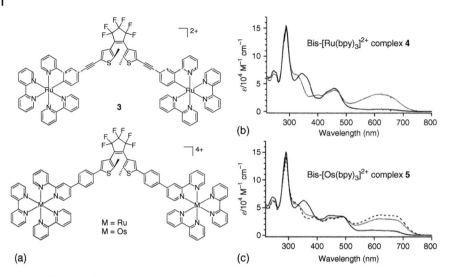

Figure 12.3 (a) Diruthenium complexes **3–5** based on dithienylethene. (b) UV–vis spectra of complex **4o** (the open form) (–) and a solution of **4o/4c** (the closed form) in the photostationary state (· · ·). Source: Reproduced with permission [24]. Copyright 2004 American Chemical Society. (c) UV–vis spectra of complex **5o** (–) and a solution of **5o/5c** in the photostationary state (· · ·). The UV–vis spectrum of pure **5c** (---) was constructed by extrapolation to 100% photoconversion. Source: Reproduced with permission [24]. Copyright 2004 American Chemical Society.

coupling is present in the open form. Upon photoisomerization to the closed DTE form, the closed isomer exhibited a noticeable IVCT transition with a H_{ab} of 200 cm^{-1} (0.025 eV).

In order to improve the efficiency of the photocyclization process, Adamo and colleagues incorporated the conventional noncyclometalated [Ru(bpy)$_3$]$^{2+}$ (bpy = 2,2′-bipyridine) as the sensitizer to realize the visible light-triggered ring closure of the DTE unit [25]. They synthesized a DTE-bridged diruthenium complex **4** with two [Ru(bpy)$_3$]$^{2+}$ termini that showed near 100% efficiency of the photochemical ring-closing process upon either direct UV irradiation of the intraligand (IL) transition of the switching unit or by irradiation into the metal-to-ligand charge-transfer (MLCT) band in the visible region and subsequent energy transfer from the excited metal center to the switching unit [25] (Figure 12.3b). As an extension of this work, De Cola and coworkers reported the photoisomerization behavior of bis-[Os(bpy)$_3$]$^{2+}$ complex **5** connected by the DTE unit [24]. The ring closure efficiency became lower after replacing the Ru(II) centers with Os(II) (Figure 12.3c). The author believed that the energy of the Os-based ^{3}MLCT state was lower than that of the ^{3}IL state of the switching unit, which prevented the energy transfer from ^{3}MLCT to the ^{3}IL state. Though **4** and **5** have been shown to undergo photocyclization, no electronic coupling has been reported for them in either open or closed forms.

The above works are mainly focused on ruthenium and osmium metal complexes. The radical cations of bis(triarylamine) derivatives with a π-conjugated bridge are

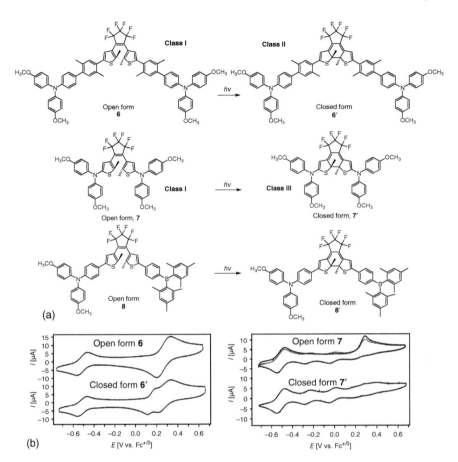

Figure 12.4 (a) Isomerization of triarylamine-terminated dithienylethene derivatives **6–8**. (b) Cyclic voltammograms measured in acetonitrile solutions of **6** (left) and **7** (right). Source: Reproduced with permission [28]. Copyright 2011 American Chemical Society.

another type of well-known MV compounds [26, 27]. Triarylamines bearing electron-rich groups, e.g. the methoxy group, show reversible electrochemistry, and the IVCT bands of bis(triarylamine) monocations are easy to detect. Wenger and coworkers reported in 2011 that the DTE-bridged molecule **6** has appended triaryamine units and *p*-xylene spacers between the amine units and the DTE bridge (Figure 12.4a) [28]. The open form of **6** displays a redox wave at +0.27 V vs. Fc$^{+/0}$, which is assigned to the oxidations of two triarylamine units (Figure 12.4b). The corresponding monocation is a Robin–Day class I system with the odd electron being localized on one triarylamine component. Upon photoisomerization to the closed isomer **6′**, a potential splitting (ΔE) of 150 mV is observed between the two amine oxidation waves of the close former, suggesting the presence of significant mutual interaction between two triarylamine units. In this case, partial charge delocalization occurs in the molecule as a class II MV system with a H_{ab} of 476 cm^{-1}.

Significantly stronger electronic coupling has been observed for MV systems in which the nitrogen centers of tertiary amines are attached directly to bridging thiophene units [29–31]. In this sense, compound **7**, with the nitrogen atoms being directly attached to the DTE spacer, was prepared. Compound **7** shows the similar amine oxidation potential (+0.29 V vs. Fc$^{+/0}$) as **6** (Figure 12.4b). The closed isomer **7′** shows a larger potential splitting between the two amine oxidation waves ($\Delta E = 230$ mV) than **6′** ($\Delta E = 150$ mV), suggestive of the enhanced coupling in **7′**. In this case, the photocyclization of the DTE core induces a changeover from class I (**7**) to class III (**7′**) MV behavior. A related donor-acceptor type of compound **8** was also reported, which contained the triarylamine unit on one end and the electron-deficient triarylboryl group on the other end of the DTE bridge [32]. A distinct N → B charger transfer (CT) transition was observed for the open isomer **8**. For the closed isomer **8′**, however, the CT band could not be identified. The addition of F$^-$ led to distinct absorption spectral changes of both **8** and **8′**.

During the studies on the effect of the carbon bridge length and the electronic properties of metal end-caps on the degree of electronic coupling, Akita and coworkers synthesized two dinuclear DTE derivatives **9** and **10** with iron or ruthenium metal centers connected by the C ≡ C spacers [33, 34] (Figure 12.5a). These two

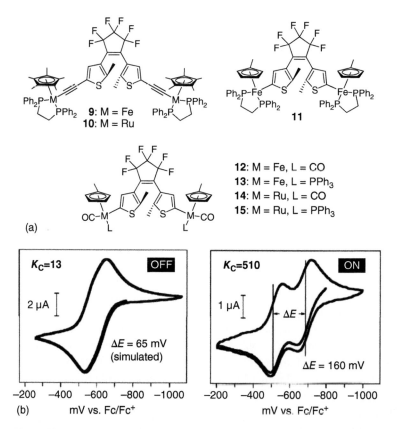

Figure 12.5 (a) Molecular structures of complexes **9–15**. (b) Cyclic voltammograms of **9** in the OFF (left) and ON (right) state in CH$_2$Cl$_2$. Source: Reproduced with permission [33]. Copyright 2007 The Royal Society of Chemistry.

organometallic molecules could be switched "on" and "off" by UV and visible light irradiation, respectively. The photoisomerization of the di-iron complex **9** from the open to the closed form increases the K_C from 13 to 510 (Figure 12.5b). The switching factor (SF = $K_{C,closed}/K_{C,open}$) of **9** is 39. The large difference in the K_C values for the two isomeric forms is ascribed to the different conjugation patterns of the bridging parts. The Hush analysis yielded a H_{ab} of 0.047 eV for the closed form and of 0 eV for the open form of **9**. However, the photoisomerization conversion of **9** is very slow (>1 hour) and the content of the closed isomer is dependent on solvents. In contrast, the photoisomerization conversion of the Ru derivative **10** is much faster than the Fe derivative **9**, while the *SF* of **10** (4.2) is smaller than that of **9**. It was also found that the ruthenium system exhibited dual photo- and electrochromism.

The *SF* value of the above system could be significantly improved by shortening the bridge. Following the studies of **9** and **10**, Akita's group reported the DTE-bridged dinuclear metal complexes **11–15** by directly connecting the iron or ruthenium redox centers to the thiophene units of DTE [35, 36]. Compared with the acetylide-linked complexes **9** and **10**, the K_C ($4.5 \times 10^3 \sim 7.5 \times 10^4$) and *SF* values ($8.8 \times 10^2 \sim 5.4 \times 10^3$) of **11–15** were significantly improved, demonstrating that the shortening of the Fe–Fe distance could enhance the metal–metal communication. A general trend is that K_C increases as the ancillary ligands on Fe(II) and Ru(II) become more electron-donating. In addition, Ru(II) complexes are generally more stable and show more efficient photochemical cyclization than their Fe(II) analogs.

Rigaut, Lagrost, and coworkers employed a similar ethynyl-disubstituted DTE unit as the bridging ligand to connect two [Ru(dppe)$_2$Cl]$^+$ (dppe = 1,2-bis[diphenylphosphino]ethane) moieties (**16** in Figure 12.6a) [38]. Compound **16** is characterized by multicolor electrochromism, electrocyclization at low voltages, and photo- and

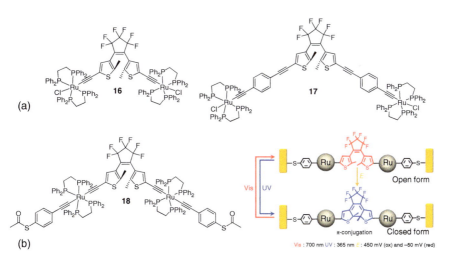

Figure 12.6 (a) Chemical structures of photoswitchable complexes **16** and **17** comprised of ruthenium alkynyl and DTE ligand. (b) Schematic representation of the molecular isomerization of **18** bridging a nanogap electrode under external controls. Source: [37]/Springer Nature/Licensed under CC BY 3.0.

electrotunable electronic communication. In a related study, Humphrey and co-workers succeeded in the synthesis of a more complicated six-state switching system by elongating compound **16** to give the dinuclear ruthenium alkynyl complex **17** with additional phenylethynyl linkers [39]. The ruthenium redox centers, the DTE bridge, and the terminal ruthenium acetylide ligand can respond to electrochemical, photochemical, and proton stimuli (isomerization between acetylide and vinylidene), respectively. The open and closed isomers of each couple can be interconverted by UV and visible light irradiation. In addition, Chen and coworkers reported the thioacetyl-terminated diruthenium alkynyl complex **18** with a DTE bridge [37] (Figure 12.6b). This complex was used to bridge a nanogap electrode to form a switchable molecular device by the *in situ* deprotection of the thioacetyl groups. The stepwise control of molecular isomerization could be repeatedly and reversibly completed with the use of orthogonal photo- and electrochemical stimuli to achieve the controllable switching of conductivity.

A series of binuclear ruthenium vinyl complexes bridged by either dithienyl perhydro- or perfluorocyclopentene units were synthesized by Liu and coworkers, as represented by complexes **19H** and **19F** in Figure 12.7a [40]. These complexes underwent cyclization and cycloreversion under irradiation with UV and visible light, respectively. The replacement of the hexafluoropentene backbone of DTE by an ordinary cyclopentene unit had no significant impact on the comproportionation constants ($K_C = 349$). The open isomer of **19F** displayed a broad oxidation wave at +0.57 V vs. Ag/Ag$^+$. Upon UV irradiation, two new well-separated waves appeared at +0.02 and +0.17 V, respectively (Figure 12.7b).

In addition to the typical structures constituting two redox termini with a switchable bridge, systems containing two or more photochromic switches have been the focus of a number of research groups. When two identical DTE–acetylide ligands are connected to a gold, platinum, or ruthenium center to afford metal component-bridged bis-DTE complexes, stepwise ring-closing and opening processes have been observed upon light irradiation [41–43]. A triphenylamine dimer **20** with two DTE moieties and a 1,3-butadiene central linker was reported by Pu

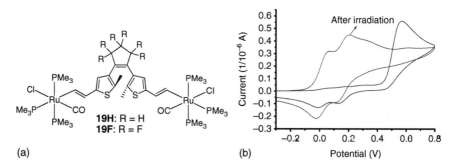

Figure 12.7 (a) Structure of diruthenium complexes **19H** and **19F**. (b) Cyclic voltammograms of **19F** in CH_2Cl_2 containing nBu_4NPF_6, before (solid line) and after (dashed line) irradiation with 302 nm light at the scan of 100 mV/s. Source: Reproduced with permission [40]. Copyright 2009 American Chemical Society.

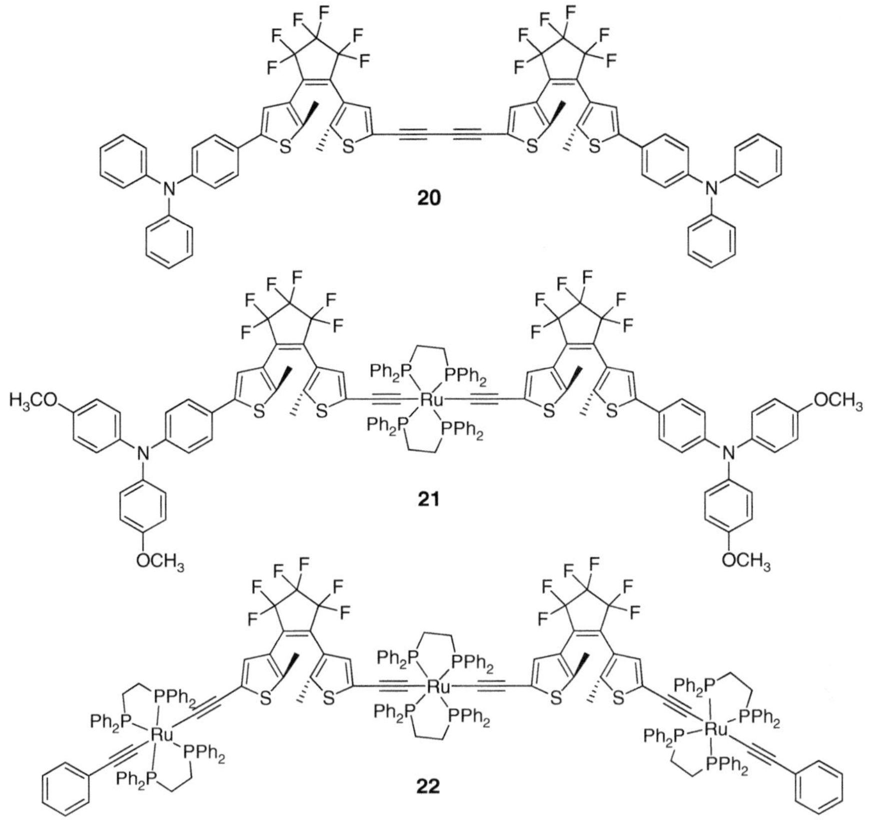

Figure 12.8 Molecular structures of complexes **20–22** containing two DTE units.

and coworkers [44] (Figure 12.8). Compound **20** displayed a reversible redox wave at +1.05 V vs. Ag/AgCl, which was attributed to the one-step two-electron process of the two noncommunicating triarylamine centers in the open isomer. Upon UV irradiation, the electronic communication between two triarylamine centers had progressive enhancement as a result of the gradual increase in the π-conjugation along the molecular backbone during the stepwise ring-closing conversion of two DTE moieties. At the same time, **20** exhibited eight switchable states through stepwise photochromic and reversible redox processes. This implied that the electronic interaction between two triarylamine centers could be well mediated through stepwise photochromic reactions.

Chen *et al.* prepared a monoruthenium complex **21** with two triarylamine-DTE-acetylide ligands to photochemically modulate the degree of electronic communication in inorganic–organic MV systems [45]. Complex **21** exhibited two separated anodic waves at +0.60 and +0.86 V, due to the oxidations of ruthenium and triarylamines, respectively. Upon irradiation at 312 nm, three reversible oxidation waves were found at +0.28, +0.37, and +0.66 V, ascribed to the oxidations of the ring-closed DTE, triarylamine, and ruthenium, respectively. The ring-closed isomer of **21** showed two moderately intense IVCT bands at 926 and 1143 nm, due to the charge transfer transitions from triarylamine and Ru(II) to the oxidized DTE, respectively. Moreover, the charge delocalization along the molecular backbone can be efficiently regulated through stepwise photocyclization. Another related compound **22** with three ruthenium centers and two DTE units has been presented by the Rigaut group [46]. The full or stepwise ring-closing of the DTE units in the same molecule was achieved through the combination of electrochemical and photo stimuli.

In 2012, He and Wenger synthesized a donor–bridge–acceptor molecule **23** composed of a phenothiazine (PTZ) electron donor, a DTE bridge, and a $[Ru(bpy)_3]^{2+}$ electron acceptor to study intramolecular electron transfer from PTZ to Ru(III) across the DTE bridge [47] (Figure 12.9). In the open form, the direct excitation of the $[Ru(bpy)_3]^{2+}$ unit resulted in rapid energy transfer to the DTE unit. Subsequently, the triplet-excited DTE underwent rapid bimolecular electron transfer with methylviologen present in the solution, followed by intramolecular electron transfer with PTZ. In the closed state **23′**, PTZ$^+$ cannot be observed any more. The authors believed that most of the excitation energy quickly converged to the local singlet-excited state of DTE, followed by rapid photoisomerization to the open form.

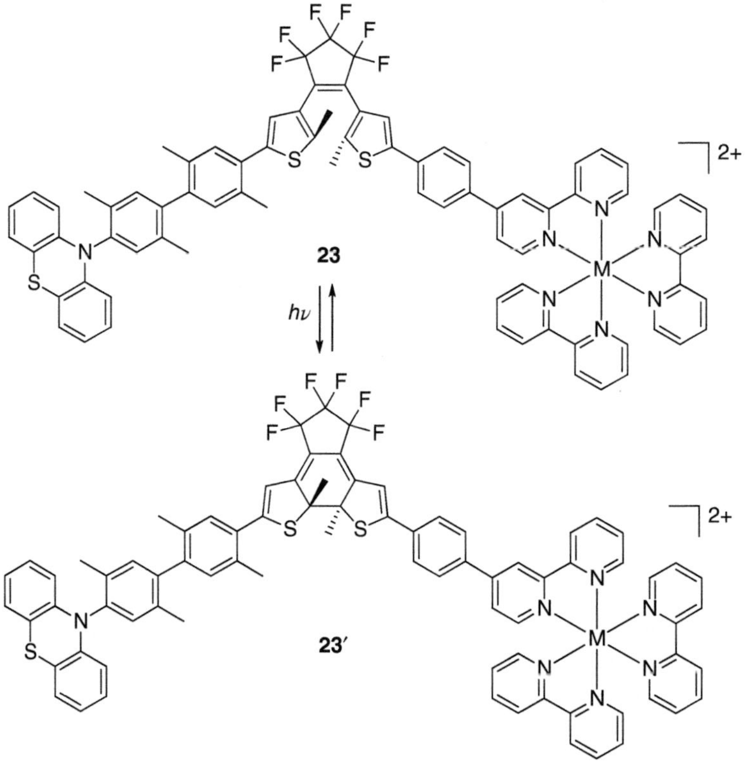

Figure 12.9 Isomerization of the ruthenium–DTE–phenothiazine molecule **23** (M = Ru).

Among photochromic molecules, the *E–Z* photoisomerization reactions of azobenzene and alkene derivatives have attracted intense interest [48–50]. The reversible *trans/cis* isomerization of azobenzene is usually induced by alternating irradiation with UV and visible light, which drives the *trans* to *cis* and *cis* to *trans* isomerization through the azo π–π* excitation and azo n–π* excitation, respectively [51]. Nishihara *et al.* synthesized the ferrocene-terminated azobenzene derivative **24** to study the photoisomerization behavior and Fe–Fe electronic interaction (Figure 12.10a) [53]. Unfortunately, because of the low stability of the

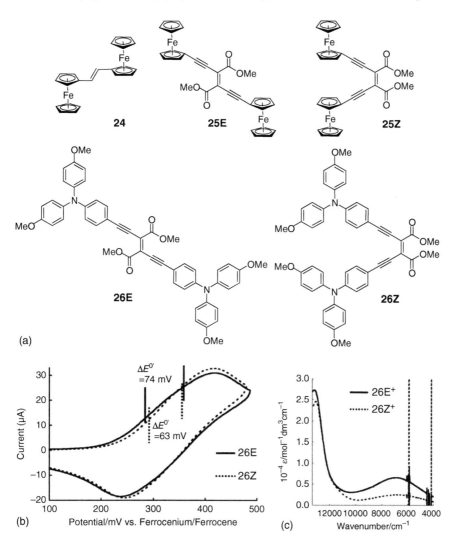

Figure 12.10 (a) Chemical structures of **24–26**. (b) Cyclic voltammograms of **26E** and **26Z** in CH_2Cl_2 at a sweep rate of 100 mV s^{-1}. Reproduced with permission [52]. Copyright 2015 Wiley-VCH. (c) Overlay of the electronic spectra of **26E$^+$** and **26Z$^+$** in dichloromethane. Source: Reproduced with permission [52]. Copyright 2015 Wiley-VCH.

cis-isomer, the study on the electronic interaction in the MV state of *cis*-azoferrocene was not successful.

Ethynylene, which can be regarded as the fundamental building block of polydiacetylene, was further used as the photoswitchable linker because of its highly extended π-system, reversible and efficient E/Z photochromism, and good thermal stability of both isomers [54]. The properties of ethynylethene derivatives are superior to those of azobenzene derivatives in terms of photochromic behavior. The photoinduced *E–Z* isomerization of the diethynylethene-linked di-ferrocene (**25**) or bis-triarylamine (**26**) units has been investigated [52, 55] (Figure 12.10a). Irradiation of **25E** in dichloromethane with visible light led to a photostationary state composed of 89% of **25Z** and 11% of **25E** [55]. The potential splitting of the two ferrocene units is 48 mV for the *Z*-form and 70 mV for the *E*-form, respectively, indicating that the electronic interaction is stronger in the *E*-form. The authors found that the triarylamine-based system **26** exhibited more favorable properties for the investigation of photoswitchable MV than the ferrocene-based molecule **25** [52]. The potential splitting between the two redox potentials is also larger in the *E*-form of **26** (74 mV for **26E** and 63 mV for **26Z**) (Figure 12.10b). The ethynylethene-bridged triarylamine complex showed an intense and well-resolved IVCT band. Analysis of the IVCT band with Hush theory gave a H_{ab} of 534 cm^{-1} for **26E** and 529 cm^{-1} for **26Z**, respectively (Figure 12.10c).

Compared with other photoswitchable units, dihydropyrene (DHP) derivatives often offer enhanced thermal stability of the two interconverting isomers and heightened fatigue resistance [56–58]. Under irradiation with visible light, DHP can be readily isomerized to the opening of the central carbon–carbon bond to form the cyclophanediene (CPD) form. The reverse transformation is triggered by irradiation with UV light or thermally. Another feature of the DHP framework is its extended π-conjugation. Because of these characteristics, DHP is also used to develop MV systems.

In 2008, Nishihara and coworkers reported the first example of photoswitchable DHP-bridged compound **27** (Figure 12.11) [59, 60]. They connected two ferrocene moieties to dimethyldihydropyrene through a π-conjugated ethynyl moiety. The DHP unit of **27** could be converted to CPD by illumination with yellow light (578 nm). The CPD form is significantly less π-conjugated than the DHP isomer. As a result, the ferrocene–ferrocene interaction in the DHP form ($K_C = 57$) is stronger than that in the CPD form ($K_C = 4$). In addition, a redox-assisted ring-closing reaction via the oxidations of the ferrocene moieties is also possible. In contrast,

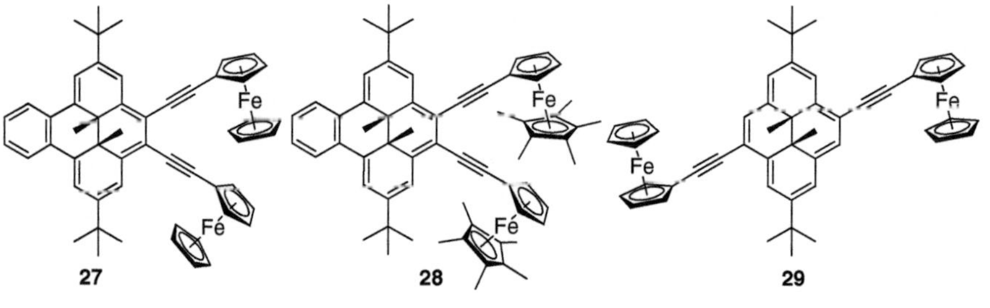

Figure 12.11 Photoswitchable MV compounds **27–29** comprising of ferrocene redox centers and dimethyldihydropyrene ligands.

complex **28** with pentamethylferrocene termini exhibited no photoisomerization process, probably because of the stronger donor ability of pentamethylferrocene (vs. ferrocene) to assist in the quenching of the photoexcited state. In addition, complex **28** did not show good electronic communication either between the two pentamethylferrocene moieties (K_C = 7.7). In 2013, these authors reported compound **29**, which contained two ferrocene groups at the 4- and 9-positions of dimethyldihydropyrene through ethynylene linkers [61]. Compound **29** showed intense electronic communication between the two ferrocene groups, which is promoted by the highly extended π-orbital of the diethynyl-substituted DHP moiety.

In addition to ferrocene derivatives, the Nishihara group and the Vila group reported Ru complexes connected to DHP via phenyl or ethynyl linkers (Figure 12.12) [62, 63]. Unfortunately, it appeared that the use of the conjugated phenyl linker in **30** led to very slow photoisomerization; whereas the use

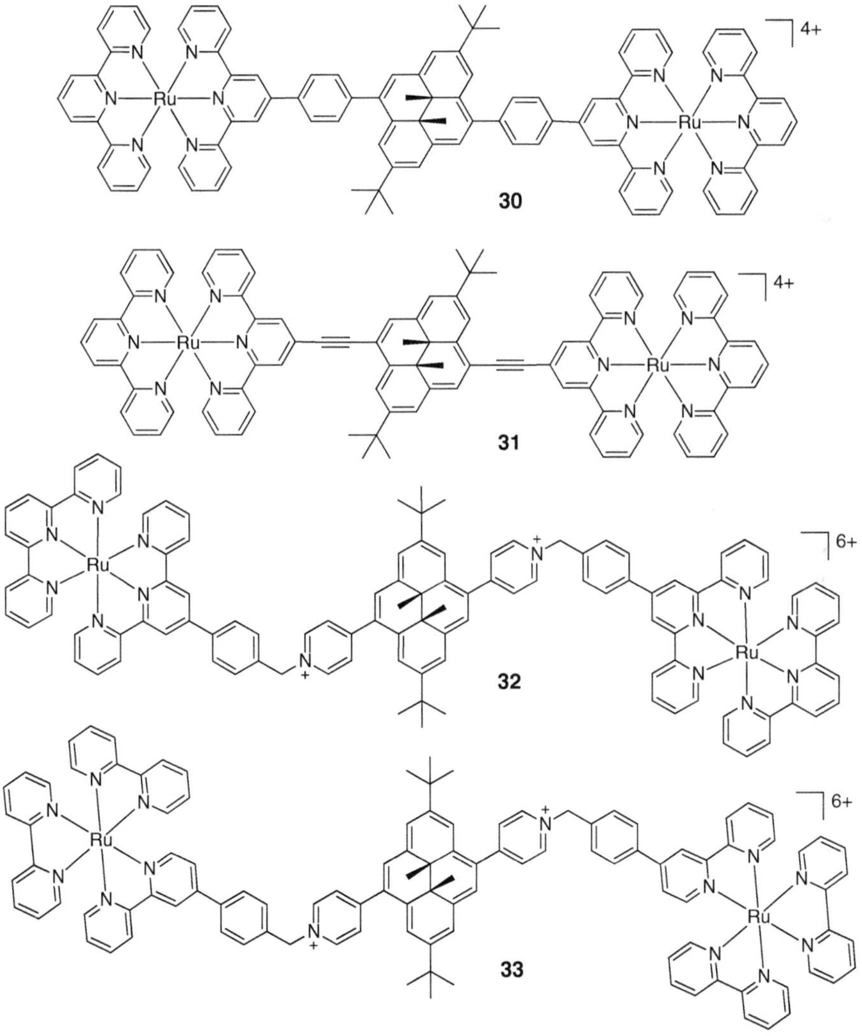

Figure 12.12 Diruthenium complexes **30–33** based on dimethyldihydropyrene.

of the ethynyl linker in **31** completely prevented the photoconversion of the DHP core. In contrast, Cobo, Royal, and coworkers demonstrated that the ring-opening efficiency for the conversion of DHP to CPD could be significantly enhanced by connecting the ruthenium components to the DHP core with a nonconjugated methylene pyridinium linker (**32,33**) [64, 65]. However, none of these complexes show efficient metal–metal communication.

12.3 Anion-Responsive Compounds

Anions are ubiquitous in nature, and they are common chemical species used in supramolecular chemistry and responsive systems [66]. The combination of anions with MV compounds can alter the involved electron transfer when the substrate contains a suitable recognition. For instance, urea is well-known to coordinate with halide anions by hydrogen bonds [67]. In the neutral (fully protonated) state, urea can provide two NH protons for multiple hydrogen-bonding interactions and is often used in anion recognition.

A urea-bridged cyclometalated diruthenium complex **34** was reported for studies on anion-responsive electronic coupling (Figure 12.13a) [68]. Complex **34** displayed two reversible Ru(II/III) waves at +0.39 and +0.49 V vs. Ag/AgCl (Figure 12.13b). According to the IVCT analysis, the MV state of **34** displayed a medium degree of electronic coupling with a H_{ab} of 480 cm^{-1} (Figure 12.13c). In the presence of Br$^-$ or Cl$^-$, the degree of the metal–metal electronic coupling could be slightly enhanced (Br$^-$: H_{ab} = 550 cm^{-1}; Cl$^-$: H_{ab} = 530 cm^{-1}). The introduction of anions was believed to form a more rigid bridge structure with the urea unit through hydrogen bonds. However, upon treatment of **34** with relatively strong basic anions such as $H_2PO_4^-$, F^-, and OAc^-, the redox waves of the diruthenium complex became highly irreversible, possibly due to proton transfer or deprotonation.

Kikuchi, Tahara, and coworkers reported a series of bistriarylamine compounds **35a**–**35c** with a urea bridge and different substituents on the terminal phenyl groups [69]. The influence of supporting electrolytes was examined on the potential splitting of these compounds. The electrochemical potential splitting of conventional MV compounds generally increases in the presence of large counterions because they form noncoordinating ion pairs with enhanced electrostatic interactions between redox components. Compounds **35a**–**35c**, however, display an opposite trend. For example, when changing supporting electrolytes from the smaller PF_6^- to the larger $BArF_4^-$, the potential split of **35b** decreased by 37 mV. This behavior was explained by the hydrogen-bonding interaction between the urea substrate and the smaller PF_6^- anion. In the presence of PF_6^-, the MV states of **35a** and **35b** show an IVCT band with H_{ab} of 700 and 830 cm^{-1}, respectively. No electronic-coupling data was reported for **35c** because of its low solubility.

Organoboron systems have attracted attention because of their intriguing electronic and photophysical properties [70, 71]. The electronic structures of organoboron compounds can be distinctly altered by fluoride bonding [72]. Wenger and coworkers synthesized a donor–bridge–acceptor molecule **36**, composed of a

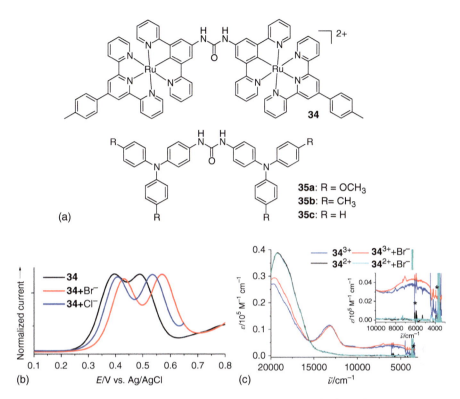

Figure 12.13 (a) The urea-bridged compounds **34** and **35**. (b) Differential pulse voltammograms of **34** in the absence or presence of Bu$_4$NBr or Bu$_4$NCl. Source: Reproduced with permission [68]. Copyright 2017 The Royal Society of Chemistry. (c) A comparison of the absorption spectra of **34**$^{2+}$ and **34**$^{3+}$ as a function of wavenumbers in the absence or presence of 5 equiv. of Br$^-$ in CH$_2$Cl$_2$. Source: Reproduced with permission [68]. Copyright 2017 The Royal Society of Chemistry.

triarylamine donor, 2,5-diboryl-1,4-phenylene bridge, and [Ru(bpy)$_3$]$^{2+}$ acceptor (Figure 12.14a) [74]. They investigated the effect of fluoride anion on the photoinduced electron transfer (PET) between donor and acceptor. Upon the addition of fluoride anion, the PET process from triarylamine to the [Ru(bpy)$_3$]$^{2+}$ component of **36** occurs more than two orders of magnitude more slowly, with the electron transfer rate k_{ET} changing from >10^8 s^{-1} to <10^6 s^{-1}. This suggests that the electron-tunneling barrier is raised by the fluoride binding to the bridge, though the driving force is essentially unaffected.

In order to examine the influence of organoboron on the charge distribution in MV systems, Abe, Tahara, and coworkers introduced the ion-responsive organoboron unit into the bistriarylamine or bisferrocene compound to construct the unsymmetrical acceptor–donor–donor (A–D–D) triad **37** and **38** [73, 75] (Figure 12.14b). The chemically oxidized MV state A–(D–D)$^{•+}$ of **37** exhibited a strong IVCT band derived from the (bistriarylamine)$^{•+}$ moiety in the near infrared (NIR) region (6610 cm^{-1}) [75]. Compared with tetrakis(4-methoxyphenyl)benzidine (MeO-TPD), the covalent attachment of the organoboron group at one end of the bistriarylamine resulted in a

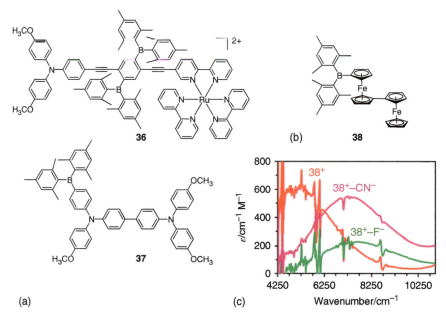

Figure 12.14 (a) Donor–acceptor compounds **36** and **37**. (b) Structures of A–D–D conjugate **38**. (c) NIR spectra of **38**$^+$, **38**$^+$-CN$^-$, and **38**$^+$-F$^-$. Source: Reproduced with permission [73]. Copyright 2015 Wiley-VCH.

slight blueshift of the IVCT band by 350 cm^{-1}. Interestingly, the direct photoexcitation of the neutral triad **37** at 400 nm led to the emergence of the IVCT band in the excited charge-separated MV state A$^{\bullet-}$–D–D$^{\bullet+}$ within 10 ps.

In the case of the singly oxidized state of the bisferrocene triad **38**, it exhibited a broad IVCT absorption band at 4950 cm^{-1} with a H_{ab} of 426 cm^{-1} [73]. In the presence of CN$^-$ and F$^-$, the IVCT band shifted to 7250 cm^{-1} and 7650 cm^{-1}, respectively. The significant blueshift of the IVCT band was attributed to the change in the degree of positive charge delocalization by the bound anions in the so-called zwitterionic MV state (CN$^-$)A-(D-D)$^{\bullet+}$ or (F$^-$)A-(D-D)$^{\bullet+}$. In addition, the addition of CN$^-$ or F$^-$ to the MV state of **38** led to the decrease of H_{ab} by 39 and 112 cm^{-1}, respectively, indicating the distinct effect of external anions on the degree of electronic coupling (Figure 12.14c).

12.4 Proton-Responsive Compounds

Protons can also be utilized as an external stimulus to modulate or switch electron-transfer processes, just like photons and anions. The electronic communication could be tuned by the addition of acid or base in MV complexes containing protonation/deprotonation sites. For instance, Launay and coworkers reported a dinuclear pentaammineruthenium complex **39** bridged by 4,4-azopyridine and studied the properties of the corresponding MV system in 1991 (Figure 12.15) [76]. The bridging ligand could be transformed between the oxidized azo and the

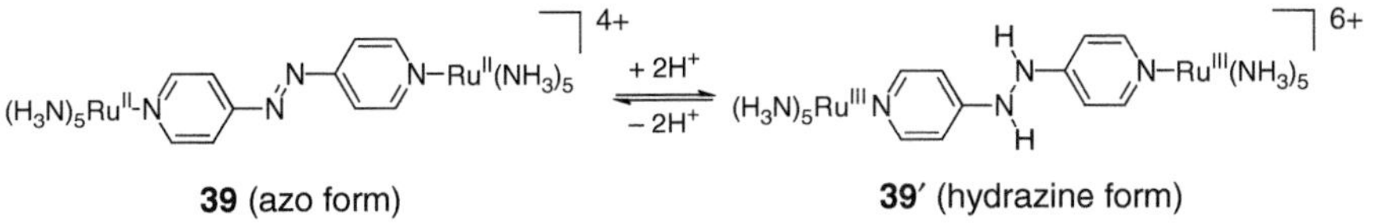

$(H_3N)_5Ru^{II}-N$ ... $-Ru^{II}(NH_3)_5$]$^{4+}$ $\xrightleftharpoons[-2H^+]{+2H^+}$ $(H_3N)_5Ru^{III}N$ ... $N-Ru^{III}(NH_3)_5$]$^{6+}$

39 (azo form)   **39′** (hydrazine form)

Figure 12.15 Protonation and deprotonation of complex **39**.

reduced hydrazine form under different pH conditions. When **39** (with two Ru^{II} sites) was singly oxidized to the MV state, an IVCT band was detected, but it was unresolved with the nearby MLCT band and could not be quantitatively analyzed. The acidification of **39** resulted in a pH-induced intramolecular redox reaction, yielding the hydrazine form **39′** with two Ru^{III} sites. The single reduction of **39′** led to the appearance of a very weak IVCT band, suggesting that the intermetallic electronic interaction in the hydrazine form is weaker than that in the azo form.

Benzimidazole possesses a dissociative imino N–H proton, which can be used to perturb the electronic communication of metal complexes [77]. Therefore, the deprotonation/protonation of metal complexes with benzimidazole moieties can act as proton-responsive MV systems. Haga and coworkers synthesized two dinuclear complexes [M(bpy)$_2$(bpbimH$_2$)M(bpy)$_2$](ClO$_4$)$_4$ (**40**, M = Ru; **41**, M = Os), in which bpbimH$_2$ is the bridging ligand 2,2′-bis(2-pyridyl)bibenzimidazole (Figure 12.16a) [78]. The MV state of the protonated dinuclear complex showed an IVCT band at 7300 cm^{-1} for **40** with H_{ab} of 60–80 cm^{-1} and 9100 cm^{-1} for **41**

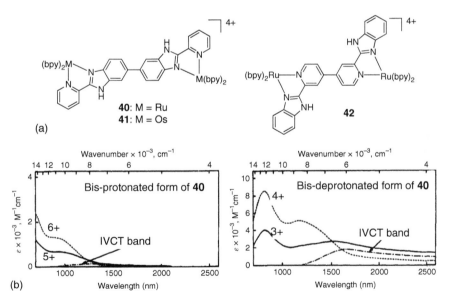

Figure 12.16 (a) Diruthenium complexes **40–42** with benzimidazole derivatives. (b) Near-infrared spectra of di-protonated form and di-deprotonated form of **40** in CH$_3$CN obtained by controlled oxidative electrolysis. The IVCT band is calculated from the direct subtraction of half of the [Ru(III)-Ru(III)] spectrum from the [Ru(II)-Ru(III)] spectrum. Source: Reproduced with permission [78]. Copyright 1991 American Chemical Society.

with H_{ab} of 30–40 cm^{-1}, respectively. When the bridging ligand was deprotonated, the H_{ab} was enhanced to 240–300 cm^{-1} and 140–170 cm^{-1} for the Ru and Os complexes, respectively. The degree of metal–metal interaction of the deprotonated dinuclear complexes becomes four to six times larger than that of the protonated complexes (Figure 12.16b). In 1996, they also reported another bridging ligand, 2,2′-bis(benzimidazol-2-yl)-4,4-bipyridine (bbbpyH$_2$), which was the structural isomer of bpbimH$_2$ [79]. In contrast, the complex **42** with bbbpyH$_2$ showed an opposite trend with respect to **41**. The deprotonation of **42** led to the decrease of the electronic interactions in the MV state and the IVCT band was no longer observed, which was likely caused by the smaller orbital mixing between the Ru(II) dπ and the bridging ligand π^* orbitals. The bridging ligand bbbpyH$_2$ was later used to examine the proton-switched electron-transfer reactions in dendrimer-type tetranuclear RuOs$_3$ complexes [80].

Complexes **40–42** are *tris*-bidentate metal complexes. Following these studies, proton-responsive bis-tridentate metal complexes containing the benzimidazole moieties have been disclosed. For instance, Haga and co-workers reported the diruthenium complex **43** with the bridging ligand consisting of two pyridylbenzimidazole groups linked by a central pyrimidine unit (Figure 12.17) [81]. In this structure, the pyrimidine-bridging group works as a strong π-acceptor and the benzimidazole groups act as π-donors. The protonated form of **43** displayed two Ru(II/III) waves at +0.91 and +1.02 V vs. the ferrocenium/ferrocene (Fc$^+$/Fc) couple with a potential difference ΔE of 110 mV. In contrast, the di-deprotonated form showed a smaller ΔE of 80 mV, suggesting that the deprotonation of the benzimidazole moieties weakened the electronic communication. The H_{ab} value of the MV complex in the protonated form was calculated to be 250 cm^{-1}, but no

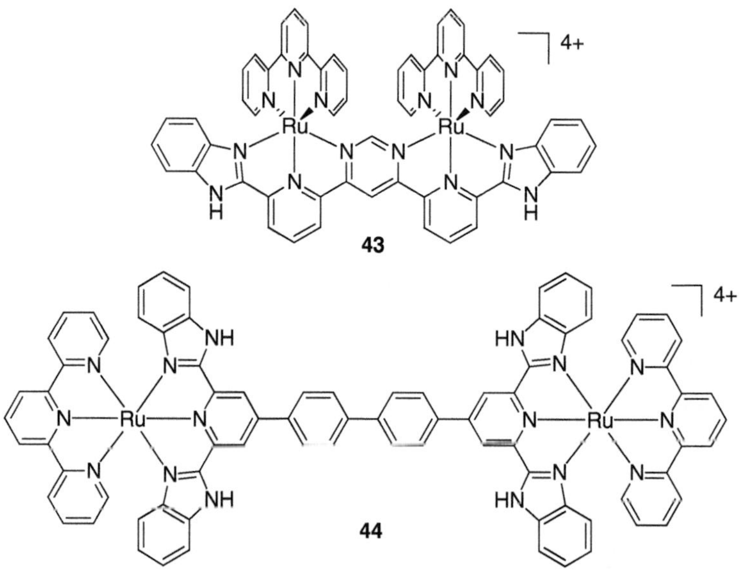

Figure 12.17 Molecular structures of complexes **43** and **44**.

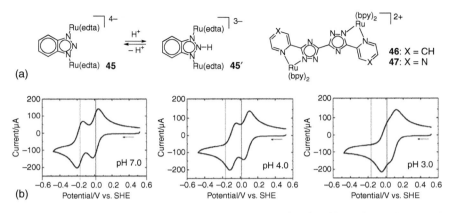

Figure 12.18 (a) Chemical structures of triazole-containing **45–47** and isomerization of **45**. edta = ethylenediaminetetraacetate. (b) Cyclic voltammograms of **45** in an aqueous solution at several pH. Source: Reproduced with permission [83]. Copyright 2001 Elsevier Science B.V.

IVCT band was observed in the deprotonated form. In 2003, Haga et al. reported the linear bis-tridentate diruthenium complex **44** with a bis(benzimidazolyl)pyridine derivative as the bridging ligand [82]. Complex **44** showed only one reversible Ru(II/III) wave at +0.54 V vs. Fc$^+$/Fc in DMF in the presence of HClO$_4$, suggesting that there is little or relatively weak interaction between the two long-separated Ru centers. However, proton-responsive changes in the absorption spectrum and the Ru(II/III) potentials of **44** were observed.

Rocha and Toma have utilized benzotriazole (btaH) as the bridging ligand to prepare the symmetric diruthenium complex **45** (Figure 12.18a) [83]. Upon protonation of the bridging ligand, the electrochemical separation between the two metal-centered redox couples decreased from 195 mV for **45** ($K_c = 2 \times 10^3$) to 75 mV for **45'** ($K_c = 20$) (Figure 12.18b). When the MV state was investigated at pH > 3.3, the complex belongs to the class III system. When pH is less than 3, the complex belonged to the class II MV system. However, higher acidic conditions (pH < 2.5) caused the decomposition of the complex.

Vos et al. studied the electronic communication of the dinuclear ruthenium(II) complexes **46** and **47** with a bis-1,2,4-triazole unit as the bridging ligand [84, 85] (Figure 12.18a). These two complexes showed a potential splitting of 180 mV for **46** and 170 mV for **47**, respectively, implying the presence of electronic interaction between two ruthenium centers. An IVCT transition has also been observed in the singly oxidized states of **46** and **47**, though further analysis was not available [84]. Upon the successive two-step protonation on the uncoordinated nitrogen atoms of the two triazole rings, the splitting of the redox waves of both compounds diminished. In the single-protonation state, the ΔE decreased to 110 mV for **46** and 60 mV for **47**. For the fully protonated species, only a single two-electron oxidation wave was observed for both compounds.

In addition, Sasaki et al. reported a dinuclear ruthenium complex **48** bridged by the 2,5-dimercapto-1,3,4-thiadiazolate (DMcT) ligand, in which the two metal centers

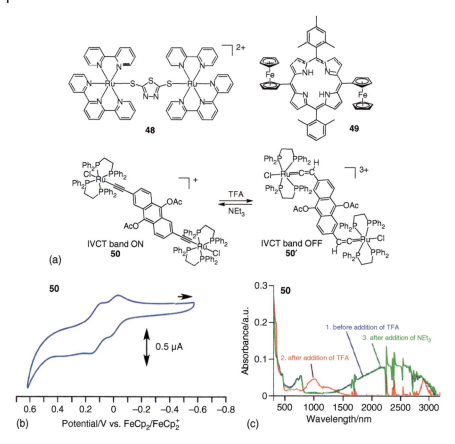

Figure 12.19 (a) Structures of compounds **48–50** and isomerization of **50**. (b) Cyclic voltammogram of **50** in N-methylpyrrolidone. Source: Reproduced with permission [86]. Copyright 2019 The Royal Society of Chemistry. (c) UV–Vis–NIR spectra of protonation and deprotonation reactions with TFA and NEt$_3$ of **50** recorded in CH$_2$Cl$_2$. Source: Reproduced with permission [86]. Copyright 2019 The Royal Society of Chemistry.

are connected by two thiolate sulfur atoms (Figure 12.19a) [87]. In addition to the role of bridging ligand, the DMcT unit can also be utilized as a proton acceptor. Complex **48** showed two reversible Ru(II/III) waves at +0.54 and +0.76 V vs. Ag/AgCl in CH$_3$CN ($K_c = 5.3 \times 10^3$). In the MV state, a broad IVCT band was observed at 1822 nm with a H_{ab} of 550 cm^{-1}. After the addition of TsOH, the electronic communication was switched off by protonation of the bridging ligand.

Porphyrins can respond to external stimuli such as protons by forming protonated porphyrins [88]. In general, protonation will occur at the pyrrolic nitrogen atoms of porphyrins, which can influence the energy of the molecular orbitals by adding two positive charges. Kalimuthu and coworkers synthesized bis(ferrocenyl)bis(mesityl)porphyrin **49** as a stimuli-responsive compound [89]. An IVCT transition between two ferrocene redox centers was observed at 829 nm upon the single oxidation of **49**. The switching of IVCT was achieved by using hydrogen chloride and ammonia vapor to protonate and deprotonate the porphyrin core.

Recently, Akita and coworkers presented the dinuclear Ru(dppe)$_2$ complex **50**, which is bridged by the diethynylated diacetoxyanthracene unit [86] (Figure 12.19a). Two reversible redox waves were observed for **48** at +0.01 and +0.11 V vs. Fc$^+$/Fc (K_C = 76) (Figure 12.19b), with the emergence of an intense IVCT band at 2392 nm in the MV state. The ON/OFF switching of the IVCT band was realized by treating with trifluoroacetic acid (TFA) or NEt$_3$ (Figure 12.19c).

12.5 Conclusion and Outlook

MV compounds have attracted intense interest in terms of fundamental electron-transfer research and potential applications in molecular electronics. The introduction of stimuli-responsive units further intensifies the interest and importance of these compounds. This chapter describes various switchable MV and related systems, in which the electron-transfer process can be switched by externally applied signals. Among them, photoswitchable systems have been mostly studied, in which electronic communication can be reversibly modulated through photoisomerizable units. Electronic communication can also be tuned by adding proton or chemical species as the external stimuli. In terms of the redox-active sites, metal complexes (particularly ruthenium complexes), ferrocene derivatives, and triarylamine derivatives are often used in MV systems. These compounds not only add additional switching functionality to the electron-transfer systems but also are potentially useful in sensing and electronic devices.

Acknowledgement

Financial support is acknowledged from the National Natural Science Foundation (21872154, 21925112, 22090021, and 21975264), Beijing Natural Science Foundation (Grant 2191003), and Youth Innovation Promotion Association, Chinese Academy of Sciences.

References

1 Creutz, C. and Taube, H. (1969). Direct approach to measuring the Franck-Condon barrier to electron transfer between metal ions. *J. Am. Chem. Soc.* 91: 3988–3989.
2 Hankache, J. and Wenger, O.S. (2011). Organic mixed valence. *Chem. Rev.* 111: 5138–5178.
3 Launay, J.-P. (2020). Mixed-valent compounds and their properties – recent developments. *Eur. J. Inorg. Chem.* 2020: 329–341.
4 Wenger, O.S. (2012). Photoswitchable mixed valence. *Chem. Soc. Rev.* 41: 3772–3779.

5 Tahara, K. and Abe, M. (2020). Stimuli-responsive mixed-valence architectures: synthetic design and interplay between mobile and introduced charges. *Chem. Lett.* 49: 485–492.

6 Mrinalini, M. and Prasanthkumar, S. (2019). Recent advances on stimuli-responsive smart materials and their applications. *ChemPlusChem* 84: 1103–1121.

7 Aki, M. (2011). Photochromic organometallics, a stimuli-responsive system: an approach to smart chemical systems. *Organometallics* 30: 43–51.

8 Tuktarov, A.R., Akhmetov, A.R., Khuzin, A.A., and Dzhemilev, U.M. (2018). Synthesis and properties of energy-rich methanofullerenes containing norbornadiene and quadricyclane moieties. *J. Organomet. Chem.* 83: 4160–4166.

9 Jacovella, U., Carrascosa, E., Buntine, J.T. et al. (2020). Photo- and collision-induced isomerization of a charge-tagged norbornadiene-quadricyclane system. *J. Phys. Chem. Lett.* 11: 6045–6050.

10 Harvey, E.C., Feringa, B.L., Vos, J.G. et al. (2015). Transition metal functionalized photo- and redox-switchable diarylethene based molecular switches. *Coord. Chem. Rev.* 282–283: 77–86.

11 Wong, C.-L., Cheng, Y.-H., Poon, C.-T., and Yam, V.W.-W. (2020). Synthesis, photophysical, photochromic, and photomodulated resistive memory studies of dithienylethene-containing copper(I) diimine complexes. *Inorg. Chem.* 59: 14785–14795.

12 Beharry, A.A. and Woolley, G.A. (2011). Azobenzene photoswitches for biomolecules. *Chem. Soc. Rev.* 40: 4422–4437.

13 Szymański, W., Beierle, J.M., Kistemaker, H.A.V. et al. (2013). Reversible photocontrol of biological systems by the incorporation of molecular photoswitches. *Chem. Rev.* 113: 6114–6178.

14 Cameron, D. and Eisler, S. (2018). Photoswitchable double bonds: synthetic strategies for tunability and versatility. *J. Phys. Org. Chem.* 31: e3858.

15 Syamala, M.S., Devanathan, S., and Ramamurthy, V. (1986). Modification of the photochemical behaviour of organic molecules by cyclodextrin: geometric isomerization of stilbenes and alkyl cinnamates. *J. Photochem.* 34: 219–229.

16 Cobo, S., Lafolet, F., Saint-Aman, E. et al. (2015). Reactivity of a pyridinium-substituted dimethyldihydropyrene switch under aerobic conditions: self-sensitized photo-oxygenation and thermal release of singlet oxygen. *Chem. Commun.* 51: 13886–13889.

17 Mitchell, R.H., Ward, T.R., Chen, Y. et al. (2003). Synthesis and photochromic properties of molecules containing [*e*]-annelated dihydropyrenes. Two and three way π-switches based on the dimethyldihydropyrene–metacyclophanediene valence isomerization. *J. Am. Chem. Soc.* 125: 2974–2988.

18 Laine, P., Marvaud, V., Gourdon, A. et al. (1996). Electron transfer through norbornadiene and quadricyclane moieties as a model for molecular switching. *Inorg. Chem.* 35: 711–714.

19 Shao, J.-Y., Gong, Z.-L., and Zhong, Y.-W. (2018). Bridged cyclometalated diruthenium complexes for fundamental electron transfer studies and multi-stage redox switching. *Dalton Trans.* 47: 23–29.

20 Fraysse, S., Coudret, C., and Launay, J.-P. (2000). Synthesis and properties of dinuclear complexes with a photochromic bridge: an intervalence electron transfer switching "on" and "off". *Eur. J. Inorg. Chem.* 2000: 1581–1590.

21 Matsuda, K. and Irie, M. (2004). Diarylethene as a photoswitching unit. *J. Photoch. Photobio. C* 5: 169–182.
22 Tian, H. and Yang, S. (2004). Recent progresses on diarylethene based photochromic switches. *Chem. Soc. Rev.* 33: 85–97.
23 Perrier, A., Maurel, F., and Jacquemin, D. (2012). Single molecule multiphotochromism with diarylethenes. *Acc. Chem. Res.* 45: 1173–1182.
24 Jukes, R.T.F., Adamo, V., Hartl, F. et al. (2004). Photochromic dithienylethene derivatives containing Ru(II) or Os(II) metal units. Sensitized photocyclization from a triplet state. *Inorg. Chem.* 43: 2779–2792.
25 Adamo, V. and Belser, P. (2003). Molecular switches containing transition metals. *Chimia* 57: 169–172.
26 Shen, J.-J., Shao, J.-Y., Zhu, X., and Zhong, Y.-W. (2016). Amine-amine electronic coupling through a dibenzo[a,e]pentalene bridge. *Org. Lett.* 18: 256–259.
27 Nie, H.-J., Yao, C.-J., Shao, J.-Y. et al. (2014). Oligotriarylamines with a pyrene core: a multicenter strategy for enhancing radical cation and dication stability and tuning spin distribution. *Chem. Eur. J.* 20: 17454–17465.
28 He, B. and Wenger, O.S. (2011). Photoswitchable organic mixed valence in dithienylcyclopentene systems with tertiary amine redox centers. *J. Am. Chem. Soc.* 133: 17027–17036.
29 Odom, S.A., Lancaster, K., Beverina, L. et al. (2007). Bis[bis-(4-alkoxyphenyl) amino] derivatives of dithienylethene, bithiophene, dithienothiophene and dithienopyrrole: palladium-catalysed synthesis and highly delocalised radical cations. *Chem. Eur. J.* 13: 9637–9646.
30 Noll, G., Avola, M., Lynch, M., and Daub, J. (2007). Comparison of alternant and nonalternant aromatic bridge systems with respect to their ET-properties. *J. Phys. Chem. C* 111: 3197–3204.
31 Lacroix, J.C., Chane-Ching, K.I., Maquere, F., and Maurel, F. (2006). Intrachain electron transfer in conducting oligomers and polymers: the mixed valence approach. *J. Am. Chem. Soc.* 128: 7264–7276.
32 Mengel, A.K.C., He, B., and Wenger, O.S. (2012). A triarylamine–triarylborane dyad with a photochromic dithienylethene bridge. *J. Organomet. Chem.* 77: 6545–6552.
33 Tanaka, Y., Inagaki, A., and Akita, M. (2007). A photoswitchable molecular wire with the dithienylethene (DTE) linker, (dppe)(η^5-C_5Me_5)Fe–C[triple bond, length as m-dash]C–DTE–C[triple bond, length as m-dash]C–Fe(η^5-C_5Me_5)(dppe). *Chem. Commun.* 43: 1169–1171.
34 Tanaka, Y., Ishisaka, T., Inagaki, A. et al. (2010). Photochromic organometallics with a dithienylethene (DTE) bridge, [Y–C≡C-DTE-C≡C-Y] (Y={MCp*(dppe)}): photoswitchable molecular wire (M=Fe) versus dual photo- and electrochromism (M=Ru). *Chem. Eur. J.* 16: 4762–4776.
35 Motoyama, K., Koike, T., and Akita, M. (2008). Remarkable switching behavior of bimodally stimuli-responsive photochromic dithienylethenes with redox-active organometallic attachments. *Chem. Commun.* 44: 5812–5814.
36 Motoyama, K., Li, H., Koike, T. et al. (2011). Photo- and electro-chromic organometallics with dithienylethene (DTE) linker, L_2CpM-DTE-MCpL_2: dually stimuli-responsive molecular switch. *Dalton Trans.* 40: 10643–10657.

37 Meng, F., Hervault, Y.-M., Shao, Q. et al. (2014). Orthogonally modulated molecular transport junctions for resettable electronic logic gates. *Nat. Commun.* 5: 3023.

38 Liu, Y., Lagrost, C., Costuas, K. et al. (2008). A multifunctional organometallic switch with carbon-rich ruthenium and diarylethene units. *Chem. Commun.* 44: 6117–6119.

39 Green, K.A., Cifuentes, M.P., Corkery, T.C. et al. (2009). Switching the cubic nonlinear optical properties of an electro-, halo-, and photochromic ruthenium alkynyl complex across six states. *Angew. Chem. Int. Ed.* 48: 7867–7870.

40 Lin, Y., Yuan, J., Hu, M. et al. (2009). Syntheses and properties of binuclear ruthenium vinyl complexes with dithienylethene units as multifunction switches. *Organometallics* 28: 6402–6409.

41 Li, B., Wu, Y.H., Wen, H.M. et al. (2012). Gold(I)-Coordination triggered multistep and multiple photochromic reactions in multi-dithienylethene (DTE) systems. *Inorg. Chem.* 51: 1933–1942.

42 Li, B., Wen, H.M., Wang, J.Y. et al. (2013). Modulating stepwise photochromism in platinum(II) complexes with dual dithienyletheneeacetylides by a progressive red shift of ring-closure absorption. *Inorg. Chem.* 52: 12511–12520.

43 Li, B., Wang, J.Y., Wen, H.M. et al. (2012). Redox-modulated stepwise photochromism in a ruthenium complex with dual dithienylethene-acetylides. *J. Am. Chem. Soc.* 134: 16059–16067.

44 Zhang, D., Fan, C., Zheng, C., and Pu, S. (2017). A new dithienylethene dimer with terminal tertiary amine redox centers: Electrochemical, UV–vis–NIR spectral and electronic transfer charges induced by a stepwise photochromic process. *Dyes Pigm.* 136: 669–677.

45 Zhang, D.-B., Wang, J.-Y., Jacquemin, D., and Chen, Z.-N. (2016). Spectroscopic and electrochemical properties of ruthenium complexes with photochromic triarylamine–dithienylethene–acetylide ligands. *Inorg. Chem. Front.* 3: 1432–1443.

46 Hervault, Y.-M., Ndiaye, C.M., Norel, L. et al. (2012). Controlling the stepwise closing of identical DTE photochromic units with electrochemical and optical stimuli. *Org. Lett.* 14: 4454–4457.

47 He, B. and Wenger, O.S. (2012). Ruthenium-phenothiazine electron transfer dyad with a photoswitchable dithienylethene bridge: flash-quench studies with methylviologen. *Inorg. Chem.* 51: 4335–4342.

48 Jiang, W., Wang, G., He, Y. et al. (2005). Photo-switched wettability on an electrostatic self-assembly azobenzene monolayer. *Chem. Commun.* 41: 3550–3552.

49 Yu, H. (2014). Recent advances in photoresponsive liquid-crystalline polymers containing azobenzene chromophores. *J. Mater. Chem. C* 2: 3047–3054.

50 Gobbi, L., Seiler, P., and Diederich, F. (1999). A novel three-way chromophoric molecular switch: pH and light controllable switching cycles. *Angew. Chem. Int. Ed.* 38: 674–678.

51 Lentes, P., Stadler, E., Röhricht, F. et al. (2019). Nitrogen bridged diazocines: photochromes switching within the near-infrared region with high quantum yields in organic solvents and in water. *J. Am. Chem. Soc.* 141: 13592–13600.

52 Sakamoto, R., Kume, S., and Nishihara, H. (2008). Visible-light photochromism of triarylamine- or ferrocene-bound diethynylethenes that switches electronic communication between redox sites and luminescence. *Chem. Eur. J.* 14: 6978–6986.

53 Kurihara, M., Matsuda, T., Hirooka, A. et al. (2000). Novel photoisomerization of azoferrocene with a low-energy MLCT band and significant change of the redox behavior between the *cis*- and *trans*-Isomers. *J. Am. Chem. Soc.* 122: 12373–12374.

54 Moonen, N.N.P., Boudon, C., Gisselbrecht, J.-P. et al. (2002). Cyanoethynylethenes: a class of powerful electron acceptors for molecular scaffolding. *Angew. Chem. Int. Ed.* 41: 3044–3047.

55 Sakamoto, R., Murata, M., and Nishihara, H. (2006). Visible-light photochromism of bis(ferrocenylethynyl)ethenes switches electronic communication between ferrocene sites. *Angew. Chem. Int. Ed.* 45: 4793–4795.

56 Roldan, D., Cobo, S., Lafolet, F. et al. (2015). A multi-addressable switch based on the dimethyldihydropyrene photochrome with remarkable proton-triggered photo-opening efficiency. *Chem. Eur. J.* 21: 455–467.

57 Bakkar, A., Cobo, S., Lafolet, F. et al. (2016). A redox- and photo-responsive quadri-state switch based on dimethyldihydropyrene-appended cobalt complexes. *J. Mater. Chem. C* 4: 1139–1143.

58 Zhang, P., Brkic, Z., Berg, D.J. et al. (2011). Cobalt complexes containing dimethyldihydropyrene-substituted cyclobutadiene ligands. *Organometallics* 30: 5396–5407.

59 Muratsugu, S., Kume, S., and Nishihara, H. (2008). Redox-assisted ring closing reaction of the photogenerated cyclophanediene form of bis(ferrocenyl) dimethyldihydropyrene with interferrocene electronic communication switching. *J. Am. Chem. Soc.* 130: 7204–7205.

60 Muratsugu, S., Kishida, M., Sakamoto, R., and Nishihara, H. (2013). Comparative study of photochromic ferrocene-conjugated dimethyldihydropyrene derivatives. *Chem. Eur. J.* 19: 17314–17327.

61 Kishida, M., Muratsugu, S., Sakamoto, R. et al. (2013). Efficient electronic communication in 4,9-bis(ferrocenylethynyl)dimethyldihydropyrene. *Chem. Lett.* 42: 361–362.

62 Muratsugu, S. and Nishihara, H. (2014). π-Conjugation modification of photochromic and redox-active dimethyldihydropyrene by phenyl- and ethynyl-terpyridines and Ru(bis-terpyridine) complexes. *New J. Chem.* 38: 6114–6124.

63 Vilà, N., Royal, G., Loiseau, F., and Deronzier, A. (2011). Photochromic and redox properties of bisterpyridine ruthenium complexes based on dimethyldihydropyrene units as bridging ligands. *Inorg. Chem.* 50: 10581–10591.

64 Jacquet, M., Lafolet, F., Cobo, S. et al. (2017). Efficient photoswitch system combining a dimethyldihydropyrene pyridinium core and ruthenium(II) bis-terpyridine entities. *Inorg. Chem.* 56: 4357–4368.

65 Jacquet, M., Uriarte, L.M., Lafolet, F. et al. (2020). All visible light switch based on the dimethyldihydropyrene photochromic core. *J. Phys. Chem. Lett.* 11: 2682–2688.

66 Beer, P.D. and Gale, P.A. (2001). Anion recognition and sensing: the state of the art and future perspectives. *Angew. Chem. Int. Ed.* 40: 486–516.

67 Bu, J., Lilienthal, N.D., Woods, J.E. et al. (2005). Electrochemically controlled hydrogen bonding. Nitrobenzenes as simple redox-dependent receptors for arylureas. *J. Am. Chem. Soc.* 127: 6423–6429.

68 Gong, Z.-L., Deng, L.-Y., Zhong, Y.-W., and Yao, J. (2017). Anion-regulated electronic communication in a cyclometalated diruthenium complex with a urea bridge. *Phys. Chem. Chem. Phys.* 19: 8902–8907.

69 Tahara, K., Nakakita, T., Starikova, A.A. et al. (2019). Small anion-assisted electrochemical potential splitting in a new series of bistriarylamine derivatives: organic mixed valency across a urea bridge and zwitterionization. *Beilstein J. Org. Chem.* 15: 2277–2286.

70 Yamaguchi, S., Akiyama, S., and Tamao, K. (2001). Colorimetric fluoride ion sensing by boron-containing π-electron systems. *J. Am. Chem. Soc.* 123: 11372–11375.

71 Siewert, I., Fitzpatrick, P., Broomsgrove, A.E.J. et al. (2011). Probing the influence of steric bulk on anion binding by triarylboranes: comparative studies of FcB(o-Tol)$_2$, FcB(o-Xyl)$_2$ and FcBMes$_2$. *Dalton Trans.* 40: 10345–10350.

72 Hudnall, T.W., Chiu, C.-W., and Gabbaï, F.P. (2009). Fluoride ion recognition by chelating and cationic boranes. *Acc. Chem. Res.* 42: 388–397.

73 Tahara, K., Terashita, N., Tokunaga, K. et al. (2019). Zwitterionic mixed valence: internalizing counteranions into a biferrocenium framework toward molecular expression of half-cells in quantum cellular automata. *Chem. Eur. J.* 25: 13728–13738.

74 Chen, J. and Wenger, O.S. (2015). Fluoride binding to an organoboron wire controls photoinduced electron transfer. *Chem. Sci.* 6: 3582–3592.

75 Tahara, K., Koyama, H., Fujitsuka, M. et al. (2019). Charge-separated mixed valency in an unsymmetrical acceptor–donor–donor triad based on diarylboryl and triarylamine units. *J. Organomet. Chem.* 84: 8910–8920.

76 Launay, J.P., Tourrel-Pagis, M., Lipskier, J.F. et al. (1991). Control of intramolecular electron transfer by a chemical reaction. The 4,4′-azopyridine/1,2-bis(4-pyridyl)hydrazine system. *Inorg. Chem.* 30: 1033–1038.

77 Wang, H., Shao, J.-Y., Duan, R. et al. (2021). Synthesis and electronic coupling studies of cyclometalated diruthenium complexes bridged by 3,3′,5,5′-tetrakis (benzimidazol-2-yl)-biphenyl. *Dalton Trans.* 50: 4219–4230.

78 Haga, M., Ano, T., Kano, K., and Yamabe, S. (1991). Proton-induced switching of metal-metal interactions in dinuclear ruthenium and osmium complexes bridged by 2,2′-bis(2-pyridyl)bibenzimidazole. *Inorg. Chem.* 30: 3843–3849.

79 Haga, M., Ali, M.M., Koseki, S. et al. (1996). Proton-induced tuning of electrochemical and photophysical properties in mononuclear and dinuclear ruthenium complexes containing 2,2′-bis(benzimidazol-2-yl)-4,4′-bipyridine: synthesis, molecular structure, and mixed-valence state and excited-state properties. *Inorg. Chem.* 35: 3335–3347.

80 Haga, M., Ali, M.M., and Arakawa, R. (1996). Proton-induced switching of electron transfer pathways in dendrimer-type tetranuclear RuOs$_3$ complexes. *Angew. Chem. Int. Ed.* 35: 76–78.

81 Kobayashi, K., Ishikubo, M., Kanaizuka, K. et al. (2011). Proton-induced tuning of metal–metal communication in rack-type dinuclear Ru complexes containing benzimidazolyl moieties. *Chem. Eur. J.* 17: 6954–6963.

82 Haga, M., Takasugi, T., Tomie, A. et al. (2003). Molecular design of a proton-induced molecular switch based on rod-shaped Ru dinuclear complexes with bis-tridentate 2,6-bis(benzimidazol-2-yl)pyridine derivatives. *Dalton Trans.* 32: 2069–2084.

83 Rocha, R.C. and Toma, H.E. (2001). Proton-induced switching and control of intramolecular electron transfer on a benzotriazole-bridged symmetric mixed-valence ruthenium complex. *Inorg. Chem. Commun.* 4: 230–236.

84 Pietro, C.D., Serroni, S., Campagna, S. et al. (2002). Proton controlled intramolecular communication in dinuclear ruthenium(II) polypyridine complexes. *Inorg. Chem.* 41: 2871–2878.

85 Fanni, S., Pietro, C.D., Serroni, S. et al. (2000). Ni(0) catalysed homo-coupling reactions: a novel route towards the synthesis of multinuclear ruthenium polypyridine complexes featuring made-to-order properties. *Inorg. Chem. Commun.* 3: 42–44.

86 Oyama, Y., Kawano, R., Tanaka, Y., and Akita, M. (2019). Dinuclear ruthenium acetylide complexes with diethynylated anthrahydroquinone and anthraquinone frameworks: a multi-stimuli-responsive organometallic switch. *Dalton Trans.* 48: 7432–7441.

87 Tannai, H., Tsuge, K., and Sasaki, Y. (2005). Switching of the electronic communication between two {Ru(trpy)(bpy)} (trpy = 2,2′:6′,2′-terpyridine and bpy = 2,2′-bipyridine) centers by protonation on the bridging dimercaptothiadiazolato ligand. *Inorg. Chem.* 44: 5206–5208.

88 Presselt, M., Dehaen, W., Maes, W. et al. (2015). Quantum chemical insights into the dependence of porphyrin basicity on the meso-aryl substituents: thermodynamics, buckling, reaction sites and molecular flexibility. *Phys. Chem. Chem. Phys.* 17: 14096–14106.

89 Sundharamurthi, S., Sudha, K., Karthikaikumar, S. et al. (2018). Switching of inter-valence charge transfer in stimuli-responsive bis(ferrocenyl)porphyrin. *New J. Chem.* 42: 4742–4747.

13

Mixed Valency in Extended Materials

Harrison S. Moore, Eleanor R. Kearns, Martin P. van Koeverden, and Deanna M. D'Alessandro

School of Chemistry, The University of Sydney, Sydney, NSW 2006, Australia

13.1 Introduction

13.1.1 Fundamental Aspects of Mixed Valency in the Solid State

Mixed valency arises when redox-active species in a structure exist in more than one formal oxidation state [1–3]. Depending on the structure of the material and the relative redox potentials for each of these moieties, charge transfer (CT), that is the movement of an electron (or hole), may occur between the sites. For metal complexes, these electrons usually arise from loosely bound individual d-electrons, while in organic systems, stable radical states provide the mobile electrons (or holes) for CT [4].

Prussian blue, $Fe_4[Fe(CN)_6]_3$ discovered in the 1700s, is one of the earliest examples of a mixed-valence (MV) compound [2, 5]. The characteristic blue color arises from electron transfer between Fe(II) and Fe(III) centers present in the structure due to the absorption of light in the red region of the electromagnetic spectrum (Figure 13.1). In reality, the formal degree of CT may be some fraction of 1, but for the purposes of our discussion here, we can consider the intervalence charge transfer (IVCT) as arising from the transfer of an electron from Fe(II) to Fe(III). When Prussian blue is fully reduced, the deep blue color is lost, and the resulting compound is called Prussian white. In the latter case, only Fe(II) ions remain in the structure and CT between Fe(II) ions is absent. Similarly, fully oxidizing Prussian blue to Prussian yellow $Fe_3[Fe(CN)_3]$ precludes CT between Fe(III) ions, and the yellow color arises from the $[Fe^{III}(CN)_6]^{3-}$ absorption at 425 nm [6]. Prussian green is the intermediate state between Prussian blue and yellow and exhibits a corresponding intermediate color. The redox behavior of Prussian blue highlights the importance of different oxidation states in mixed valency.

In addition to multiple accessible redox states, CT is only favorable when the different valence sites reside in relatively similar local chemical environments [2], thus minimizing the reorganization energy required for CT to occur. For example, CT is favorable in Prussian blue, because both Fe(II) and Fe(III) centers reside

Mixed-Valence Systems: Fundamentals, Synthesis, Electron Transfer, and Applications, First Edition.
Edited by Yu-Wu Zhong, Chun Y. Liu, and Jeffrey R. Reimers.
© 2023 WILEY-VCH GmbH. Published 2023 by WILEY-VCH GmbH.

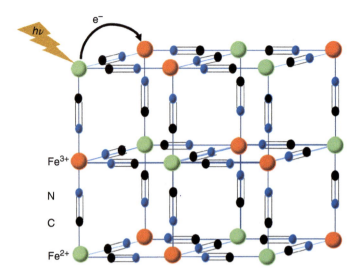

Figure 13.1 Structure of Prussian blue and its associated CT. Interstitial charge-balancing counter ions are omitted for clarity. Green = Fe(II), brown = Fe(III), blue = N, and black = C.

in octahedral coordination environments formed by CN^- ligands, so CT does not require large distortions to the coordination spheres of Fe(II) or Fe(III). Contrastingly, CT does not occur in $[Ga^I(Ga^{III}Cl_4)]$, which contains Ga(III) tetrahedrally coordinated by Cl^-, and Ga(I) is surrounded by an uneven dodecahedron of Cl^- ligands because CT would require significant changes in metal coordination environments [1].

For electron transfer between sites in an MV compound (MVC), orbital mixing must be present. This can occur via direct overlap of the orbitals on the moieties involved, or via a bridging ligand by mixing of the relevant molecular orbitals, known as through-bond IVCT [1]. IVCT can also occur through space, where CT relies on π-orbital overlap to provide a path for the transferring electrons. Significantly, the donor and acceptor orbitals have positive and negative regions of electron density [7]. When the orbitals overlap **constructively** (positive/positive or negative/negative) is CT favored or "symmetry allowed"; if the orbitals overlap **destructively** (positive/negative), CT is not favorable and is "symmetry disallowed". As a result, both the angle of orbital approach and the interorbital distance affect the rate of electron transfer. When IVCT occurs in solution, the measured rate is the mean rate of electron transfer, as it includes all angles of approach and interorbital distances at which CT occurs. IVCT in the solid state removes this ambiguity, allowing for the direct computation of CT parameters for a specific angle and distance.

13.1.2 Quantum Mechanical Considerations in Mixed Valency and IVCT

Theoretically, IVCT can be modeled using classical Marcus–Hush theory [1, 8]. In this elegant two-state model, the donor and acceptor are regarded as two noninteracting spheres. IVCT is assumed to be an adiabatic process in which an electron can adjust smoothly to the change in reaction coordinates between the two states (i.e. the donor and acceptor), with only one eigenfunction required to describe the electron transfer process. Independently, Marcus and Hush derived models to describe IVCT [8–10]. Traditional Hush theory enables visualization of CT reactions along a single reaction coordinate. In this model, it is assumed that a change in redox state of the donor or acceptor leads to a change in the bonding environment – this contribution constitutes the activation energy for CT.

The Franck–Condon principle, which, in turn, is derived from the Born–Oppenheimer approximation, states that electronic transitions are more likely to occur when the donor and acceptor are in similar vibrational states [7, 11, 12]. Classically, this is because nuclei are heavier than electrons so that electronic transitions will occur faster than vibrational transitions. Quantum mechanically, the intensity of an electronic transition is given by the square of the overlap integrals of the donor and acceptor vibronic states. When the Franck–Condon principle is applied, nuclei are considered fixed, allowing electronic transitions to be visualized as vertical movements between the donor and acceptor curves in a potential energy diagram (see Chapter 1, Section 1.6 for further discussion). In this case, CT can be represented as two overlapping parabolas that represent cross sections of the potential energy surfaces for the electron transfer [2, 7, 8, 13].

13.1.3 Marcus–Hush Theory and the Quantification of CT

Several important parameters are defined by Marcus–Hush theory [8, 9, 14, 15] (see Chapter 1 for further discussion). The vertical reorganization energy (λ) describes the energy required to excite an electron vertically from the minimum of the donor curve to meet the acceptor curve and is equivalent to the optical activation energy (E_{IT}). H_{ab} quantifies the electronic delocalization between the donor and acceptor moieties [7, 8]. The H_{ab} parameter is an important factor in CT and indicates the amount of mixing between donor and acceptor wavefunctions. An increasing H_{ab} indicates a greater degree of electronic delocalization between the valence states, and concomitantly lower amount of thermal energy required for CT to occur. As the charge becomes more delocalized, the donor and acceptor wavefunctions/potential energy surfaces become more mixed. This changes the relationship between the activation energy and the reaction coordinate. Considering a physical system, increasing electronic delocalization decreases the unique oxidation state character of each site, as well as reducing the magnitude of supramolecular environment and solvent dipole rearrangements required.

13.1.4 Classifications of Mixed Valency

As described in Chapter 1, MV systems can be described in three different classes according to the Robin–Day scheme, depending on the magnitude of H_{ab} and the relationship to λ [2]. In class I systems, $H_{ab} = 0$ is indicative of an absence of orbital mixing [16–18], which can result from large spatial separation between donor and acceptor sites or because the sites exist in very different bonding environments. As a result, only the properties of the individual fragments are observed so that the donor and acceptor are crystallographically distinguished and possess integer redox states. As the degree of orbital mixing increases, so too does H_{ab} [16–18]. For H_{ab} less than $\lambda/2$, the system is classified as class II. In class II systems, the donor and acceptor sites remain crystallographically distinguishable, and therefore still possess integer oxidation states. Class II systems exhibit perturbed properties of the individual components and may also display some properties of the mixed state system. Further increasing H_{ab} (so that $H_{ab} > \lambda/2$) eliminates the thermal activation barrier to electron transfer [16–18], so the system is considered a class III compound. Class III systems will still undergo optical excitations at $h\nu = 2H_{ab}$; however, these occur within the molecular orbital manifold due to the spatial delocalization of the charge. Crystallographically, the donor and acceptor are indistinguishable, and exhibit noninteger oxidation states. In these systems, only the properties of the MVC are observed.

13.1.5 Organic Mixed Valency

Mixed valency is a term often dedicated to inorganic chemistry, taking 30 years after the conception of mixed valency for organic systems to be denoted as MV [19, 20]. This may be in part because, in organic MV systems, it is often difficult to distinguish which organic groups are the redox centers taking part in CT, in contrast to inorganic MV systems (see Chapter 9, Section 9.1) [19]. Despite this, organic systems are perhaps more suited to the study of mixed valency as their optical transitions are not overlapped with other excitation modes. However, when considering organic mixed valency, it is important to distinguish MVCs from donor–acceptor (DA) compounds, each of which behaves differently on excitation. Upon excitation in MVCs, a new electronic ground state is obtained, while in DA compounds, an excited electronic state is obtained. A thorough review by Lambert and coworkers elegantly highlights these distinctions [19].

Much like inorganic MVCs, organic MVCs can undergo through-bond or through-space CT [19]. Through-bond CT is mediated by a bridging unit between the donor and acceptor sites via either a hopping or superexchange mechanism. In contrast, through-space CT is mediated through cofacially aligned overlapping donor and acceptor π-orbitals only via a superexchange mechanism (see Chapter 9, Section 9.5 for further details). In all cases, IVCT in organic MVCs exhibits a strong dependence on the spatial separation between the donor and acceptor functionalities. Where hopping processes dominate, the rate of CT, k, exhibits an inverse relationship with distance, $k = 1/r$, which has been observed experimentally by Amthor and Lambert [21]. In contrast, superexchange mechanisms exhibit an exponential decrease in CT rate with increasing spatial separation between donor and acceptor sites [12].

The desire for organic electronic devices drives research into organic MVCs; common motifs observed in the study of organic mixed valency include cyclophanes [22], 1,2-phenylene derivatives [23], and metal–organic frameworks (MOFs) [24].

13.2 Electron Transfer in Extended MV Materials

13.2.1 Introduction to Extended Materials

MOFs are a subset of coordination polymers consisting of metal-containing nodes connected by organic linkers [25, 26]. MOFs are strictly two- or three-dimensional (2D or 3D) nanoporous materials, with a large percentage of the framework volume consisting of pores or channels. MOFs are highly designable materials and can incorporate a range of behaviors specific to their component groups, coordination complexes, and coordination polymers [27], including conductivity [28–31], CT [32], luminescence [33–35], ferromagnetism [36–38], redox activity [39], spin crossover, and selective encapsulation and interactions with guests. MOFs are often synthesized under mild conditions relative to conditions used for other inorganic solids, such as zeolites [40]. Additionally, MOFs are insoluble in most solvents and can be modified postsynthetically. Consequently, MOFs have been developed and investigated at length for their synergistic properties, particularly in combination with their high surface areas. Some applications include nanowires, catalysis, porous sensors, kinetic separation, gas capture, and storage. Postsynthetic modification (PSM) of MOFs is an increasingly relevant technique employed by MOF chemists to further tune the properties of frameworks [41]. There are four main categories of PSM, broadly defined by the component of the framework to be added, substituted, or removed. Relevant to this review is the ability to modify the oxidation states of ligands and metals in some framework structures. For example, Tang et al. oxidized Fe(II) sites in MIL-100(Fe^{II}) to produce a MV system by heating the original framework under vacuum [42].

As IVCT interactions are entirely dependent on orbital interactions, CT can occur through one of two avenues: through-bond and through-space, as described previously [43]. Ligand design and framework structure are therefore extremely important factors to consider when attempting to produce a MV framework, as the spatial relationship between MV functional groups is integral to the mechanism and extent of coupling [31]. Engineering through-bond CT in extended materials is readily achieved by judicious selection of metals and ligands to ensure close matching in energy levels and efficient orbital overlap [44]. In contrast, through-space IVCT in supramolecular systems is an uncommon phenomenon, appearing in only a handful of examples [45–47]. The requirements for through-space IVCT to occur are restrictive to the extent that the structure of a material must be carefully considered and designed to produce the desired supramolecular orbital overlaps between the MV moieties. In particular, the crystal packing arrangements of metals and ligands in MOFs can sometimes be difficult to control, precluding the supramolecular interactions required for through-space CT to be predictably engineered.

13.2.2 Organic-Based Mixed Valency in Extended Frameworks

13.2.2.1 Thiazolo[5,4-*d*]thiazole-Based Compounds

The redox-active and electron-deficient thiazolo[5,4-*d*]thiazole (TzTz) moiety has been extensively incorporated into conjugated organic polymers to promote high charge mobility [48–51]. Despite the more recent use of MOF precursor ligands [52], TzTz-based frameworks represent some of the largest recent steps forward in the development and characterization of extended CT materials. Significantly, TzTz-based MOFs were the first to exhibit through-space IVCT, in addition to being used as a platform to develop new methods of quantifying mixed valency in solid-state materials, by determining the coupling constant H_{ab}.

Hua et al. synthesized and characterized the novel framework [Zn$_2$(BPPTzTz)$_2$(tdc)$_2$] (BPPTzTz = 2,5-bis(4-(pyridine-4-yl)phenyl)thiazolo[5,4-d]thiazole, tdc = thiophene-2,5-dicarboxylate) [53]. This framework exhibits an IVCT band in the near infrared (NIR) region upon single electron reduction of the neutral cofacially stacked (BPPTzTz)$_2$ units to the cofacial MV radical dimer (BPPTzTz$^{0/-\bullet}$)$_2$ (Figure 13.2). Through-space IVCT was favorable in the reduced framework due to the cofacial orientation of the TzTz units in the MV (BPPTzTz$^{0/-\bullet}$)$_2$ dimers, in addition to the short ligand-to-ligand distance of 3.80 Å. In agreement with Marcus–Hush theory [54], the relationship between distance and the degree of CT was confirmed by synthesizing the framework [Zn$_4$(BDPPTzTz)$_2$(tdc)$_2$] (BDPPTzTz = 2,5-bis(4-(pyridine-4-yl)-3,5-dimethylphenyl)thiazolo[5,4-d]thiazole). In this compound, the bulky 3,5-dimethylphenyl groups increased the TzTz separation to 8.93 Å, precluding any cofacial interaction and preventing the through-space IVCT between TzTz groups from occurring.

Ding et al. synthesized two structurally related frameworks [M$_2$(BPPTzTz)$_2$(sdc)$_2$] (M = Zn, Cd; sdc = selenophene-2,5-dicarboxylate) [24]. While cofacial stacking of the TzTz units occurred in both systems, the separation of 3.76 Å in the Zn system

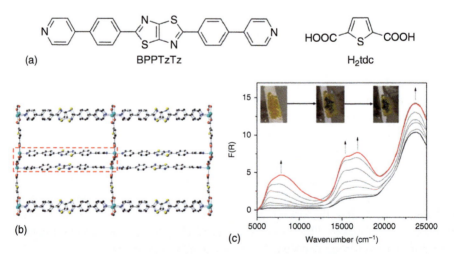

Figure 13.2 (a) Structures of the ligands in [Zn$_2$(BPPTzTz)$_2$(tdc)$_2$]. (b) X-ray crystal structure of [Zn$_2$(BPPTzTz)$_2$(tdc)$_2$] showing cofacial alignment of the BPPTzTz units (red dashed box). Green = Zn, yellow = S, red = O, blue = N, and black = C. H atoms not shown for clarity. (c) Solid-state UV-vis-NIR spectroelectrochemistry of [Zn$_2$(BPPTzTz)$_2$(tdc)$_2$] from 0 (black trace) to −1.75 V (red trace) vs. Ag/Ag$^+$; inset shows marked color change that occurs during framework reduction. Source: Hua et al. 2018 [53]/Reproduced with permission from American Chemical Society.

was slightly reduced compared to the Cd system (3.79 Å), which could cause a lower degree of electronic coupling between TzTz units in the latter. A degree of cofacial slippage occurs in the Cd system, where the TzTz units become misaligned and thus the π-orbital overlap is reduced. This likely contributes to the observed decrease in coupling. A new method for calculating the coupling constant H_{ab} in solid-state materials was also developed using these compounds as a platform. The molar absorption coefficients (ε) of π → π* bands in the neutral framework were determined through transmission measurements on KBr pellets containing a variety of dilutions of the bulk material, termed Kubelka–Munk analysis [55]. This value was then used to calculate the ε values of the IVCT bands in the spectra of the reduced materials from, which the value of H_{ab} could be calculated.

Building upon studies of the aforementioned [Zn$_2$(BPPTzTz)$_2$(tdc)$_2$] framework [53], Doheny et al. synthesized a novel framework [Cd(BPPTzTz)(tdc)]·2DMF, which was structurally related to the Zn compound [56]. As with the Zn compound, through-space IVCT was similarly observed between cofacial TzTz units in the Cd compound, likely due to the equivalent interligand distances of 3.786(6) and 3.782(8) Å in the Cd and Zn compounds, respectively. Subsequently, H_{ab} was determined in both the Cd and Zn compounds. While the previously described Kubelka–Munk analysis was performed on the Zn system, Doheny et al. noted that in this technique, significant error could be introduced due to absorption of atmospheric water by KBr. To overcome these limitations, the authors employed single-crystal UV-vis-NIR spectroscopy measurements developed earlier by Krausz and coworkers [57, 58]. This new method involved quantifying the ε value of an absorption band in the neutral framework, thereby eliminating errors introduced by using a KBr matrix. The same conversion was performed to determine ε for the IVCT band in the diffuse reflectance spectrum, allowing the value of H_{ab} to be determined. Importantly, the values of H_{ab} for the Zn framework were relatively close when calculated using either Kubelka–Munk analysis or single-crystal absorption spectroscopy. Given the values of H_{ab} and ν_{max}, the relationship of $2H_{ab} \ll \nu_{max}$ was determined, and the Cd system was characterized as a Robin–Day class II MV system.

13.2.2.2 Tetrathiafulvalene (TTF)-Based Compounds

Tetrathiafulvalene (TTF) based compounds have significant applications in the synthesis of DA CT systems because they act as effective electron donors [59]. Such systems rely on an easily accessible and reversible one-electron oxidation of the system to TTF$^{+\bullet}$, the radical cation form, an intrinsically MV structure [60]. IVCT in TTF systems requires a π-stacked (TTF$_2$)$^{+\bullet}$ pair [60, 61] via a through-space interaction and is, thus, rarely seen in discrete systems. As with other through-space CT relationships, the distance between TTF molecules strongly impacts the degree of electron delocalization. A correlation between S⋯S distance and conductivity in TTF-based systems has been firmly established [62].

Using the four-connecting TTF-based ligand tetrathiafulvalene tetrabenzoate (TTFTB), Leong et al. synthesized a series of TTF-based MOFs with the formula [M$_2$(TTFTB)(H$_2$O)$_2$] (M = Zn, Mn, Co, Cd) [61] (Figure 13.3). The TTF group was

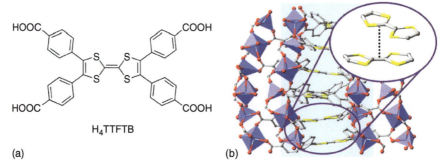

Figure 13.3 (a) Structure of the tetrathiafulvalene tetrabenzoic acid (H_4TTFTB) ligands. (b) X-ray crystal structure of the [M_2(TTFTB)(H_2O)$_2$] (M = Zn, Mn, Co, Cd) family of frameworks showing cofacial through-space interaction between TTF units. Source: Leong et al. 2018 [61]/Reproduced with permission of Elsevier.

shown to exist in the radical (TTF$_2$)$^{+\bullet}$ MV state in the as-synthesized material, which produced an IVCT band in the NIR region. It was speculated that the presence of the radical TTF$^{+\bullet}$ ions in the framework was due to the low oxidation potential of TTF^0, which occurs below 0.17 V vs. Fc/Fc$^+$. Postsynthetic framework doping with I_2 increased the electrical conductivity of the Zn compound by ca. 60%, ascribed to the oxidation of TTF^0 to TTF$^{+\bullet}$ providing a greater concentration of charge carriers.

Hu et al. produced a series of isomorphic novel lanthanide MOFs with the formula [M_4(TTF-DC)$_6$(DMF)$_4$(H_2O)$_2$]·4DMF (M = Gd, Tb, Dy, and Er; TTF-DC = tetrathiafulvalene dicarboxylate) [63]. Upon chemical oxidation of the frameworks with I_2, significant changes in the absorbance spectrum and electronic properties of the materials were observed. An IVCT band was observed in the NIR region for all the oxidized frameworks, occurring between neutral and radical TTF pairs. A change in the ultraviolet absorption profile of the oxidized compounds was attributed to the presence of the oxidized form of TTF ligands. The conductivities of the frameworks increased by three orders of magnitude upon oxidation, likely due to the additional charge mobility afforded by (TTF$_2$)$^{+\bullet}$ coupling. Significant changes in magnetic properties were also observed after the oxidation of the frameworks, likely due to spin contributions from the radical TTF$^{+\bullet}$ species.

13.2.2.3 Tetraoxolene-Based Compounds

For several decades, tetraoxolene ligands derived from 2,5-dihydroxy-1,4-benzoquinone (H_2dhbq) and the 3,6-disubstituted analogues (termed anilic acids, H_2Xan, X = F, Cl, Br, I, CN, NO$_2$, etc.) have been used to produce a range of interesting molecular complexes and 1-, 2-, and 3D frameworks [64, 65]. These ligands exist in three accessible oxidation states (Figure 13.4): a diamagnetic quinoid dianion (dhbq^{2-} or Xan^{2-}), a paramagnetic semiquinoid trianion (dhbq$^{3-\bullet}$ or Xan$^{3-\bullet}$), and an aromatic tetraanion (dhbq^{4-} or Xan^{4-}), all of which are capable of coordinating to metal centers [65–67]. These ligands can produce CT materials via strong electronic interactions between metal (especially with Fe) and the ligands [68]. In addition,

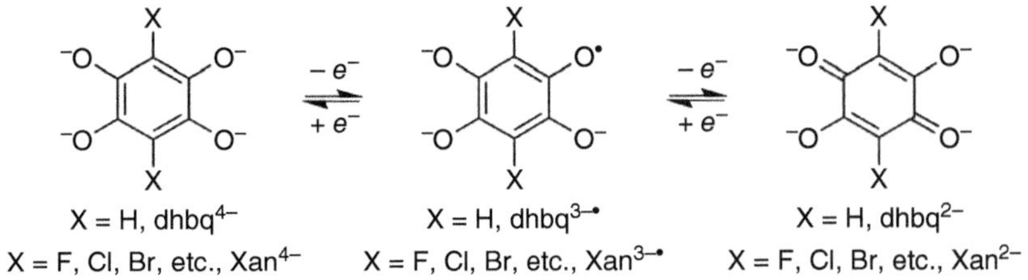

X = H, dhbq^{4-}
X = F, Cl, Br, etc., Xan^{4-}

X = H, dhbq$^{3-•}$
X = F, Cl, Br, etc., Xan$^{3-•}$

X = H, dhbq^{2-}
X = F, Cl, Br, etc., Xan^{2-}

Figure 13.4 Redox states of the tetraoxolene ligands derived from 2,5-dihydroxy-1,4-benzoquinone (H$_2$dhbq) and 3,6-disubstituted analogues (H$_2$Xan).

magnetic ordering is often observed in these systems, facilitated by strong direct coupling between paramagnetic transition metals and the radical ligands [69, 70].

The 3D framework (NBu$_4$)$_2$[Fe$_2$(dhbq)$_3$] (NBu$_4^+$ = tetra-n-butylammonium) first synthesized in 2011 by Abrahams and coworkers [71] was later found to exhibit ligand-centered mixed valency [70]. The compound critically contained the Fe(III)(dhbq)$_3$ moiety, where Fe(III) centers are bridged by two dhbq$^{3-•}$ and one dhbq^{2-} ions; electron transfer between the Fe(II) precursor and dhbq^{2-} gave rise to the Fe(III) and dhbq$^{3-•}$ valence distribution. This led to moderate electronic coupling within the dhbq^{2-}/$^{3-•}$ manifold and high-temperature ferromagnetic interactions with a net ferrimagnetic behavior at low temperature. Analogously, a reaction between 3,6-dichloro-2,5-dihydroxy-1,4-benzoquinone (chloranilic acid, H$_2$Clan) and an Fe(II) precursor afforded the 2D honeycomb framework (H$_2$NMe$_2$)$_2$[Fe$_2$(Clan)$_3$]·6DMF·2H$_2$O (H$_2$NMe$_2^+$ = dimethylammonium) (Figure 13.5) [72]. Again, spontaneous electron transfer between Fe(II) and Clan^{2-} gave rise to an Fe(III) framework containing MV chloranilate ligands. This led to semiconductor behavior [73], while long-range ferrimagnetic ordering results from strong direct antiferromagnetic coupling between Fe(III) and the radical bridging Clan$^{3-•}$ ligands [72, 73].

The Long group from Berkeley has made significant strides in the synthesis and characterization of 2D tetraoxolene honeycomb frameworks. Ziebel et al. conducted detailed electrochemical and spectroscopic analyses on the aforementioned 2D honeycomb framework (H$_2$NMe$_2$)$_2$[Fe$_2$(Clan)$_3$] [74]. This framework displayed high charge mobility due to the arrangement of chloranilate ligands around Fe centers, leading to π-d overlap throughout the structure of the material. Additionally, the Fe(III) sites could be partially reduced, resulting in both MV iron and chloranilate sites present in the structure. When used as a lithium-ion battery (LIB) cathode, introduction of Fe-based and ligand-based mixed valency occurred, with the high charge capacity of the LIB devices attributed to involvement of both metal- and ligand-centered redox processes. Another 2D system (H$_2$NMe$_2$)$_4$[Fe$_3$(Clan)$_3$(SO$_4$)$_2$] was synthesized, although it exhibited notably different connectivity compared to (H$_2$NMe$_2$)$_2$[Fe$_2$(Clan)$_3$]. Spectroscopic analyses revealed (H$_2$NMe$_2$)$_4$[Fe$_3$(Clan)$_3$(SO$_4$)$_2$] primarily contains reduced Fe(II) sites and incomplete metal–ligand charge transfer (MLCT), preventing long-range charge transport. Both (H$_2$NMe$_2$)$_4$[Fe$_3$(Clan)$_3$(SO$_4$)$_2$] and the partially reduced (H$_2$NMe$_2$)$_2$[Fe$_2$(Clan)$_3$] exhibited similar IVCT bands, suggesting a similar CT

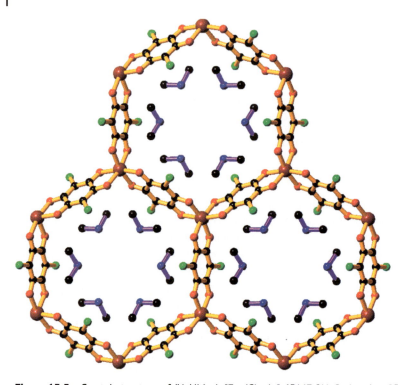

Figure 13.5 Crystal structure of $(H_2NMe_2)_2[Fe_2(Clan)_3] \cdot 6DMF \cdot 2H_2O$ showing 2D honeycomb topology and location of $H_2NMe_2^+$ cations (purple) in the honeycomb channels. Maroon = Fe, green = Cl, red = O, blue = N, and black = C. H atoms and solvents (DMF and H_2O) in honeycomb channels not shown for clarity.

mechanism occurs in both systems. The authors noted that the interaction observed in $(H_2NMe_2)_4[Fe_3(Clan)_3(SO_4)_2]$ may represent a localized electron transition not capable of facilitating long-range charge transport.

To better understand the effect of metal–ligand covalency upon the properties of MV tetraoxolene honeycomb frameworks, Ziebel et al. synthesized a series of honeycomb frameworks $(H_2NMe_2)_2[M_2(Clan)_3]$ (M = Ti, V) and $(H_2NMe_2)_{1.5}[Cr_2(Clan)_3]$ [75], analogous to the aforementioned $(H_2NMe_2)_2[Fe_2(Clan)_3] \cdot 6DMF \cdot 2H_2O$ framework [72]. Clear differences in magnetic and electronic behavior between the different frameworks were observed. Electronically, all three frameworks exhibit some degree of IVCT; however, the V framework exhibits the most delocalized CT relationship, followed by the Ti and Cr systems, with the electrical conductivity proportional to the degree of charge delocalization. Electron delocalization in the V framework is higher than the Ti and Cr frameworks due to the more covalent character of the V–Clan orbitals. Surprisingly, none of the three frameworks exhibited magnetic ordering. The Ti framework showed features of antiferromagnetic interactions between ligands, but magnetic ordering did not occur likely due to magnetic frustration. The lack of ordering in the Cr framework was attributed to the localized electronic structure of the material. In contrast, the highly covalent interaction between V and chloranilate ligands in the V framework resulted in low spin density on ligands, precluding any significant long-range magnetic ordering.

Ziebel et al. also synthesized two similar 2D honeycomb chloranilate frameworks using Mo and Nb nodes, with the formulas $(H_2NMe_2)_2[Nb_2(Clan)_3]$ and $Mo_2(Clan)_3$ [76]. Magnetic coupling in the Nb system leads to a framework that is diamagnetic above 5 K and exhibits partially localized ligand valence states. This was evidenced by the presence of high-energy IVCT bands (indicating a class II Robin–Day material), as well as vibrational modes in the IR spectrum suggesting valence localization of $Clan^{3-\bullet}$ ions, as well as valence delocalization of $Clan^{3-\bullet}$ and $Clan^{4-}$ ions. The seemingly contradictory observation in the infrared spectrum likely indicates an electronic structure where some ligands are valence trapped and others are more delocalized. An asymmetric IVCT band in the Mo compound indicated charge delocalization across the Mo ions is high. This material also displayed a relatively high electrical conductivity, with a value of 1.8×10^{-2} S cm^{-1} at 298 K. Both the asymmetric IVCT band and high electrical conductivity point toward a strongly coupled MV system, consistent with a class III Robin–Day material.

This work establishes the increased covalency between second-row transition metals and chloranilate ligands, in comparison to analogous frameworks with first-row transition metals. The primary result of this covalency is increased magnetic and electronic interactions throughout framework structures. These interactions are not isolated from each other however and may compete. While excellent charge mobility is present in the Mo framework, the Nb framework is dominated by strong magnetic interactions, precluding charge transport of the same magnitude. It is notable that such interactions can significantly impact the degree of coupling between MV centers, stressing the distinction between the behavior of MV pairs in discrete and network materials.

Hybrid density functional theory (DFT) calculations were performed on three 2D honeycomb structures incorporating Fe and Mn by Tyminska and Kepenekian [77]. This work determined that Fe honeycomb systems exhibited greater electron delocalization than equivalent Mn systems due to larger valence band dispersion and smaller band gaps between the oxidation states of metal centers. This relationship between band gap magnitude and conductivity could explain the high degree of conductivity observed in vanadium systems.

Since chloranilate ligands have dominated in the synthesis of MV tetraoxolene frameworks, Murase et al. synthesized a novel Fe honeycomb framework $(NEt_4)_2[Fe_2(Fan)_3]$ (H_2Fan=3,6-difluoro-2,5,-dihydroxy-1,4-benzoquinone, fluoranilic acid; NEt_4^+ = tetraethylammonium) (Figure 13.6a) [78]. Importantly, the fluoranilate framework is isostructural to the previously synthesized $(NEt_4)_2[Fe_2(Clan)_3]$ framework [79], which allowed an understanding of ligand substituent effects on spectroscopic and physical properties to be obtained. The fluoranilate framework shows greater electron delocalization than the isostructural chloranilate framework, in addition to comparable conductivity to previously reported 2D chloranilate frameworks, such as $(H_2NMe_2)_2[Fe_2(Clan)_3]$. $(NEt_4)_2[Fe_2(Fan)_3]$ shows a greater electrical conductivity than $(NEt_4)_2[Fe_2(Clan)_3]$ analogue as

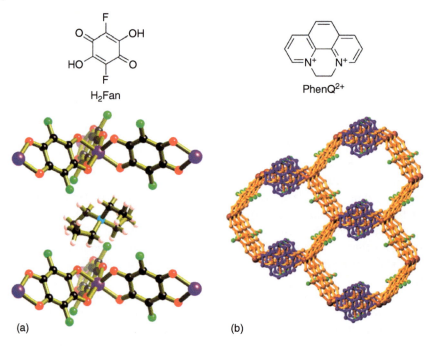

Figure 13.6 (a) Structure of the (NEt$_4$)$_2$[Fe$_2$(Fan)$_3$] framework showing NEt$_4^+$ cations located between the MV [Fe$_2$(Fan)$_3$]$^{2-}$ layers. Source: Murase et al. 2020 [78]/Reproduced with permission of American Chemical Society. (b) Structure of the (PhenQ)[Fe$_2$(Clan)$_3$]·solvent framework showing CT interactions of electron-deficient PhenQ^{2+} cations (purple) sandwiched between MV [Fe$_2$(Clan)$_3$]$^{2-}$ layers. Source: van Koeverden et al. 2020 [81]/Reproduced with permission of American Chemical Society.

well as more strongly coupled intervalence interaction. Comparisons between the frameworks discussed above were made with isostructural 3D systems (NBu$_4$)$_2$[Fe$_2$(dhbq)$_3$] and (NBu$_4$)$_2$[Fe$_2$(Fan)$_3$] [70, 80]. From this comparison, a correlation between lower v_{max} values of IVCT transitions and higher conductivities was established.

Notably, redox-inactive alkylammonium-based countercations have dominated in the synthesis of the aforementioned MV tetraoxolene frameworks. To elucidate new mechanisms to control the electronic structure and properties in these materials, van Koeverden et al. synthesized the 2D MV framework (PhenQ)[Fe$_2$(Clan)$_3$]·solvent (Clan^{n-} = 3,6-dichloro-2,5,-dihydroxy-1,4-benzoquinone) (Figure 13.6b), incorporating an electron-deficient dication 5,6-dihydropyrazino[1,2,3,4-*lmn*][1, 10]-phenanthrolindiium (PhenQ^{2+}) [81]. CT interactions between the electron-poor cation and ligands in the framework produced partial localization of radical trianion Clan$^{3-•}$ valence state. This resulted in a weak-to-moderately coupled IVCT transition within the MV ligand manifold, characterized as Class II by the Robin–Day classification scheme. Due to the ligand-based mixed valency, the framework also exhibited semiconductor behavior measuring between 3.5–5.4×10^{-4} S cm^{-1}

at 300 K. Long-range magnetic ordering was also observed with ferromagnetic coupling at high temperatures and net ferrimagnetic ordering at low temperatures. This work displays the degree of control possible over the CT properties of tetraoxolene frameworks.

13.2.2.4 Naphthalenediimide (NDI)-Based Compounds

Naphthalenediimides are widely used functional groups in solid-state chemistry, often used as a charge acceptor in DA systems. Additionally, NDI undergoes a one-electron reversible reduction to give a stable radical form [82]. Given the planar geometry of the NDI system, it is an ideal target for cofacial MV systems [23]. For example, Ding et al. synthesized a novel cadmium framework with the formula [Cd(DPNDI)(tdc)] (DPNDI = N,N′-di(4-pyridyl)-1,4,5,8-naphthalenediimide; H_2tdc = thiophene-2,5-dicaboxylic acid) [83]. This compound contains cofacially stacked DPNDI units at a distance of 3.3–3.5 Å (Figure 13.7). The radical reduced form of the DPNDI dimer, $(DPNDI)_2^{-\bullet}$, was generated by photoexcitation, where the radical state was stabilized by a DA relationship with tdc^{2-} ligands. Through-space IVCT was observed, which occurs between cofacial DPNDI pairs in the photoexcited state. There is an indication of charge delocalization between NDI units, which stabilized the radical anion state. As a result of the delocalization exhibited by the

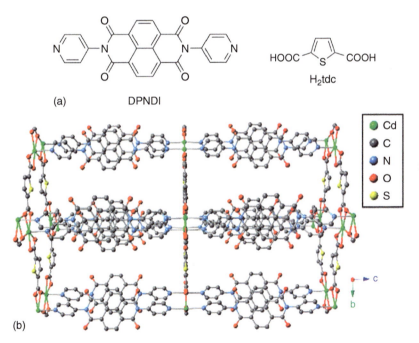

Figure 13.7 (a) Structure of the ligands in the [Cd(DPNDI)(tdc)] framework. (b) Crystal structure of the [Cd(DPNDI)(tdc)] framework showing cofacial stacking of the DPNDI units. Source: Ding et al. 2021 [83]/Reproduced with permission of Royal Society of Chemistry.

material in the photoexcited state, the system was investigated as a photocathode for the reduction of CO_2, performing with a 78% Faradaic efficiency.

13.2.2.5 Phenalenyl-Based Compounds

Phenalenyl is an odd-electron aromatic system comprising three six-membered rings with a resonance-stabilized radical delocalized over the entire π-system [84]. The phenalenyl core can be incorporated into larger organic molecules, including ligands for framework synthesis. The 1,3-diazaphenalenyl (DAP) radical is isoelectronic with the phenalenyl radical [84], so, consequently, Koo et al. synthesized a new ligand, 2,5,8-tri(4-pyridyl)-1,3-diazaphenalene (TPDAP) containing the redox-active DAP core [85]. Deprotonation of TPDAP and subsequent one-electron oxidation produces a neutral radical species. This radical can undergo reversible oxidation and reduction, and was used to produce cadmium-based frameworks [86].

In 2018, Ha et al. synthesized a ligand containing two redox-active DAP centers linked by a phenylene bridge, 5,5′,8,8′-tetra(4-pyridyl)-2,2′-(1,4-phenylene)bis-1H-perimidine (H_2TPP) (Figure 13.8a) [87]. The neutral diradical form of the ligand TPP•• contains a stabilized radical on each DAP core, with the neutral diradical state able to undergo two one-electron oxidations and two one-electron reductions. Two of the oxidation states TPP$^{-•}$ and TPP$^{+•}$ comprise MV DAP cores, which exhibit intramolecular IVCT between the redox-active DAP moieties mediated by the phenylene bridge. A broad IVCT band was observed in NIR spectrum for the MV TPP$^{+•}$ state, but no sufficiently intense band was observed in the TPP$^{-•}$ species. The IVCT in the cationic TPP$^{+•}$ species was characterized as Robin–Day class II mixed valency based on the shape of the NIR band. Subsequently, the framework $[Cd_2(NO_3)_4(H_2TPP)(DMF)_2]\cdot 0.5DMF\cdot 5H_2O$ (DMF = N,N'-dimethylformamide) was synthesized using this new ligand (Figure 13.8b). Cyclic voltammetry demonstrated the framework exhibits multiple accessible redox states, presumably including the MV states of the ligand. However, spectroscopy was not performed on the oxidized and reduced forms of the framework.

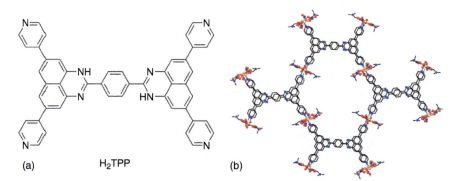

Figure 13.8 (a) Structure of the 5,5′,8,8′-tetra(4-pyridyl)-2,2′-(1,4-phenylene)bis-1H-perimidine (H_2TPP) ligand. (b) X-ray crystal structure of the $[Cd_2(NO_3)_4(H_2TPP)(DMF)_2]\cdot 0.5DMF\cdot 5H_2O$ framework. Yellow = Cd, red = O, blue = N, and black = C. H atoms and guest solvents not shown for clarity.

13.2.2.6 Covalent-Organic Frameworks (COFs)

COFs are a class of organic polymer material that have many features in common with MOFs, including permanent nanoporosity, 2D and 3D interconnected topologies, and crystallinity [88, 89]. Notably, COFs can incorporate redox-active functionality [90], allowing the development of 2D graphene-like COFs, with conductive layers linked to other layers through π-stacking interactions [91]. Thus far, there are only three COFs in the literature that have incorporated some form of mixed valency.

Cai et al. prepared a TTF-based COF by using imine linkage of 1,4-diaminobenzene with a tetraldehyde derivative of TTF to produce extended 2D sheets [92]. This material is an early example of through-space IVCT in an extended material, exhibiting very similar CT behavior to the TTF-based frameworks synthesized by Leong et al. [61]. While few efforts were made to quantify the IVCT band observed in this material, the conductivity could be tuned by I_2 doping, similar to that used by Leong et al. in $[Zn_2(TTFTB)(H_2O)_2]$ [61], ascribed to increasing delocalization upon framework oxidation.

Two new materials, both incorporating redox-active triphenylamine (TPA) groups, demonstrate the impact of topology on the degree and mechanisms of conductivity and IVCT. The earliest system, COF_{3PA-TT}, is a 2D framework consisting of flat sheets of tris(4-aminophenyl)amine (TAPA) and thieno[3,2-b]thiophene-2,5-dicarbaldehyde (TTDA) (Figure 13.9) [93]. The system was not characterized structurally, which obscures much of the possible analysis of CT. Upon electrochemical oxidation, the material displayed absorption in the NIR region, characterized as IVCT between neutral TPA groups and cationic radical TPA groups. There is significant ambiguity as to the nature of the CT mechanism in this system. On one hand, π-stacking is an established pathway to through-space IVCT between planar, redox-active organic groups [24, 53, 56]. COFs with similar ligands and bond geometries to COF_{3PA-TT} have produced materials with cofacial interactions [94, 95], indicating that COF_{3PA-TT} may also exhibit such a structural relationship, and thus, through-space IVCT may occur. On the other hand, there are a number of literature examples of TPA groups exhibiting through-bond IVCT [96], including the following work by Wang and coworkers [97]. The authors suggested that through-bond IVCT may be mediated by TTDA bridging groups, as there is precedent for imine linkers improving conductivity in extended materials [94]. Without further structural characterization and computational modeling, no definitive conclusion could be drawn on the exact nature of this transition. The framework also demonstrated applicability as an electrochromic sensor, exhibiting a reversible oxidation, corresponding to the formation of the MV state. A quasi-solid-state device was constructed to exploit this property.

The second system of note $COF_{TPDA-PDA}$ made use of N,N,N',N'-tetrakis (4-aminophenyl)-1,4-benzenediamine (TPDA) and terephthalaldehyde (PDA) as building blocks (Figure 13.9), for a 2D extended material [97]. TPDA can be considered as a dimer of aminophenyl units bridged by phenylene bridge. Both the ligand and the new framework exhibited IVCT consistent with a Robin–Day class III MV system. The electron delocalization between triarylamine groups is

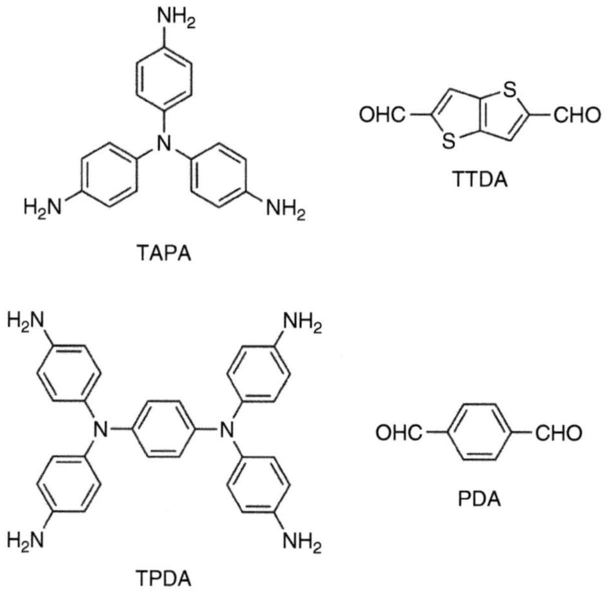

Figure 13.9 Arylamine and dialdehyde ligands used to prepare MV COFs.

notably much higher than the IVCT phenomenon observed in COF$_{3PA\text{-}TT}$, where the CT mechanism is hindered by either bridge distance or TAPA separation in a through-space interaction. The new framework undergoes two reversible redox processes, with all three oxidation states having distinct colors, demonstrating the potential as an electrochromic material. This property was exploited to construct a working flip/flop logic gate.

13.2.3 Metal-Based Mixed Valency

13.2.3.1 First-Row Transition Metals

First-row transition metals are widely incorporated into coordination polymers and MOFs. Many first-row metals employed for reticular synthesis have multiple accessible oxidation states within polymers, which leads to mixed valency under the right conditions. However, the extent of coupling between metal centers in these systems is often limited due to the large differences in coordination geometry between oxidation sites as well as the limited size of the first-row transition metal d orbitals, which precludes orbital overlap in most structures. Consequently, many frameworks employing first-row transition metals with mixed valency have been developed with catalysis and photocatalysis in mind.

Copper MV copper(I/II) systems have been commonly observed in the literature [98]. Cluster arrangements are frequently observed in extended systems containing multiple copper nodes [99, 100]. One limitation upon the use of Cu(I/II) mixed valency is that Cu(I) and Cu(II) typically occupy very different bonding environments in secondary building units (SBUs). IVCT is therefore often unfavorable due to the large coordination environment reorganization required upon electron transfer, which causes the compounds to contain uncoupled arrangements of MV clusters [17]. Despite this, MV Cu systems have found application in catalysis, which is discussed in depth below.

Tomar et al. synthesized two novel MV Cu frameworks incorporating Cu(II) paddlewheel SBUs, with the formulas [Cu^{I}Cu^{II}$_2$L$_2$(H$_2$O)(pyridine)$_2$](NO$_3$)·4DMF· 3H$_2$O and [Cu^{I}Cu^{II}$_2$L$_2$(pyridine)$_3$](NO$_3$)·DMF·4H$_2$O (H$_2$L = 4,4′-(methylenebis (3,5-dimethyl-1H-pyrazole-4,1-diyl))dibenzoic acid) [101]. Both frameworks contain Cu(II) ions in close proximity, with Cu(I) centers present outside the SBU. While MV systems are possible within the copper paddlewheel SBUs [102, 103], neither of these aforementioned novel frameworks appear to have any degree of coupling between copper centers.

Liu et al. synthesized a novel framework [(Cu^I$_4$Cu^{II}$_4$L$_4$)·3H$_2$O] (L = [3-(naphthalene-1-carbonyl)-thioureido] acetic acid) [104]. The SBUs contain eight Cu centers, with four central Cu(I) centers bonded to sulfur and nitrogen (Figure 13.10). There is also evidence of Cu–Cu interactions within the SBU. The outer four Cu(II) centers are bound in a nearly square–pyramidal coordination environment, each bonded to one carboxylate, two halves of a carboxylate, and a nitrogen ion. The bonding in these clusters is very complex; however, there does not appear to be any direct route for through-bond electronic coupling between metal centers. The framework was modified by exposure to ammonia vapor, producing a new material with improved proton conductivity, although the structure of this new material was not determined.

Gao et al. synthesized a novel copper framework, [Cu$_4$Cl(ptca)$_5$·5H$_2$O] (Hpcta = (4-pyridyl)thiazole-4-carboxylic acid) [105]. This framework contains four crystallographically distinct Cu sites, including two Cu(I) and two Cu(II) ions. The coordination environment of each site was distinct, with the Cu(I) ions displaying slightly different square–planar geometries, and the Cu(II) ions coordinated in distorted square–planar geometries. While Cu-based mixed valency was established by examination of the Cu 2p X-ray photoelectron spectrum (XPS), due to the spatial positions of the Cu centers in the asymmetric unit, no coupling or CT was observed.

Iron Iron is a frequently used metal for extended materials, including Hoffman-type and Hoffman-like ferro/ferricyanide coordination polymers [106, 107], and MOFs containing iron-based paddlewheel [108] and oxo-centered cluster SBUs [109, 110]. There is a rich history of mixed valency in materials containing group eight metals, including one of the first extended materials, Prussian blue being identified as containing MV iron centers [2, 5].

Xie et al. synthesized a novel iron framework, Fe$_2$(H$_{0.67}$BDT)$_3$·17(H$_2$O)·0.5(^{i}PrOH), where H$_2$BDT = 5,5′-(1,4-phenylene)bis(1H-tetrazole) (Figure 13.11a) [111]. The as-synthesized framework contains only valence-pure Fe(II) sites, with infinite 1D Fe–N–N chains forming the pillars of the structure. Upon exposure to air for several days, the framework darkens from orange–red to red and then black, suggesting the oxidation of Fe(II) facilitated CT in the structure, although the Fe oxidation

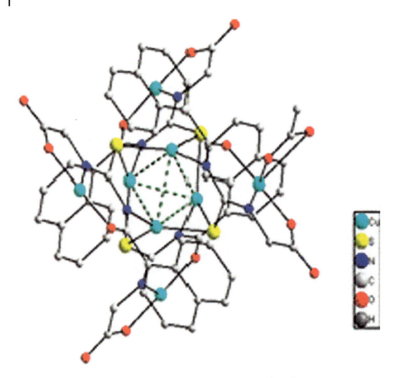

Figure 13.10 Structure of the SBU found in MV [(Cu$^I_4$Cu$^{II}_4$L$_4$)·3H$_2$O] (L = [3-(naphthalene-1-carbonyl)-thioureido] acetic acid). Source: Liu et al. 2018 [104]/Reproduced with permission of American Chemical Society.

state could not be unambiguously assigned by ^{57}Fe Mössbauer spectroscopy. An IVCT band in the visible-to-NIR region was observed in the oxidized compound (Figure 13.11b), and the system characterized as a Robin–Day class II material due to valence localization. Additionally, the single-crystal conductivity of the system could be tuned with exposure to oxygen, with conductivity increasing up to 5 orders of magnitude for the most oxidized structure.

Park et al. synthesized a family of MV iron-based triazolate frameworks [Fe(tri)$_2$](BF$_4$)$_x$ (Htri = 1,2,3-triazole; x = 0.09, 0.22, and 0.33) via chemical oxidation of the valence-pure Fe(II) framework [Fe(tri)$_2$] [112]. All three MV frameworks exhibit an IVCT band; however, the full band shapes could not be observed due to the low wavelengths of the IVCT transitions. ^{57}Fe Mössbauer spectroscopy at 290 K indicated all three frameworks contained valence averaged Fe(II)/Fe(III), a strong indication of class III Robin–Day mixed valency [17]. All three frameworks were electrically conductive; however, [Fe(tri)$_2$](BF$_4$)$_{0.33}$ exhibits an unusually high electronic conductivity value due to charge delocalization in the MV system.

Sobczak and Katrusiak postsynthetically modified a MV perovskite framework α-(H$_2$NMe$_2$)[Fe^{II}Fe^{III}(HCOO)$_6$] (HCOO$^-$ = formate) [113], previously synthesized by Zhao et al. [114]. PSM was achieved by compressing the framework in various liquid media, producing several different phases. Notably, a new framework

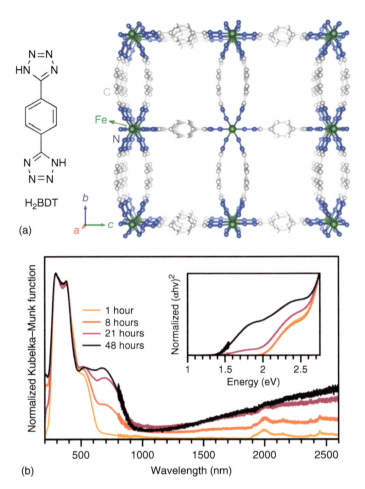

Figure 13.11 (a) Ligand and crystal structure of $Fe_2(H_{0.67}BDT)_3 \cdot 17(H_2O) \cdot 0.5(^iPrOH)$. (b) Diffuse reflectance UV-vis-NIR spectra of $Fe_2(H_{0.67}BDT)_3 \cdot 17(H_2O) \cdot 0.5(^iPrOH)$ at different periods of exposure to air. Source: Xie et al. 2018 [111]/Reproduced with permission of American Chemical Society.

$(H_2NMe_2)_3[Fe^{II}{}_3Fe^{III}(HCOO)_{12}] \cdot CO_2$ resulted when the original framework was compressed at 1.15 GPa in a mixture of methanol and ethanol. The transition is believed to occur due to dissolution of $(H_2NMe_2)[Fe^{II}Fe^{III}(HCOO)_6]$ in methanol/ethanol under high pressures, in combination with decomposition of $HCOO^-$ ligands. Additionally, a new phase of $(H_2NMe_2)[Fe^{II}Fe^{III}(HCOO)_6]$ was obtained by compression at 1.4 GPa in Daphne 7474 oil and isopropanol. The phase transition is reversible and results in the monoclinic space group $P\,2_1/c$ from the original trigonal space group $P\,\bar{3}1c$.

Furthering the studies on Fe–formate networks, Collings et al. synthesized a MV perovskite framework with the formula $[(H_2NMe_2)_3(H_2O)][Fe^{II}{}_3Fe^{III}(HCOO)_{12}]$ [115]. ^{57}Fe Mössbauer spectroscopy revealed the 3 : 1 ratio of Fe(II) to Fe(III) sites. Variable-temperature and pressure single-crystal X-ray diffraction (SC-XRD) studies were performed on the system, where VT SC-XRD revealed a shift from dynamic to the static disorder of $H_2NMe_2^+$ at low temperatures, and VP SC-XRD revealed the existence of a high-pressure phase of the material.

Liyanage et al. and Fu et al. synthesized two related oxo-centered iron frameworks [116, 117]. Both frameworks incorporated the conjugate base of 5-bromonicotinic acid (HBNA), giving 2D and 3D frameworks, respectively, and have the same formula [$Fe^{III}{}_2Fe^{II}(\mu_3$-$O)(BNA)_6$]. Both frameworks contain trimeric oxo-centered Fe clusters, which include six carboxylates and three nitrogens, all attached to the BNA ligand. This coordination environment of iron is well documented in the literature [118, 119]. Notably, the Fe(II) sites exist in a similar octahedral coordination environment to the Fe(III) sites, with notably different bond lengths and angles. No investigation of CT was performed.

Cobalt Li et al. synthesized a novel dicarboxyferrocene-based MOF, incorporating Co(II/III) mixed valency [120]. The framework [$Co^{II}{}_2Co^{III}(DFc)_2(OH)_3$]·$H_2O$ (DFc = 1,1′-dicarboxyferrocene) consists of octahedral $Co^{II}O_6$ and $Co^{III}O_6$ centers formed into puckered 2D sheets of $Co^{II}{}_2Co^{III}(ROO)_2(OH)_3$, which are then decorated with pendant DFc groups. The new material was investigated as a cathode material for LIBs because the accessible Fe(II)/Fe(III) and Co(II)/Co(III) couples give rise to a number of framework redox states. Upon oxidation of the material, PF_6^- anions are reversibly incorporated into the framework. When used as a LIB cathode, the devices displayed high-energy density, operating potential, rate performance, and stability, making it a promising platform for further development of dual-ion batteries with MOFs.

A highly selective gaseous ammonia sensor was developed by Zhang et al. using a novel MV cobalt framework with the formula [$Co(H_{0.27}L)$]·$4H_2O$·$0.5DMF$ (H_3L = tris-(4-tetrazolyl-phenyl)amine) [121]. In this compound, ammonia was selectively encapsulated by the framework within the pores due to hydrogen bonding between the guest and the tetrazole ligands. The adsorption of ammonia molecules is believed to alter the coordination environment of Co(II) and Co(III) ions within the framework, resulting in easily discernable color switching (Figure 13.12), making the framework a highly effective ammonia sensor.

Vanadium Polyoxovanadates (POVs) are a family of vanadium oxoclusters and a subclass of polyoxometalates (POMs), which are anionic molecular clusters formed from condensed oxyanions of early transition metals (M = Mo^{VI}, W^{VI}, V^V, Nb^V, Ta^V) [122]. MV POVs are popular targets, as the ratio of V(IV) and V(V) centers can be synthetically controlled [123]. Moreover, POMs can undergo facile redox processes with retention of the overall structure [122], allowing the electronic and magnetic properties to be modulated [124]. POVs are also common motif in vanadium MOFs [125–127].

Chen et al. synthesized the framework $(H_3DETA)_3(DETA)\{Na\{[V^{III}(DETA)]_2[W^{VI}{}_8V^{IV}{}_4O_{36}(PO_4)]\}_2\}_2$·$H_2O$ (DETA = diethylenetriamine) [128]. The framework incorporates POM nodes containing W and V. The system contains two uncoupled MV sites of V, one within and one outside the POM cluster. The system was

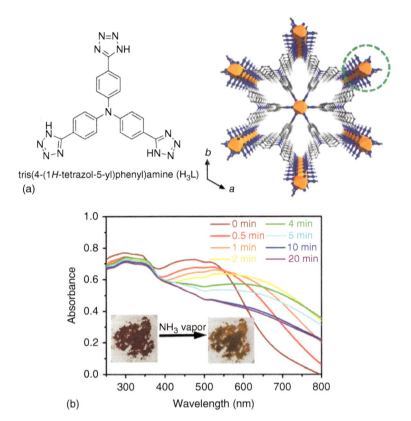

Figure 13.12 (a) Crystal structure of the [Co(H$_{0.27}$L)]·4H$_2$O·0.5DMF (H$_3$L = tris-(4-tetrazolyl-phenyl)amine) framework showing coordination of the tetrazole groups to the Co centers. Brown polyhedral = Co, blue = N, and gray = C. H atoms and guest solvent not shown for clarity. (b) Diffuse reflectance UV-vis spectra of [Co(H$_{0.27}$L)]·4H$_2$O·0.5DMF (H$_3$L = tris-(4-tetrazolyl-phenyl)amine) after different periods of exposure to ammonia vapor, including the visual color change. Source: Zhang et al. 2018 [121]/Reproduced with permission of American Chemical Society.

investigated for magnetic properties, where it was discovered to be paramagnetic at low temperatures and exhibited antiferromagnetic exchange at higher temperatures.

Chromium Park et al. synthesized a novel chromium triazolate framework [Cr(tri)$_2$](CF$_3$SO$_3$)$_{0.33}$ [129]. The single-crystal structure was observed to contain one CF$_3$SO$_3^-$ anion per pore, indicating the presence of Cr(II) and Cr(III) ions within the material. The material was designed to promote π–d conjugation between metal centers and linkers, which lead to high charge delocalization and magnetic ordering at high temperatures due to an interaction known as "double exchange". This interaction occurs when electrons are delocalized throughout the material and consequently promotes spin alignment across the structure. An IVCT band was observed in the NIR and mid-IR regions at notably low energy, indicating a high degree of electronic delocalization over the structure. Ferromagnetic ordering was observed below 225 K, the highest magnetic ordering temperature observed in a coordination solid or MOF.

13.2.3.2 Other Metals

Molybdenum Chen et al. prepared a framework incorporating polyoxomolybdates (POMos) linked with imidazole [130]. This follows earlier work by Martían-Zarza et al. into POMo compounds [131]. Chen et al. then produced a MV species of this Mo-MOF via photoreduction of the POMo clusters, which resulted in a slight color change. The modified framework contained oxygen vacancies resulting from the photoreduction of the POMo sites. The authors made use of this property by testing the framework as a surface-enhanced Raman spectroscopy (SERS) substrate. The efficacy of CT between substrate and sample in SERS is shown to enhance measurement signals [132].

Lanthanoids Jin et al. synthesized a novel cerium framework $[Ce^{III}_2Ce^{IV}_2(bdc)_6O(DMF)(H_2O)_4·5H_2O]$ (H_2bdc = benzenedicarboxylic acid) named CSUST-1 [133]. The framework consists of MV Ce(III) and Ce(IV) centers linked by bdc^{2-} groups, leading to high oxygen vacancies, which were identified through structural analysis. As the Ce(III) and Ce(IV) centers are not in close proximity to one another, IVCT behavior is unlikely in this system. CSUST-1 was subsequently incorporated into a composite material with activated carbon nanotubes. The composites acted as efficient separator coating materials for lithium–sulfur batteries, as they facilitated high sulfur loading due to the high concentration of oxygen vacancies [134].

Another MV Ce-MOF was developed by Xiong et al. [135]. This framework was found to exhibit oxidase-like activity. Later, the framework was discovered to exhibit uricase-like catalytic behavior able to efficiently catalyze the oxidation of uric acid to allantoin [136].

Lanthanoid MOFs are known for narrow emission and color purity of luminescence. Emissions tend to be sensitive to the chemical environment, making them ideal for sensing pollutants. In an attempt to exploit this property, Wang et al. synthesized a novel framework $[Tb_2(HCOO)(cyhex)·2H_2O]$ (H_6cyhex = 1,2,3,4,5,6-cyclohexanehexacarboxylic acid), which contained Tb(III) and Tb(IV) centers [137]. The characteristic UV-excited luminescence of this material occurs in the visible region, with bands at 489, 545, 489, and 621 nm. The luminescence mechanism arises from a ligand-to-metal charge transfer (LMCT) interaction, not as a direct result of Tb-based mixed valency. The material was used for selective sensing of small molecule pharmaceutical compounds, where it displayed stability after multiple cycles.

13.2.3.3 Catalysis in Uncoupled MV Systems

Catalysis is one of the most prevalent applications of MOFs due to their high porosity and highly variable structures and designs [138]. Two of the most popular strategies for catalysis in extended materials include promoting coordinatively unsaturated metal sites (CUS) and creating lattice defects. Both techniques are useful for creating active catalytic sites within frameworks [139]. Mixed valency offers an effective route to these properties in some materials, as the following systems will show.

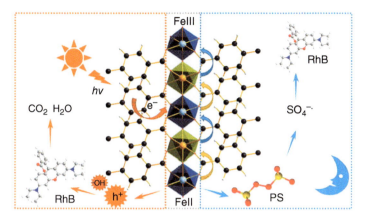

Figure 13.13 Schematic representation of the catalytic behavior of MIL-53(Fe$^{II/III}$), including photocatalytic degradation of Rhodamine B (RhB) and catalytic activation of persulfate (PS) for pollutant degradation. Source: Reproduced with permission from Chen et al. 2019 [141]. Copyright © 2019 American Chemical Society.

Iron-Based Catalysts MIL-53(Fe) has gained recent attention for its applicability as a photocatalyst, particularly the MV MIL-53(Fe$^{II/III}$) system [140]. In the scope of this review, two works made use of this system for catalytic applications. The first work found that the ratio of Fe(II) to Fe(III) in MIL-53(Fe) could be altered by post-synthetically heating the framework at different temperature under vacuum [141]. The material was tested at different Fe(II)/Fe(III) ratios to determine the most effective catalytic behavior in the degradation of Rhodamine B (Figure 13.13). The catalytic behavior was directly correlated to the prevalence of mixed valency within the structure, as Fe(III) ions introduced holes and charge mobility into the system. The catalytic activity was increased by increasing the ratio of Fe(II) to Fe(III), but the integrity of the 1D Fe chains began to degrade at higher Fe(II) content, leading to partial collapse of the structure.

The second work produced MIL-53(Fe$^{II/III}$) through *in situ* chemical reduction of MIL-53(Fe) with ethylene glycol (EG) during the solvothermal reaction [142]. Varying the ratio of EG to DMF could be used to tune the ratio Fe(II) to Fe(III) ions in the structure. MV MIL-53(Fe^{II}/Fe^{III}) has also been tested as a nitrogenase-like catalyst in the nitrogen fixation process, i.e. converting N$_2$ to ammonia [142]. The MV frameworks performed better in this process compared to valence-pure MIL-53(Fe^{II}).

MIL-100(Fe) is another Fe system that contains CUS as a result of mixed valency, which can be achieved by partial reduction of the framework [143]. Two recent works have made use of MIL-100(Fe$^{II/III}$) as a highly efficient catalyst [42, 144]. Similar to MIL-53(Fe$^{II/III}$), MIL-100(Fe$^{II/III}$) is a more effective catalyst than the valence-pure Fe(II) framework due to the presence of additional CUS and higher charge mobility within the MV framework. Tang et al. investigated the system for use as a Fenton-like catalyst, which displayed a 100% degradation of sulfamethazine over 180 minutes [42]. Similarly, Rostamnia and Alamgholiloo found that MIL-100(Fe$^{II/III}$) was an effective catalyst for the aza-Michael reaction giving a yield of 94% in the synthesis of 3-(phenylamino)propanoate from methyl acrylate and aniline [144].

Ahmad et al. produced a MV framework from MIL-88(Fe) by thermal activation under vacuum [145]. Similar to other MIL(Fe) frameworks, this produced a number of CUS which were employed in photo-Fenton catalysis. The new material was deposited on ultrathin Ti_3C_2 nanosheets and used as a photocatalyst in the production of water from hydrogen peroxide. The MV system showed significantly higher Fenton yield than MIL-88B(Fe).

MOFs have also been used to template materials with similar properties to those above, where Xu and coworkers produced several MV hematite derivatives for CUS-enhanced Fenton photocatalysis [146].

POM-Based Catalysts $[Ni(4,4'-bpy)_2]_2[V^{IV}_7V^V_9O_{38}Cl]\cdot(4,4'-bpy)\cdot 6H_2O$, named NENU-MV-1 [147], is a MV POV-based MOF, that was subsequently used by Wang et al. to catalytically oxidize cyclohexane [148]. The authors determined a relationship between particle size and catalytic efficacy, with smaller particles producing higher yields. The catalytic mechanism may involve metal-assisted radical generation. The framework showed excellent stability and low solubility in the solvent system, making it a promising target for future catalyst development.

Tian et al. synthesized a novel MV POV-based framework $[Cu^I(bbi)]_2\{[Cu^I(bbi)]_2 V^{IV}_2V^V_8O_{26}\}\cdot 2H_2O$ (bbi = 1,1'-(1,4-butanediyl)bis(imidazole)), called NENU-MV-5 [149], with comparable catalytic behavior to NENU-MV-1. NENU-MV-5 was used for the catalytic cleavage of β-O-4 lignin compounds. It was determined that the catalytic mechanism involved the production of $O_2^{-}\cdot$ radicals via the oxidation of O_2 by V^{IV} ions. V^V ions are also believed to contribute to the activation of Cα–H bonds of the lignin, further contributing to the catalytic behavior of this material.

Cobalt-Based Catalysts Zhang et al. also discovered a MV framework that behaved as a catalyst via radical stabilization [150]. The framework $[Co^{II}Co^{III}_2(\mu_3\text{-}O)(bdc)_3(tpt)]$ (tpt = 2,4,6-tri(4-pyridyl)-1,3,5-triazine) was used to selectively oxidize cyclohexene. The framework was believed to produce *tert*-butoxy and hydroxyl radicals when *tert*-butyl hydroperoxide and oxygen were used as terminal oxidants, respectively. The proposed mechanism is unique to the $[Co^{II}Co^{III}_2(\mu_3\text{-}O)]$ cluster.

Zhang et al. produced a MV Co framework with the formula $[Co_3(\mu_3\text{-}OH)(pydc)(H_2O)_3]\cdot 3DMF\cdot 8H_2O$ (H_6pydc = 5,5',5''-(pyridine-2,4,6-triyl)tri-isophthalic acid) [151]. The framework contains MV triangular$[Co^{II}_2Co^{III}(\mu_3\text{-}OH)(COO)_6(H_2O)_3]$ clusters linked by pydc ligands. As the clusters contained three quasi-open metal sites (occupied by water ligands), PSM was performed by substituting the labile H_2O ligands for the planar ligand tris(3-pyridyl)-1,3,5-benzene (tpb). The addition of this additional linker greatly improved the stability of the framework by providing further structural support to the framework, which was assisted by π–π stacking between the newly introduced pyridyl groups. With this additional linker, the framework retained pore size when the pores were activated, whereas the original framework shrunk significantly. Both materials were then tested for selective CO_2 adsorption, where the newly modified framework adsorbed 43 cm³ g⁻¹ at 298 K, whereas the unmodified framework did not adsorb CO_2 by any measure at this temperature.

Copper-Based Catalysts Xu and coworkers synthesized a novel Cu MV system $[Cu^I{}_3Cu^{II}{}_3(pbt)_3Cl_2]\cdot ClO_4\cdot H_2O$ (H_2pbt = 5′-(pyridin-2-yl)-2H,4′H-3,3′-bi(1,2,4-triazole)), which contains clusters of three Cu(I) and three Cu(II) ions [152]. The compound was used to photocatalytically degrade methylene blue and methyl orange dyes. The proposed mechanism involves the production of electron-hole pairs and charge localization in the framework (via LMCT), after which OH• radicals are generated through adsorption of O_2 or OH^- onto the surface of the framework. The mixed valency present in the framework is believed to reduce the band gap between the conduction and valence bands in the material, thereby improving catalytic performance.

Gao et al. synthesized a novel Cu(I)/Cu(II) framework $[(Cu_4I_4)_{2.5}[Cu_3(\mu_4\text{-O})(\mu_3\text{-I})(pmc)_3(dabco)_3]\cdot 2.5DMF\cdot 2MeCN]$ (dabco = 1,4-diazabicyclo[2.2.2]octane; pmc = pyrimidine-5-carboxylic acid), which incorporates both $Cu_3OI(CO_2)_3$ and Cu_4I_4 moieties [153]. This system was employed for the photoreduction of CO_2 to CH_4. The catalytic behavior of this system involves both copper clusters. The Cu_4I_4 clusters are responsible for photoelectron generation, and the $Cu_3OI(CO_2)_3$ clusters act as photoelectron collectors and active sites for CO_2 reduction. The mixed valency of this system directly contributes to the catalytic behavior. The Cu_4I_4 clusters likely generate photoelectrons by virtue of electron-rich Cu(I) ions, and the catalytic mechanism of $Cu_3OI(CO_2)_3$ clusters involves the reduction of Cu(II) ions. The versatility of Cu in different oxidation states, including coordination environments and electron transfer behavior, is exemplified by this work.

Another photocatalyst incorporating MV Cu was developed by Zhao et al., with the formula $[Cu^I{}_2Cu^{II}(tpt)_2(L)]\cdot 15H_2O$ (L = 3,3′-disulfonyl-4,4′-biphenyldicarboxylate) [154]. This system was investigated as a dual catalyst for both water splitting (hydrogen evolution) and the degradation of Rhodamine B. The system compared favorably to monovalent transition metal systems synthesized using similar reagents for two reasons. Firstly, the presence of both Cu(I) and Cu(II) ions in the structure afforded two different coordination geometries with triangular and octahedral coordination, respectively. This allowed the structure to adopt a pillared structure, increasing dimensionality, which, in turn, increased catalytic performance due to better separation of electron-hole pairs and faster CT.

Li et al. developed a framework with the formula $(NH_4)\{Cu^{II}{}_3\cdot [Cu^{II}Cu^I{}_6(OH)_6(Ad)_6]_2\}\cdot (H_2O)_x$ (Ad = adenine) [155]. This system contains MV Cu clusters and was employed for the kinetic separation of acetylene (C_2H_2) from ethylene (C_2H_4). The central clusters of this framework comprise planar heptanuclear $[Cu^{II}Cu^I{}_6(OH)_6]$ clusters that contain several open metal sites by virtue of the Cu(I) ions. As a result, the framework is capable of binding a high density of hydrocarbons and is able to separate C_2H_2 from C_2H_4 with a high level of selectivity.

MOF Nanozymes MOFs can be used to emulate enzyme-like catalytic behavior [156, 157] due to their designable structures and pores, large surface areas [158], as well as the properties already explored in this work, including open metal sites and CT behavior. Many nanozyme MOFs rely on redox-active metals and tend to perform better with mixed valency or mixed metals.

Luo et al. synthesized a novel Ce–MOF, called Ce-BPyDC, containing MV Ce(III/IV) centers [158]. The framework exhibits both oxidase-like and peroxidase-like activity. The oxidase-like mechanism is believed to directly involve the MV Ce centers, via the reduction of the Ce(IV) ions to Ce(III), and subsequent oxidation back to Ce(IV). The peroxidase-like mechanism involves the generation of OH• radicals, also by virtue of the MV metal centers.

Ma et al. developed an electrochemiluminescence (ECL) sensor from a MV Ce framework, which was used to detect the insecticide imidacloprid in plant-derived foods [159]. Despite issues with the sensitivity of the system, the strategy used to develop the ECL sensor will undoubtedly be useful for generating more effective systems.

Zhao and coworkers used MV MIL-53(Fe) to emulate nitrogenase catalytic activity [142]. This system performed favorably in this reaction compared to the valence-pure MIL-53(Fe^{III}) system.

13.3 Conclusion

Throughout this work, we have observed a unique relationship between the design of a framework, i.e. the chosen topology, clusters, linkers and their spatial and orbital relationships, and mixed valency. As is the case for many of the systems discussed within this work, MV MOFs frequently exhibit emergent properties either because of or for reasons relating to mixed valency. Perhaps the best example of this relationship is the prevalence of conductive and semiconductive extended systems exhibiting mixed valency. Through-bond and through-space IVCT can only occur where charge mobility is sufficient for electrons to transfer efficiently between the redox-active centers. As a result, these systems frequently exhibit high charge mobility throughout their entire structures, leading to semiconductivity or conductivity.

Coupled MV systems can facilitate photocatalysis by providing pathways for other CT relationships (such as MLCT and LMCT) to occur, thereby facilitating photoactivation of species. Many MV frameworks have multiple accessible oxidation states, which can often be accessed reversibly. This property of MV frameworks can lead to porous electrode materials, which can reversibly store high concentrations of ions, and are extremely useful for battery development. While not always directly related to mixed valency, magnetic properties often arise in these systems where the mixed valency involves organic radicals which facilitate magnetic interactions between paramagnetic metal centers. Mechanisms, such as double exchange, involve a direct relationship between charge mobility and spin orientation. Mixed valency still represents a useful target when designing systems where CT or conductivity are not desirable or important. Mixed valency allows easy introduction of CUS, which in conjunction with nanoporous structures can create an ideal environment for efficient catalysis to occur.

The growing focus on organic MV and through-space IVCT interactions in extended materials has allowed a more in-depth exploration of these types of mechanisms and opened up a range of possibilities with regard to viable ligands and tunability of properties, such as conductivity, catalytic activity, and magnetic behavior.

References

1 Brown, D.B. (1980). *Mixed-Valence Compounds: Theory and Applications in Chemistry, Physics, Geology, and Biology*. Dordrecht: Springer Netherlands.
2 Day, P., Hush, N.S., and Clark, R.J.H. (2008). Mixed valence: origins and developments. *Phil. Trans. R. Soc. A* 366: 5–14.
3 Wheland, R.C. (1976). Correlation of electrical conductivity in charge-transfer complexes with redox potentials, steric factors, and heavy atom effects. *J. Am. Chem. Soc.* 98: 3926–3930.
4 D'Alessandro, D.M., Kanga, J.R.R., and Caddy, J.S. (2011). Towards conducting metal-organic frameworks. *Aust. J. Chem.* 64: 718–722.
5 Karyakin, A.A. (2001). Prussian blue and its analogues: electrochemistry and analytical applications. *Electroanalysis* 13: 813–819.
6 Mortimer, R.J. (1999). *Spectroelectrochemistry, applications*. In: *Encyclopedia of Spectroscopy and Spectrometry* (ed. J.C. Lindon, G.E. Tranter and J.L. Holmes), 2161–2174. Oxford: Elsevier.
7 Fletcher, S. (2010). The theory of electron transfer. *J. Solid State Electrochem.* 14: 705–739.
8 Allen, G.C. and Hush, N.S. (1967). *Intervalence-transfer absorption. Part 1. Qualitative evidence for intervalence-transfer absorption in inorganic systems in solution and in the solid state*. In: *Progress in Inorganic Chemistry*, vol. 8 (ed. F.A. Cotton), 357–389. New York: Wiley.
9 Hush, N.S. (1961). Adiabatic theory of outer sphere electron-transfer reactions in solution. *Trans. Faraday Soc.* 57: 557–580.
10 Marcus, R.A. (1963). On the theory of oxidation—reduction reactions involving electron transfer. V. Comparison and properties of electrochemical and chemical rate constants1. *J. Phys. Chem.* 67: 853–857.
11 Wong, K.Y., Schatz, P.N., and Piepho, S.B. (1979). Vibronic coupling model for mixed-valence compounds. Comparisons and predictions. *J. Am. Chem. Soc.* 101: 2793–2803.
12 Newton, M.D. (1991). Quantum chemical probes of electron-transfer kinetics: the nature of donor-acceptor interactions. *Chem. Rev.* 91: 767–792.
13 Hush, N.S. (1967). *Intervalence-transfer absorption. Part 2. Theoretical considerations and spectroscopic data*. In: *Progress in Inorganic Chemistry*, vol. 8 (ed. F.A. Cotton), 391–444. New York: Wiley.
14 Marcus, R.A. (1956). On the theory of oxidation-reduction reactions involving electron transfer. I. *J. Chem. Phys.* 24: 966–978.

15 Marcus, R.A. (1964). Chemical and electrochemical electron-transfer theory. *Annu. Rev. Phys. Chem.* 15: 155–196.
16 Creutz, C. (1983). *Mixed valence complexes of d^5-d^6 metal centers*. In: *Progress in Inorganic Chemistry*, vol. 30 (ed. S.J. Lippard), 1–73. New York, Wiley.
17 Robin, M.B. and Day, P. (1968). *Mixed valence chemistry-a survey and classi fication*. In: *Advances in Inorganic Chemistry and Radiochemistry*, vol. 10 (ed. H.J. Eeméus and A.G. Sharpe), 247–422. New York City: Academic Press.
18 Parthey, M. and Kaupp, M. (2014). Quantum-chemical insights into mixed-valence systems: within and beyond the Robin–Day scheme. *Chem. Soc. Rev.* 43: 5067–5088.
19 Heckmann, A. and Lambert, C. (2012). Organic mixed-valence compounds: a playground for electrons and holes. *Angew. Chem. Int. Ed.* 51: 326–392.
20 Matsunaga, Y. and Takayanagi, K. (1980). Electrical and optical properties of the 5,6:11,12-bis(epidithio)naphthacene-bromine system (TTT–Br_n). Formation of organic mixed-valence compounds (TTT)(TTT$^+$)Br$^-$ and (TTT$^+$)(TTT^{2+})(Br$^-$)$_3$. *Bull. Chem. Soc. Jpn.* 53: 2796–2799.
21 Amthor, S. and Lambert, C. (2006). [2.2]paracyclophane-bridged mixed-valence compounds: application of a generalized Mulliken–Hush three-level model. *J. Phys. Chem. A* 110: 1177–1189.
22 Kaupp, M., Gückel, S., Renz, M. et al. (2016). Electron transfer pathways in mixed-valence paracyclophane-bridged bis-triarylamine radical cations. *J. Comput. Chem.* 37: 93–102.
23 Wu, Y., Frasconi, M., Gardner, D.M. et al. (2014). Electron delocalization in a rigid cofacial naphthalene-1,8:4,5-bis(dicarboximide) dimer. *Angew. Chem. Int. Ed.* 53: 9476–9481.
24 Ding, B., Hua, C., Kepert, C.J., and D'Alessandro, D.M. (2019). Influence of structure–activity relationships on through-space intervalence charge transfer in metal–organic frameworks with cofacial redox-active units. *Chem. Sci.* 10: 1392–1400.
25 Zhou, H.-C.J. and Kitagawa, S. (2014). Metal–organic frameworks (MOFs). *Chem. Soc. Rev.* 43: 5415–5418.
26 Rowsell, J.L.C. and Yaghi, O.M. (2004). Metal–organic frameworks: a new class of porous materials. *Microporous Mesoporous Mater.* 73: 3–14.
27 Janiak, C. (2003). Engineering coordination polymers towards applications. *Dalton Trans.* 32: 2781–2804.
28 Gómez-García, C.J., Coronado, E., Laukhin, V., and Galán-Mascarós, J.R. (2000). Coexistence of ferromagnetism and metallic conductivity in a molecule-based layered compound. *Nature* 408: 447–449.
29 Takaishi, S., Hosoda, M., Kajiwara, T. et al. (2009). Electroconductive porous coordination polymer Cu[Cu(pdt)$_2$] composed of donor and acceptor building units. *Inorg. Chem.* 48: 9048–9050.
30 Sun, L., Miyakai, T., Seki, S., and Dincă, M. (2013). Mn$_2$(2,5-disulfhydrylbenzene-1,4-dicarboxylate): a microporous metal–organic framework with infinite (−Mn−S−)$_\infty$ chains and high intrinsic charge mobility. *J. Am. Chem. Soc.* 135: 8185–8188.

31 Sun, L., Campbell, M.G., and Dincă, M. (2016). Electrically conductive porous metal-organic frameworks. *Angew. Chem. Int. Ed.* 55: 3566–3579.
32 Stavila, V., Talin, A.A., and Allendorf, M.D. (2014). MOF-based electronic and opto-electronic devices. *Chem. Soc. Rev.* 43: 5994–6010.
33 Chen, Z.-F., Xiong, R.-G., Zhang, J. et al. (2000). The first chiral 2-D molecular triangular grid. *J. Chem. Soc., Dalton Trans.* 4010–4012.
34 Dai, J.-C., Wu, X.-T., Fu, Z.-Y. et al. (2002). Synthesis, structure, and fluorescence of the novel cadmium(II)–trimesate coordination polymers with different coordination architectures. *Inorg. Chem.* 41: 1391–1396.
35 Fu, Z.-Y., Wu, X.-T., Dai, J.-C. et al. (2002). The structure and fluorescence properties of two novel mixed-ligand supramolecular frameworks with different structural motifs. *Eur. J. Inorg. Chem.* 2002: 2730–2735.
36 Kurmoo, M. and Kepert, C.J. (1998). Hard magnets based on transition metal complexes with the dicyanamide anion, $\{N(CN)_2\}^-$. *New J. Chem.* 22: 1515–1524.
37 Riou-Cavellec, M., Albinet, C., Livage, C. et al. (2002). Ferromagnetism of the hybrid open framework $M[M_3(btc)_3]\cdot 5H_2O$ (M = Fe, Co) or MIL-45. *Solid State Sci.* 4: 267–270.
38 Ouellette, W., Galan-Mascaros, J.R., Dunbar, K.R., and Zubieta, J. (2006). Hydrothermal synthesis and structure of a three-dimensional cobalt(II) triazolate magnet. *Inorg. Chem.* 45: 1909–1911.
39 Talin, A.A., Centrone, A., Ford, A.C. et al. (2014). Tunable electrical conductivity in metal-organic framework thin-film devices. *Science* 343: 66–69.
40 Kitagawa, S., Kitaura, R., and Noro, S.-i. (2004). Functional porous coordination polymers. *Angew. Chem. Int. Ed.* 43: 2334–2375.
41 Mandal, S., Natarajan, S., Mani, P., and Pankajakshan, A. (2021). Post-synthetic modification of metal-organic frameworks toward applications. *Adv. Funct. Mater.* 31: 2006291.
42 Tang, J.T. and Wang, J.L. (2018). Metal organic framework with coordinatively unsaturated sites as efficient Fenton-like catalyst for enhanced degradation of sulfamethazine. *Environ. Sci. Technol.* 52: 5367–5377.
43 Hoffmann, R. (1971). Interaction of orbitals through space and through bonds. *Acc. Chem. Res.* 4: 1–9.
44 Xie, L.S., Skorupskii, G., and Dincă, M. (2020). Electrically conductive metal–organic frameworks. *Chem. Rev.* 120: 8536–8580.
45 Dinolfo, P.H., Lee, S.J., Coropceanu, V. et al. (2005). Borderline class II/III ligand-centered mixed valency in a porphyrinic molecular rectangle. *Inorg. Chem.* 44: 5789–5797.
46 Dinolfo, P.H., Williams, M.E., Stern, C.L., and Hupp, J.T. (2004). Rhenium-based molecular rectangles as frameworks for ligand-centered mixed valency and optical electron transfer. *J. Am. Chem. Soc.* 126: 12989–13001.
47 Dinolfo, P.H. and Hupp, J.T. (2004). Tetra-rhenium molecular rectangles as organizational motifs for the investigation of ligand-centered mixed valency: three examples of full delocalization. *J. Am. Chem. Soc.* 126: 16814–16819.

48 Osaka, I., Sauvé, G., Zhang, R. et al. (2007). Novel thiophene-thiazolothiazole copolymers for organic field-effect transistors. *Adv. Mater.* 19: 4160–4165.

49 Osaka, I., Zhang, R., Sauvé, G. et al. (2009). High-lamellar ordering and amorphous-like π-network in short-chain thiazolothiazole–thiophene copolymers lead to high mobilities. *J. Am. Chem. Soc.* 131: 2521–2529.

50 Mishra, S.P., Palai, A.K., Kumar, A. et al. (2010). Highly air-stable thieno[3,2-*b*]thiophene-thiophene-thiazolo[5,4-*d*]thiazole-based polymers for light-emitting diodes. *Macromol. Chem. Phys.* 211: 1890–1899.

51 Jung, I.H., Yu, J., Jeong, E. et al. (2010). Synthesis and photovoltaic properties of cyclopentadithiophene-based low-bandgap copolymers that contain electron-withdrawing thiazole derivatives. *Chem. Eur. J.* 16: 3743–3752.

52 Rizzuto, F.J., Faust, T.B., Chan, B. et al. (2014). Experimental and computational studies of a multi-electron donor-acceptor ligand containing the thiazolo[5,4-*d*]thiazole core and its incorporation into a metal-organic framework. *Chem. Eur. J.* 20: 17597–17605.

53 Hua, C., Doheny, P.W., Ding, B.W. et al. (2018). Through-space intervalence charge transfer as a mechanism for charge delocalization in metal-organic frameworks. *J. Am. Chem. Soc.* 140: 6622–6630.

54 Hush, N.S. (1985). Distance dependence of electron transfer rates. *Coord. Chem. Rev.* 64: 135–157.

55 Kubelka, P. and Munk, F. (1931). An article on optics of paint layers. *Z. Tech. Phys.* 12: 593–609.

56 Doheny, P.W., Clegg, J.K., Tuna, F. et al. (2020). Quantification of the mixed-valence and intervalence charge transfer properties of a cofacial metal-organic framework via single crystal electronic absorption spectroscopy. *Chem. Sci.* 11: 5213–5220.

57 Krausz, E. (1998). Low temperature polarised absorption spectroscopy of microcrystals made easy. *Aus. Opti. Soc. News* 12: 21–24.

58 Krausz, E. (1993). A single-beam approach to the absorption spectroscopy of microcrystals. *Aust. J. Chem.* 46: 1041–1054.

59 Jérome, D. (2004). Organic conductors: from charge density wave TTF–TCNQ to superconducting $(TMTSF)_2PF_6$. *Chem. Rev.* 104: 5565–5592.

60 Saad, A., Barriere, F., Levillain, E. et al. (2010). Persistent mixed-valence $[(TTF)_2]^{+\bullet}$ dyad of a chiral bis(binaphthol)-tetrathiafulvalene (TTF) derivative. *Chem. Eur. J.* 16: 8020–8028.

61 Leong, C.F., Wang, C.H., Ling, C.D., and D'Alessandro, D.M. (2018). A spectroscopic and electrochemical investigation of a tetrathiafulvalene series of metal-organic frameworks. *Polyhedron* 154: 334–342.

62 Park, S.S., Hontz, E.R., Sun, L. et al. (2015). Cation-dependent intrinsic electrical conductivity in isostructural tetrathiafulvalene-based microporous metal–organic frameworks. *J. Am. Chem. Soc.* 137: 1774–1777.

63 Hu, J.J., Li, Y.G., Wen, H.R. et al. (2021). A family of lanthanide metal-organic frameworks based on a redox-active tetrathiafulvalene-dicarboxylate ligand showing slow relaxation of magnetisation and electronic conductivity. *Dalton Trans.* 50: 14714–14723.

64 Kitagawa, S. and Kawata, S. (2002). Coordination compounds of 1,4-dihydroxybenzoquinone and its homologues. Structures and properties. *Coord. Chem. Rev.* 224: 11–34.

65 Guo, D. and McCusker, J.K. (2007). Spin exchange effects on the physicochemical properties of tetraoxolene-bridged bimetallic complexes. *Inorg. Chem.* 46: 3257–3274.

66 Dei, A., Gatteschi, D., Pardi, L., and Russo, U. (1991). Tetraoxolene radical stabilization by the interaction with transition-metal ions. *Inorg. Chem.* 30: 2589–2594.

67 Min, K.S., DiPasquale, A.G., Golen, J.A. et al. (2007). Synthesis, structure, and magnetic properties of valence ambiguous dinuclear antiferromagnetically coupled cobalt and ferromagnetically coupled iron complexes containing the chloranilate(2−) and the significantly stronger coupling chloranilate(•3−) radical trianion. *J. Am. Chem. Soc.* 129: 2360–2368.

68 Mercuri, M.L., Congiu, F., Concas, G., and Sahadevan, S.A. (2017). Recent advances on anilato-based molecular materials with magnetic and/or conducting properties. *Magnetochemistry* 3: 17.

69 Ward, M.D. (1996). A dinuclear ruthenium(ii) complex with the dianion of 2,5-dihydroxy-1,4-benzoquinone as bridging ligand. Redox, spectroscopic, and mixed-valence properties. *Inorg. Chem.* 35: 1712–1714.

70 Darago, L.E., Aubrey, M.L., Yu, C.J. et al. (2015). Electronic conductivity, ferrimagnetic ordering, and reductive insertion mediated by organic mixed-valence in a ferric semiquinoid metal–organic framework. *J. Am. Chem. Soc.* 137: 15703–15711.

71 Abrahams, B.F., Hudson, T.A., McCormick, L.J., and Robson, R. (2011). Coordination polymers of 2,5-dihydroxybenzoquinone and chloranilic acid with the (10,3)-*a* topology. *Cryst. Growth Des.* 11: 2717–2720.

72 Jeon, I.-R., Negru, B., Van Duyne, R.P., and Harris, T.D. (2015). A 2D semiquinone radical-containing microporous magnet with solvent-induced switching from T_c = 26 to 80 K. *J. Am. Chem. Soc.* 137: 15699–15702.

73 DeGayner, J.A., Jeon, I.-R., Sun, L. et al. (2017). 2D conductive iron-quinoid magnets ordering up to T_c = 105 K via heterogenous redox chemistry. *J. Am. Chem. Soc.* 139: 4175–4184.

74 Ziebel, M.E., Gaggioli, C.A., Turkiewicz, A.B. et al. (2020). Effects of covalency on anionic redox chemistry in semiquinoid-based metal-organic frameworks. *J. Am. Chem. Soc.* 142: 2653–2664.

75 Ziebel, M.E., Darago, L.E., and Long, J.R. (2018). Control of electronic structure and conductivity in two-dimensional metal–semiquinoid frameworks of titanium, vanadium, and chromium. *J. Am. Chem. Soc.* 140: 3040–3051.

76 Ziebel, M.E., Ondry, J.C., and Long, J.R. (2020). Two-dimensional, conductive niobium and molybdenum metal-organic frameworks. *Chem. Sci.* 11: 6690–6700.

77 Tyminska, N. and Kepenekian, M. (2021). Interplay between electronic, magnetic, and transport properties in metal organic-radical frameworks. *J. Phys. Chem. C* 125: 11225–11234.

78 Murase, R., Commons, C.J., Hudson, T.A. et al. (2020). Effects of mixed valency in an Fe-based framework: coexistence of slow magnetic relaxation, semiconductivity, and redox activity. *Inorg. Chem.* 59: 3619–3630.

79 Kingsbury, C.J., Abrahams, B.F., D'Alessandro, D.M. et al. (2017). Role of NEt_4^+ in orienting and locking together $[M_2lig_3]^{2-}$ (6,3) sheets (H_2lig = chloranilic or fluoranilic acid) to generate spacious channels perpendicular to the sheets. *Cryst. Growth Des.* 17: 1465–1470.

80 Murase, R., Abrahams, B.F., D'Alessandro, D.M. et al. (2017). Mixed valency in a 3D semiconducting iron–fluoranilate coordination polymer. *Inorg. Chem.* 56: 9025–9035.

81 van Koeverden, M.P., Abrahams, B.F., D'Alessandro, D.M. et al. (2020). Tuning charge-state localization in a semiconductive iron(III)-chloranilate framework magnet using a redox-active cation. *Chem. Mater.* 32: 7551–7563.

82 Bhosale, S.V., Jani, C.H., and Langford, S.J. (2008). Chemistry of naphthalene diimides. *Chem. Soc. Rev.* 37: 331–342.

83 Ding, B.W., Chan, B., Proschogo, N. et al. (2021). A cofacial metal-organic framework based photocathode for carbon dioxide reduction. *Chem. Sci.* 12: 3608–3614.

84 Morita, Y., Aoki, T., Fukui, K. et al. (2002). A new trend in phenalenyl chemistry: a persistent neutral radical, 2,5,8-tri-*tert*-butyl-1,3-diazaphenalenyl, and the excited triplet state of the gable *syn*-dimer in the crystal of column motif. *Angew. Chem. Int. Ed.* 41: 1793–1796.

85 Koo, J.Y., Yakiyama, Y., Kim, J. et al. (2015). Redox-active diazaphenalenyl-based molecule and neutral radical formation. *Chem. Lett.* 44: 1131–1133.

86 Koo, J.Y., Yakiyama, Y., Lee, G.R. et al. (2016). Selective formation of conductive network by radical-induced oxidation. *J. Am. Chem. Soc.* 138: 1776–1779.

87 Ha, J.Y., Koo, J.Y., Ohtsu, H. et al. (2018). An organic mixed-valence ligand for multistate redox-active coordination networks. *Angew. Chem. Int. Ed.* 57: 4717–4721.

88 Cote, A.P. (2005). Porous, crystalline, covalent organic frameworks. *Science* 310: 1166–1170.

89 El-Kaderi, H.M., Hunt, J.R., Mendoza-Cortes, J.L. et al. (2007). Designed synthesis of 3D covalent organic frameworks. *Science* 316: 268–272.

90 Dogru, M. and Bein, T. (2014). On the road towards electroactive covalent organic frameworks. *Chem. Commun.* 50: 5531–5546.

91 Liu, X.-H., Guan, C.-Z., Wang, D., and Wan, L.-J. (2014). Graphene-like single-layered covalent organic frameworks: synthesis strategies and application prospects. *Adv. Mater.* 26: 6912–6920.

92 Cai, S.-L., Zhang, Y.-B., Pun, A.B. et al. (2014). Tunable electrical conductivity in oriented thin films of tetrathiafulvalene-based covalent organic framework. *Chem. Sci.* 5: 4693–4700.

93 Hao, Q., Li, Z.-J., Lu, C. et al. (2019). Oriented two-dimensional covalent organic framework films for near-infrared electrochromic application. *J. Am. Chem. Soc.* 141: 19831–19838.

94 Bisbey, R.P. and Dichtel, W.R. (2017). Covalent organic frameworks as a platform for multidimensional polymerization. *ACS Cent. Sci.* 3: 533–543.

95 Xu, H., Gao, J., and Jiang, D. (2015). Stable, crystalline, porous, covalent organic frameworks as a platform for chiral organocatalysts. *Nat. Chem.* 7: 905–912.

96 Reitzenstein, D., Quast, T., Kanal, F. et al. (2010). Synthesis and electron transfer characteristics of a neutral, low-band-gap, mixed-valence polyradical. *Chem. Mater.* 22: 6641–6655.

97 Hao, Q., Li, Z.-J., Bai, B. et al. (2021). A covalent organic framework film for three-state near-infrared electrochromism and a molecular logic gate. *Angew. Chem. Int. Ed.* 60: 12498–12503.

98 Dunaj-Jurčo, M., Ondrejovič, G., Melník, M., and Garaj, J. (1988). Mixed-valence copper(I)-copper(II) compounds: analysis and classification of crystallographic data. *Coord. Chem. Rev.* 83: 1–28.

99 He, Y., Li, B., O'Keeffe, M., and Chen, B. (2014). Multifunctional metal–organic frameworks constructed from *meta*-benzenedicarboxylate units. *Chem. Soc. Rev.* 43: 5618–5656.

100 Lu, H., Zhu, Y., Chen, N. et al. (2011). Ligand-directed assembly of a series of complexes bearing thiourea-based carboxylates. *Cryst. Growth Des.* 11: 5241–5252.

101 Tomar, K., Verma, A., and Bharadwaj, P.K. (2018). Exploiting dimensional variability in Cu paddle-wheel secondary building unit based mixed valence Cu(II)/Cu(I) frameworks from a bispyrazole ligand by solvent/pH variation. *Cryst. Growth Des.* 18: 2397–2404.

102 Nijem, N., Bluhm, H., Ng, M.L. et al. (2014). Cu^{1+} in HKUST-1: selective gas adsorption in the presence of water. *Chem. Commun.* 50: 10144–10147.

103 Guo, Y., Feng, C., Wang, S. et al. (2020). Construction of planar-type defect-engineered metal–organic frameworks with both mixed-valence sites and copper-ion vacancies for photocatalysis. *J. Mater. Chem. A* 8: 24477–24485.

104 Liu, R.L., Zhao, L.L., Yu, S.H. et al. (2018). Enhancing proton conductivity of a 3D metal-organic framework by attaching guest NH_3 molecules. *Inorg. Chem.* 57: 11560–11568.

105 Gao, X., Fu, A.-Y., Liu, B. et al. (2019). Unique topology analysis by TOPOSPRO for a metal–organic framework with multiple coordination centers. *Inorg. Chem.* 58: 3099–3106.

106 Gao, J., Cong, J., Wu, Y. et al. (2018). Bimetallic Hofmann-type metal–organic framework nanoparticles for efficient electrocatalysis of oxygen evolution reaction. *ACS Appl. Energy Mater.* 1: 5140–5144.

107 Niel, V., Martinez-Agudo, J.M., Muñoz, M.C. et al. (2001). Cooperative spin crossover behavior in cyanide-bridged Fe(II)–M(II) bimetallic 3D Hofmann-like networks (M = Ni, Pd, and Pt). *Inorg. Chem.* 40: 3838–3839.

108 Xie, L., Liu, S., Gao, C. et al. (2007). Mixed-valence iron(II, III) trimesates with open frameworks modulated by solvents. *Inorg. Chem.* 46: 7782–7788.

109 Feng, D., Wang, K., Wei, Z. et al. (2014). Kinetically tuned dimensional augmentation as a versatile synthetic route towards robust metal–organic frameworks. *Nat. Commun.* 5: 5723.

110 Wei, Y.-S., Shen, J.-Q., Liao, P.-Q. et al. (2016). Synthesis and stabilization of a hypothetical porous framework based on a classic flexible metal carboxylate cluster. *Dalton Trans.* 45: 4269–4273.

111 Xie, L.S., Sun, L., Wan, R.M. et al. (2018). Tunable mixed-valence doping toward record electrical conductivity in a three-dimensional metal-organic framework. *J. Am. Chem. Soc.* 140: 7411–7414.

112 Park, J.G., Aubrey, M.L., Oktawiec, J. et al. (2018). Charge delocalization and bulk electronic conductivity in the mixed-valence metal-organic framework Fe(1,2,3-triazolate)$_2$(BF$_4$)$_x$. *J. Am. Chem. Soc.* 140: 8526–8534.

113 Sobczak, S. and Katrusiak, A. (2019). Environment-controlled postsynthetic modifications of iron formate frameworks. *Inorg. Chem.* 58: 11773–11781.

114 Zhao, J.-P., Hu, B.-W., Lloret, F. et al. (2010). Magnetic behavior control in niccolite structural metal formate frameworks [NH$_2$(CH$_3$)$_2$][Fe^{III}M^{II}(HCOO)$_6$] (M = Fe, Mn, and Co) by varying the divalent metal ions. *Inorg. Chem.* 49: 10390–10399.

115 Collings, I.E., Saines, P.J., Mikolasek, M. et al. (2020). Static disorder in a perovskite mixed-valence metal-organic framework. *CrystEngComm* 22: 2859–2865.

116 Liyanage, R., Yang, Q., Fang, M. et al. (2018). A μ_3-oxo-centered mixed-valence triiron coordination polymer constructed by 5-bromonicotinato ligands. *Inorg. Chem. Commun.* 92: 121–124.

117 Fu, L., Yang, Q., Li, D., and Lu, J.Y. (2020). New 3D metal–organic framework isomer containing trinuclear oxo-centered mixed valence iron. *Mendeleev Commun.* 30: 589–591.

118 Woehler, S.E., Wittebort, R.J., Oh, S.M. et al. (1986). Solid-state deuterium NMR, iron-57 Mössbauer, and X-ray structural characteristics of μ_3-oxo-bridged mixed-valence [Fe$_3$O(O$_2$CCH$_3$)$_6$(4-Me-py)$_3$](C$_6$H$_6$): Dynamics of the benzene solvate molecules influencing intramolecular electron transfer. *J. Am. Chem. Soc.* 108: 2938–2946.

119 Oh, S.M., Henderickson, D.N., Hassett, K.L., and Davis, R.E. (1985). Valence-detrapping modes for electron transfer in the solid state of mixed-valence, oxo-centered, trinuclear iron acetate complexes: X-ray structure and physical data for [Fe$_3$O(O$_2$CCH$_3$)$_6$(4-Et-py)$_3$](4-Et-py). *J. Am. Chem. Soc.* 107: 8009–8018.

120 Li, C., Yang, H.Y., Xie, J. et al. (2020). Ferrocene-based mixed-valence metal-organic framework as an efficient and stable cathode for lithium-ion-based dual-ion battery. *ACS Appl. Mater. Interfaces* 12: 32719–32725.

121 Zhang, J.D., Ouyang, J., Ye, Y.X. et al. (2018). Mixed-valence cobalt(II/III) metal-organic framework for ammonia sensing with naked-eye color switching. *ACS Appl. Mater. Interfaces* 10: 27465–27471.

122 Pope, M.T. and Müller, A. (1991). Polyoxometalate chemistry: an old field with new dimensions in several disciplines. *Angew. Chem. Int. Ed. Engl.* 30: 34–48.

123 Müller, A., Penk, M., Rohlfing, R. et al. (1990). Topologically interesting cages for negative ions with extremely high "coordination number": An unusual property of V-O clusters. *Angew. Chem. Int. Ed.* 29: 926–927.
124 Müller, A., Sessoli, R., Krickemeyer, E. et al. (1997). Polyoxovanadates: high-nuclearity spin clusters with interesting host–guest systems and different electron populations. Synthesis, spin organization, magnetochemistry, and spectroscopic studies. *Inorg. Chem.* 36: 5239–5250.
125 Zhang, C.-D., Liu, S.-X., Gao, B. et al. (2007). Hybrid materials based on metal–organic coordination complexes and cage-like polyoxovanadate clusters: synthesis, characterization and magnetic properties. *Polyhedron* 26: 1514–1522.
126 Dong, B.-X., Peng, J., Gómez-García, C.J. et al. (2007). High-dimensional assembly depending on polyoxoanion templates, metal ion coordination geometries, and a flexible bis(imidazole) ligand. *Inorg. Chem.* 46: 5933–5941.
127 Lu, B.-B., Yang, J., Liu, Y.-Y., and Ma, J.-F. (2017). A polyoxovanadate–resorcin[4]arene-based porous metal–organic framework as an efficient multifunctional catalyst for the cycloaddition of CO_2 with epoxides and the selective oxidation of sulfides. *Inorg. Chem.* 56: 11710–11720.
128 Chen, W.H., Lai, Y.Z., Hu, Z.B. et al. (2018). A new redox- based and stepwise synthetic strategy lead to an unprecedented mixed-valence Keggin-type tungstovanadophosphate (W^{VI}/V^{IV}) bi-capped by vanadium(V^{III})-complexes. *New J. Chem.* 42: 8738–8744.
129 Park, J.G., Collins, B.A., Darago, L.E. et al. (2021). Magnetic ordering through itinerant ferromagnetism in a metal-organic framework. *Nat. Chem.* 13: 594–598.
130 Chen, Z.Y., Su, L.J., Ma, X.H. et al. (2021). A mixed valence state Mo-based metal-organic framework from photoactivation as a surface-enhanced raman scattering substrate. *New J. Chem.* 45: 5121–5126.
131 Martían-Zarza, P., Arrieta, J.M., Muñoz-Roca, M.C., and Gili, P. (1993). Synthesis and characterization of new octamolybdates containing imidazole, 1-methyl- or 2-methyl-imidazole co-ordinatively bound to molybdenum. *J. Chem. Soc., Dalton Trans.* 1551–1557.
132 Zhang, Q., Li, X., Ma, Q. et al. (2017). A metallic molybdenum dioxide with high stability for surface enhanced Raman spectroscopy. *Nat. Commun.* 8: 14903.
133 Jin, H.G., Wang, M.Y., Wen, J.X. et al. (2021). Oxygen vacancy-rich mixed-valence cerium MOF: an efficient separator coating to high-performance lithium-sulfur batteries. *ACS Appl. Mater. Interfaces* 13: 3899–3910.
134 Peng, H.-J., Huang, J.-Q., Cheng, X.-B., and Zhang, Q. (2017). Review on high-loading and high-energy lithium–sulfur batteries. *Adv. Energy Mater.* 7: 1700260.
135 Xiong, Y., Chen, S., Ye, F. et al. (2015). Synthesis of a mixed valence state Ce-MOF as an oxidase mimetic for the colorimetric detection of biothiols. *Chem. Commun.* 51: 4635–4638.

136 Liu, D., Yang, P., Wang, F. et al. (2021). Study on performance of mimic uricase and its application in enzyme-free analysis. *Anal. Bioanal.Chem.* 413: 6571–6580.

137 Wang, C.-Y., Yu, B., Fu, H. et al. (2019). A mixed valence Tb(III)/Tb(IV) metal–organic framework: crystal structure, luminescence property and selective detection of naproxen. *Polyhedron* 159: 298–307.

138 Lee, J., Farha, O.K., Roberts, J. et al. (2009). Metal–organic framework materials as catalysts. *Chem. Soc. Rev.* 38: 1450–1459.

139 Rogge, S.M.J., Bavykina, A., Hajek, J. et al. (2017). Metal–organic and covalent organic frameworks as single-site catalysts. *Chem. Soc. Rev.* 46: 3134–3184.

140 Gao, Y., Yu, G., Liu, K. et al. (2017). Integrated adsorption and visible-light photodegradation of aqueous clofibric acid and carbamazepine by a Fe-based metal-organic framework. *Chem. Eng. J.* 330: 157–165.

141 Chen, H., Liu, Y., Cai, T. et al. (2019). Boosting photocatalytic performance in mixed-valence MIL-53(Fe) by changing Fe^{II}/Fe^{III} ratio. *ACS Appl. Mater. Interfaces* 11: 28791–28800.

142 Zhao, Z.F., Yang, D., Ren, H.J. et al. (2020). Nitrogenase-inspired mixed-valence MIL-53(Fe^{II}/Fe^{III}) for photocatalytic nitrogen fixation. *Chem. Eng. J.* 400.

143 Yoon, J.W., Seo, Y.-K., Hwang, Y.K. et al. (2010). Controlled reducibility of a metal-organic framework with coordinatively unsaturated sites for preferential gas sorption. *Angew. Chem. Int. Ed.* 49: 5949–5952.

144 Rostamnia, S. and Alamgholiloo, H. (2018). Synthesis and catalytic application of mixed valence iron (Fe^{II}/Fe^{III})-based OMS-MIL-100(Fe) as an efficient green catalyst for the aza-Michael reaction. *Catal. Lett.* 148: 2918–2928.

145 Ahmad, M., Quan, X., Chen, S., and Yu, H.T. (2020). Tuning Lewis acidity of MIL-88B-Fe with mix-valence coordinatively unsaturated iron centers on ultrathin Ti_3C_2 nanosheets for efficient photo fenton reaction. *Appl. Catal., B* 264: 118534.

146 Xu, W.K., Xue, W.J., Huang, H.L. et al. (2021). Morphology controlled synthesis of α-Fe_2O_{3-x} with benzimidazole-modified Fe-MOFs for enhanced photo-Fenton-like catalysis. *Appl. Catal., B* 291.

147 Liu, S., Xie, L., Gao, B. et al. (2005). An organic-inorganic hybrid material constructed from a three-dimensional coordination complex cationic framework and entrapped hexadecavanadate clusters. *Chem. Commun.* 41: 5023–5025.

148 Wang, S., Sun, Z.X., Zou, X.Y. et al. (2019). Enhancing catalytic aerobic oxidation performance of cyclohexane via size regulation of mixed-valence $\{V_{16}\}$ cluster-based metal-organic frameworks. *New J. Chem.* 43: 14527–14535.

149 Tian, H.R., Liu, Y.W., Zhang, Z. et al. (2020). A multicentre synergistic polyoxometalate-based metal-organic framework for one-step selective oxidative cleavage of β-O-4 lignin model compounds. *Green Chem.* 22: 248–255.

150 Zhang, T., Hu, Y.Q., Han, T. et al. (2018). Redox-active cobalt(II/III) metal-organic framework for selective oxidation of cyclohexene. *ACS Appl. Mater. Interfaces* 10: 15786–15792.

151 Zhang, Q.Q., Liu, X.F., Ma, L. et al. (2018). Remoulding a MOF's pores by auxiliary ligand introduction for stability improvement and highly selective CO_2-capture. *Chem. Commun.* 54: 12029–12032.

152 Xu, Z.Q., Li, Q.Q., He, X.L. et al. (2018). Construction of mixed-valence Cu(I)/Cu(II) 3-D framework and its photocatalytic activities. *Polyhedron* 151: 478–482.

153 Gao, Y.J., Zhang, L., Gu, Y.M. et al. (2020). Formation of a mixed-valence Cu(I)/Cu(II) metal–organic framework with the full light spectrum and high selectivity of CO_2 photoreduction into CH_4. *Chem. Sci.* 11: 10143–10148.

154 Zhao, Y., Liu, Z.Y., Wang, X.G. et al. (2020). Transition metal ion-directed coordination polymers with mixed ligands: Synthesis, structure, and photocatalytic activity for hydrogen production and Rhodamine B degradation. *Z. Anorg. Allg. Chem.* 646: 1765–1773.

155 Li, J., Jiang, L., Chen, S. et al. (2019). Metal–organic framework containing planar metal-binding sites: efficiently and cost-effectively enhancing the kinetic separation of C_2H_2/C_2H_4. *J. Am. Chem. Soc.* 141: 3807–3811.

156 Nath, I., Chakraborty, J., and Verpoort, F. (2016). Metal organic frameworks mimicking natural enzymes: a structural and functional analogy. *Chem. Soc. Rev.* 45: 4127–4170.

157 Ma, L., Jiang, F., Fan, X. et al. (2020). Metal–organic-framework-engineered enzyme-mimetic catalysts. *Adv. Mater.* 32: 2003065.

158 Luo, L.P., Huang, L.J., Liu, X.N. et al. (2019). Mixed-valence Ce-bpydc metal–organic framework with dual enzyme-like activities for colorimetric biosensing. *Inorg. Chem.* 58: 11382–11388.

159 Ma, X., Pang, C., Li, S. et al. (2021). Biomimetic synthesis of ultrafine mixed-valence metal–organic framework nanowires and their application in electrochemiluminescence sensing. *ACS Appl. Mater. Interfaces* 13: 41987–41996.

14

Near-Infrared Electrochromism Based on Intervalence Charge Transfer

Ying Han[1,*], Xiaohua Cheng[1,*], Yu-Wu Zhong[2,3], and Bin-Bin Cui[1]

[1]*Beijing Institute of Technology (BIT), Advanced Research Institute of Multidisciplinary Science, School of Chemistry and Chemical Engineering, 5 South Zhongguancun Street, Haidian District, Beijing 100081, China*
[2]*Institute of Chemistry, Chinese Academy of Sciences, Beijing National Laboratory for Molecular Sciences, CAS Key Laboratory of Photochemistry, 2 Bei Yi Jie, Zhong Guan Cun, Haidian District, Beijing 100081, China*
[3]*University of Chinese Academy of Sciences, School of Chemical Sciences, No.19(A) Yuquan Road, Shijingshan District, Beijing 100081, China*

14.1 Introduction

Electrochromism is the phenomenon in which the optical properties, including reflectivity, transmittance, absorptivity, etc. of a material display stable and reversible changes under the external applied electric field, giving rise to reversible changes in color and transparency in appearance [1–7]. It usually traces the research origin of electrochromism to the pioneering work in 1969 in which Deb first reported the seminal colorless/blue electrochromism based on tungsten oxide (WO_3) films [8]. He also developed the first electrochromic device (ECD) of a thin film in the early 1970s [9]. Electrochromic materials (ECMs) have received a lot of attention over the past decades as they are widely used in "smart windows," [10, 11] antiglare rearview mirrors [12], switchable photodetectors [13], and electronic display screens [14]. Lampert [15] and Granqvist [16] et al. first proposed the concept of "smart windows" based on electrochromic films. The near-infrared (NIR) electrochromism typically refers to the reversible change of absorption spectra occurring in the NIR region (750–2500 nm) [6, 7, 17, 18].

Generally, ECMs in practical applications need to have the following properties: (i) low operating voltages; (ii) reversible color changes and fast response time; (iii) good chemical and thermodynamic stability; (iv) stable electrochemical redox reversibility; and (v) easy to assemble into devices [19]. In addition, due to a certain optical state-keeping function, ECMs have been applied to optical memory storage and molecular logic operation with electrical input signals and optical

---

* Y. Han, X. Cheng contributed equally to this work.

Mixed-Valence Systems: Fundamentals, Synthesis, Electron Transfer, and Applications, First Edition.
Edited by Yu-Wu Zhong, Chun Y. Liu, and Jeffrey R. Reimers.
© 2023 WILEY-VCH GmbH. Published 2023 by WILEY-VCH GmbH.

output signals [20, 21]. The research and development of efficient NIR ECMs and devices have important scientific values and application prospects. NIR ECMs, which are based on reversible modulation of the NIR absorption, have been used in "smart windows" to regulate the temperature of buildings and aircraft [5, 22]. This will effectively reduce energy consumption in maintaining comfortable indoor temperature, which is of great significance to achieve the global "carbon neutrality" and "low-carbon" targets. In light of the minimum signal transmission attenuation in the NIR region, NIR ECMs can be used in optical fiber communications [23].

ECMs are the cores of ECDs [24]. Currently, electrochromism is mainly realized with transition metal oxides (WO_3 [25, 26], NiO [27], V_2O_5 [28], Co_3O_4 [29], etc.), organic small molecules (e.g. viologen and triphenylamine (TPA) derivatives) [6, 30], metal complexes (e.g. polypyridine metal complexes) [31, 32], and conducting polymers (e.g. polyaniline and polythiophene) [33]. According to the material components, ECMs can be sorted into inorganic, organic, and inorganic–organic hybrid materials. Molecular ECMs typically consist of electron "donor-bridge-acceptor" (D-B-A) structures, and their electrochromism characteristics originate from the charge transfer between two or more redox centers [34, 35]. Most of these D-B-A-type ECMs are pseudo-mixed-valence compounds [36], and the intervalence charge transfer (IVCT) between different redox sites in the singly oxidized or reduced states leads to the corresponding changes in NIR spectral absorptions [4, 7]. Here, we mainly focus and discuss the NIR electrochromism based on mixed-valence compounds. We summarize and discuss NIR ECMs from the perspectives of classifications, electrochromism mechanisms, applications in ECDs, and basic physical parameters for a systematic survey of their research status. At the end of this chapter, a prospective outlook for the development of NIR ECMs in the future is given.

14.2 Near-Infrared Electrochromic Materials

In addition to the common transition metal oxides, organic small molecules, and metal complexes, the emerging NIR ECMs based on organic or organic–inorganic hybrid conducting polymers, electropolymerized films [18, 32], and covalence-organic frameworks (COFs) [21, 37] are mainly composed of mixed-valence systems in the singly oxidized or reduced states. It is worth mentioning that not all mixed-valence compounds have the function of NIR electrochromism because the characteristic of IVCT depends on the degree of intervalence electron delocalization in these compounds [31, 36]. For the organic and organic–inorganic hybrid mixed-valence compounds, the nature of the NIR transition is clearer than that of the transition metal oxides based ECMs with multiple oxidation states.

Organic and organic–inorganic hybrid mixed-valence compounds are model systems for studying intramolecular electron transfer processes, and the discovery of the "Creutz–Taube" ion has greatly promoted the development of mixed-valence chemistry [38, 39]. According to the degree of electron coupling, mixed-valence compounds are classified as Robin–Day class I (no electron coupling), class II (moderate coupling), and class III systems in which electrons are completely delocalized. In addition, Meyer et al. [40] proposed an intermediate II–III system taking the solvent effect of electron delocalization and molecular configuration changes of intramolecular electron transfer into account [41]. In fact, most mixed-valence compounds in the Robin–Day class II form the significant NIR electrochromism [42]. Herein, inorganic, organic, and inorganic–organic hybrid NIR ECMs are discussed as follows.

14.2.1 Inorganic NIR Electrochromic Materials

Inorganic ECMs include cathodic and anodic materials. The cathodic ECMs represented mainly by WO_3 have modulation functions in the NIR region, and the spectral modulation range of anodic ECMs is mainly in the visible light region [22]. Although the exact electrochromic mechanism of the transition metal oxides is unclear, it is for sure that the IVCT between different oxidation states of metals is involved. Most research articles attribute the electrochromic mechanism of WO_3 to the injection/extraction of charges and ions [10, 26], which lead to the optical modulation from the visible region to the NIR region. The electrochromic mechanism of WO_3 film is usually described as follows: [43]

$$WO_3 \text{ (bleached)} + xM^+ + xe^- \leftrightarrow M_xWO_3 \text{ (colored)} \qquad (14.1)$$

where M^+ is H^+ or alkali metal ions. The double injection/extraction of electrons and ions in WO_3 leads to the reversible switching of the colored and bleached state of WO_3 films.

Among the inorganic ECMs, electrochromism from ordinary transition metal oxides, such as the amorphous WO_3 film, is simple in the visible light region [44]. Inorganic ECMs in the NIR region can be obtained from the modification or nanocrystallization of some transition metal oxides and most of them are cathodic ECMs [45]. For instance, constructing a special stereoscopic structure of WO_3 is one of the most reliable methods to improve the electrochromism. Recently, Li et al. fabricated a monolayer hollow spherical WO_3 film by magnetron sputtering method with the colloidal crystal template (Figure 14.1a) [46]. The hollow spherical WO_3 film with a large specific surface area, short ion diffusion distance, and weak scattering shows a high transmittance modulation in the NIR range (78.8% at ~1000 nm, Figure 14.1b), a fast-switching time (coloration time: 2.41 s, bleaching time: 1.28 s, Figure 14.1c), and a high coloring efficiency at 1000 nm (102.9 cm [2]·C^{-1}). The hollow spherical mode is an effective method to regulate the incident light of WO_3 films, especially in the NIR band. The response times of the hollow spherical WO_3 film are faster than those of the dense WO_3 film, indicating a faster reaction kinetics.

Thin ECM films are commonly deposited on an indium tin oxide (ITO). However, this transparent conductive electrode has low NIR transmittance. With high transparency, high mechanical flexibility, and good conductivity, silver nanowire (Ag NW) electrodes have great potential to replace ITO. Especially, Ag NW electrodes have particularly high transmittance of NIR light. Using an electrochemical deposition method, Yan et al. deposited WO_3 films on Ag NW electrodes

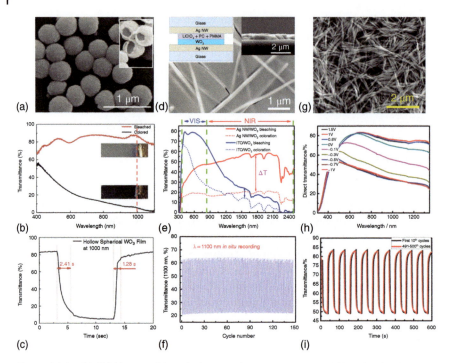

Figure 14.1 (a) SEM Image, (b) optical transmittance spectra, and (c) switching response curve of the hollow spherical WO$_3$ film. Source: Zhang et al. [46]/Reproduced with permission Elsevier. (d) SEM image, (e) optical transmittance spectra, and (f) *in situ* optical transmittance evolutions on cycling test of the WO$_3$/Ag NW film. Source: Zhou et al. [47]/Reproduced with permission from Springer Nature. (g) SEM image, (h) optical transmittance spectra, and (i) stability and reversibility test after running 500 cycles of the W$_{18}$O$_{49}$ nanoneedle film. Source: Li et al. [25]. Reproduced with Permission from The Royal Society of Chemistry.

(Figure 14.1d) [47]. In addition to being excellent materials of optical modulators in the visible region, WO$_3$/Ag NW films show better modulating ability than the previous WO$_3$/ITO film (Figure 14.1e). The modulation of WO$_3$/Ag NW in the NIR region amounts to 57.74%, which is much larger than that of WO$_3$/ITO (43.09%). Furthermore, these WO$_3$/Ag NW films not only modulate NIR light distinctly but also exhibit excellent cycling stability (Figure 14.1f).

Due to the unique optical performance of needle-like W$_{18}$O$_{49}$ nanocrystals, they were selected to fabricate as optical devices to modulate NIR light (Figure 14.1g) [25]. The needle-like W$_{18}$O$_{49}$ nanocrystals have intense and wide NIR absorptions (780–2500 nm; Figure 14.1h). The W$_{18}$O$_{49}$ nanocrystals were assembled onto ITO glass by a layer-by-layer (LBL) method, making the resulting films more tunable for electrochromic investigations. The nanostructured W$_{18}$O$_{49}$ films exhibit high contrast, fast-switching response, high coloration efficiencies (150 cm^2·C^{-1} at 650 nm and 255 cm^2·C^{-1} at 1300 nm), long-term redox switching stability (retention of 98% of contrast ratio after 500 cycles, Figure 14.1i), and wide electrochromic spectrum covering the visible and NIR regions.

14.2.2 Organic NIR Electrochromic Materials

Viologen and TPA derivatives are representative organic materials that have been modified to show IVCT for controllable NIR electrochromism, exhibiting good modifiability in structural and tunability in color [48]. Organic conducting polymers (OCPs) are applied to NIR electrochromism due to their unique advantages, such as excellent film forming performance, variety of species, easy processing, and so on [49]. The representative NIR electrochromic OCP mainly includes polyaniline (PAN), polyselenophene (PSP), polypyrrole (PPy), and their derivatives. Organic NIR ECMs are usually coated on transparent ITO to fabricate NIR ECDs.

14.2.2.1 Viologen Derivatives

N,N-substituted bipyridine (viologen, **1^{2+}**) has shown excellent electrochromic characteristics in the visible region as a functional organic material [50, 51]. It exhibits three reversible redox states, including a colorless dication, a brilliant violet–blue radical cation, and a yellow–brown neutral form [52]. The charge transfer between nitrogen atoms at different valence states of viologen gives rise to valuable electrochromic properties with diverse colors [53]. However, ECDs with viologen have poor stability in the long term [54, 55]. Adding a bridging group between two pyridyls was proven to be a very effective optimization strategy to obtain modulated NIR electrochromism, as demonstrated by the viologen derivatives **2^{2+}**–**5^{2+}** (Figure 14.2) [30, 56].

The four viologens with conjugation-extended and thiophene derivative structure, thiophene viologen (TV, **2^{2+}**), 3,4-ethylenedioxythiophene viologen (ETV, **3^{2+}**), thieno[3,2-b]thiophene viologen (TTV, **4^{2+}**), and 2,2-bithiophene viologen (BTV, **5^{2+}**), have been successfully obtained by Chang et al. [30] The incorporation of thiophene prolongs the effective conjugation length of the viologen derivative, extending their electrochromic response to the NIR region. Moreover, significant changes in the visible and NIR range of the absorption spectra were observed. For instance, new absorption bands around 724 nm continuously grow up to 1192 and 1436 nm for **5^{2+}** upon applying reductive potentials, resulting in distinct color changes (Figure 14.3a). The transmittance change ($\Delta T\%$) at 724 nm for the ECDs based on the gels of **5^{2+}** (10%, w/w) is 80%. The obvious transmittance in the NIR region has also changed significantly ($\Delta T\%$ is 69% at 1436 nm). The ECDs of **5^{2+}** display response times (90% of the full transmittance change) of 2.9 and 8.0 s at 724 and 1436 nm, respectively (Figure 14.3b). The longer response time of **5^{2+}** compared with **2^{2+}** and **3^{2+}** with single thiophene bridges can be attributed to the larger double thiophene bridges that slowed the charge and mass transport. The transmittance changes of the ECD of **5^{2+}** over 500 continuous cycles demonstrate its good performance stability in the application of visible and NIR regions.

With benzene and tetraphenyl ethane as the core, two viologen analogues 4, 4, 4, 4-(benzene-1, 2, 4, 5-tetrayl)-tetrakis(1-butylpyridin-1-ium) (BTTBP, **6^{4+}**) and 1, 1, 2, 2-tetrakis(4-[1-butyl-pyridin-4-yl]-phenyl)ethane (TBPPE, **7^{4+}**) were prepared [57]. Compounds **6^{4+}** and **7^{4+}** show excellent electrochromic performance and extensive absorptions extending to the NIR region. Particularly, compound **7^{4+}** presents outstanding electrochromic behavior, such as excellent optical contrast, obvious color

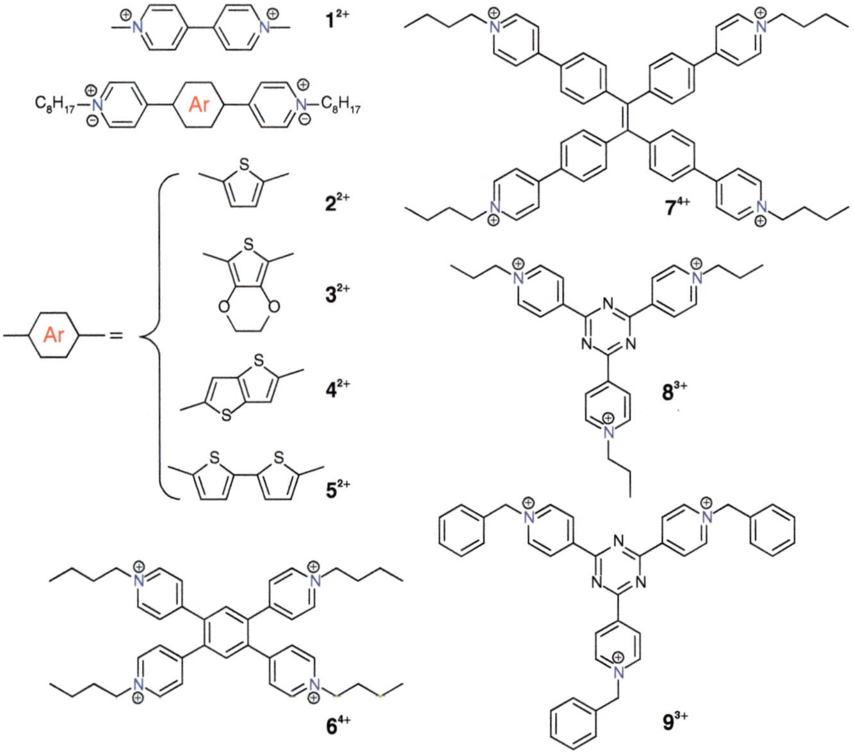

Figure 14.2 Chemical structures of viologen and its derivatives.

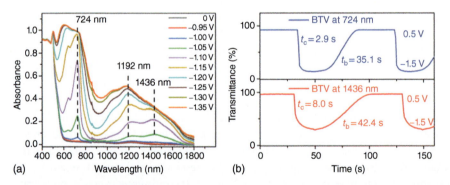

Figure 14.3 (a) Absorption spectra of the ECDs containing gels (10%, w/w) of 5^{2+} upon applying different potentials. (b) Electrochromic switching properties of the ECDs of 5^{2+} at two different wavelengths. Source: Chang et al. [30]/Reproduced with Permission from The Royal Society of Chemistry.

change, great cycling stability, and high coloration efficiency in the NIR region. These materials and strategies offer guidance for preparing high-performance ECMs that can be applied to energy-saving smart windows.

Wu et al. synthesized the viologen derivatives 2, 4, 6-tri(pyridyl-4-propyl)-1, 3, 5-triazine bromide (TPPT, 8^{3+}) and 2, 4, 6-tri(pyridyl-4-benzyl)-1, 3, 5-triazine bromide (TPBT, 9^{3+}) with triazine-centered structure [58]. Moreover, the compounds 8^{3+} and 9^{3+} present outstanding dual-band electrochromic behaviors. Among them, the ΔT% values of the 8^{3+}-based ECD are 66.63% and 50.71% at 887 nm and 600 nm, respectively. The ΔT% values in the switching of the 9^{3+}-based ECD are 59.79% and 66.90% at 900 nm and 600 nm, respectively. The dual-band ECDs were fabricated based on the gel of 8^{3+} or 9^{3+} and have high coloration efficiency in the NIR region. Additionally, this ECD was certified to work at a low voltage of 0.6 V. This work lays a foundation for designing viologen derivative materials with dual frequency electrochromic properties.

14.2.2.2 Triphenylamine Derivatives

TPA with propeller-shaped structure is well known as an excellent ECM. TPA has a strong ability in donating an electron and great capability in transporting holes due to their low ion potential. TPA can be reversibly oxidized with the parapositions protection of phenyl rings, and its absorption spectra in the NIR region change significantly. In recent years, the most studied organic mixed-valence system is the para-substituted triaromatic amine because of its simple synthesis and good stability of the resulting aminium cationic radical. Some examples are shown in Figure 14.4.

The TPA derivative **10** with an anthracene bridge was synthesized and characterized [59]. The IVCT transitions are verified in the NIR region when a single electron was oxidized. The results show that the position of the amine substituent on the anthracene bridge plays a decisive role in the degree of amine–amine electron coupling. Lambert and his workers investigated a series of bridged bis-triarylamines molecules **11–13** [60]. In compounds **11–13**, the electron interaction decreases with the increasing N–N distance. The corresponding cationic radicals, however, all show intense IVCT absorptions in the NIR region due to intramolecular charge transfer.

Thiophene polymers have many applications in the field of photoelectric materials, so the charge transfer of thiophene polymers has attracted much attention. Lacroix and collaborators studied the TPA systems **14** containing multiple thiophene bridging units and simulated the charge delocalization of the corresponding cationic radicals with theoretical calculations [61]. The results suggest that, instead of the long-range charge transfer between TPA termini, the possibility of charge transfer from the polythiophene bridge to the terminal TPA units increases with increasing bridge length.

Corrente et al. synthesized a novel organic mixed-valence system with excellent Vis–NIR electrochromic properties in which two or three amino redox centers were bridged by dibenzofulvene (DBF) unit (**15–18**) [62]. For **15** and **16**, the weak absorption was observed at 502 and 460 nm, respectively. The two absorption bands with the characteristics of HOMO → LUMO transition were categorized as

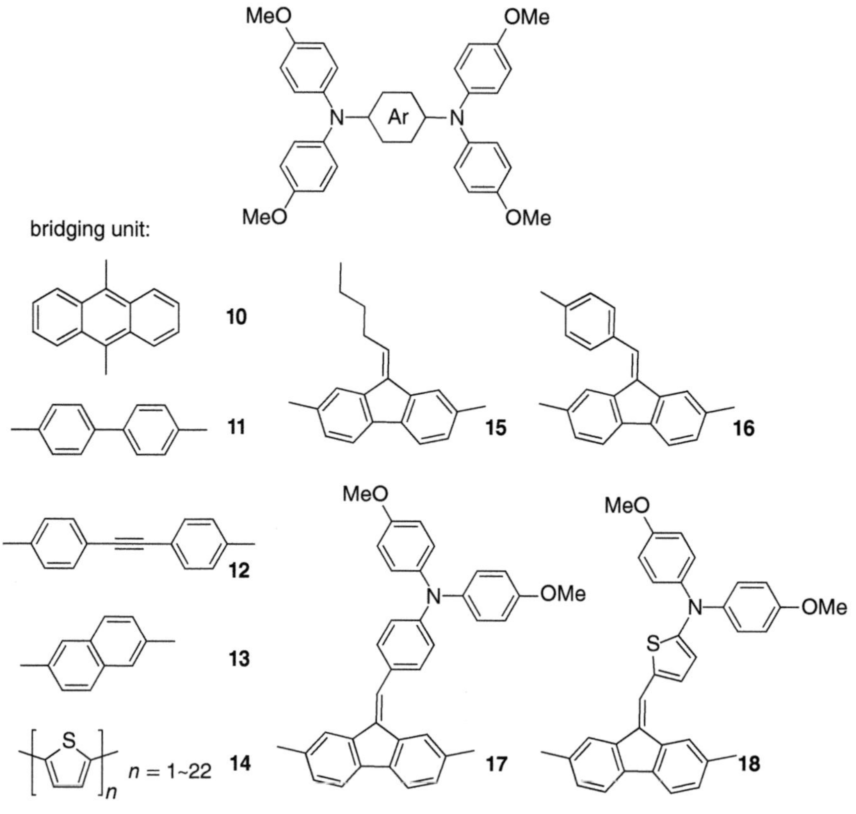

Figure 14.4 Chemical structures of TPA derivatives.

the charge transfer from the diarylamines to the DBF moieties. In addition, two reversible oxidations of **15** in +0.3 and 1.2 V vs. AgCl/Ag were observed in the potential window of cyclic voltammograms (Figure 14.5a). The oxidation of **15–18** causes remarkable changes in absorption spectra in the visible or NIR range. As the oxidation potential increased, new bands appeared in the 500–700 nm for all compounds, resulting in obvious color changes. The absorption bands of molecule **15** bleached at higher voltages, and a secondary structure band was observed at 600–1000 nm since the ionized species were formed, resulting in the solution appearing black (Figure 14.5b). The devices based on **15** show $\Delta T\%$ of over 40% at both 1370 and 910 nm.

In addition, a new luminescent agent based on dithieno[3,2-b : 2,3-d]pyrrole-based (DTP) was synthesized by adding two TPA units to the core of rigid conjugated DTP core [63]. This compound exhibits two consecutive reversible redox processes when voltages are 0.08 and 0.47 V and several new absorption bands during the process of the first two-electron oxidation. The new absorption band appears and continually increased at 600 nm, with the simultaneous appearance of strong absorptions in the NIR region. Finally, the behavior of electrochromic switching in solution and the phenomenon of reversible mechanochromic luminescence were exhibited.

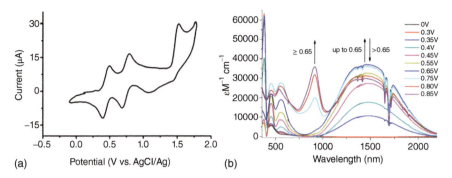

Figure 14.5 (a) Cyclic voltammetry of **15** (100 mV/s) at 10^{-3} mol/L in CH_2Cl_2/$TBAPF_6$ (0.1 M) vs. Ag/AgCl. (b) Spectroelectrochemistry of **15** in CH_2Cl_2 (0.2 mM) and $TBAPF_6$ (0.1 M). Source: Corrente et al. [62]/Reproduced with Permission from American Chemical Society.

14.2.2.3 Organic Conducting Polymers

Conductive polymers are promising ECMs because of their good machining properties, great optical contrast, polychromatic properties, fast-switching characteristics, and excellent band-gap tunability. During the process of doping and dedoping, conductive polymers usually exhibit strong NIR absorption bands due to the formation of polarons and bipolarons. Some representative conducting polymers for electrochromic studies are shown in Figure 14.6.

The Stille coupling reaction of 1,4-dibromobenzene and tributyl(2-selenophenyl) stannane was utilized to synthesize 1,4-di(selenophene-2-yl)-benzene (DSB) based on selenophene monomer. In CH_2Cl_2, the corresponding polymer (PDSB, **Poly-1**) was prepared using tetrabutylammonium hexafluorophosphate ($TBAPF_6$) as the

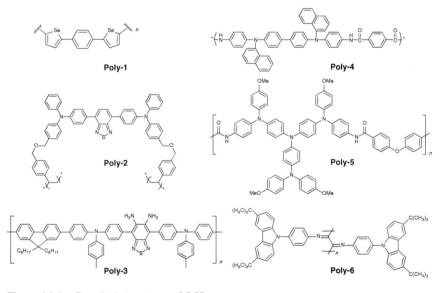

Figure 14.6 Chemical structures of OCPs.

electrolyte [64]. As shown in Figure 14.7a, the NIR absorptions drastically increased with the increase of applied potential. The polaron band of radical cation and the bipolaron band of dication were observed at 608 nm and 1250 nm, respectively. Spectroelectrochemistry analysis and kinetic studies of **Poly-1** revealed a rapid (0.6 s) transmittance change (35%) at 1250 nm, indicating that **Poly-1** is a suitable NIR-ECM.

A monomer consisting of a benzothiadiazole core flanked by two TPAs and two styrene pendant moieties and the corresponding polymer **Poly-2** were prepared [65]. When a potential greater than the oxidation potential was applied, the absorption at 475 nm decreased and a new wide absorption was formed in the NIR region. The color changes from yellow to gray as the potential switches between the neutral and oxidized states (Figure 14.7b). The change of applied potential can also affect the fluorescence intensity of **poly-2**. The low band-gap-conjugated polymer **poly-3** contains a 5,6-diaminobenzothiadiazole bridging unit and has an intense absorption band at 1200 nm of NIR region in the oxidized state [66]. The color of **poly-3** changes from orange in the neutral state to slate gray in the oxidized state (Figure 14.7c).

The aromatic polyamide **poly-4** was obtained by the polymer condensation reaction of diamine and binary acid substrates [67]. As the applied voltage increased from 0 to 0.8 V, the absorption of **poly-4** film gradually decreased at 366 nm, and new absorption appeared at 480 and 1420 nm of NIR region. Furthermore, the original absorption bands gradually decreased and a new absorption peak appeared near 940 nm when the potential increased to 1.05 V (Figure 14.7d). The color changes of **poly-4** in different oxidized states were observed from yellow to brown and to atroceruleous finally. Similarly, the solution-processable NIR electrochromic aromatic polyamide **poly-5** consisting of starburst triarylamine units displays multiple redox states and multiwavelength electrochromism in the visible to NIR regions (Figure 14.7e) [33]. Chen et al. synthesized the electrochromic poly(phenyl isocyanide) helix cross-conjugated polyisocyanides with strong donor (carbazole) and weak acceptor (phenyl isocyanide) units, as represented by **poly-6** [68]. With the increase of applied potential from 0 to 1.10 V, the maximum absorption intensity of **poly-6** at 450 nm decreased (Figure 14.7f). Meanwhile, a new absorption peak appears at 830 nm, and the peak intensity improved gradually.

In addition to the polymeric materials, the device structure and electrolyte are important in the research of ECDs. Song et al. constructed an integrated infrared ECD with a highly flexible monolithic by using a gold-plated nylon porous membrane [69]. The unique structure gives an outstanding electrochromic performance in the NIR region with great flexibility and excellent stability compared with the traditional sandwich configuration.

In addition, the flexible camphorsulfonic acid (CSA) doped PAN films were fabricated via electrochemical deposition on the surface of ductile Au/nylon 66 porous substrate, which showed excellent electrochromic performance in the NIR to infrared region [70]. The results show that poly(ether ether ketone) (PEEK) film has a good pore structure and high porosity, and can be used as the electrolyte material of infrared variable emittance devices (IR-VEDs) [71]. Compared with the conventional gel electrolytes, the porous PEEK film provides a high liquid

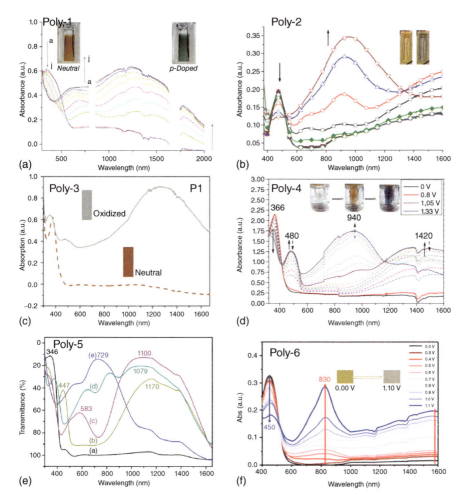

Figure 14.7 (a) Spectroelectrochemistry of **poly-1**/ITO in 0.1 M TBAPF$_6$/CH$_2$Cl$_2$ with applied potentials from 0 to 1.0 V vs. Ag/Ag$^+$. Source: Aydemir et al. [64] Reproduced with permission from Elsevier. (b) Spectroelectrochemistry of **poly-2**/ITO with applied potentials. Source: Wałęsa-Chorab et al. [65]/Reproduced with permission from American Chemical Society. (c) UV/Vis/NIR absorption spectra of **poly-3**/ITO in its neutral (dashed line) and oxidized states (solid line). Source: Qian et al. [66]/Reproduced with permission from Royal Society of Chemistry (d) Electronic absorption spectra of the **poly-4** thin film. Source: Wang et al. [67]/Reproduced with permission from ELSEVIER. (e) Electrochromic behavior of **poly-5** at applied potentials of 0, 0.55, 0.80, 1.10, 1.45 V vs. Ag/AgCl (from curves a to e). Source: Reproduced with permission [33]. Copyright 2011 American Chemical Society. (f) Spectroelectrochemistry of **poly-6**/ITO film. Source: Zhai et al. [68]/Reproduced with permission from American Chemical Society.

electrolyte load and maintains great membrane integrity; thus, the new ECD with high conductivity and a simple process was obtained. The resulting IR-VED displays excellent and comprehensive performance, including the superior regulation capability of infrared emittance, fast-switching time, and great stability, which has broad application prospects in the field of dynamic thermal control.

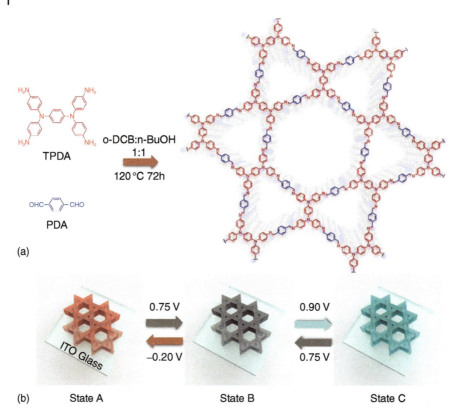

Figure 14.8 (a) Chemical structures of TPDA, PDA, and COF$_{TPDA-PDA}$. (b) Color change of three-state near-infrared electrochromic behavior of COF$_{TPDA-PDA}$ thin film. Source: Hao at al. [72]/Reproduced with permission from John Wiley & Sons

14.2.2.4 Covalence-Organic Framework (COF)

Hao et al. synthesized a COF material with Kagome structure and indicated its three-state NIR electrochromic performance (Figure 14.8a) [72]. The obtained COF$_{TPDA-PDA}$ film is formed by using hexagonal-shaped nanosheets with high crystallinity and presents three different and reversible color changes at diverse applied potentials. The three reversible color changes at different potentials are plum, gray, and light blue, respectively (Figure 14.8b). The film shows distinct changes in the NIR region of absorption spectra due to the strong IVCT interaction between the conjugated TPA. Moreover, thanks to their unique structure with highly ordered porosity and the π–π stacking, the fast response time and long retention time were verified and the excellent electrochromic properties of the NIR region were illustrated. This COF film has great potential in the application of optical memory logic gates.

Yu et al. synthesized dark purple electrochromic COF material (EC-COF-1) with layered structure via the reaction of the N,N,N,N-tetrakis(p-aminophenyl)-p-benzenediamine (TPBD) and the 2,1,3-benzothiadiazole-4,7-dicarboxaldehyde (BTDD) (Figure 14.9a) [73]. The dual-band bleaching peaks of a sandwiched device

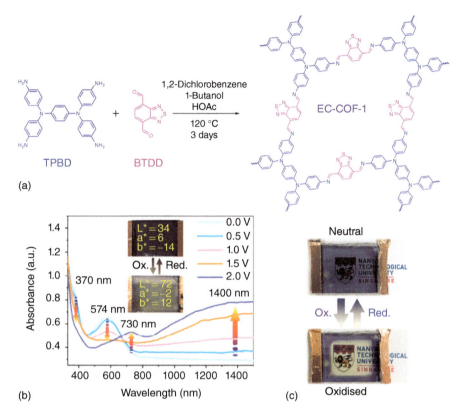

Figure 14.9 (a) Synthetic route of EC-COF-1. Source: [73]/Springer Nature/Licensed under CC BY 4.0(b) Changes in the optical spectrum of EC-COF-1 film as a function of applied potential. (c) Color switching of the EC-COF-1 EC device. Source: [73] Xu et al. (2021)/Reproduced with permission from Springer Nature.

based on EC-COF-1 at 370 and 574 nm were observed and become transparent gradually upon the application of potential, and finally, the induced absorption peak at 1400 nm appears (Figure 14.9b). Additionally, the color of the prepared two-dimensional (2-D) COF film changes from opaque black to transparent when the voltage increases from 1.8 to +2.0 V (Figure 14.9c).

Bessinger et al. developed the full-organic and porous COF films with highly efficient and fast-switching electrochromic, which has good superposition with thienoisoindigo moiety [34]. The redox processes of COF are completely reversible and have fast electron transfer kinetics. The first oxidation ($-100 \sim +100$ mV) generates spectral characteristics at 1000 nm, and the second oxidation produces a stronger absorption peak at 900 nm. The material has an excellent electrochromic efficiency at 880 nm and maintains a 95% of the origin electrochromic response after 100 oxidation/reduction cycles. The high-performance combination of high color rendering efficiency and fast switching provides great opportunities for the application of COF materials in ECMs.

14.2.3 Organic–Inorganic Hybrid NIR Electrochromic Materials

Organometallic complexes and their conducting polymers are important parts of organic–inorganic hybrid NIR ECMs [4]. Generally, they have two or more redox centers, and these centers can be either metal complexes (Ru, Fe, Mo, etc.) or organic triarylamines [19]. IVCT between two redox centers has weak or moderate electron coupling, which is responsible for the NIR electrochromism of these materials. The charge coupling strength is related to the ligand field and the coordination strength of the ligand on the central metal [7, 31]. NIR electrochromic films of metal complexes have been fabricated by the methods of spin coating, inkjet printing, molecular self-assembly, or electropolymerization, and these films are potentially useful in smart windows, optical communication and storage, and other NIR optical devices [74].

14.2.3.1 Metal Complexes

Transition metal complexes have a particular value in the NIR field. First, they usually exhibit reversible electrochemical behavior. Second, due to the strong IVCT transitions, some complexes have intense absorptions in the NIR region. Third, the electrochemical properties and absorption spectra of metal complexes can be finely tuned by the modification of coligands or ligand substituents. Finally, metal complexes can easily be functionalized, allowing them to be incorporated into films by polymerization or adsorption on metal oxide surfaces [4]. Some recent and classical intervalence metal complexes for applications in NIR electrochromism are displayed in Figure 14.10.

Metal complexes, represented by Ru(II)-dioxolane complexes, have variable optical attenuation and information storage owing to their electrochromic ability in the visible and NIR ranges. Creutz and Taube first designed and synthesized a mixed-valence bimetallic ruthenium complex $[(NH_3)_5Ru(pz)Ru(NH_3)_5]^{5+}$ (19^{5+}) with pyrazine as the bridging ligand (so-called the Creutz–Taube ion) [75, 76]. In addition, Bignozz and coworkers synthesized asymmetric binuclear complexes containing a carboxylic acid group on the terminal pyridine ligand [77]. In these complexes, the two ruthenium components are connected by a cyanide bridge. Upon oxidation of the Ru(II)-amine component, an IVCT transition appears between 763 and 1149 nm.

Ward et al. found that the NIR transitions were more pronounced when two or more chromophores were connected by conjugated bridging ligands [78]. Complex 20^{n+} is based on a "back-to-back" bis-catecholate ligand. Four redox processes were observed due to the two redox interconversions among the catecholate, semiquinone, and quinone redox states at each terminus. A related complex 21^{n+} is functionalized with the units of peripheral carboxylic acid, enabling it to be attached to the SnO_2:Sb surface. In particular, dye oxidations cause an absorbance reduction in the region of 600 to 750 nm, leading to an obvious color change from blue to pink, and an absorbance increment occurs in the NIR region (1000–1500 nm) due to which the adsorption complex oxygenated from 21^+ to 21^{3+} [79, 80].

Yao et al. prepared bis-cyclometalated diruthenium complex 22^{n+} with a bicyclic metallized bridging ligand 1, 3, 6, 8-tetra(2-pyridyl)pyrene [81]. This complex

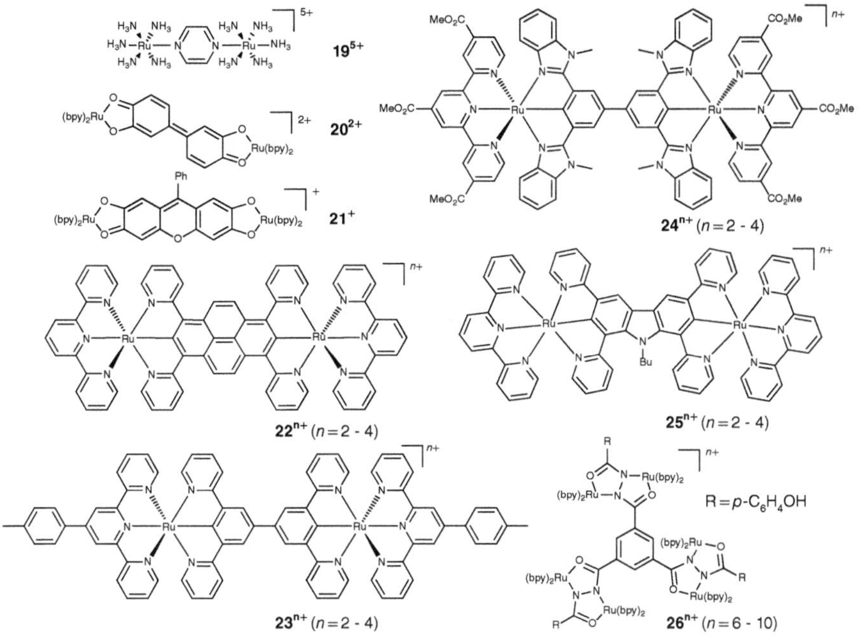

Figure 14.10 Chemical structures of intervalence complexes.

exhibited a distinct IVCT band in the mixed-valence state due to the electronic coupling between individual metal centers. The degree of the electronic coupling of 22^{n+} is slightly stronger with respect to the classical cyclometalated diruthenium complex 23^{n+} reported by Launay and Collin et al. in 1994 [82]. In addition, structurally related diruthenium complexes, e.g. 24^{n+} and 25^{n+}, with intense IVCT bands in the mixed-valence state have been reported [83, 84]. All of these complexes are potential candidates as NIR ECMs. Besides, Wang et al. also synthesized the dendritic dinuclear ruthenium complex 26^{5+} [85]. The hydroxyl groups in this complex enabled it to form thin films on an electrode by chemical cross linking with isocyanate materials. The optical attenuation of the cross-linked films at 1550 nm is 5.4 dB and the switching time is 2 s.

Furthermore, some other transition metal complexes have been studied in the NIR electrochromism field. For example, Enemark reported the mononuclear oxo-Mo(V) complex, which can undergo reversible Mo(VI/V) and Mo(V/IV) redox interconversion at easily accessible potentials [86]. Ward et al. reported the binuclear complex 27^{n+} [87], which shows intense NIR absorptions at 1100 and 1000 nm at +1 and +2 redox states, respectively (Figure 14.11a).

Recently, the bis-cyclometalated complex 28^{n+} bridged by 1,2,4,5-tetra(2-pyridyl)-benzene was reported [88]. It shows two sequential redox couples with a potential difference (ΔE°) of 430 mV between two Ru(II/III) half-wave potentials at +0.12 and +0.55 V. At the mixed-valence state (+3), it displays an intense IVCT band at round 1160 nm (Figure 14.11b). Besides, Wang et al. presented a series of NIR electrochromic binuclear and trinuclear Ni and Pt metallodithioene complexes [90].

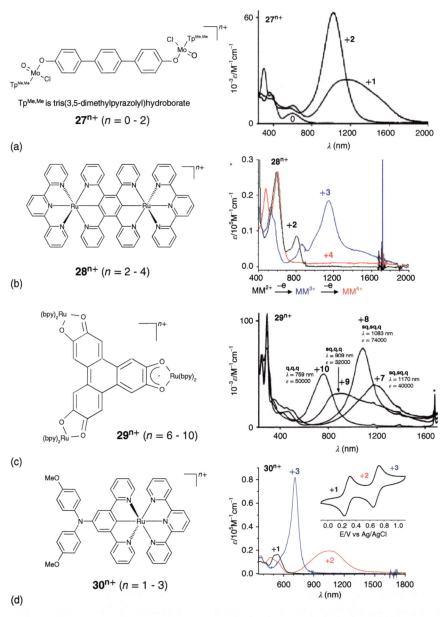

Figure 14.11 (a) Chemical structure and electronic absorption spectra of **27^{n+}**. Source: Bayly et al. [87]/Reproduced with Permission from The Royal Society of Chemistry. (b) Chemical structure and absorption spectra of **28^{n+}**. Source: Yao et al. [88]/Reproduced with Permission from American Chemical Society. (c) Chemical structure and absorption spectra of **29^{n+}**. Source: Barthram et al. [89]/Reproduced with Permission from The Royal Society of Chemistry. (d) Chemical structure, cyclic voltammetric analysis, and absorption spectra of **30^{n+}**. Source: Yao et al. [17]/Reproduced with Permission from The Royal Society of Chemistry.

The trinuclear complex **29^{n+}** based on the hexahydroxytriphenylene bridging ligand exhibits polyelectrochromic properties as a result of the redox interconversions among four stable redox states of the bridging ligand (Figure 14.11c) [89]. In addition, Yao et al. reported a 1,4-benzene-bridged **30^{n+}** of triarylamine and cyclometalated ruthenium covalent hybrid (Figure 14.11d) [17]. It presents two continuous reversible couples at +0.27 and +0.68 V vs. Ag/AgCl, belonging to the N/N$^{•+}$ and Ru$^{II/III}$ processes, respectively. Strong charge transfer transition at about 1050 nm is observed at the singly oxidized state.

The cyclometalated diruthenium complex **31^{n+}** and triruthenium complex **32^{n+}** with a triarylamine bridge were recently reported (Figure 14.12a,b) [18, 91]. Complex **31^{n+}** shows three independent redox processes and an intense absorption band at 750, 1170, and 1680 nm is observed at the +3, +4, and +5 redox states, respectively (Figure 14.12c–e). In comparison, the star-shaped complex **32^{n+}** shows more complicated four-step NIR spectral changes [91].

14.2.3.2 Conducting Polymers of Metal Complexes

The attractive NIR electrochromic behavior of transition metal complexes stimulates further studies on the thin films of these materials. Electropolymerization

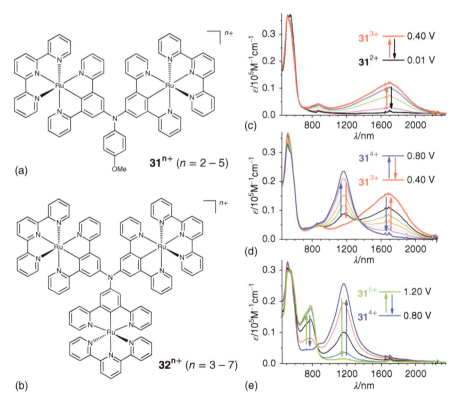

Figure 14.12 (a,b) Chemical structures of complexes **31^{n+}** and **32^{n+}**. (c–e) Absorption spectral changes of **31^{n+}** at different redox states. Source: Cui et al. [18], John Wiley & Sons.

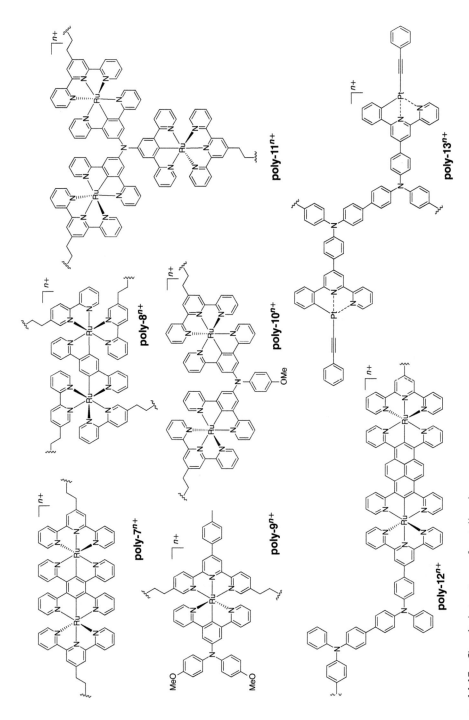

Figure 14.13 Chemical structures of metallopolymers.

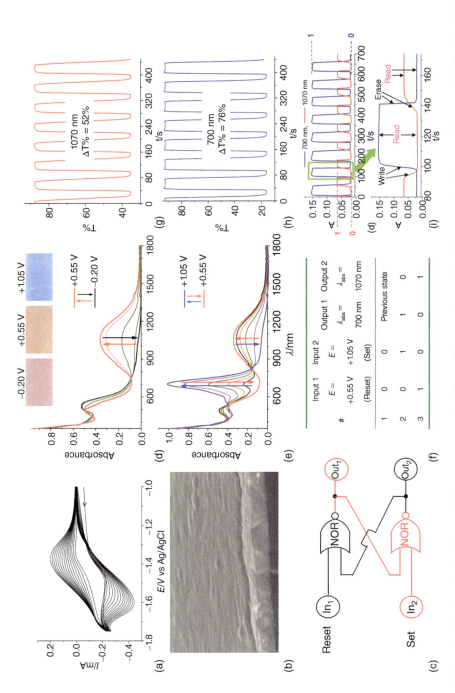

Figure 14.14 (a) Reductive electropolymerization to prepare **poly-9**$^{n+}$ on ITO glass electrode by 15 repeated cyclic potential scans at 100 mV/s. (b) SEM cross-sectional image of **poly-9**$^{n+}$/ITO film. Source: Cui et al. [93]/Reproduced with permission Royal Society of Chemistry. (c) Set/reset flip-flop circuit built with two cross-coupled NOR logic gates. (d,e) Absorption spectral and color changes upon (d) single and (e) double oxidation during spectroelectrochemical measurements. The applied potentials are referenced vs. Ag/AgCl. (f) Truth table for a set/reset flip-flop logic gate. (g,h) Transmittance changes of **poly-9**$^{n+}$/ITO film. (i) Absorbance (A) switching monitored at 1070 nm (red curve) and 700 nm (blue curve).

is a feasible and practical method to prepare metallopolymeric films with good film quality. The structures of some representative NIR electrochromic metallopolymers prepared by electropolymerization are displayed in Figure 14.13. For instance, Yao et al. prepared the adherent metallopolymeric films of **Poly-7^{n+}** by reductive electropolymerization of a 1,2,4,5-tetra(2-pyridyl)-benzene bridging vinyl-functionalized bis-cyclometalated ruthenium complex [7]. Due to the IVCT band switching of mixed-valence bimetal units, the prepared films exhibited NIR electrochromism at 1165 nm. The adherent metallopolymeric films of **Poly-8^{n+}** with tris-bidentate cyclometalated ruthenium components have also been successfully synthesized, which showed NIR electrochromism at 1300 nm [92].

The thin film of **poly-9^{n+}** was prepared from the reductive electropolymerization of a vinyl-functionalized ruthenium-amine complex (Figure 14.14a) [93]. The film surface is flat and about 50 nm thick (Figure 14.14b). During the single oxidation step, the contrast ratio is 52% at 1070 nm (Figure 14.14d,g). In the double oxidation step, the contrast ratio achieved 76% at 700 nm (Figure 14.14e,h). The **poly-9^{2+}** and **poly-9^{3+}** states of the thin film can be used to mimic the set/reset flip-flop operation with electrochemical potentials as the input signals and the absorbance at 1070 and 700 nm as the output signals (Figure 14.14c,f,i). In a similar fashion, the previously discussed amine-bridged diruthenium complex **31^{n+}** and triruthenium complex **32^{n+}** were functionalized with vinyl groups on the terminal terpyridine ligand and subjected to reductive electropolymerization to give **poly-10^{n+}** and **poly-11^{n+}**, respectively [18, 91]. These thin films show multistep electrochromism in the NIR region.

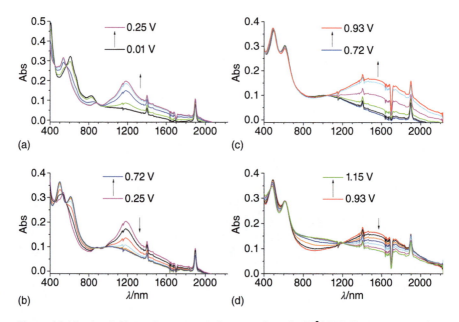

Figure 14.15 (a–d) Absorption spectral changes of a **poly-12^{2+}**/ITO film upon stepwise oxidations. The applied potentials are referenced vs. Ag/AgCl. Source: Yao et al. [32]/Reproduced with Permission from American Chemical Society.

Yao et al. prepared **poly-12^{n+}** polymer films [32], which exhibit NIR electrochromism with five stages with good contrast ratio and coloration efficiency by switching the IVCT bands of the organic (1200 nm) and inorganic (1500 nm) mixed-valence units, respectively (Figure 14.15). Qiu and Zhou et al. synthesized the **poly-13^{n+}** homogeneous films with controlled thickness by oxidation electropolymerization of cyclometalated Pt complex (Figure 14.13) [94]. The metal polymer films have stable reversible redox behavior and promising dual-wavelength electrochromic behavior at 1520 and 773 nm.

Rubner et al. first used the mononuclear ruthenium complex polymers as emitters in the solid-state devices in 1997 [95]. Wang et al. later synthesized and characterized the NIR electrochromic of electroluminescent polymers containing binuclear ruthenium complexes, as represented by **poly-14^{n+}** shown in Figure 14.16a [96]. This polymer displays strong absorption at 1600 nm when oxidized to the mixed-valence state (Figure 14.16b). Higuchi et al. synthesized dual-redox metallo-supramolecular polyFe-N and polyRu-N polymers with a zigzag structure by complexing Fe(II) or

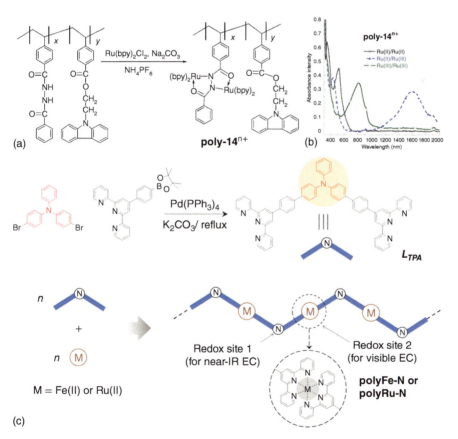

Figure 14.16 (a) Preparation of **poly-14^{n+}** and (b) absorption spectra of **poly-14^{n+}** spin coated on ITO at three oxidation states. Source: Wang et al. [96]/Reproduced with Permission from American Chemical Society. (c) Preparations of dual-redox metallo-supramolecular polymers polyFe-N and polyRu-N. Source: Mondal et al. [97]/Reproduced with Permission from American Chemical Society.

Ru(II) ion and a redox-active triarylamine bridging ligand L_{TPA} (Figure 14.16c) [97]. These films show dual-wavelength electrochromism with good reversibility and high contrast ratio (up to 80% in the NIR region).

14.2.3.3 Monolayer and Multilayer Assembled Films

Cui et al. reported a monolayer molecular switch capable of three-state switching via electrochemical inputs and NIR optical outputs [20]. The ruthenium complex 33^{n+} with three carboxylic groups was used to form a monolayer film on ITO surfaces (Figure 14.17a). Although it is a monolayer scale, the film shows a significant NIR absorbance signal (Figure 14.17b). In particular, during the singly oxidized state (+2 sate; state B), an IVCT band is observed at 1020 nm, which is absent at +1 and +3 redox states. In addition, the monolayer film exhibits a long optical retention time at each easy oxidation state, making it attractive for potential applications as memory devices.

In addition to monolayer films, metal complex thin films can be fabricated by LBL assembly with controllable thickness and composition [99, 100]. The LBL assembly can be assisted by the electrostatic interaction [101], hydrogen bonds [102], covalent bonds [103], and metal–ligand coordination [104]. LBL techniques have been used by different research groups to prepare electrochromic films [105–108]. Li et al. synthesized a cyclometalated diruthenium complex 34^{n+} with carboxylic units on

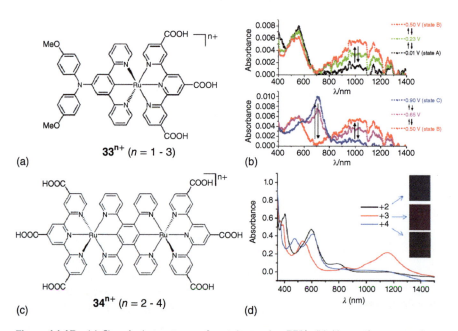

Figure 14.17 (a) Chemical structures of metal complex 33^{n+}. (b) Absorption spectral changes of the monolayer of 33^{n+} during spectroelectrochemical measurements. The applied potentials are referenced vs. Ag/AgCl. Source: Cui et al. [20]/Reproduced with Permission from American Chemical Society. (c) Chemical structures of metal complex 34^{n+}. (d) Absorption spectra and the color changes of the multilayered film of 34^{n+}. Source: Li et al. [98]/Reproduced with permission from Springer Nature.

terminal ligands (Figure 14.17c) [98]. The multilayer films were prepared on TiO_2 or SnO_2 substrates by the LBL coordinative assembly of **34^{n+}** with Zr (IV) ions. The best contrast ratio achieved 56% at 1150 nm with good cyclic stability. Furthermore, a star-shaped tri-Ru complex was synthesized and used to prepare multilayer films on Sb-doped SnO_2 nanocrystalline substrates by LBL assembly [109]. The contrast ratio of the two-layer film at 1550 nm is 36%, which is the best overall performance. The response time of LBL films was found to be significantly shortened with respect to that of electropolymerized films of **poly-11^{n+}**.

14.3 Potential Applications of NIR Electrochromic Materials

14.3.1 Smart Windows

As global climate change and energy crisis pose major threats, an increasing number of countries are promoting "carbon neutrality" as a national strategy, proposing a carbon-free future. Smart windows based on NIR ECMs can selectively absorb or reflect the thermal radiation of NIR lights, reducing the large amount of energy consumed by office buildings and civil houses to remain cool in summer and warm in winter. NIR ECMs can be used in smart windows, but it requires that they can easily be deposited to conducting glasses to fabricate ECDs. As mentioned in the previous sections, spin coating and electropolymerization provide convenient ways to prepare electrochromic films.

14.3.2 Molecular Logic Gates and Optical Storage

The rapid development of the information industry requires high-density and fast response information storage materials and information calculation methods [110]. Stimuli-responsive molecules, ions, and polymer materials can be used to mimic the simple or complex logic gate function with various input and output signals [18, 21, 111]. NIR ECMs with spectral responses can be applied as molecular logic gates by utilizing the material's response in the NIR light region as output signals and electrochemical voltages as the input signals. Furthermore, NIR ECMs can store optical information as a result of the state-retention effect. For instance, thin films of **poly-9^{n+}** and **poly-10^{n+}** have been successfully used to simulate flip-flop and flip-flap-flop molecular logic gates with optical storage functions.

14.3.3 Optical Communication

Variable optical attenuator (VOA) is a significant optical passive device in optical fiber communications, which can control the signal in real time by attenuating the transmitted optical power in the NIR region, especially at wavelengths around 850, 1310, and 1550 nm. Microelectromechanical system-based technology is a common method to fabricate VOA. The next-generation VOAs are considered to be flexible and lightweight devices with short response time but without moving parts. ECDs can modulate the NIR optical signals by applying a low voltage, providing a candidate for this purpose. Multiple NIR ECMs, such as conducting polymers and transition metal oxides, are conceived as potential applications for VOAs. For instance, the cross-linked thin films of the dendritic mixed-valence binuclear ruthenium complex exhibit promising electrochromic phenomena in the NIR region, with an optical attenuation of 5.4 dB/μm at 1550 nm, and a switching time of two seconds [86].

14.3.4 Military Camouflage

The NIR detection at 750–1500 nm mainly depends on NIR night vision devices and identifies the target by the difference between the green vegetation reflection value and the target. In military warfare, the working frequency band of military detectors is not limited to the infrared range. In this sense, the electrochromic camouflage system needs to meet the wavelength requirements of all compatible integrated equipment, including visible, NIR, far infrared, and radar waves. A good electrochromic camouflage device should display an independent color and infrared radiation changes to hide an army. Additionally, in order to fit the military multispectral detection development, microwave and millimeter wave absorption properties should also be considered. Compared with inorganic materials, conductive polymers exhibit multicolor variation, radar absorption properties, and infrared modulation features, allowing them to have potential applications in military camouflage [2]. However, some critical issues remain to be solved. For instance, the electrochromic VOA is difficult to achieve a microsecond response time required for practical applications.

14.4 Summary and Outlook

This chapter highlights recent progress in the development of NIR ECMs showing IVCT switching. Intramolecular charge transfer mechanisms of these materials are clarified. Extensive electron D-B-A-type compounds, including nanomaterials of transition metal oxides, metal complexes organic small molecules, conductive polymers, and emerging COF materials, have been developed to show promising NIR electrochromism, demonstrating their potential applications in the field of "smart windows," thermal modulation, optical fiber telecommunication, optical logic gates, and memory devices. Among them, NIR-electrochromic transition metal complexes, in particular, mixed-valence ruthenium complexes are of great values. The corresponding metallopolymeric films are characterized by low switching potential, good contrast ratio, and long-term stability. These films can readily be prepared by electropolymerization, spin coating, or LBL assembly.

Despite of the current NIR ECMs have achieved innovative results and applications, there are some urgent challenges to be addressed. From the material perspective, nonprecious materials are desirable, considering the high cost and low abundance of ruthenium. In addition, most of the reported electrochromic performances belong to the behavior of a film-loaded electrode in an electrochemical cell. Few of them have been tested in a true solid device, which is required for practical uses. The electrochromic performance of a single electrode should be very different when it is integrated into a solid device, considering that the ion transport is more sluggish and the surface area could be larger in the latter. In this sense, the research on the fabrication of ECDs with short switching time and good color uniformity should be another focus in the future. Another direction that could be considered is to combine NIR electrochromism with other advanced technologies, such as storage and energy harvesting technology, allowing these materials for multifunctional applications.

Acknowledgment

This work was supported by funding from the National Natural Science Foundation (21872154, 21925112, 21703008, and 22075022) and the "Cultivate Creative Talents Project" of Beijing Institute of Technology (BIT).

References

1 Granqvist, C.G. (2012). Oxide electrochromics: an introduction to devices and materials. *Sol. Energ. Mat. Sol. C.* 99: 1–13.
2 Niu, J., Wang, Y., Zou, X. et al. (2021). Infrared electrochromic materials, devices and applications. *Appl. Mater. Today* 24: 101073.
3 Bai, Z., Li, R., Li, K. et al. (2020). Transparent metal-organic framework-based gel electrolytes for generalized assembly of quasi-solid-state electrochromic devices. *ACS Appl. Mater. Interfaces* 12: 42955–42961.
4 Ward, M.D. (2005). Near-infrared electrochromic materials for optical attenuation based on transition-metal coordination complexes. *J. Solid State Electrochem.* 9: 778–787.
5 Thakur, V.K., Ding, G., Ma, J. et al. (2012). Hybrid materials and polymer electrolytes for electrochromic device applications. *Adv. Mater.* 24: 4071–4096.
6 Chuang, Y.W., Yen, H.J., Wu, J.H., and Liou, G.S. (2014). Colorless triphenylamine-based aliphatic thermoset epoxy for multicolored and near-infrared electrochromic applications. *ACS Appl. Mater. Interfaces* 6: 3594–3599.
7 Yao, C.-J., Zhong, Y.-W., Nie, H.-J. et al. (2011). Near-IR electrochromism in electropolymerized films of a biscyclometalated ruthenium complex bridged by 1,2,4,5-tetra(2-pyridyl)benzene. *J. Am. Chem. Soc.* 133: 20720–20723.
8 Deb, S.K. (1969). A Novel Electrophotographic System. *J. Appl. Opt. Suppl.* 3: 193.
9 Deb, S.K. (2006). Optical and photoelectric properties and colour centres in thin films of tungsten oxide. *Philos. Mag.* 27: 801–822.

10 Lee, S.H., Deshpande, R., Parilla, P.A. et al. (2006). Crystalline WO_3 nanoparticles for highly improved electrochromic applications. *Adv. Mater.* 18: 763–766.

11 Gu, H., Guo, C., Zhang, S. et al. (2008). Highly efficient, near-infrared and visible light modulated electrochromic devices based on polyoxometalates and $W18O49$ nanowires. *ACS Nano* 12: 559–567.

12 Zhang, S., Chen, S., Yang, F. et al. (2019). A facile preparation of SiO_2/PEDOT core/shell nanoparticle composite film for electrochromic device. *J. Mater. Sci.-Mater. El.* 30: 3994–4005.

13 Lan, Z. and Zhu, F. (2021). Electrically switchable color-selective organic photodetectors for full-color imaging. *ACS Nano* 8: 13674–13682.

14 Gong, H., Zhou, K., Zhang, Q. et al. (2020). A self-patterning multicolor electrochromic device driven by horizontal redistribution of ions. *Sol. Energ. Mat. Sol. C.* 215: 110642.

15 Lampert, C.M. (1982). Electrochromic materials and devices for energy efficient windows. *Sol. Energ. Mater.* 11: 162.

16 Svensson, J.G.C.G.J. (1984). Electrochromic coatings for smart windows. *Proc. SPIE.* 30: 502.

17 Yao, C.-J., Zheng, R.H., Shi, Q. et al. (2012). 1,4-Benzene-bridged covalent hybrid of triarylamine and cyclometalated ruthenium: a new type of organic-inorganic mixed-valent system. *Chem. Commun.* 48: 5680–5682.

18 Cui, B.-B., Tang, J.-H., Yao, J., and Zhong, Y.-W. (2015). A molecular platform for multistate near-infrared electrochromism and flip-flop, flip-flap-flop, and ternary memory. *Angew. Chem. Int. Ed.* 54: 9192–9197.

19 Zhong, Y.-W., Yao, C.-J., and Nie, H.-J. (2013). Electropolymerized films of vinyl-substituted polypyridine complexes: Synthesis, characterization, and applications. *Coord. Chem. Rev.* 257: 1357–1372.

20 Cui, B.-B., Zhong, Y.-W., and Yao, J. (2015). Three-state near-infrared electrochromism at the molecular scale. *J. Am. Chem. Soc.* 137: 4058–4061.

21 Hao, Q., Li, Z.J., Lu, C. et al. (2019). Oriented two-dimensional covalent organic framework films for near-infrared electrochromic application. *J. Am. Chem. Soc.* 141: 19831–19838.

22 Gillaspie, D.T., Tenent, R.C., and Dillon, A.C. (2010). Metal-oxide films for electrochromic applications: present technology and future directions. *J. Mater. Chem.* 20: 9585.

23 Dyer, A.L., Grenier, C.R.G., and Reynolds, J.R. (2007). A Poly(3,4-alkylenedioxythiophene) Electrochromic Variable Optical Attenuator with Near-Infrared Reflectivity Tuned Independently of the Visible Region. *Adv. Funct. Mater.* 17: 1480–1486.

24 Granqvist, C.G. (1995). *Handbook of Inorganic Electrochromic Materials.* Amsterdam: The Netherlands.

25 Li, G., Zhang, S., Guo, C., and Liu, S. (2016). Absorption and electrochromic modulation of near-infrared light: realized by tungsten suboxide. *Nanoscale* 8: 9861–9868.

26 Chen, H., Xu, N., Deng, S. et al. (2007). Electrochromic properties of WO_3 nanowire films and mechanism responsible for the near infrared absorption. *J. Appl. Phys.* 101: 114303.
27 Diao, X., Liu, X., and Zhong, X. (2021). Electrochromic devices based on tungsten oxide and nickel oxide: a review. *J. Inorg. Mater.* 36: 128.
28 Margoni, M.M., Mathuri, S., Ramamurthi, K. et al. (2018). Hydrothermally grown nano and microstructured V_2O_5 thin films for electrochromic application. *Appl. Surf. Sci.* 449: 193–202.
29 Dhas, C.R., Venkatesh, R., Sivakumar, R. et al. (2016). Fast electrochromic response of porous-structured cobalt oxide (Co_3O_4) thin films by novel nebulizer spray pyrolysis technique. *Ionics* 22: 1911–1926.
30 Chang, M., Chen, W., Xue, H. et al. (2020). Conjugation-extended viologens with thiophene derivative bridges: near-infrared electrochromism, electrofluorochromism, and smart window applications. *J. Mater. Chem. C* 8: 16129–16142.
31 Yao, C.-J., Nie, H.-J., Yang, W.-W. et al. (2014). Strongly coupled cyclometalated ruthenium-triarylamine hybrids: tuning electrochemical properties, intervalence charge transfer, and spin distribution by substituent effects. *Chem. Eur. J.* 20: 17466–17477.
32 Yao, C.-J., Zhong, Y.-W., and Yao, J. (2013). Five-stage near-infrared electrochromism in electropolymerized films composed of alternating cyclometalated bisruthenium and bis-triarylamine segments. *Inorg. Chem.* 52: 10000–10008.
33 Yen, H.-J., Lin, H.-Y., and Liou, G.-S. (2011). Novel starburst triarylamine-containing electroactive aramids with highly stable electrochromism in near-infrared and visible light regions. *Chem. Mater.* 23: 1874–1882.
34 Bessinger, D., Muggli, K., Beetz, M. et al. (2021). Fast-switching vis-IR electrochromic covalent organic frameworks. *J. Am. Chem. Soc.* 143: 7351–7357.
35 Gu, H., Wang, K., Wu, Z. et al. (2021). Stable low-bandgap isoindigo-bisEDOT copolymer with superior electrochromic performance in NIR window. *Electrochim. Acta* 399: 139418.
36 Nie, H.-J., Shao, J.-Y., Yao, C.-J., and Zhong, Y.-W. (2014). Organic-inorganic mixed-valence systems with strongly-coupled triarylamine and cyclometalated osmium. *Chem. Commun.* 50: 10082–10085.
37 Yu, F., Liu, W., Ke, S.W. et al. (2020). Electrochromic two-dimensional covalent organic framework with a reversible dark-to-transparent switch. *Nat. Commun.* 11: 5534.
38 Creutz, C. and Taube, H. (1969). A Direct Approach to Measuring the Franck-Condon Barrier to Electron Transfer between Metal Ions. *J. Am. Chem. Soc.* 91: 3988.
39 Creutz, C. and Taube, H. (1973). Binuclear complexes of ruthenium ammines. *J. Am. Chem. Soc.* 95: 1068.
40 Demadis, K.D., Hartshorn, C.M., and Meyer, T.J. (2001). The localized-to-delocalized transition in mixed-valence chemistry. *Chem. Rev.* 101: 2655.

41 Hush, N.S.P. (1985). Distance dependence of electon transfer rates. *Coord. Chem. Rev.* 64: 135.

42 Hush, N.S. (1967). Intervalence-transfer absorption. Part 2. Theoretical considerations and spectroscopic data. *Prog. Inorg. Chem.* 8: 391.

43 Zhang, W., Li, H., Hopmann, E., and Elezzabi, A.Y. (2020). Nanostructured inorganic electrochromic materials for light applications. *Nanophotonics* 10: 825–850.

44 Zhu, L.-L., Huang, Y.-E., Gong, L.-K. et al. (2020). Ligand control of room-temperature phosphorescence violating Kasha's rule in hybrid organic–inorganic metal halides. *Chem. Mater.* 32: 1454–1460.

45 Wei, D., Scherer, M.R., Bower, C. et al. (2012). A nanostructured electrochromic supercapacitor. *Nano Lett.* 12: 1857–1862.

46 Zhang, X., Dou, S., Li, W. et al. (2019). Preparation of monolayer hollow spherical tungsten oxide films with enhanced near infrared electrochromic performances. *Electrochim. Acta* 297: 223–229.

47 Zhou, K., Wang, H., Zhang, S. et al. (2017). Electrochromic modulation of near-infrared light by WO_3 films deposited on silver nanowire substrates. *J. Mater. Sci.* 52: 12783–12794.

48 Zhuang, Y., Zhao, W., Wang, L. et al. (2020). Soluble triarylamine functionalized symmetric viologen for all-solid-state electrochromic supercapacitors. *Sci. China Chem.* 63: 1632–1644.

49 Beaujuge, P.M. and Reynolds, J.R. (2010). Color control in π-conjugated organic polymers for use in electrochromic devices. *Chem. Rev.* 110: 268–320.

50 Ding, J., Zheng, C., Wang, L. et al. (2019). Viologen-inspired functional materials: synthetic strategies and applications. *J. Mater. Chem. A* 7: 23337–23360.

51 Michaelis, L. and Hill, E.S. (1933). The viologen indicators. *J. Gen. Physiol.* 16: 859–873.

52 Shah, K.W., Wang, S.-X., Soo, D.X.Y., and Xu, J. (2019). Viologen-based electrochromic materials: from small molecules, polymers and composites to their applications. *Polymers* 11: 1839.

53 Striepe, L. and Baumgartner, T. (2017). Viologens and their application as functional materials. *Chem- Eur. J.* 23: 16924–16940.

54 Tarábek, J., Kolivoška, V., Gál, M. et al. (2015). Impact of the extended 1,1′-bipyridinium structure on the electron transfer and π-dimer formation. Spectroelectrochemical and computational study. *J. Phys. Chem. C* 119: 18056–18065.

55 Funston, A., Kirby, J.P., Miller, J.R. et al. (2005). One-electron reduction of an "extended viologen" p-phenylene-bis-4,4′-(1-aryl-2,6-diphenylpyridinium) dication. *J. Phys. Chem. A.* 109: 10862–10869.

56 Madasamy, K., Velayutham, D., Suryanarayanan, V. et al. (2019). Viologen-based electrochromic materials and devices. *J. Mater. Chem. C.* 7: 4622–4637.

57 Huang, Z.-J., Li, F., Xie, J.-P. et al. (2021). Electrochromic materials based on tetra-substituted viologen analogues with broad absorption and good cycling stability. *Sol. Energ. Mat. Sol. C.* 223: 110968.

58 Wu, N., Ma, L., Zhao, S., and Xiao, D. (2019). Novel triazine-centered viologen analogues for dual-band electrochromic devices. *Sol. Energ. Mat. Sol. C.* 195: 114–121.
59 Feng, J.S., Shao, J., Gong, Z., and Zhong, Y. (2016). Amine-amine electronic coupling through an anthracene bridge. *Chin. J. Org. Chem.* 36: 2407.
60 Lambert, C. and Nöll, G. (1999). The class II/III transition in triarylamine redox systems. *J. Am. Chem. Soc.* 121: 8434–8442.
61 Lacroix, J.C., Chane-Ching, K.I., Maquère, F., and Maurel, F. (2006). Intrachain electron transfer in conducting oligomers and polymers: the mixed valence approach. *J. Am. Chem. Soc.* 128: 7264–7276.
62 Corrente, G.A., Fabiano, E., Manni, F. et al. (2018). Colorless to all-black full-NIR high-contrast switching in solid electrochromic films prepared with organic mixed valence systems based on dibenzofulvene derivatives. *Chem. Mater.* 30: 5610–5620.
63 Zhang, J., Chen, Z., Yang, L. et al. (2017). Dithionopyrrole compound with twisted triphenylamine termini: reversible near-infrared electrochromic and mechanochromic dual-responsive characteristics. *Dyes Pigm.* 136: 168–174.
64 Aydemir, K., Tarkuc, S., Durmus, A. et al. (2008). Synthesis, characterization and electrochromic properties of a near infrared active conducting polymer of 1,4-di(selenophen-2-yl)-benzene. *Polymer* 49: 2029–2032.
65 Wałęsa-Chorab, M. and Skene, W.G. (2017). Visible-to-NIR electrochromic device prepared from a thermally polymerizable electroactive organic monomer. *ACS Appl. Mater. Interfaces* 9: 21524–21531.
66 Qian, G., Abu, H., and Wang, Z.Y. (2011). A precursor strategy for the synthesis of low band-gap polymers: an efficient route to a series of near-infrared electrochromic polymers. *J. Mater. Chem.* 21: 7678–7685.
67 Wang, S., Wu, X., Zhang, X. et al. (2017). Synthesis, fluorescence, electrochromic properties of aromatic polyamide with triarylamine unit serving as functional group. *Eur. Polym. J.* 93: 368–381.
68 Zhai, Y., Wang, Y., Zhu, X. et al. (2021). Carbazole-functionalized poly(phenyl isocyanide)s: synergistic electrochromic behaviors in the visible light near-infrared region. *Macromolecules* 54: 5249–5259.
69 Song, S., Xu, G., Wang, B. et al. (2021). Highly-flexible monolithic integrated infrared electrochromic device based on polyaniline conducting polymer. *Synth. Met.* 278: 116822.
70 Zhang, L., Xia, G., Li, X. et al. (2019). Fabrication of the infrared variable emissivity electrochromic film based on polyaniline conducting polymer. *Synth. Met.* 248: 88–93.
71 Li, X., Zhang, L., Wang, B. et al. (2020). Highly-conductance porous poly(ether ether ketone) electrolyte membranes for flexible electrochromic devices with variable infrared emittance. *Electrochim. Acta* 332: 135357.
72 Hao, Q., Li, Z.-J., Bai, B. et al. (2021). A covalent organic framework film for three-state near-infrared electrochromism and a molecular logic gate. *Angew. Chem. Int. Ed.* 60: 12498–12503.

73 Xu, J. and Bu, X.-H. (2021). Electrochromic two-dimensional covalent organic framework with a reversible dark-to-transparent switch. *Chem. Res. Chin. U.* 37: 185–186.

74 Xiong, S., Yin, S., Wang, Y. et al. (2017). Organic/inorganic electrochromic nanocomposites with various interfacial interactions: a review. *Mater. Sci. Eng., B* 221: 41–53.

75 Creutz, C. and Taube, H. (1973). Binuclear complexes of ruthenium ammines. *J. Am. Chem. Soc.* 95: 1086–1094.

76 Creutz, C. and Taube, H. (1969). Direct approach to measuring the Franck-Condon barrier to electron transfer between metal ions. *J. Am. Chem. Soc.* 91: 3988–3989.

77 Biancardo, M., Schwab, P.F., Argazzi, R., and Bignozzi, C.A. (2003). Electrochromic devices based on binuclear mixed valence compounds adsorbed on nanocrystalline semiconductors. *Inorg. Chem.* 42: 3966–3968.

78 Joulié, L.F., Schatz, E., Ward, M.D. et al. (1994). Electrochemical control of bridging ligand conformation in a binuclear complex-A possible basis for a molecular switch. *J. Chem. Soc., Dalton Trans.* 6: 799–804.

79 Biancardo, M., Schwab, P.F., and Bignozzi, C.A. (2003). Electrochromic behaviour of polynuclear ruthenium complexes on nanocrystalline SnO_2. *Collect. Czech. Chem. Commun.* 68: 1710–1722.

80 Meacham, A.P., Wijayantha, K., Peter, L.M., and Ward, M.D. (2004). Polyelectrochromic behaviour in the visible and near-infrared region of a window based on a dinuclear ruthenium–dioxolene complex adsorbed onto a nanocrystalline SnO_2 electrode. *Inorg. Chem. Commun.* 7: 65–68.

81 Yao, C.-J., Sui, L.-Z., Xie, H.-Y. et al. (2010). Electronic coupling between two cyclometalated ruthenium centers bridged by 1, 3, 6, 8-tetra (2-pyridyl) pyrene (tppyr). *Inorg. Chem.* 49: 8347–8350.

82 Beley, M., Chodorowski-Kimmes, S., Collin, J.P. et al. (1994). Pronounced electronic coupling in rigidly connected N, C, N-coordinated diruthenium complexes over a distance of up to 20 Å. *Angew. Chem. Inter. Ed.* 33: 1775–1778.

83 Shao, J.-Y., Yang, W.-W., Yao, J., and Zhong, Y.-W. (2012). Biscyclometalated ruthenium complexes bridged by 3,3′,5,5′-tetrakis(N-methylbenzimidazol-2-yl) biphenyl: synthesis and spectroscopic and electronic coupling studies. *Inorg. Chem.* 51: 4343–4351.

84 Wang, L., Yang, W.-W., Zhong, Y.-W., and Yao, J. (2013). Enhancing the electronic coupling in a cyclometalated bisruthenium complex by using the 1,3,6,8-tetra(pyridin-2-yl)carbazole bridge. *Dalton Trans.* 42: 5611–5564.

85 Qi, Y. and Wang, Z.Y. (2003). Dendritic mixed-valence dinuclear ruthenium complexes for optical attenuation at telecommunication wavelengths. *Macromolecules* 36: 3146–3151.

86 Cleland, W. Jr., Barnhart, K.M., Yamanouchi, K. et al. (1987). Syntheses, structures, and spectroscopic properties of six-coordinate mononuclear oxo-molybdenum (V) complexes stabilized by the hydrotris (3, 5-dimethyl-1-pyrazolyl) borate ligand. *Inorg. Chem.* 26: 1017–1025.

87 Bayly, S.R., Humphrey, E.R., Paredes, C.G. et al. (2001). Electronic and magnetic metal–metal interactions in dinuclear oxomolybdenum (V) complexes across bis-phenolate bridging ligands with different spacers between the phenolate termini: ligand-centred vs. metal-centred redox activity. *J. Chem. Soc. Dalton Trans.* 9: 1401–1414.

88 Yao, C.-J., Zhong, Y.-W., and Yao, J. (2011). Charge delocalization in a cyclometalated bisruthenium complex bridged by a noninnocent 1, 2, 4, 5-tetra (2-pyridyl) benzene ligand. *J. Am. Chem. Soc.* 133: 15697–15706.

89 Barthram, A., Cleary, R., and Ward, M. (1998). A new redox-tunable near-IR dye based on a trinuclear ruthenium (II) complex of hexahydroxytriphenylene. *Chem. Commun.* 24: 2695–2696.

90 Chen, X., Qiao, W., Liu, B. et al. (2017). Synthesis and near infrared electrochromic properties of metallodithiolene complexes. *Sci. China Chem.* 60: 77–83.

91 Tang, J.-H., He, Y.-Q., Shao, J.-Y. et al. (2016). Multistate redox switching and near-infrared electrochromism based on a star-shaped triruthenium complex with a triarylamine core. *Sci. Rep.* 6: 35253.

92 Nie, H.-J. and Zhong, Y.-W. (2014). Near-infrared electrochromism in electropolymerized metallopolymeric films of a phen-1,4-diyl-bridged diruthenium complex. *Inorg. Chem.* 53: 11316–11322.

93 Cui, B.-B., Yao, C.-J., Yao, J., and Zhong, Y.-W. (2014). Electropolymerized films as a molecular platform for volatile memory devices with two near-infrared outputs and long retention time. *Chem. Sci.* 5: 932–941.

94 Pi, Q., Bi, D., Qiu, D. et al. (2021). A dual-wavelength electrochromic film based on a Pt(II) complex for optical modulation at telecommunication wavelengths and dark solid-state display devices. *J. Mater. Chem. C* 9: 8994–9000.

95 Lee, J.-K., Yoo, D., and Rubner, M.F. (1997). Synthesis and characterization of an electroluminescent polyester containing the Ru (II) complex. *Chem. Mater.* 9: 1710–1712.

96 Wang, S., Li, X., Xun, S. et al. (2006). Near-infrared electrochromic and electroluminescent polymers containing pendant ruthenium complex groups. *Macromolecules* 39: 7502–7507.

97 Mondal, S., Chandra Santra, D., Ninomiya, Y. et al. (2020). Dual-redox system of metallo-supramolecular polymers for visible-to-near-IR modulable electrochromism and durable device fabrication. *ACS Appl. Mater. Interfaces* 12: 58277–58286.

98 Li, Z.-J., Yao, C.-J., and Zhong, Y.-W. (2019). Near-infrared electrochromism of multilayer films of a cyclometalated diruthenium complex prepared by layer-by-layer deposition on metal oxide substrates. *Sci. China Chem.* 2: 1675–1685.

99 Richardson, J.J., Cui, J., Bjornmalm, M. et al. (2016). Innovation in layer-by-layer assembly. *Chem. Rev.* 116: 14828–14867.

100 Xiao, F.-X., Pagliaro, M., Xu, Y.-J., and Liu, B. (2016). Layer-by-layer assembly of versatile nanoarchitectures with diverse dimensionality: a new perspective for rational construction of multilayer assemblies. *Chem. Soc. Rev.* 45: 3088–3121.

101 Zhang, L. and Sun, J. (2009). Layer-by-layer deposition of polyelectrolyte complexes for the fabrication of foam coatings with high loading capacity. *Chem. Commun.* 26: 3901–3903.

102 Yang, S.Y. and Rubner, M.F. (2002). Micropatterning of polymer thin films with pH-sensitive and cross-linkable hydrogen-bonded polyelectrolyte multilayers. *J. Am. Chem. Soc.* 124: 2100–2101.

103 Li, M., Ishihara, S., Akada, M. et al. (2011). Electrochemical-coupling layer-by-layer (ECC–LbL) assembly. *J. Am. Chem. Soc.* 133: 7348–7351.

104 Sakamoto, R., Wu, K.-H., Matsuoka, R. et al. (2015). π-Conjugated bis(terpyridine) metal complex molecular wires. *Chem. Soc. Rev.* 44: 7698–7714.

105 DeLongchamp, D.M., Kastantin, M., and Hammond, P.T. (2003). High-contrast electrochromism from layer-by-layer polymer films. *Chem. Mater.* 15: 1575–1586.

106 Bucur, C.B., Sui, Z., and Schlenoff, J.B. (2006). Ideal mixing in polyelectrolyte complexes and multilayers: entropy driven assembly. *J. Am. Chem. Soc.* 128: 13690–13691.

107 Cui, M., Ng, W.S., Wang, X. et al. (2015). Enhanced electrochromism with rapid growth layer-by-layer assembly of polyelectrolyte complexes. *Adv. Funct. Mater.* 25: 401–408.

108 Elool Dov, N., Shankar, S., Cohen, D. et al. (2017). Electrochromic metallo-organic nanoscale films: fabrication, color range, and devices. *J. Am. Chem. Soc.* 139: 11471–11481.

109 Li, Z.-J., Tang, J.-H., Shao, J.-Y., and Zhong, Y.-W. (2020). Near-infrared electrochromism of multilayer films of an NC N-pincer *tri*-ruthenium (II) complex. *Eur. J. Inorg. Chem.* 2020: 2882–2888.

110 Tang, J.H., Yao, C.J., Cui, B.B., and Zhong, Y.-W. (2016). Ruthenium-amine conjugated organometallic materials for multistate near-IR electrochromism and information storage. *Chem. Rec.* 16: 754–767.

111 Lu, Q., Cai, W., Zhang, X. et al. (2018). Multifunctional polymers for electrochromic, memory device, explosive detection and photodetector: donor-acceptor conjugated isoindigo derivatives with strong fluorescence. *Eur. Polym. J.* 108: 124–137.

15

Manipulation of Metal-to-Metal Charge Transfer Toward Switchable Functions

Wen Wen[1,2], Yin-Shan Meng[1], and Tao Liu[1]

[1]*Dalian University of Technology, State Key Laboratory of Fine Chemicals, 2 Linggong Rd., Dalian 116024, China*
[2]*Yantai University, China and College of Chemistry & Chemical Engineering, 30 Qingquan Rd., Yantai 264005, China*

15.1 Introduction

In recent years, switchable phenomena have attracted lots of research interest in molecular materials because of their ability to switch electronic states, which have potential applications in displays, sensing, data storage, and molecular electronics [1–4]. In coordination chemistry, the combination of metals with ligands allows the tailoring of metal complexes for various switchable behaviors; it is possible to switch a molecule's architecture, magnetic moment, spin state, charge distribution, and other physical and chemical characteristics [5–8]. Molecule-based materials, which can switch electronic states via applications of external stimuli, include spin crossover (SCO) complexes [9, 10], valence tautomeric complexes [11, 12], and heterometallic metal-to-metal charge transfer (MMCT) complexes [13, 14]. Especially, mixed valence complexes, including electron-transfer-coupled spin transitions (ETCST) or charge-transfer-induced spin transitions (CTIST), play crucial roles in energy, biology, catalysis, materials, and other fields [15, 16]. Unlike SCO, where the spin transition happens at a metal center, valence tautomerism (VT) and MMCT are based on a charge transfer between the metal ion and ligand or between two metal ions. Charge transfer is one of the most fundamental phenomena in chemistry where the exchange of one electron between two sites, from a donor D to an acceptor A, takes place. In most situations, the switch between ground D–A and excited D$^+$–A$^-$ can be mediated through light absorption.

As shown in Figure 15.1, when a redox-active metal ion is bound to a redox-active ligand, two different charge distributions may be observed. This reversible switching between one charge distribution and another through an external stimulus is called VT, which corresponds to a charge transfer between ligand and metal. VT is very

Mixed-Valence Systems: Fundamentals, Synthesis, Electron Transfer, and Applications, First Edition.
Edited by Yu-Wu Zhong, Chun Y. Liu, and Jeffrey R. Reimers.
© 2023 WILEY-VCH GmbH. Published 2023 by WILEY-VCH GmbH.

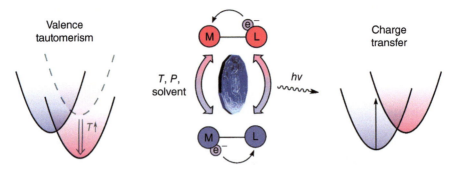

Figure 15.1 Valence tautomeric transitions and charge-transfer excitations are two sides of the same coin. Source: Nadurata et al. [17], Royal Society of Chemistry.

interesting because of its potential to induce reversible magnetic and optical property conversions via thermal- and photoinduced electron transfer between redox-active metal and ligand centers. The photoinduced excited metastable state D^+–A^- is often long-lived in low temperatures, and this is called light-induced VT (LIVT) [18, 19]. In VT systems, the charge distribution in a molecule depends upon the mixing of valence orbitals of metal ions and ligands. There are two extreme cases for a general D–A pair: one involving fully localized redox states (D^+A^-) and another involving fully delocalized redox states ($D^{0.5+}A^{0.5-}$). According to the Robin–Day's classification [20], mixed-valence (MV) complexes can be classified into three classes depending on the strength of intermetallic interactions: no electronic interaction with localized electrons for Class I; moderate electronic interaction with partially localized valence electrons for Class II; and strong electronic interaction with delocalized valence electrons for Class III. Later, Meyer and co-workers proposed a new category on the borderline between Class II and III systems that exhibits intermediate (IM) behavior [21, 22]. The energy of the absorption of class II–III complexes is solvent-independent, but the charge is localized [23].

Other than the VT with the metal-to-ligand charge-transfer interactions, the MMCT provides two valence isomers with markedly different electronic configurations. It has been established that the MMCT can be manipulated by external stimuli, such as light, heat, pressure, and electric fields [24–28]. For example, a cyanide-bridged MV (Fe–CN–Co) chromophore exhibits MMCT, yielding paramagnetic (Fe^{III}_{LS}–CN–Co^{II}_{HS}) (LS = low spin; HS = high spin) and diamagnetic (Fe^{II}_{LS}–CN–Co^{III}_{LS}) configurations. This MMCT accompanied the spin-state change of the Co ion and is called a CTIST. The CTIST can generate bistability, which has thermodynamically stable phases at certain temperatures, and the two phases show significant variations in magnetic properties [14, 29, 30]. One of the challenges in this field is designing and preparing materials with dynamic multifunction through the CTIST process. External stimuli such as heat, light, magnetic and electric fields, and mechanical forces can switch the electronic states or phases. Here, we will summarize the cyanide-bridged MMCT systems and their switchable functions with some typical examples in recent years.

15.2 Switchable Cyanide-Bridged MMCT Systems

It is well known that cyanide-bridged complexes contribute to an essential class of switchable molecular materials. The cyanide ion has a very strong affinity for transition metal ions, which is due to the negatively charged cyanide carbon atom. The cyanide ions easily coordinate with a variety of metal ions, yielding polycyanometallates in which metal ions have versatile oxidation states and different coordination numbers from 2 to 8 [31–36]. On the other hand, the σ-bonding nature of the cyanide nitrogen atom enables the cyanide group to work as a bridging ligand with other transition metal ions. Benefiting from the bridging binding ability of the cyanide group, the polycyanometallate can be purposefully used to link functional units, resulting in homo- and heterometallic assemblies with various types of sophisticated structures ranging from discrete clusters to three-dimensional (3-D) networks [37–39]. The assembled systems exhibit versatile physical properties such as magnetic bistability, ferroelectricity, conductivity, second harmonic generation, and luminescence [40–45].

One of the most extensively studied cyanide-bridged systems is a series of Prussian blue analogs (PBAs), in which transition metal ions are bridged by $[M(CN)_6]^{n-}$ units. Although Prussian blue compounds have long been used in painting and dyeing for more than hundreds of years, it was in 1996 that Hashimoto and co-workers reported the FeCo PBA of $K_{0.2}Co_{1.4}[Fe(CN)_6]\cdot 6.9H_2O$ exhibiting thermo- and photoinduced MMCT accompanied with a remarkable diamagnetic and ferromagnetic transitions [46]. Since then, researchers have begun to pay attention to their attractive switchable properties. The FeCo PBAs were constituted by the Fe–CN–Co linkages and had the general formula of $M_xCo_y[Fe(CN)_6]\cdot nH_2O$ (M = Na^+, K^+, Cs^+, Rb^+), of which linkage has electronic states interconverted between diamagnetic Fe^{II}_{LS}–CN–Co^{III}_{LS} and paramagnetic Fe^{III}_{LS}–CN–Co^{II}_{HS} states. This CTIST involves an intermetallic one-electron transfer process, making the FeCo PBAs ideal candidates to access magnetically bistable materials that are potentially applicable in information storage and molecular switches [14, 47]. In addition, super-exchange interactions originated from delocalized electrons can be propagated between the metal centers even through the diamagnetic $[Fe^{II}(CN)_4]^{2-}$ unit, demonstrating that the cyanide bridge can efficiently transfer both the electrons and magnetic exchange coupling interactions. It, therefore, offers a convenient way to build functional materials to switch the long-range magnetic ordering and other physical properties by external stimuli such as heat, light, pressure, electric/magnetic fields, and guest molecules. The MMCT property of FeCo PBAs depends highly on the stoichiometry of alkaline and cobalt ions [48, 49]. Previous studies revealed that increasing the number of iron vacancies leads to increasing the number of water molecules on the cobalt site, causing the positively shifted redox potential of the cobalt ions [50]. Although the MMCT property can be tuned by modifying the number of alkaline metal ions and vacancies, the uniformity and exact structures are still difficult to access and are very sensitive to synthetic conditions.

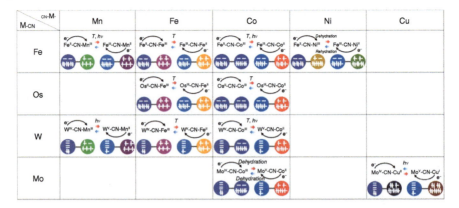

Figure 15.2 Schematic illustration of various external stimuli-tuned MMCT and MMCT-coupled spin transition with M—C ≡ N—M' linkages. The notation "*T*" and "*hv*" represent thermal treatment and light irradiation, respectively. Source: Meng et al. [66], John Wiley & Sons.

On the other hand, molecular PBAs have become a hot research topic in the last two decades, mainly because of the development of molecular and crystal engineering methods that provide an effective way to fine-tune the local coordination sphere and the dimensionality by using suitable polycyanometallate and organic ligands. The structurally well-defined molecular PBAs allow us to understand the MMCT mechanism by comprehensively analyzing their structures and functions. Moreover, MMCT functional units can also be integrated to build systems with collective functions. Notably, multiple properties can be conveniently switched by light with high temporal and spatial resolution, which is appealing for designing photoswitchable molecular materials with multifunctionality. To date, versatile polycyanometallates and organic ligands have been designed and used to tune the redox potentials of metal ions suitable for the MMCT property in switchable molecular PBAs, including Fe–Co [32, 51], Fe–Mn [52, 53], Fe–Fe [54], Fe–Ni [55, 56], Os–Fe [57], W–Fe [58], Os–Co [59], W–Co [60], W–Fe–Co [61], Mo–Co [62], Mo–Cu [63, 64], and W–Mn [65], (Figure 15.2). More interestingly, the photoinduced spin-state changes in octacyanometallates, namely $[M^{IV}(CN)_8]^{4-}$ (M = Mo and W), have also been reported [67, 68]. The MMCT-actuated functions, such as magnetic ordering, molecular nanomagnet, ferroelectricity, mechanical deformation, and photochromism properties, have been deeply investigated [61, 69–71]. In this section, we discuss the effects of chemical modifications, solvent molecules, counterions, and intermolecular interactions, including hydrogen bond (HB) interactions, CH··· π/π ···π interactions, and short contact interactions, in tuning the MMCT properties of switchable $\{Fe_2Co_2\}$ squares as examples, as most of the representative MMCT complexes have been nicely summarized in the literature.

In 2004, Dunbar and co-workers reported a discrete MMCT cluster (**1**: $\{[Co(tmphen)_2]_3[Fe(CN)_6]_2\}$) by reacting $[Fe(CN)_6]^{4-}$ with $[Co(tmphen)_2]^{2+}$ (tmphen = = 3,4,7,8-tetramethyl-1,10-phenanthroline) [72]. A magnetic study revealed that the $\{Fe_2Co_3\}$ cluster exhibits thermally induced intramolecular one-electron transfer and has the electronic configuration of [(Co$^{II}_{HS}$)$_3$(Fe$^{III}_{LS}$)$_2$] in the high-temperature phase and of [(Co$^{II}_{HS}$)$_2$Co$^{III}_{LS}$Fe$^{II}_{LS}$Fe$^{III}_{LS}$] in the low-temperature phase. Interestingly, the red-colored {Fe$_2$Co$_3$} cluster can transform into a blue-colored phase when exposed to humidity and undergoes another type of ET between [(Co$^{II}_{HS}$)$_3$(Fe$^{III}_{LS}$)$_2$] and [(Co$^{II}_{HS}$)$_2$Co$^{III}_{LS}$Fe$^{II}_{LS}$Fe$^{III}_{LS}$] configurations. Later, Dunbar, Clérac, Mathonière, and co-workers found that the blue-colored phase can exhibit significant changes in magnetic susceptibility after white-light irradiation, indicative of the photoinduced CTIST behavior [73]. Compound **1** is considered the first example of molecular PBAs exhibiting the MMCT function at the molecular level, which opens a new avenue for studying switchable molecular PBAs.

After that, considerable MV and heterometallic MMCT complexes with a metal number ranging from 2 to 15 have been synthesized [13, 24, 71, 73–76]. As aforementioned, the MMCT property depends on the relative redox potentials of the iron and cobalt centers. The {Fe$_2$Co$_2$} system provides an excellent platform to investigate the influences of chemical modifications, guest molecules, and intermolecular interactions, such as crystal packing, π–π interaction, and HB interaction on the MMCT behavior, as the local coordination sphere can be easily and systematically modified. We focus on the {Fe$_2$Co$_2$} squares to show how researchers manipulate the MMCT property. The typical {Fe$_2$Co$_2$} square is composed of two [Tp^{R}Fe^{III}(CN)$_3$]$^-$ units and two Co^{II} ions with the coordination of two bidentate N-ligands, in which iron and cobalt centers reside in alternate corners and are bridged by cyanides to each other. At the beginning of the twenty-first century, Oshio and co-workers first reported two cyanide-bridged Fe–Co molecular squares, [Fe$^{II}_2$Co$^{II}_2$(μ-CN)$_4$(bpy)$_8$](PF$_6$)$_4$·3CHCl$_3$·2CH$_3$CN (**2**) and [Fe$^{II}_2$Co$^{III}_2$(μ-CN)$_4$(bpy)$_8$](PF$_6$)$_6$·2CHCl$_3$·4CH$_3$NO$_2$ (**3**) (bpy = 2,2′-bipyridine) [77]. Magnetic susceptibility measurement revealed that Fe^{II} and Co^{III} ions in **3** were diamagnetic. The HS Co^{II} ions in **2** were in the HS state and weakly coupled through the LS Fe^{II} ions. However, no MMCT behavior was observed. It was until 10 years later that Holmes, Clérac, Mathonière, and co-workers reported the MMCT property in [{(Tp*)Fe^{III}(CN)$_3$}$_2${Co^{II}(bpy)$_2$}$_2$][OTf]$_2$·4DMF·2H$_2$O (**4**, Tp*: tris[3,5-dimethyl]pyrazolyl borate; OTf: trifluoromethanesulfonate) [78]. The temperature-dependent magnetic susceptibility of compound **4** showed a broad thermal hysteresis of 18 K, indicating that the charge-transfer-induced spin transition is of the first order. Light-monitored magnetization and surface reflectivity spectra suggested the complete photoinduced CTIST from the diamagnetic {(Fe$^{II}_{LS}$)$_2$(Co$^{III}_{LS}$)$_2$} state to the paramagnetic {(Fe$^{III}_{LS}$)$_2$(Co$^{II}_{HS}$)$_2$} state. It is noted that the photoexcited state of **4** has a relatively long lifetime; the relaxation time was estimated to be three minutes at 120 K.

At the same time, Oshio and co-workers also reported a series of Fe–Co molecular squares, [Co$_2$Fe$_2$(CN)$_6$(Tp*)$_2$(dtbbpy)$_4$](PF$_6$)$_2$·2MeOH (**5**), [Co$_2$Fe$_2$(CN)$_6$(Tp*)$_2$(bpy)$_4$](PF$_6$)$_2$·2MeOH (**6**), and [Co$_2$Fe$_2$(CN)$_6$(Tp)$_2$(dtbbpy)$_4$](PF$_6$)$_2$·4H$_2$O (**7**) (Tp: hydrotris(pyrazol-1-yl)borate, Tp*: hydrotris(3,5-dimethylpyrazol-1-yl)borate, bpy: 2,2′-bipyridine, dtbbpy =4,4′-di-*tert*-butyl-2,2′-bipyridine), by modifying the capping ligands of the iron site and substituent groups on bipyridine of cobalt sites [79]. The chemical modifications significantly changed their intervalence electron-transfer behavior. Compounds **6** and **7** maintained the electronic configurations of $\{(Fe^{III}_{LS})_2(Co^{II}_{HS})_2\}$ and $\{(Fe^{II}_{LS})_2(Co^{III}_{LS})_2\}$, respectively, throughout the entire temperature range measured. Although compound **5** exhibited a two-step CTIST centered at transition temperatures of 275 and 310 K in solid. Interestingly, the synchrotron X-ray diffraction analysis revealed that the IM phase **5** exhibits the positional ordering of $\{(Fe^{III}_{LS})_2(Co^{II}_{HS})_2\}$ and $\{(Fe^{II}_{LS})_2(Co^{III}_{LS})_2\}$ squares with the 2 : 2 ratio. And the LS phase of **5** can be photoexcited by 808-nm light, which finally relaxes to the initial state at 80 K. By analyzing the cyclic voltammetry data of three compounds, the authors suggested that introducing electron-donating groups on the Tp ligands and bpy ligands can effectively decrease the redox potentials of the iron and cobalt sites. This hypothesis is verified by considerable switchable $\{Fe_2Co_2\}$ squares, despite that some exhibit opposite results. Zhang and co-workers reported two structurally related $\{Fe_2Co_2\}$ squares (**8**, $\{[(Tp^{Me})Fe(CN)_3]_2[Co(bpy)_2]_2[(TpMe)Fe(CN)_3]_2\}\cdot 12H_2O$; **9**, $\{[(Tp^{Me})Fe(CN)_3]_2[Co(bpy)_2]_2[BPh_4]_2\}\cdot 6MeCN$), demonstrating the same conclusion that stronger σ donation of Tp derivatives can energetically stabilize the LS state of Fe^{II} species, leading to the lower transition temperature of electron transfer [80]. Recently, Nihei has also discussed the correlation between electronic states of molecular PBAs and the redox properties of the polycyanidometallates [81]. The authors found that different values between the redox potentials of the iron and cobalt species in CTIST-active $\{Fe_2Co_2\}$ squares are around −0.8 V, which suggests that the combination of iron- and cobalt-based units with the difference in redox potential values around −0.8 V will result in new $\{Fe_2Co_2\}$ complexes, exhibiting switchable CTIST behavior at room temperature.

It should be stressed that most of the redox potentials of iron and cobalt species here are based on the cyclic voltammetry results in solution. The conclusion on the substituent effect would apply only when $\{Fe_2Co_2\}$ squares share the same crystal packing and counterions. This suggests that the intermolecular interactions also strongly influence the MMCT properties. Zhang and co-workers have discussed the intermolecular interactions by analyzing the intermolecular $\pi\cdots\pi$ interactions and short-contact interactions [80]. These interactions are necessary for abrupt and hysteretic electron-transfer transitions. However, they found no obvious relation between the transition temperatures and distances of contact interactions. Intermolecular interactions seem to have a small perturbation on their intrinsic electron-transfer properties. Although recent studies have shown that the variations of intermolecular interactions and the co-crystalized solvent molecules can also alter the intervalence of electron-transfer behavior. In 2018, Liu, Jiao, and co-workers reported a square complex $[(Tp^{Pz})Fe^{II}(CN)_3]_2[Co^{III}_2(dpq)_4]\cdot 2ClO_4\cdot 2CH_3OH\cdot 4H_2O$ (**10**·$2CH_3OH\cdot 4H_2O$, Tp^{Pz} = tetrakis(pyrazolyl)borate, dpq = pyrazino[2, 3-f] (**1, 10**) phenanthroline) [82]. The solvated phase of **10**·$2CH_3OH\cdot 4H_2O$ exhibits a thermally induced reversible intermetallic charge transfer in the mother liquor. When it was subjected to desolation in the air, it transformed into a red-colored paramagnetic phase $[(Tp^{Pz})Fe^{II}(CN)_3]_2[Co^{III}_2(dpq)_4]\cdot 2ClO_4$ (**10**) that showed no CTIST property. More interestingly, the desolvated **10** can undergo another single-crystal-to-single-crystal transformation when exposed to water vapor at 100 °C, resulting in a green-colored polymorph **10a**. In contrast to **10**, **10a** exhibited reversible thermally induced charge transfer, which can be further photoswitched by alternating 532- and 808-nm light irradiations. Structural analysis revealed that the average distances of the π···π and C–H···π interactions showed a successive converge upon two-step phase transition, which provided an additional driving force for the successive two-step single-crystal-to-single-crystal transformation. For their different CTIST properties, the authors suggested that the HB interactions that are destroyed by the loss-of-solvent molecules may have a substantial effect on the Fe-CN–bonding interactions. The variation of intermolecular π···π interactions also modified the coordination environment of the cobalt sites, both of which will affect the redox potentials of metal ions and subsequent electron-transfer properties. To further probe the role of intermolecular interactions, Liu, Jiao, and co-workers synthesized a series of solvent-free {Fe$_2$Co$_2$} squares sharing the same cationic tetranuclear {[(Tp^{Pz})Fe(CN)$_3$]$_2$[Co(dpq)$_2$]$_2$}$^{2+}$ (**11**, [BF$_4$]$^-$; **12**, [PF$_6$]$^-$; **13**, OTf; **14**, [(Tp^{Pz})Fe(CN)$_3$]$^-$) [83]. The authors found that anions with different sizes can alter the intermolecular π···π interactions between dpq ligands coordinated to the cobalt ions, resulting in changes in the distortion degree of cobalt coordination spheres and, consequently, their electron-transfer behavior. Magnetostructural analyses on the compounds **10–14** suggested that a large distortion of the cobalt coordination sphere favors the paramagnetic {Fe^{III}$_{LS}$(μ-CN)Co^{II}$_{HS}$} state and the lower transition temperature (Figure 15.3). This is supported by the DFT calculations based on the LS structures of **11** and **12**, revealing that the smaller energy gap can induce the electron transfer from the Fe-based occupied orbitals to Co-based unoccupied orbitals more easily, which affords the lower transition temperature. In addition, the authors try to correlate the distortion degree of the cobalt coordination sphere with the stability of the {Fe^{II}$_{LS}$(μ-CN)Co^{III}$_{LS}$} states. They performed the DFT calculations on a modeled structure of **11** by simultaneously changing the four Co — N ≡ C bond angles. The results showed that the HOMO–LUMO energy gap

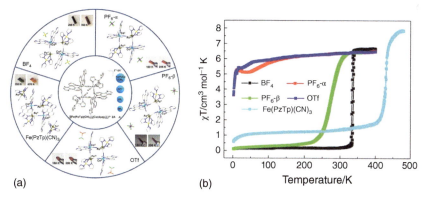

Figure 15.3 (a) Crystal structures of **11–14** with different counter anions. (b) Temperature-dependence susceptibility of **11–14**. Source: Jiao et al. [83], John Wiley & Sons.

gradually increased with the increasing distortion degree of the cobalt coordination sphere, indicating that the corresponding $\{Fe^{II}_{LS}(\mu\text{-}CN)Co^{III}_{LS}\}$ state becomes more stable. These results demonstrate that intermolecular $\pi\cdots\pi$ interactions not only play an important role in determining the occurrence of MMCT but also affect the transition temperature of the electron-transfer process. Despite their significant influences on the MMCT property, it should be pointed out that the intermolecular interactions indirectly play a part. Manipulating the electron-transfer behavior through fine-tuned intermolecular interactions is still challenging and unpredictable. In addition to this, the photoresponsive property of MMCT complexes is proposed to be related to the intermolecular interactions, as the short contact interactions may affect the stability of the photoexcited state. Nevertheless, this assumption is still under debate and needs further exploration.

Recently, Li, Lescouëzec, and co-workers studied the application of moderate pressure to convert a paramagnetic $Fe^{III}_2Co^{II}_2$ square complex into a molecular switch, exhibiting a full diamagnetic to paramagnetic transition [25]. The $\{Fe_2Co_2\}$ square with the formula $\{[Fe(Tp)(CN)_3]_2[Co(vbik)_2]_2\}^{2+}$ (**15**, Tp = hydrotris(pyrazolyl)borate; vbik = bis(1-vinylimidazol-2-yl)ketone) was reported earlier, where it was possible to crystallize in its $\{Fe^{III}_2Co^{II}_2\}^{2+}$ state when the temperature was set at 35 °C (phase 1) or in its $\{Fe^{II}_2Co^{III}_2\}^{2+}$ state when the temperature was set at 5 °C (phase 2). Phase 1 was paramagnetic, while phase 2 showed a steep thermally induced CTIST, which underlined the impact of solid-state interactions on CTIST. As shown in Figure 15.4, the application of pressure on phase 1 recovered the phase transition with a rare behavior: the higher the pressure, the broader the hysteresis. Further measurements and theoretical analyses revealed the increased intermolecular interactions mediated by the ligand were responsible for the enhancement of elastic interactions and the pressure-enhanced bistability, suggesting that $\{Fe_2Co_2\}$ complexes could be very sensitive piezo devices for the potential use as sensors.

As mentioned above, the nitrogen atom of terminal cyanide groups in polycyanometallates has an affinity for electron-deficient units and can be used as a

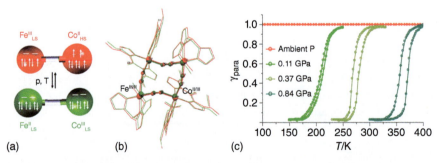

Figure 15.4 (a) Scheme of the electronic-state changes occurring in the Fe–CN–Co pair of **15**. (b) Perspective view of the square complex at 10^{-5} and 0.85 GPa of **15**. (c) Thermal variation of the paramagnetic $\{Fe^{III}_2Co^{II}_2\}$ molar fraction under various hydrostatic pressures of **15**. Source: Li et al. [25], John Wiley & Sons.

Lewis base. This provides another way to tune the redox potential of the polycyanometallate unit and the electron-transfer property. This has been previously exemplified by the acid titration experiment on complex **5** in butyronitrile. As the amount of trifluoroacetic acid was increased, the charge-transfer band, characteristic of $\{(Fe^{III}_{LS})_2(Co^{II}_{HS})_2\}$ species, decreased in intensity. And a new absorption band, characteristic of $\{(Fe^{II}_{LS})_2(Co^{III}_{LS})_2\}$ species, appeared at the same time. The reverse spectral changes were achieved by adding triethylamine, indicating a protonation-actuated CTIST. Such a pH-dependent electron-transfer property can be explained by the fact that the protonation of the terminal cyanide nitrogen atoms results in a more positive shift of the redox potential of the iron site, thereby making the transition from $\{(Fe^{III}_{LS})_2(Co^{II}_{HS})_2\}$ to $\{(Fe^{II}_{LS})_2(Co^{III}_{LS})_2\}$ more easily. Later, Marínez and co-workers carefully studied the stepwise protonation behavior of a MV square complex $[\{Co^{III}\{(Me)_2(\mu\text{-}ET)cyclen\}\}_2\{(\mu\text{-}NC)_2Fe^{II}(CN)_4\}_2]^{2-}$ (**16**) [84]. The acid titration experiment revealed a three-step protonation of the square complex. Cyclic voltammetry measurements also revealed a positively shifted redox potential of the iron ion, demonstrating that the protonation can effectively influence the CTIST property through cyanide groups. Recently, a cyanide-bridged square-shaped $\{Fe_2Co_2\}$ tetranuclear complex, $[\{Co(MeTPyA)(\mu_2\text{-}NC)_2Fe(bbp)(CN)\}_2]\cdot 3H_2O$ (**17**, MeTPyA = tris((3,5-dimethylpyrazol-1-yl)methyl)amine, and H_2bbp = bis(2-benzimidazolyl)-pyridine), was also reported by Gogoi and co-workers [85]. UV–Vis, electrochemical, and ^{1}H NMR studies verified the reversible intramolecular CTIST that could be triggered upon adding either acid or base.

The terminal cyanide groups of polycyanometallates can also act as acceptors of HBs. This may alter the spin state of Fe–Co clusters depending on the strength of the HB interactions. This HB interaction is usually found between solvent molecules and MMCT clusters. The adsorption or absorption of solvent molecules will lead to the break or formation of the HBs, resulting in the different CTIST behavior of cyanide-bridged Fe–Co complexes. The HB interactions can also be introduced on purpose by adding exogenetic HB donor ligands. Recently, Oshio and co-workers reported a HB donor–acceptor system composed of a CTIST $\{Fe_2Co_2\}$ unit acting as an HB acceptor and 4-cyanophenol acting as an HB donor. The co-crystallization of the two components yielded a square complex $[Co_2Fe_2(bpy^*)_4(CN)_6(Tp^*)_2](PF_6)_2\cdot 2CP\cdot 8BN$ (**18**, bpy* = 4,4′-dimethyl-2,2′-bipyridine, CP = 4-cyanophenol, BN = benzonitrile) [86]. Complex **18** has two crystallographically independent square cations (Sq.A and Sq.B) and four CP molecules (Figure 15.5). A magnetic study revealed a three-step spin transition with four stable phases. X-ray structural analyses showed that the HB interactions for the HS phases are substantially weaker than in the LS phases, suggesting that the intervalence electron-transfer behavior is strongly coupled with the alternation of the strength of HB interactions. This assumption is further verified by checking the distances of HB interactions in two IM states; for that, the distances between nitrogen atoms of cyanide groups and oxygen atoms of CP molecules decrease in a sequential way when the spin states of Sq. A and Sq. B moieties transform from the HS to LS states. The stepped CTIST behavior accompanied by the alternation of HB interactions can be explained

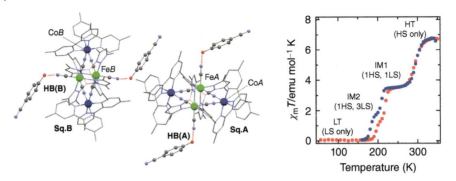

Figure 15.5 Crystal structures and temperature-dependent susceptibility of **18**. Source: Nihei et al. [86], John Wiley & Sons.

by the fact that the HB interaction between HB-acceptor cyanide groups and HB donor withdraws the d-electron density from the iron center, leading to the stabilization of the iron d-orbitals and the positively shifted iron oxidation potential. This result provides a new strategy to achieve multistability by controlling the HB interactions through external stimuli. The same group reported the construction of new aggregates with controllable dimensionality through directional HBs from a stimuli-responsive tetranuclear complex [87]. The two-terminal cyanide nitrogen atoms of the tetranuclear cation $[Co_2Fe_2(CN)_6(tp^*)_2(bpy^*)_4]^{2+}$ (**19^{2+}**, tp* = hydrotris(3,5-dimethylpyrazol-1-yl)borate; bpy* = 4,4′-dimethyl-2,2′-bipyridine) acts as a hydrogen-bond acceptor (HBA) with a linear bridging mode. The co-crystallization of **19^{2+}** with the H$_2$Q (p-hydroquinone) molecules afforded the formation of a one-dimensional aggregate, exhibiting one-step thermal CTIST behavior. The hydrogen-bonded honeycomb network obtained from the co-crystallization of **19^{2+}** with H$_3$PG (phloroglucinol) molecules showed stepwise thermal CTIST behavior. These dimensionally controlled assemblies could be useful to construct controlled molecular arrangements having multiple stable phases.

15.3 Cyanide-Bridged MMCT Complexes Showing Switchable Functional Properties

As mentioned above, the polycyanometallate-based complexes constitute a large family of magnetic materials that can combine versatile physical properties. This part will briefly review the typical MMCT complexes constructed by polycyanometallates, which show the switching of magnetic, conductivity, and dielectric constant changes, thermal expansion, and photochromism properties.

15.3.1 Modulating Molecular Nanomagnet Behavior

Molecular nanomagnets have been pursued since the study of the famous Mn$_{12}$ molecule in 1993 [88]. Magnet-like behavior stems from a large-spin ground state and uniaxial magnetic anisotropy. The corresponding magnetic complexes, known as single-molecule magnets and single-chain magnets, are promising candidates for high-density information storage. In addition, manipulating the molecular nanomagnet behavior by external stimuli is essential to make them useful for practicable applications. The photoinduced spin transition provides an attractive alternative to engineering molecular magnets. Recent examples of Fe^{II}/Fe^{III} spin-crossover complexes have been reported to exhibit phototunable nanomagnet behavior [89–91]. The MMCT process involves concomitant spin state and magnetic anisotropy changes. Therefore, it can switch the on/off of the magnetic bistable state. The most famous systems showing photomagnet behavior are the Co–Fe Prussian blue analogs. Recent progress has also been made in phototunable nanomagnets with low dimensionality [66, 92].

An example showing photoinduced single-chain magnet property is a cyanide-bridged $Fe^{III}{}_2Co^{II}$ double zigzag chain (**20**, [Fe(bpy)(CN)$_4$]$_2$Co(4,4'-bipyridine)}·4H$_2$O, bpy = 2,2'-bipyridine) [93]. A magnetic study revealed the reversible but incomplete MMCT behavior with a thermal hysteresis. After light irradiation to the LS phase, the spin topology changed from the discrete clusters to the one-dimensional because of the photoinduced transformation from diamagnetic $Fe^{II}{}_{LS}(\mu\text{-CN})Co^{III}{}_{LS}$ to metastable paramagnetic $Fe^{III}{}_{LS}(\mu\text{-CN})Co^{II}{}_{HS}$ units. Alternating current (AC) magnetic susceptibility measurements revealed the slow magnetic relaxation behavior. However, due to the interchain antiferromagnetic interaction, the complex formed an antiferromagnetically ordered phase below 3.8 K. To obtain a well-isolated magnetic chain, a bulkier building block [Fe(pzTp)(CN)$_3$]$^-$ (pzTp = tetrakis(pyrazolyl)borate) and monodentate 4-styrylpyridine ligand were used to build the chain (**21**, {[Fe(pzTp)(CN)$_3$]$_2$Co(4-styrylpyridine)$_2$}·2H$_2$O·2CH$_3$OH) [94]. Interestingly, the charge transfer occurred between the cobalt sites and one of the two iron sites, resulting in a regular 1-D structure, in which 1-D chains were sufficiently separated by bulky ligands. The author proposed that the different intermolecular π···π contacts and HB interactions involving the iron sites may be responsible for the directional charge transfer. The AC magnetic susceptibility study on a 532 nm light-irradiated complex ruled out the antiferromagnetic ordering, confirming the single-chain magnet behavior. Later, the photoinduced bidirectional MMCT behavior was adopted into the well-isolated double zigzag chain {[Fe(bpy)(CN)$_4$]$_2$[Co(phpy)$_2$]}·2H$_2$O (**22**, phpy = 4-phenylpyridine) to control the single-chain magnet behavior [32]. By alternatively applying 808 and 532 nm laser irradiations, the single-chain magnet behavior can be switched in the on–off–on sequences.

Oshio and co-workers designed a chiral cyanobridged Fe/Co chain-coordination polymer with a square-wave structure to achieve phototunable nanomagnets. The freshly prepared complex {[Co(II)([R]-pabn)][Fe(III)(tp)(CN)$_3$](BF$_4$)}·MeOH·2H$_2$O (**23**, (R)-pabn = (R)-N(2),N(2')-bis(pyridin-2-ylmethyl)-1,1'-binaphtyl-2,2'-diamine) could transform into two new phases with one and three water molecules in the lattice under nitrogen and air atmospheres, respectively [28]. Temperature-dependent magnetic susceptibility studies showed that all three phases could exhibit hysteretic phase transitions. The comparison of the three phases demonstrated how the HB interactions influenced the metal sites' redox potential and the thermally induced CTIST behavior. The 808 nm laser irradiation on the stable phase that contains one water molecule resulted in the ferromagnetic chain. The AC susceptibility measurements showed two individual relaxation processes: the major one could be attributed to the slow magnetic relaxation of the photoinduced single-chain magnet; the minor one was probably caused by a degree of fragmentation in the metastable HS phase (Figure 15.6).

In addition to the one-dimensional (1-D) cases, the discrete charge-transfer clusters can also exhibit photomagnet behavior. Oshio and co-workers proposed an extended cyanide-bridged square system that may exhibit enhanced magnetic anisotropy and a ground spin state to address this goal. By replacing the two bidentate ligands with one tridentate ligand and leaving one vacant coordination site on the cobalt ions of the [Fe_2Co_2] squares, two additional building blocks were linked to the system, forming an extended hexanuclear cluster (**24**, [$Co_2Fe_4(bimpy)_2(CN)_6(\mu\text{-}CN)_6(pztp)_4$]·2(1-PrOH)·$4H_2O$, bimpy = 2,6-bis(benzimidazol-2-yl)pyridine) [75]. The reversible CTIST behavior was revealed between the cobalt and iron sites within the square core. The photomagnetic study revealed a sharp increase of $\chi_m T$ values after light irradiation, suggesting intramolecular ferromagnetic interactions. In addition, the X-ray diffraction experiment demonstrated the photoinduced transformation before and after light irradiation at 20 K. When a small external field of 500 Oe was applied, the AC signals showed frequency and temperature dependences of the peak maxima, demonstrating that the phototunable single-molecule magnet can also be achieved in discrete charge-transfer molecules. Recently, Sato et al. reported a novel V-shaped trinuclear cyanide-bridged {Fe_2Co} complex (**25**, {[$Fe(Tp)(CN)_3$]$_2$($Co[dpa]_2$)}·$2H_2O$, dpa = 2,2'-dipyridylamine), in which an unexpected reverse effect is observed in the $S = \frac{1}{2}$ {$Fe^{II}_{LS}Co^{III}_{LS}\text{-}Fe^{III}_{LS}$} ground state for the first time [95]. The complex exhibited the light-switchable Fe–Co charge transfer with repeatable on/off switching by alternating light irradiation at 560 and 785 nm. Intriguingly, the field-induced slow magnetic relaxation behavior was observed in the $S = \frac{1}{2}$ ground state with the possible phonon-bottlenecked direct relaxation process. In this work, a new way to modulate slow–fast magnetic relaxation with photoinduced CTIST was discovered to promote future research on the photo effect in slow magnetic relaxation complexes.

15.3.2 Modulating Molecular Electric Dipole

It has been established that significant changes in the dielectric constant are caused by the motions of polar molecules or groups. In contrast to the displacement and order–disorder mechanisms, the movement of the electron will change the charge distribution and electric dipole by breaking the charge symmetry during MMCT. In this way, both the magnetic and electric changes can be effectively and reversibly controlled through external stimuli. Research on changing charge-transfer-coupled electric properties focused on Fe–Co Prussian blue analogs. The typical examples are $Na_{0.5}Co^{II}_{1.25}[Fe^{III}(CN)_6]\cdot4.8H_2O$ (**26**) and $Na_{0.38}Co^{II}_{1.31}[Fe^{III}(CN)_6]\cdot5.4H_2O$ (**27**) which showed an abrupt change in conductivity at the temperature where

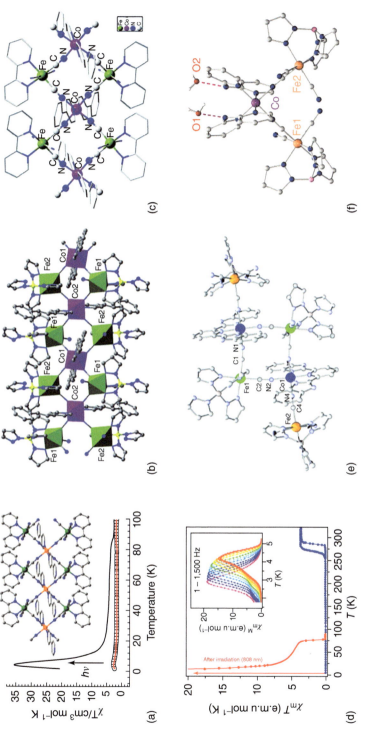

Figure 15.6 (a) Crystal structures and temperature-dependent susceptibilities of 1D double zigzag chain **20**. Source: Liu et al. [93], Reproduced with Permission from American Chemical Society. (b) Crystal structures of 1D double zigzag chain **21**. Source: Dong et al. [94], John Wiley & Sons. (c) Crystal structures of 1D chain **22**. Source: Jiang, et al. [32], Reproduced with Permission from Royal Society of Chemistry. (d) Light-induced magnetic susceptibility data of **23**. Inset: Low-temperature frequency dependence of the out-of-phase magnetic response, indicative of a slow magnetic relaxation. Source: Hoshino et al. [28], Reproduced with Permission from Springer Nature. (e) Crystal structures of hexanuclear complex **24**. Source: Nihei et al. [75], John Wiley & Sons. (f) Crystal structures of V-shaped trinuclear complex **25**. Source: Li et al. [95], John Wiley & Sons.

the CTIST occurred [42]. Their analog $\{Rb_{0.5}Co^{III}Co^{II}_{0.25}[Fe^{II}(CN)_6]\}\cdot 5.9H_2O$ (**28**) exhibited no CTIST effect but abrupt conductivity transition by applying a higher electric field, indicating that Fe–Co Prussian blue can show conductivity switching by varying the temperature and by applying an electric field. Later, Ohkoshi and co-workers observed the coexistence of ferroelectricity and ferromagnetism in $Rb^{I}_{0.82}Mn^{II}_{0.20}Mn^{III}_{0.80}[Fe^{II}(CN)_6]_{0.80}[Fe^{III}(CN)_6]_{0.14}\cdot H_2O$ (**29**) where the ferroelectricity is caused by the mixing of Fe^{II}, Fe^{III}, Fe vacancies, Mn^{II}, and Jahn–Teller distorted Mn^{III} centers [52]. The ferromagnetism is mainly because of the parallel ordering of the magnetic moments on the Mn^{III} ions. In 2009, further research was conducted on similar types of Prussian blue analogs $Rb_{0.8}Mn[Fe(CN)_6]_{0.93}\cdot 1.62H_2O$ (**30**) to search for electrically switchable devices [53]. This cubic architecture exhibited a charge-transfer phase change from the high-temperature (HT) and low-temperature (LT) phases with a large thermal hysteresis loop. It was worth noting that the reverse transition from the LT to HT phase could not be induced by an electric field but by the heating process over the hysteresis range. The new type of electric-field-induced phase transition was related to a paraelectric–ferroelectric transition based on the fact that dielectric anomaly peaks of the real part were observed. The electric field could trigger such a phase transition and benefit by stabilizing the LT phase (Figure 15.7).

Compared with 3-D Prussian blue networks, an MMCT complex with low dimensionality is more desirable for understanding the phase-transition mechanism because of its regular composition and defined structure. As mentioned above, complex **23** is a [CoFe] cyanide-bridged chiral chain complex that displays thermal- and light-induced CTIST, coupled with conductivity switching and light-induced SCM properties. The conductivity property transformed from insulator in the LT phase to semiconductor in the HT phase was direct evidence of interchain dipolar switching induced by the IVCT. Given that MMCT can cause charge redistribution, a drastic change in molecular electric dipole is expected within the centrosymmetric discrete charge-transfer molecules. Therefore, a linear trinuclear cluster $\{[FeTp(CN)_3]Co(Meim)_4\}\cdot 6H_2O$ (**31**, Meim = N-methylimidazole), exhibiting the thermally reversible and photoinduced electron transfer, was reported by Liu and co-workers [96]. Accompanying the electron transfer, the trinuclear Fe_2Co cluster possessed reversible polar–nonpolar switching due to breaking the inversion center induced by charge transfer between the Co and Fe ions. The DFT-calculated results indicated that the LT phase has a permanent electric dipole of 18.4 D. In contrast, the HT phase has no permanent dipole, confirming the polarity conversion during the MMCT. Later, a similar light-induced bidirectional MMCT trinuclear $\{Fe_2Co\}$ cluster was reported with switchable polarity and magnetism [97]. The dielectric study revealed the dielectric anomaly associated with the reorientation of the molecular dipole during the MMCT process, despite the crystal showing centrosymmetric to the centrosymmetric phase transition. Recently, Meng and co-workers assembled the asymmetric $\{Fe_2Co\}$ unit into a cyanide-bridged MV chain $\{[(Tp)Fe(CN)_3]_2Co(BIT)\}\cdot 2CH_3OH$ (**32**, BIT = 3,4-bis-[1H-imidazole-1-yl]thiophene), which showed reversible multiphase transitions accompanied by photoswitchable single-chain magnet properties and a

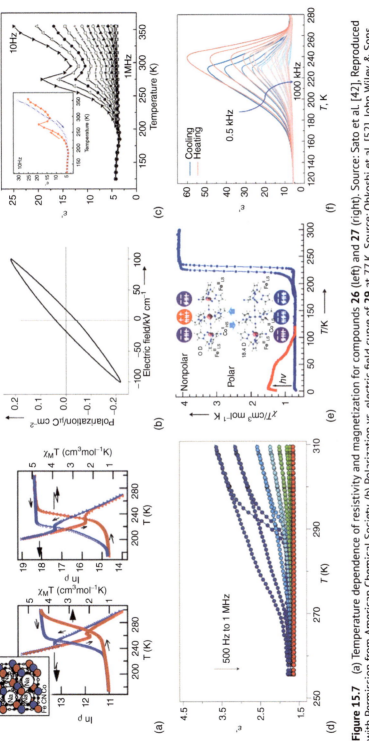

Figure 15.7 (a) Temperature dependence of resistivity and magnetization for compounds **26** (left) and **27** (right). Source: Sato et al. [42], Reproduced with Permission from American Chemical Society. (b) Polarization vs. electric field curve of **29** at 77 K. Source: Ohkoshi et al. [52], John Wiley & Sons. (c) Temperature dependence of the real part of the complex dielectric permittivity of **30**. Source: Mahfoud et al. [53], Reproduced with Permission from American Chemical Society. (d) Dielectric constant (ε') vs. frequency plots of **23**. Source: Hoshino et al. [28], Reproduced with Permission from Springer Nature. (e) Temperature-dependent magnetic susceptibilities and polar–nonpolar transformation of **31**. Source: Liu et al. [96], John Wiley & Sons. (f) Temperature-dependent ε' with different frequencies for **32**. Source: Yao et al. [70], John Wiley & Sons.

dielectric anomaly [70]. However, the selective ET compound showed a remarkable dielectric constant change that is uncommon for Class II MV complexes, most likely because the total dipoles do not cancel out. Furthermore, when complex **32** was rapidly cooled to 30 K, a single crystal-to-single crystal transformation from the HT phase (*Pnma*) to the HT^{sc} phase (supercooled phase, $P2_12_12_1$) was observed. These findings provide a new platform for multiphase transitions and multiswitches adjusted by selective metal-to-metal ET.

15.3.3 Modulating Thermal Expansion Behavior

Thermal expansion (TE), a fundamental issue in solid molecular materials, has attracted attention because relatively weak intermolecular interactions in their flexible structures sometimes produce characteristic TE behavior such as anisotropic and negative thermal expansion (NTE) [98]. The control of a given thermal expansion behavior is requested for modern applications, for example, the significant thermal expansion for micro- and nanoscale actuators and machines and zero thermal expansion for precision devices [99]. Most of the research on MMCT complexes mainly focuses on magnetization switching. Nevertheless, the charge-transfer-induced phase transition may also bring fascinating thermal expansion properties.

A typical example is the Perovskite-type material of A-site-ordered $LaCu_3Fe_4O_{12}$, which shows thermally induced charge transfer between the copper and iron ions, along with the magnetism- and conductivity-phase transitions [100]. This first-order phase transition led to an abrupt unit cell volume contraction of c. 1%. As molecule-based MMCT complexes feature a more flexible character, the structural variation and thermal expansion would be more significant. For instance, Sieklucka et al. reported that cyanide-bridged $\{Fe_6Co_3[W(CN)_8]_6\}$ (**33**) cluster exhibited reversible charge transfer from 190 to 210 K [101]. Spectroscopic, magnetic, and structural characterization indicated the coexistence of two active electron-transfer channels in one trimetallic material, which was reported for the first time. As shown in Figure 15.8b, crystallographic cell parameter measurements revealed a volumetric expansion of c. 5% (from 16 100 to 16 900 Å^3), and this positive thermal expansion occurred along all three axes. The changes in the unit cell parameters are mainly caused by the variation of the Co—N bond distances (by c. 0.16 Å, 7.4%) and the M—N and M—O bond distances (c. 0.07 Å, 3.5%) as well as the HB distance reduction (1.8–4.1%). Furthermore, the $[M(CN)_8]^{n-}$ moieties as rigid linkers remain unchanged during cluster expansion and contraction.

In 2017, Liu and co-workers found another example of cyanide-bridged compound $\{[(Tp)Fe(CN)_3]_2Co(Bib)_2\}$ (**34**, Bib = 1,4-Bis-[1H-imidazol-1-yl]benzene) that exhibited colossal PTE and NTE at different temperatures [69]. The trinuclear $\{Fe_2Co\}$ units were linked with organic ligands into the infinite two-dimensional (2-D) structures, and a charge-transfer-induced magnetic transition with hysteresis was observed between the Fe^{III}_{LS}–CN–Co^{II}_{HS} and Fe^{II}_{LS}–CN–Co^{III}_{LS} states. A detailed study revealed the volume increased abruptly with the expansion coefficient of 1498 MK^{-1} in the temperature of 180–240 K. The colossal PTE mainly

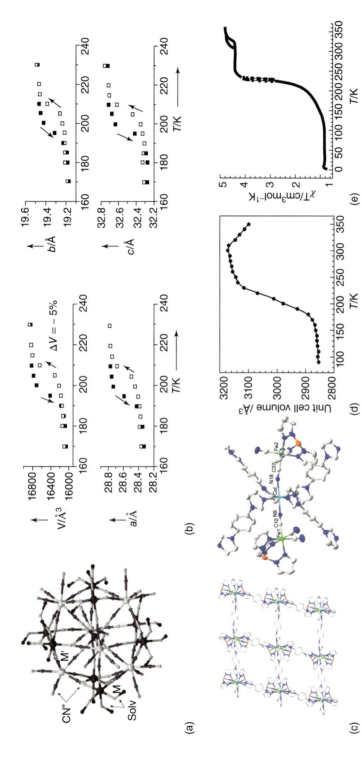

Figure 15.8 (a–b) The molecular structure and change of the crystallographic cell parameters for the phase transitions of **33**. Source: Podgajny et al. [101], John Wiley & Sons. (c–e) Crystal structure, changes in unit cell volume, and temperature-dependent susceptibilities of **34**. Source: Hu et al. [69], John Wiley & Sons.

contributed to the elongation of the coordination bond around the Co site and the interlayer π...π distances. Moreover, rotation of building blocks with the enhanced interlayer π...π interactions resulted in a negative volume expansion with a coefficient of −489 MK^{-1} as well as another magnetic transition between 300 and 350 K. The PTE and NTE behaviors were also accompanied by two hysteretic magnetic phase transitions, demonstrating the importance of the MMCT and weak bonding interactions on crystal thermodynamics in flexible materials.

15.3.4 Modulating Photochromic Behavior

Another remarkable feature of MMCT complexes is that they can exhibit different electronic states with different colors, also known as photochromism. The most well-studied photochromic compounds are organic species like azobenzene, spiropyran, naphthopyran, and diarylethene. Their detailed photochromic properties have been well-summarized in the literature [102–104]. Compared with organic species, the MMCT complexes do not involve drastic structural changes like *cis/trans* or open-/closed-ring isomerization but can also exhibit color and absorption transitions because of the rearrangement of electronic configurations. Note that the organic–inorganic hybridized materials can exhibit photochromic properties in solid form and with high fatigue resistance and a fast response rate.

A typical example is an optical property of a bimetallic compound [{(Tp)Fe(CN)$_3$}{Co(PY5Me$_2$)}](CF$_3$SO$_3$) (**35**, PY5Me$_2$ = 2,6-bis(1,1-bis[2-pyridyl]ethyl)pyridine) reported by Clerac et al. (Figure 15.9) [71]. The solvent DMF-free analog exhibited reversible and abrupt MMCT between 175 and 165 K, representing the smallest photoinduced charge-transfer complexes in solid. The temperature-dependent reflectivity exhibited a significant decrease in the band above 700 nm upon cooling, indicating the occurrence of the paramagnetic-to-diamagnetic transition. After light irradiation at 10 K, the reflectivity at 850 nm recovered nearly to the one recorded at 310 K and then relaxed to the thermal equilibrium state, which was proven by the photomagnetic measurement. In addition, a 3-D Co–W bimetallic compound [{Co^{II}(4-methylpyridine)(pyrimidine)}$_2${Co^{II}(H$_2$O)$_2$}{W^V(CN)$_8$}$_2$]·4H$_2$O (**36**) was reported to show thermally induced charge transfer between the Co$^{II}_{HS}$–NC–W^V and Co$^{III}_{LS}$–NC–W^{IV} with a relatively large hysteresis of 69 K as well as drastic color changes [60]. Furthermore, the Co–W complex underwent photoinduced charge transfer, exhibiting photochromic behavior at low temperatures.

However, one drawback is that most of the charge transfer complexes can only exhibit photochromic properties at cryogenic temperatures because of the short lifetime of the metastable states. Prolonging their lifetimes remains a considerable challenge for future applications. A few Prussian blue networks showing photoinduced charge transfer around room temperature were found. In 2002, Hashimoto et al. found that {Na$_{0.68}$Co$_{1.20}$[Fe(CN)$_6$]}·3.7H$_2$O (**37**) exhibited thermally induced charge transfer around room temperature with a significant change in IR spectra after applying 532-nm one-shot-laser pulse on the LT phase at 295 K, indicating the transition from the LT phase to HT phase [105]. It was noteworthy that only the power density surpassed the threshold, and the photochromic change could

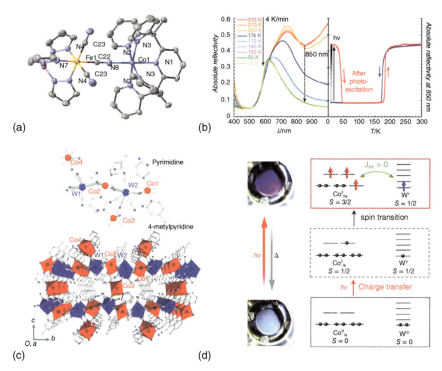

Figure 15.9 (a) Crystal structure of complex **35**. (b) Left: Surface reflectivity of **35** upon cooling. Right: Absolute reflectivity of **35** at 850 nm with cooling (blue), heating (red), and heating after 12 hours of white-light irradiation at 10 K (red). Source: Koumousi et al. [71], Reproduced with Permission from American Chemical Society. (c) Crystal structure of complex **36**. (d) Photographs of **36** before and after the irradiation and schematic illustration of the ferromagnetic ordering because of optically induced charge-transfer-induced spin transition. Source: Ozaki et al. [60], John Wiley & Sons.

be observed. Ohkoshi and co-workers also reported a rubidium–manganese hexacyanoferrate. The phase transition from Fe^{II}_{LS}–CN–Mn^{III}_{HS} to Fe^{III}_{LS}–CN–Mn^{II}_{HS} in $Rb_{0.98}Mn_{1.01}[Fe(CN)_6]\cdot 0.2H_2O$ (**38**) was observed at room temperature with one-shot-laser-pulse irradiation, which would inevitably be accompanied by photochromic behavior [106].

As mentioned above, some organic–inorganic hybridized materials composed of polymetallic polycyanometallate units can exhibit photochromic properties in solid state and at room temperature. Taking the discrete 3d-4f hexacyanoferrate $[Eu^{III}(18C6)(H_2O)_3]Fe^{III}(CN)_6\cdot 2H_2O$ (**39**, 18C6 = 18-crown-6) reported by Guo and co-workers as an example [107], this compound exhibited electron transfer with a color change from yellowish green to orange under UV–Vis light irradiation (Figure 15.10). The orange crystal could slowly reverse back in the dark. The UV–Vis and IR spectra indicated the formation of the Fe^{II} species. At the same time, the photoluminescence study suggested that the Eu^{III} species remained during light irradiation. EPR and IR studies confirmed radical formation on the crown site, suggesting electron transfer between the crown ligand and Fe^{III} sites. More recently,

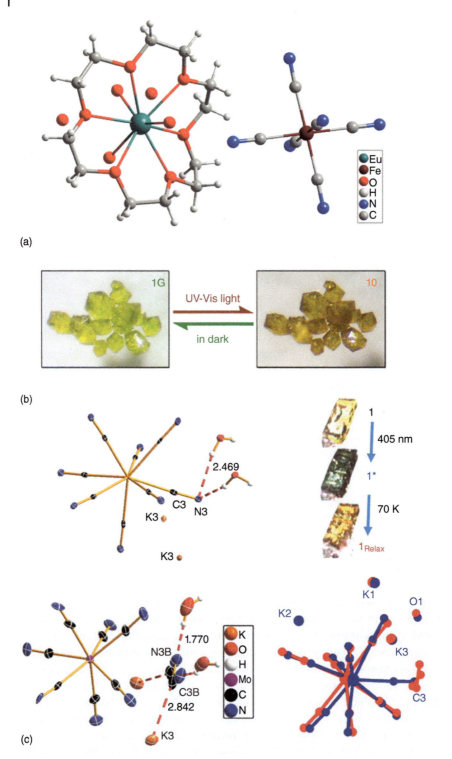

Figure 15.10 (a) Crystal structure of complex **39**. (b) Photochromism of **39** at room temperature. Source: Cai et al. [107], Reproduced with Permission from American Chemical Society. (c) Crystal structures and photographs of crystals for **40**. Source: Qi et al. [108], Reproduced with Permission from John Wiley & Sons.

even the isolated polycyanometallate-based compound $K_4Mo(CN)_8 \cdot 2H_2O$ (**40**) was reported to show photochromic and photomagnetic properties [108]. In-depth investigation by crystallographic, magnetic, optical, and theoretical calculation analyses demonstrated that reversible photomagnetism was associated not only with spin changes but also with a profound structural modification. It was proposed that the photodissociation of a Mo—CN bond can be activated in a solid at a low temperature or in a solution at room temperature. Both ways allowed us to isolate an unprecedented $[Mo^{IV}(CN)_7]^{3-}$ complex with a triplet spin ground state with significant magnetic anisotropy. The photochromic properties of the compound were also evaluated through reflectivity measurements. A new band around 650 nm appeared when exposing the compound to specific wavelengths ranging from 365 to 455 nm at 10 K. As such, the compound was irradiated at 405 nm for 30 minutes to obtain a spectrum of the stable photoexcited state at 10 K. After further heating, the photoexcited state began to relax to the unirradiated state and finally reached the diamagnetic one at 70 K. This result opens exciting perspectives in molecular magnetism for the photochemical preparation of new anisotropic building blocks.

15.4 Conclusion and Outlook

To summarize, we reviewed the recent progress of MMCT complexes, including the influencing factors on MMCT property and its switchable functionality. The manipulation of the MMCT is essential to control a material's electronic state, whose magnetic, electric, optical, and mechanical properties can be modulated simultaneously. The flexibility and response to external stimuli render the MMCT complexes versatile actuators to be used in multifunctional devices.

However, there are still some challenges that need to be solved. Firstly, the working temperature of photoinduced MMCT is still below liquid nitrogen temperature. And the electron-transfer mechanism also needs to be further explored by ultrafast spectroscopic techniques, which will shed light on improving the photo-operating temperature for MMCT. In addition, the lack of large hysteresis and coercive fields in low-dimensional photoinduced molecular magnets also needs to be overcome. Concerning the fact that 3d/4d/5d assemblies can display photoinduced charge-transfer properties, a possible way is to introduce the 4d/5d metal ions with strong spin–orbit coupling and large magnetic anisotropy. Secondly, the CTIST-induced ferroelectric transition is still a big challenge for charge-transfer-mediated electric properties. Introducing an asymmetric chiral ligand or intermolecular interaction may lead the complexes to crystalize in the polar space group and generate the possible directional electron transfer, thus helping to realize the goal. As the anomalous thermal expansion behavior is caused by the charge-transfer-coupled phase transition, the manipulation of the thermal expansion lies in the control of the transition temperature range and transition degree. Recently, Kepert and co-workers investigated the SCO-induced thermal expansion behavior in Hofmann-type metal–organic frameworks [109]. This material can exhibit tunable spin-transition behavior through the dilution method, thus representing different expansion properties. This strategy, tuning the redox potential of the metal site and intermolecular interactions, will help to control the thermal expansions.

Finally, combining MMCT with other physical properties will help to design more fascinating multifunctional materials. For instance, the coupling between MMCT and fluorescence may offer a promising alternative for manipulating the luminescence and phosphorescence properties by external stimuli of light and electric fields. We believe that the research of MMCT materials is not only limited to the demonstrated aspects aforementioned but also in many cross-fields such as photocatalysts and chemical biology, which will help researchers to develop a new generation of multifunctional materials.

References

1 Li, H.B., Tebikachew, B.E., Wiberg, C. et al. (2020). A memristive element based on an electrically controlled single-molecule reaction. *Angew. Chem. Int. Ed.* 59: 11641–11646.

2 Yang, Q., Zhong, T.T., Tu, Z.Y. et al. (2019). Design of single-molecule multiferroics for efficient ultrahigh-density nonvolatile memories. *Adv. Sci.* 6: 1801572.

3 Dei, A., Gatteschi, D., Sangregorio, C., and Sorace, L. (2004). Quinonoid metal complexes: toward molecular switches. *Acc. Chem. Res.* 37: 827–835.

4 Irie, M. (2000). Photochromism: memories and switches-introduction. *Chem. Rev.* 100: 1683–1684.

5 Bubnov, M.P., Kozhanov, K.A., Skorodumova, N.A., and Cherkasov, V.K. (2020). Photo- and thermosensitive molecular crystals. Valence tautomeric interconversion as the cause of the photomechanical effect in crystals of bis(dioxolene)cobalt complex. *Inorg. Chem.* 59: 6679–6683.

6 Shao, D., Shi, L., Shen, F.X. et al. (2019). Reversible on–off switching of the hysteretic spin crossover in a cobalt(II) complex via crystal to crystal transformation. *Inorg. Chem.* 58: 11589–11598.

7 Natterer, F.D., Yang, K., Paul, W. et al. (2017). Reading and writing single-atom magnets. *Nature* 543: 226–228.

8 Conyard, J., Cnossen, A., Browne, W.R. et al. (2014). Chemically optimizing operational efficiency of molecular rotary motors. *J. Am. Chem. Soc.* 136: 9692–9700.

9 Valverde-Munoz, F.J., Seredyuk, M., Meneses-Sanchez, M. et al. (2019). Discrimination between two memory channels by molecular alloying in a doubly bistable spin crossover qaterial. *Chem. Sci.* 10: 3807–3816.

10 Hogue, R.W., Singh, S., and Brooker, S. (2018). Spin crossover in discrete polynuclear iron(ii) complexes. *Chem. Soc. Rev.* 47: 7303–7338.

11 Li, B., Wang, X.N., Kirchon, A. et al. (2018). Sophisticated construction of electronically labile materials: a neutral, radical-rich, cobalt valence tautomeric triangle. *J. Am. Chem. Soc.* 140: 14581–14585.

12 Tezgerevska, T., Alley, K.G., and Boskovic, C. (2014). Valence tautomerism in metal complexes: stimulated and reversible intramolecular electron transfer between metal centers and organic ligands. *Coord. Chem. Rev.* 268: 23–40.

13 Jimenez, J.R., Glatz, J., Benchohra, A. et al. (2020). Electron transfer in the Cs⊂{Mn$_4$Fe$_4$} cubic switch: a soluble molecular model of the MnFe prussian-blue analogues. *Angew. Chem. Int. Ed.* 59: 8089–8093.

14 Aguila, D., Prado, Y., Koumousi, E.S. et al. (2016). Switchable Fe/Co prussian blue networks and molecular analogues. *Chem. Soc. Rev.* 45: 203–224.

15 Cammarata, M., Zerdane, S., Balducci, L. et al. (2021). Charge transfer driven by ultrafast spin transition in a CoFe prussian blue analogue. *Nat. Chem.* 13: 10–14.

16 Goujon, A., Varret, F., Escax, V. et al. (2001). Thermo-chromism and photo-chromism in a prussian blue analogue. *Polyhedron* 20: 1339–1345.

17 Nadurata, V.L. and Boskovic, C. (2021). Switching metal complexes via intramolecular electron transfer: connections with solvatochromism. *Inorg. Chem. Front.* 8: 1840–1864.

18 Schmidt, R.D., Shultz, D.A., Martin, J.D., and Boyle, P.D. (2010). Goldilocks effect in magnetic bistability: remote substituent modulation and lattice control of photoinduced valence tautomerism and light-induced thermal hysteresis. *J. Am. Chem. Soc.* 132: 6261–6273.

19 Dapporto, P., Dei, A., Poneti, G., and Sorace, L. (2008). Complete direct and reverse optically induced valence tautomeric interconversion in a cobalt–dioxolene complex. *Chem. Eur. J* 14: 10915–10918.

20 Robin, M.B. and Day, P. (1968). *Mixed Valence Chemistry-A Survey and Classification*. Oxford: University of Oxford.

21 Brunschwig, B.S., Creutz, C., and Sutin, N. (2002). Optical transitions of symmetrical mixed-valence systems in the class II–III transition regime. *Chem. Soc. Rev.* 31: 168–184.

22 Demadis, K.D., Hartshorn, C.M., and Meyer, T.J. (2001). The localized-to-delocalized transition in mixed-valence chemistry. *Chem. Rev.* 101: 2655–2686.

23 D'Alessandro, D.M. and Keene, F.R. (2006). Current trends and future challenges in the experimental, theoretical and computational analysis of intervalence charge transfer (IVCT) transitions. *Chem. Soc. Rev.* 35: 424–440.

24 Zhao, X.H., Shao, D., Chen, J.T. et al. (2021). Spin and valence isomerism in cyanide-bridged {Fe$_2^{III}$M^{II}} (M = Fe and Co) clusters. *Dalton Trans.* 50: 9768–9774.

25 Li, Y., Benchohra, A., Xu, B. et al. (2020). Pressure-induced conversion of a paramagnetic FeCo complex into a molecular magnetic switch with tuneable hysteresis. *Angew. Chem. Int. Ed.* 59: 17272–17276.

26 Nihei, M., Shiroyanagi, K., Kato, M. et al. (2019). Intramolecular electron transfers in a series of [Co$_2$Fe$_2$] tetranuclear complexes. *Inorg. Chem.* 58: 11912–11919.

27 Mondal, A., Li, Y., Seuleiman, M. et al. (2013). On/off photoswitching in a cyanide-bridged {Fe$_2$Co$_2$} magnetic molecular square. *J. Am. Chem. Soc.* 135: 1653–1656.

28 Hoshino, N., Iijima, F., Newton, G.N. et al. (2012). Three-way switching in a cyanide-bridged [CoFe] chain. *Nat. Chem.* 4: 921–926.

29 Mathonière, C. (2018). Metal-to-metal electron transfer: a powerful tool for the design of switchable coordination compounds. *Eur. J. Inorg. Chem.* 2018: 248–258.

30 Ohkoshi, S.-i. and Tokoro, H. (2012). Photomagnetism in cyano-bridged bimetal assemblies. *Acc. Chem. Res.* 45: 1749–1758.

31 Chorazy, S., Zakrzewski, J.J., Magott, M. et al. (2020). Octacyanidometallates for multifunctional molecule-based materials. *Chem. Soc. Rev.* 49: 5945–6001.

32 Jiang, W.J., Jiao, C.Q., Meng, Y.S. et al. (2018). Switching single chain magnet behavior via photoinduced bidirectional metal-to-metal charge transfer. *Chem. Sci.* 9: 617–622.

33 Peng, Y.Y., Wu, S.G., Chen, Y.C. et al. (2020). Asymmetric seven−/eight-step spin-crossover in a three-dimensional hofmann-type metal–organic framework. *Inorg. Chem. Front.* 7: 1685–1690.

34 Shi, L., Shao, D., Wei, X.-Q. et al. (2020). Enhanced single-chain magnet behavior via anisotropic exchange in a cyano-bridged Mo^{III}–Mn^{II} chain. *Angew. Chem. Int. Ed.* 59: 10379–10384.

35 Zheng, H., Meng, Y.S., Zhou, G.L. et al. (2018). Simultaneous modulation of magnetic and dielectric transition via spin-crossover-tuned spin arrangement and charge distribution. *Angew. Chem. Int. Ed.* 57: 8468–8472.

36 Zhuge, F., Wu, B., Yang, J. et al. (2010). Microscale hexagonal rods of a charge-assisted second-sphere coordination compound [Co(DABP)$_3$][Fe(CN)$_6$]. *Chem. Commun.* 46: 1121–1123.

37 Kobylarczyk, J., Kuzniak, E., Liberka, M. et al. (2020). Modular approach towards functional multimetallic coordination clusters. *Coord. Chem. Rev.* 419: 213394.

38 Pinkowicz, D., Podgajny, R., Nowicka, B. et al. (2015). Magnetic clusters based on octacyanidometallates. *Inorg. Chem. Front.* 2: 10–27.

39 Zenere, K.A., Duyker, S.G., Trzop, E. et al. (2018). Increasing spin crossover cooperativity in 2D Hofmann-type materials with guest molecule removal. *Chem. Sci.* 9: 5623–5629.

40 Jornet-Molla, V., Duan, Y., Gimenez-Saiz, C. et al. (2017). A ferroelectric iron(II) spin crossover material. *Angew. Chem. Int. Ed.* 56: 14052–14056.

41 Qiu, Y.R., Cui, L., Cai, P.Y. et al. (2020). Enhanced dielectricity coupled to spin-crossover in a one-dimensional polymer iron(ii) incorporating tetrathiafulvalene. *Chem. Sci.* 11: 6229–6235.

42 Sato, O., Kawakami, T., Kimura, M. et al. (2004). Electric-field-induced conductance switching in FeCo prussian blue analogues. *J. Am. Chem. Soc.* 126: 13176–13177.

43 Wang, C.F., Sun, M.J., Guo, Q.J. et al. (2016). Multiple Correlations between spin crossover and fluorescence in a dinuclear compound. *Chem. Commun.* 52: 14322–14325.

44 Wang, J.L., Liu, Q., Meng, Y.S. et al. (2018). Fluorescence modulation via photoinduced spin crossover switched energy transfer from fluorophores to Fe(II) ions. *Chem. Sci.* 9: 2892–2897.

45 Yadav, J., Mondal, D.J., and Konar, S. (2021). High-temperature electron transfer coupled spin transition (ETCST) with hysteresis in a discrete [Fe_2Co_2] prussian blue analogue. *Chem. Commun.* 57: 5925–5928.

46 Sato, O., Iyoda, T., Fujishima, A., and Hashimoto, K. (1996). Photoinduced magnetization of a cobalt-iron cyanide. *Science* 272: 704–705.

47 Risset, O.N., Quintero, P.A., Brinzari, T.V. et al. (2014). Light-induced changes in magnetism in a coordination polymer heterostructure, $Rb_{0.24}Co[Fe(CN)_6]_{0.74}@K_{0.10}Co[Cr(CN)_6]_{0.70} \cdot nH_2O$ and the role of the shell thickness on the properties of both core and shell. *J. Am. Chem. Soc.* 136: 15660–15669.

48 Bleuzen, A., Lomenech, C., Escax, V. et al. (2000). Photoinduced ferrimagnetic systems in prussian blue analogues $C^I_xCo_4[Fe(CN)_6]_y$ (C^I = alkali cation). 1. Conditions to observe the phenomenon. *J. Am. Chem. Soc.* 122: 6648–6652.

49 Escax, V., Bleuzen, A., Cartier dit Moulin, C. et al. (2001). Photoinduced ferrimagnetic systems in prussian blue analogues $C^I_xCo_4[Fe(CN)_6]_y$ (C^I = alkali cation). 3. Control of the photo- and thermally induced electron transfer by the [$Fe(CN)_6$] vacancies in cesium derivatives. *J. Am. Chem. Soc.* 123: 12536–12543.

50 Cafun, J.-D., Champion, G., Arrio, M.-A. et al. (2010). Photomagnetic CoFe prussian blue analogues: role of the cyanide ions as active electron transfer bridges modulated by cyanide–alkali metal ion interactions. *J. Am. Chem. Soc.* 132: 11552–11559.

51 De, S., Jiménez, J.-R., Li, Y. et al. (2016). One synthesis: two redox states. temperature-oriented crystallization of a charge transfer {Fe_2Co_2} square complex in a {$Fe^{II}_{LS}Co^{III}_{LS}$}$_2$ diamagnetic or {$Fe^{III}_{LS}Co^{II}_{HS}$}$_2$ paramagnetic state. *RSC Adv.* 6: 17456–17459.

52 Ohkoshi, S.-i., Tokoro, H., Matsuda, T. et al. (2007). Coexistence of ferroelectricity and ferromagnetism in a rubidium manganese hexacyanoferrate. *Angew. Chem. Int. Ed.* 119: 3302–3305.

53 Mahfoud, T., Molnár, G., Bonhommeau, S. et al. (2009). Electric-field-induced charge-transfer phase transition: a promising approach toward electrically switchable devices. *J. Am. Chem. Soc.* 131: 15049–15054.

54 Zhang, K., Kang, S., Yao, Z.S. et al. (2016). Charge-transfer phase transition of a cyanide-bridged Fe^{II}/Fe^{III} coordination polymer. *Angew. Chem. Int. Ed.* 55: 6047–6050.

55 Reczyński, M., Pinkowicz, D., Nakabayashi, K. et al. (2020). Room-temperature bistability in a Ni–Fe chain: electron transfer controlled by temperature, pressure, light, and humidity. *Angew. Chem. Int. Ed.* 60: 2330–2338.

56 Nowicka, B., Reczyński, M., Rams, M. et al. (2015). Hydration-switchable charge transfer in the first bimetallic assembly based on the [Ni(cyclam)]$^{3+}$ –

magnetic CN-bridged chain $\{(H_3O)[Ni^{III}(cyclam)][Fe^{II}(CN)_6]\cdot 5H_2O\}_n$. *Chem. Commun.* 51: 11485–11488.

57 Hilfiger, M.G., Chen, M., Brinzari, T.V. et al. (2010). An unprecedented charge transfer induced spin transition in an Fe–Os cluster. *Angew. Chem. Int. Ed.* 49: 1410–1413.

58 Chorazy, S., Podgajny, R., Nogaś, W. et al. (2014). Charge transfer phase transition with reversed thermal hysteresis loop in the mixed-valence $Fe_9[W(CN)_8]_6\cdot xMeOH$ cluster. *Chem. Commun.* 50: 3484–3487.

59 Avendano, C., Hilfiger, M.G., Prosvirin, A. et al. (2010). Temperature and light induced bistability in a $Co_3[Os(CN)_6]_2\cdot 6H_2O$ prussian blue analog. *J. Am. Chem. Soc.* 132: 13123–13125.

60 Ozaki, N., Tokoro, H., Hamada, Y. et al. (2012). Photoinduced magnetization with a high curie temperature and a large coercive field in a Co-W bimetallic assembly. *Adv. Funct. Mater.* 22: 2089–2093.

61 Chorazy, S., Stanek, J.J., Nogas, W. et al. (2016). Tuning of charge transfer assisted phase transition and slow magnetic relaxation functionalities in $\{Fe_{9-x}Co_x[W(CN)_8]_6\}$ ($x = 0$–9) molecular solid solution. *J. Am. Chem. Soc.* 138: 1635–1646.

62 Arimoto, Y., Ohkoshi, S.-i., Zhong, Z.J. et al. (2003). Photoinduced magnetization in a two-dimensional cobalt octacyanotungstate. *J. Am. Chem. Soc.* 125: 9240–9241.

63 Ohkoshi, S.-i., Tokoro, H., Hozumi, T. et al. (2006). Photoinduced magnetization in copper octacyanomolybdate. *J. Am. Chem. Soc.* 128: 270–277.

64 Herrera, J.M., Marvaud, V., Verdaguer, M. et al. (2004). Reversible photoinduced magnetic properties in the heptanuclear complex $[Mo^{IV}(CN)_2(CN-CuL)_6]^{8+}$: a photomagnetic high-spin molecule. *Angew. Chem. Int. Ed.* 43: 5468–5471.

65 Mathonière, C., Podgajny, R., Guionneau, P. et al. (2005). Photomagnetism in cyano-bridged hexanuclear clusters $[Mn^{II}(bpy)_2]_4[M^{IV}(CN)_8]_2\cdot xH_2O$ (M=Mo, x=14, and M=W, x=9). *Chem. Mater.* 17: 442–449.

66 Meng, Y.S., Sato, O., and Liu, T. (2018). Manipulating metal-to-metal charge transfer for materials with switchable functionality. *Angew. Chem. Int. Ed.* 57: 12216–12226.

67 Magott, M., Stefańczyk, O., Sieklucka, B., and Pinkowicz, D. (2017). Octa-cyanidotungstate(IV) coordination chains demonstrate a light-induced excited spin state trapping behavior and magnetic exchange photoswitching. *Angew. Chem. Int. Ed.* 56: 13283–13287.

68 Bridonneau, N., Long, J., Cantin, J.L. et al. (2015). First evidence of light-induced spin transition in molybdenum(iv). *Chem. Commun.* 51: 8229–8232.

69 Hu, J.X., Xu, Y., Meng, Y.S. et al. (2017). A material showing colossal positive and negative volumetric thermal expansion with hysteretic magnetic transition. *Angew. Chem. Int. Ed.* 56: 13052–13055.

70 Yao, N.T., Zhao, L., Yi, C. et al. (2021). Manipulating selective metal-to-metal electron transfer to achieve multi-phase transitions in an asymmetric [Fe_2Co]-assembled mixed-valence chain. *Angew. Chem. Int. Ed.* 61: e202115367.

71 Koumousi, E.S., Jeon Ie, R., Gao, Q. et al. (2014). Metal-to-metal electron transfer in Co/Fe prussian blue molecular analogues: the ultimate miniaturization. *J. Am. Chem. Soc.* 136: 15461–15464.

72 Berlinguette, C.P., Dragulescu-Andrasi, A., Sieber, A. et al. (2004). A charge-transfer-induced spin transition in the discrete cyanide-bridged complex {[Co(tmphen)$_2$]$_3$[Fe(CN)$_6$]$_2$}. *J. Am. Chem. Soc.* 126: 6222–6223.

73 Funck, K.E., Prosvirin, A.V., Mathoniere, C. et al. (2011). Light-induced excited spin state trapping and charge transfer in trigonal bipyramidal cyanide-bridged complexes. *Inorg. Chem.* 50: 2782–2789.

74 Wen, W., Meng, Y.S., Jiao, C.Q. et al. (2020). A mixed-valence {Fe_{13}} cluster exhibiting metal-to-metal charge-transfer-switched spin crossover. *Angew. Chem. Int. Ed.* 59: 16393–16397.

75 Nihei, M., Okamoto, Y., Sekine, Y. et al. (2012). A light-induced phase exhibiting slow magnetic relaxation in a cyanide-bridged [Fe_4Co_2] complex. *Angew. Chem. Int. Ed.* 51: 6361–6364.

76 Meng, L.Y., Deng, Y.F., Liu, S.H. et al. (2021). A smart post-synthetic route towards [Fe_2Co_2] molecular capsules with electron transfer and bidirectional switching behaviors. *Sci. China Chem.* 64: 1340–1348.

77 Oshio, H., Onodera, H., Tamada, O. et al. (2000). Cyanide-bridged Fe–Fe and Fe–Co molecular squares: structures and electrochemistry of [$Fe_4^{II}(\mu\text{-}CN)_4(bpy)_8$](PF$_6$)$_4$·4H$_2$O, [$Fe_2^{II}Co_2^{II}(\mu\text{-}CN)_4(bpy)_8$](PF$_6$)$_4$· 3CHCl$_3$·2CH$_3$CN, and [$Fe_2^{II}Co_2^{III}(\mu\text{-}CN)_4(bpy)_8$](PF$_6$)$_6$·2CHCl$_3$·4CH$_3NO_2$. *Chem. Eur. J* 6: 2523–2530.

78 Zhang, Y., Li, D., Clérac, R. et al. (2010). Reversible thermally and photoinduced electron transfer in a cyano-bridged {Fe_2Co_2} square complex. *Angew. Chem. Int. Ed.* 49: 3752–3756.

79 Nihei, M., Sekine, Y., Suganami, N. et al. (2011). Controlled intramolecular electron transfers in cyanide-bridged molecular squares by chemical modifications and external stimuli. *J. Am. Chem. Soc.* 133: 3592–3600.

80 Zhang, Y.Z., Ferko, P., Siretanu, D. et al. (2014). Thermochromic and photoresponsive cyanometalate fe/co squares: toward control of the electron transfer temperature. *J. Am. Chem. Soc.* 136: 16854–16864.

81 Nihei, M. (2020). Molecular prussian blue analogues: from bulktomolecules and low-dimensional aggregates. *Chem. Lett.* 49: 1206–1215.

82 Jiao, C.Q., Jiang, W.J., Meng, Y.S. et al. (2018). Manipulation of successive crystalline transformations to control electron transfer and switchable functions. *Natl. Sci. Rev.* 5: 507–515.

83 Jiao, C.Q., Meng, Y.S., Yu, Y. et al. (2019). Effect of intermolecular interactions on metal-to-metal charge transfer: a combined experimental and theoretical investigation. *Angew. Chem. Int. Ed.* 58: 17009–17015.

84 Alcazar, L., Aullon, G., Ferrer, M., and Martinez, M. (2016). Redox-assisted self-assembly of a water-soluble cyanido-bridged mixed valence {Co^{III}/Fe^{II}}$_2$ square. *Chem. Eur. J* 22: 15227–15230.

85 Mudoi, P.P., Choudhury, A., Li, Y. et al. (2021). Observation of protonation-induced spin state switching in a cyanido-bridged {Fe$_2$Co$_2$} molecular square. *Inorg. Chem.* 60: 17705–17714.

86 Nihei, M., Yanai, Y., Hsu, I.J. et al. (2017). A hydrogen-bonded cyanide-bridged [Co$_2$Fe$_2$] square complex exhibiting a three-step spin transition. *Angew. Chem. Int. Ed.* 56: 591–594.

87 Sekine, Y., Nihei, M., and Oshio, H. (2017). Dimensionally controlled assembly of an external stimuli-responsive [Co$_2$Fe$_2$] complex into supramolecular hydrogen-bonded networks. *Chem. Eur. J* 23: 5193–5197.

88 Sessoli, R., Gatteschi, D., Caneschi, A., and Novak, M.A. (1993). Magnetic bistability in a metal-ion cluster. *Nature* 365: 141–143.

89 Liu, T., Zheng, H., Kang, S., Shiota, Y., Hayami, S., Mito, M., Sato, O., Yoshizawa, K., Kanegawa, S. and Duan, C. (2013) A light-induced spin crossover actuated single-chain magnet. *Nat. Commun.* 4: 2826.

90 Liu, Q., Hu, J.X., Meng, Y.S. et al. (2021). Asymmetric coordination toward a photoinduced single-chain magnet showing high coercivity values. *Angew. Chem. Int. Ed.* 60: 10537–10541.

91 Ohkoshi, S., Imoto, K., Tsunobuchi, Y. et al. (2011). Light-induced spin-crossover magnet. *Nat. Chem.* 3: 564–569.

92 Huang, W., Ma, X., Sato, O., and Wu, D.Y. (2021). Controlling dynamic magnetic properties of coordination clusters via switchable electronic configuration. *Chem. Soc. Rev.* 50: 6832–6870.

93 Liu, T., Zhang, Y.-J., Kanegawa, S., and Sato, a. O. (2010). Photoinduced metal-to-metal charge transfer toward single-chain magnet. *J. Am. Chem. Soc.* 132: 8250–8251.

94 Dong, D.P., Liu, T., Kanegawa, S. et al. (2012). Photoswitchable dynamic magnetic relaxation in a well-isolated {Fe$_2$Co} double-zigzag chain. *Angew. Chem. Int. Ed.* 51: 5119–5123.

95 Li, J., Wu, S., Su, S. et al. (2020). Manipulating slow magnetic relaxation by light in a charge transfer {Fe$_2$Co} complex. *Chem. Eur. J.* 26: 3259–3263.

96 Liu, T., Dong, D.P., Kanegawa, S. et al. (2012). Reversible electron transfer in a linear {Fe$_2$Co} trinuclear complex induced by thermal treatment and photoirraditaion. *Angew. Chem. Int. Ed.* 51: 4367–4370.

97 Hu, J.X., Luo, L., Lv, X.J. et al. (2017). Light-induced bidirectional metal-to-metal charge transfer in a linear Fe$_2$Co complex. *Angew. Chem. Int. Ed.* 56: 7663–7668.

98 Liu, Z., Gao, Q., Chen, J. et al. (2018). Negative thermal expansion in molecular materials. *Chem. Commun.* 54: 5164–5176.

99 Sergeenko, A.S., Ovens, J.S., and Leznoff, D.B. (2018). Designing anisotropic cyanometallate coordination polymers with unidirectional thermal expansion (TE): 2D Zero and 1D colossal positive TE. *Chem. Commun.* 54: 1599–1602.

100 Long, Y.W., Hayashi, N., Saito, T. et al. (2009). Temperature-induced A–B intersite charge transfer in an a-site-ordered $LaCu_3Fe_4O_{12}$ perovskite. *Nature* 458: 60–63.

101 Podgajny, R., Chorazy, S., Nitek, W. et al. (2013). Co-NC-W and Fe-NC-W electron-transfer channels for thermal bistability in trimetallic $\{Fe_6Co_3[W(CN)_8]_6\}$ cyanido-bridged cluster. *Angew. Chem. Int. Ed.* 52: 896–900.

102 Bandara, H.M.D. and Burdette, S.C. (2012). Photoisomerization in different classes of azobenzene. *Chem. Soc. Rev.* 41: 1809–1825.

103 Beharry, A.A. and Woolley, G.A. (2011). Azobenzene photoswitches for biomolecules. *Chem. Soc. Rev.* 40: 4422–4437.

104 Wenger, O.S. (2012). Photoswitchable mixed valence. *Chem. Soc. Rev.* 41: 3772–3779.

105 Naonobu, S., Shin-ichi, O., Osamu, S., and Kazuhito, H. (2002). One-shot-laser-pulse-induced cooperative charge transfer accompanied by spin transition in a Co-Fe prussian blue analog at room temperature. *Chem. Lett.* 31: 486–487.

106 Tokoro, H., Matsuda, T., Hashimoto, K., and Ohkoshi, S.-i. (2005). Optical switching between bistable phases in rubidium manganese hexacyanoferrate at room temperature. *J. Appl. Phys.* 97: 10M508.

107 Cai, L.Z., Chen, Q.S., Zhang, C.J. et al. (2015). Photochromism and photomagnetism of a 3d-4f hexacyanoferrate at room temperature. *J. Am. Chem. Soc.* 137: 10882–10885.

108 Qi, X., Pillet, S., de Graaf, C. et al. (2020). Photoinduced Mo-CN bond breakage in octacyanomolybdate leading to spin triplet trapping. *Angew. Chem. Int. Ed.* 59: 3117–3121.

109 Mullaney, B. R., Goux-Capes, L., Price, D. J., Chastanet, G., Letard, J. F. and Kepert, C. J. (2017) Spin crossover-induced colossal positive and negative thermal expansion in a nanoporous coordination framework material. *Nat. Commun.* 8: 1053.

Index

a

ab initio molecular dynamics (AIMD)   110, 111
abpy and bpip bridged diruthenium complexes   140
absorption intensity   28, 253, 440
acetyl-CoA synthase activity   329
A-cluster   329, 330
alkene   366, 375
anharmonic regime   52, 53, 57, 63, 69, 74, 75, 79, 82
anion-responsive compounds   378–380
azobenzene   366, 375, 376, 480

b

basic FeS clusters   325–326
benzimidazole   381, 382
bidirectional redox active   128
2,2′-bis(2-pyridyl)bibenzimidazole   381
bis-bidentate 2,5-pyrazine-dicarboxylate (BL_1^{2-})   124
2,2′-bis(benzimidazol-2-yl)-4,4′-bipyridine ($bbbpyH_2$)   382
bis(ferrocenyl)bis(mesityl)porphyrin   384
bis(pentaammineruthenium) complex   366
bis(trianisylamine)s   312
bistriarylamine compounds   378
bistriarylamine radical cations   94
BLYP35-based pragmatic protocol   104
B3LYP hybrid functional   302
BNB molecule   75
BO approximation   48, 49, 51, 58, 64–65, 72, 75, 76, 81
bridge band   302
bridged M_2 dimers   231
bridged MV compounds   9
bridging units   121, 185, 270, 297–299, 302, 304–309

c

carbon monoxide dehydrogenase (CODH)   323, 329–331
carbon neutrality   432, 453
catalysis in uncoupled mixed-valence systems   414
C_4 butadiynediyl bridge   98
C-clusters   330
[Cd(BPPTzTz)(tdc)] · 2DMF   399
charge-delocalized excited state   181
charge-transfer-induced spin transitions (CTIST)   463–465
Chisholm   233
chloroperoxidase   331
chromium triazolate framework   413
cis/trans-configuration   280
Class I ribonucleotide reductases (RNRs)   339
CNS model   32–34, 242
cobalt   412
cobalt-based catalysts   416–417
comproportionation constant   13, 122, 133, 168, 231, 350, 351
conducting polymers of metal complexes   447–452
copper   408
copper-based catalysts   417
coupled harmonic-oscillator model   48–54
coupling integral   2, 3, 15, 16, 24, 26
coupling regimes   49–53, 57–59, 61, 63, 65, 68–70, 73, 74, 76, 81
covalently-bonded diruthenium complexes   9
covalent-organic frameworks (COFs)   407–408, 442–443
Cp′(L_2)Ru-based termini   158–163
Creutz-Taube complexes   5, 297
Creutz-Taube ion (CT ion)   121, 182, 272, 432
Cross conjugation and quantum destructive effect   257
cubane-type [4Fe-4S] clusters   325

Mixed-Valence Systems: Fundamentals, Synthesis, Electron Transfer, and Applications, First Edition.
Edited by Yu-Wu Zhong, Chun Y. Liu, and Jeffrey R. Reimers.
© 2023 WILEY-VCH GmbH. Published 2023 by WILEY-VCH GmbH.

cyanide bridged mixed-valence (Fe–CN–Co) chromophore 464
cyanide-bridged MMCT complexes
 electric dipole 474–478
 molecular nanomagnet behavior 472–483
 thermal expansion (TE) 478–480
cyanide-bridged MV complex
 advantage 270
 dinuclear 272
 electron transfer and electron coupling 271
 tetranuclear and higher nuclear 284–290
 trinuclear 276–284
cyanide–isocyanide configuration 274
cyanoacetylide bridging ligand (CCCN) 101
cyclometalated diruthenium complex 452
cyclometalated MV systems 366
cyclopenta-ring-fused rylene π-bridged bis(dianisylamine) radical cations 304
cytochrome c oxidase 121, 332, 333
cytochrome c peroxidase 331, 332

d

D-B-A experimental models 231
D-B-A model 270
D-COSMO-RS 97, 110
degenerate MV systems 299
density-functional theory (DFT) 48
deprotonated 2-thiouracil 133
dianionic 124, 126, 133
dianionic 2,5-bis (2-oxidophenyl)pyrazine (BL_3^{2-}) bridging ligand 126
dianionicoxamidato (2^-) bridged diruthenium(III) complex 136
2,5-diboryl-1,4-phenylene bridge 379
dicarboxylates 232, 233
3,6-dichloro-2,5-dihydroxy-1,4-benzoquinone 401
diethynylated diacetoxyanthracene unit 385
diethynylethene-linked di-ferrocene 376
dihydropyrenes (DHP) derivatives 376
dimanganese complexes 273
2,5-dimercapto-1,3,4-thiadiazolate (DMcT) ligand 383
dimer of dimers 9, 141, 231, 235, 243
dimetal tetracarboxylate 231, 233
dimethyldihydropyrene 366, 376, 377
dimolybdenum dimers of dimers 234
dinitroaryl radical anions 96
1,3-dinitrobenzene 96, 110, 111
dinuclear complex 100, 122, 157, 162, 163, 170, 381, 382
dinuclear cyanide-bridged mixed-valence complex 272
dinuclear iron-ethynyl complexes
 with butadiynediyl bridge 153–154
 with non-conjugated C_4-bridge core 156–157
dinuclear pentaammineruthenium complex 380
dinuclear Ru(dppe)$_2$ complex 385
dinuclear ruthenium-ethynyl complexes
 with alternating polyyndiyl and capped Ru-Ru units 165–166
 Cp′(L$_2$)Ru-based termini 158–163
 ruthenium-ethynyl termini and core units 166–168
 with Ru(dppe)$_2$X-based termini 163–165
diprotic bridging ligand 233
diruthenium(II) complexes 130
dithienylethene (DTE) 366
dithienyl perhydro/perfluorocyclopentene units 372
divinylphenylene diruthenium complex 352
donor-bridge-acceptor molecule 378
double exchange mechanism 315
$d(\delta)(M_2)$-p(π)(ligand) conjugation 235
DTE-bridged diruthenium complex 368

e

electric dipole 474–478
electrochemical methods 13–14
electrochromic martials (ECMs) 431
electrochromism, defined 431
electronic coupling 151
 and electron transfer
  conformational effects of 252–255
  distance-dependence of 247–252
  hydrogen bonds 258–260
  [M$_2$-BL-M$_2$]$^+$ systems 247
 modulation via host-guest or through-space interaction 356–361
 through hydrogen bonds
  between MV organic fragments 353–356
  between transition metal centers 350–353
electron paramagnetic resonance (EPR) spectroscopy 12–13
electron transfer 269, 365
 pathways 122
 process 121
 reactions 50
electron-transfer-coupled spin transitions (ETCST) 463
electron transporters 323
energy conversion processes 4
environmental effects 109
enzymes' biocatalystas 323
ethynylene 376
exact-exchange (EXX) admixture 95

extended mixed-valence materials
  covalent-organic frameworks (COFs) 407–408
  introduction to 397
  naphthalenediimide-based compounds 405–406
  phenalenyl-based compounds 406
  tetraoxolene-based compounds 400–405
  tetrathiafulvalene (TTF)-based compounds 399–400
  thiazolo[5,4-*d*]thiazole (TzTz)-based compounds 397–399

f

faster vibrational spectroscopic techniques 123
[FeFe]-hydrogenase 323, 326–328
Fe(II,III) MV tetranuclear cage complexes 356
ferredoxin [2Fe-2S] cofactors 349
ferridoxin 328
ferrocene-terminated azobenzene derivative 375
ferrocenyl-functionalized heterocycles 181
ferrocenyl-functionalized 5-membered heterocycles
  with group-13 elements 183
  with group-14 elements 183–184
  with group-15 elements 185–201
  with group-16 elements 201–213
  with transition metal elements 213–217
ferrocenyl-functionalized 6-membered heterocycles 217–218
ferrocenyl methylhydantoin 5-ferrocenyl-5-methylimidazolidine-2,4-dione 351
FeS clusters 325–326
field-free absorption spectrum 57
finite-field difference spectrum 57
first-derivative correction 59
first row transition metals 408–414
flavanthoxins 328
formamidinate ancillary ligands 233
fragile entanglement 63
fragment molecular orbitals (FMO) 304
Franck-Condon
  principle 3, 395
  progressions 57
  transition 2, 17
free energy surfaces (FESs) 299
functionalized dinuclear iron-ethynyl complexes 157–158

g

gauge problem of LHs 105
generalized Mulliken–Hush theory (GMH) 27–28
generalized Mulliken–Hush three-state analysis 302–303
ground-state delocalisation, Creutz-Taube ion 68–69
GS adiabatic 77

h

Hartree–Fock exchange 95
Hartree–Fock theory 94, 95
hemocyanins 333
hexa(dianisylamine)-substituted hexaarylbenzene 311
5,5′(4H,4H′)-spirobi[cyclopenta[*c*]pyrrole]2,2,′6,6′-tetrahydro cation 109
hole transport in a molecular conducting material 68
homodinuclear metal-ethynyl complexes 152
horseradish peroxidase 331
Hush formula 123, 127, 128, 134
Hush model 3, 5, 14, 245, 276
HXH bond angle 47, 76, 78
hybrid functionals 95, 100, 105, 112, 302
hydrogen bond (HB) 353
  BLs 11
  bridged MV systems 350
hydrogen-bond acceptor (HBA) 472

i

inorganic MV compounds (MVCs) 396
inorganic NIR electrochromic materials 433–434
intermediate-coupling regime 51–53, 61, 63, 73, 74
intermolecular mixed-valence 298
intervalence charge-transfer (IVCT) 97, 123, 151, 298, 393
intramolecular charge transfer (CT) reaction 299
intramolecular electron transfer 2, 121, 124, 127, 232, 244, 269, 273
inverted region 20, 52, 66
iron 409
iron-based catalysts 415–416
iron-ethynyl complexes
  dinuclear iron-ethynyl complexes
    with butadiynediyl bridge 153–154
    with non-conjugated C_4-bridge core 156–157
  functionalized dinuclear iron-ethynyl complexes 157–158
iron (Fe) nitrogenase 328
IVCT absorption 256

k
Kubelka-Munk analysis 399

l
Landau–Zener model 31, 249, 252
lanthanoids 414
Levich–Dogonadze–Marcus expression 31, 252
ligand bridged mixed valent diruthenium complex 122
Liptay equation 57
local hybrid functionals (LHs) 100, 104
localization/delocalization 95
local mixing function (LMF) 104

m
macrocyclic tetranuclear platinum(II) complex 356
magnetic protein biocompass 323, 326
Marcus-Hush theory 25, 26, 212, 242, 248, 252, 254, 259, 262, 298, 352, 395, 398
M-BL-M 34
M_2-BL-M_2 systems 239, 240, 247, 256
McConnell's theory 6
McConnell superechange expression 32–34
metal-based mixed valency
    catalysis 414–418
    first row transition metals 408–414
    metal complexes 444–447
metal-metal charge transfer (MMCT) 272
metal(M_2)-metal(M_2) coupling 239
metal-metal quadruple bonds 230, 262
metal to ligand charge transfer (MLCT) 192, 195, 239, 368, 464
metal-to-metal charge transfer (MMCT)
    cyanide-bridged MMCT complexes 472–483
    switchable cyanide bridged MMCT systems 465–472
meta-para paradigm 310
mixed-valence (MV) chemistry
    analysis of IVCT band shape 28
    attraction of 2
    BO approximation 64–65
    compounds 1
    coupled harmonic-oscillator model 48–54
    definition 1
    diversity of 6–12
    electric fields on MV optical band shapes 56–58
    electrochemical methods 13–14
    electronic coupling matrix element and the transition moments 26–27
    electron paramagnetic resonance spectroscopy 12–13
    electron transfer in 3
    Generalized Mulliken–Hush theory (GMH) 27–28
    history 4–6
    from localized to delocalized 16–17
    McConnell superexchange mechanism and the CNS model 32–34
    molecules and iconic model systems
        ground-state delocalisation, Creutz–Taube ion 68–69
        hole transport in a molecular conducting material 68
        MV excited states in a bis-metal complex 66–68
        photochemical charge separation 66
        photochemical charge separation during bacterial photosynthesis 70–73
    MV complexes as potential quantum qubits 60–63
    non-adiabatic effects 58–60
    optical analysis 14–15
    origins within chemical bonding theory 47–48
    potential energy surfaces from classical two-state model 18–20
    prominent feature of 6
    quantum description of the potential energy surfaces 20–24
    reorganization energies 24–26
    solvent control of electron transfer 17–18
    stereoisomerism
        ammonia 75–79
        aromaticity in benzene 80–81
        BNB molecule 75
        proton-transfer reactions 79
        pyridine 74–75
mixed-valence complex 34, 231, 232, 254, 272–275
mixed-valence (MV) compounds 365
mixed valence diruthenium dimers 260–262
mixed-valence iron-sulfur clusters in biological and bio-mimic systems
    basic FeS clusters 325–326
    carbon monoxide dehydrogenase 329–331
    [FeFe]-hydrogenase 326–328
    nitrogenases 328–329
mixed-valence metal complexes 323
mixed-valence multi-copper cofactors 332
mixed-valence multi-manganese cofactors 339
mixed-valence phenomena 4
mixed valence systems
    diversity of 6
    Multiiron-contained biological systems 331–332
mixed-valence (MV) systems 93, 349

mixed valency
  classifications of 395–396
  fundamental aspects of 393–394
  Marcus-Hush theory 395
  organic 396
  quantum mechanical considerations 394–395
$Mo_2(DAniF)_3(O_2CC_6H_5)$ 242
$Mo_2(DAniF)_3(S_2CC_6H_5)$ 242
$Mo_2(DArF)_4$ 230
$Mo_2(O_2CCH_3)_4$ 230
MOF nanozymes 417–418
molecular nanomagnets 472–474
molecule-based materials 463
molecules and iconic model systems
  ground-state delocalisation, Creutz-Taube ion 68–69
  hole transport in a molecular conducting material 68
  MV excited states in a bis-metal complex 66–68
  photochemical charge separation 66
  photochemical charge separation during bacterial photosynthesis 70–73
molybdenum (Mo) 414
  nitrogenase 328–329
monolayer and multilayer assembled films 452–453
monoruthenium complex 373
Mulliken charge transfer theory 5
Mulliken–Hush expression 5, 13, 15, 26, 27, 34, 241, 262, 352
Mulliken–Hush formalism 16
Mulliken–Hush treatment 93
Mulliken-Hush two-mode analysis 301–302
multi-configuration self-consistent-field (MCSCF) approaches 94
multicopper oxidase (MCO) 333, 334
multidimensional organic MV systems 315
MV copper(I/II) systems 408
MV excited states in a bis-metal complex 66–68
MV tetraoxolene frameworks 403

n

naphthalenediimide-based compounds 405–406
near-infrared electrochromic materials
  applications
    military camouflage 454
    molecular logic gates and optical storage 453
    optical communication 453–454
    smart windows 453
  inorganic NIR electrochromic materials 433–434

organic-inorganic hybrid NIR electrochromic materials 444–453
organic NIR electrochromic materials 435–443
near-infrared (NIR) electrochromism 431
neutral in cation geometry (NCG) method 240
nitrogenases 323, 328
nitrous oxide reductase 333
N,N'-dimethyl-4,4'-bipyridinium 359
N,N'-dimethylpiperidine (DMP$^+$) 108
non-adiabatic corrections 59
non-adiabatic effects 58–60
non-adiabatic weak-coupling limit 60
norbornadiene 366, 367

o

octacyanometallates 466
oligo(p-phenylene) (OPP) 306, 308
oligoacene-bridged MV systems 308
one-electron oxidation 133, 232
one-electron reduction of Ru^{III}Ru^{III} 127
organic conducting polymers 439–441
organic-inorganic hybrid NIR electrochromic materials
  conducting polymers of metal complexes 447–452
  metal complexes 444–447
  monolayer and multilayer assembled films 452–453
organic mixed-valence systems
  bridging units and electronic coupling 304–309
  charge and/or spin localization 315
  delocalization 315
  electronic coupling 299
  generalized Mulliken-Hush three-state analysis 302–303
  history of 297–298
  intramolecular charge transfe 299
  IVCT band 303
  Mulliken-Hush two-mode analysis 301–302
  redox centers 310–311
  through-bond or through-space 311–313
organic mixed valency 396
organic NIR electrochromic materials
  covalent organic framework 442–443
  organic conducting polymers 439–441
  triphenylamine derivatives 437–439
  viologen derivatives 435–437
organoboron systems 378
organometallic mixed-valence (MV) complexes 151
organometallic MV systems 103
organometallic Ru (II/III) building blocks 6

oxido/carboxylato and oxido/pyrazolato bridged diastereomeric 138
oxygen (O) centered triruthenium cluster 10
oxygen evolution complex (OEC) 323, 336
oxygen evolving center (OEC) 12, 3336

p

peroxidases 331, 332
phenalenyl-based compounds 406
phenothiazine (PTZ) electron donor 374
phenylene bridged Mo_2-Mo_2 series 247
photochemical charge separation 66
  during bacterial photosynthesis 70–73
photoinduced multielectron charge transfer process 284
photoswitchable MV compounds 365
polycyanometallates 465, 466, 470–472, 481, 483
polynuclear complexes 121
polyoxometalate-based catalysts 416
polyoxometalates (POMs) 412
polyoxovanadates (POVs) 412
porphyrins 384
potential-energy surfaces 50
proton-coupled electron transfer (PCET) 12, 328, 351
proton-responsive compounds 380–385
proton-transfer reactions 79, 328
prussian blue analogues (PBAs) 465
putative magnetic receptor 326
pyrazine (pz) bridged mixed-valent $Ru^{II}Ru^{III}$ state 124–130
pyridine 74–75

q

quadruply-bonded dimolybdenum compounds 350
quadruply-bonded M_2 complexes 229
quadruply-bonded M_2 unit 229, 230
quantitative analysis 51, 53, 73, 79, 82
quantum-chemical approaches 94
quantum conductance 355
quantum entanglement 46, 51, 60, 81
quantum information systems 60

r

redox-active centers 297–299, 302, 314, 418
redox-inactive alkylammonium-based countercations 404
reorganization energy 24–26, 299
Resonant valency 4
respiratory complex IV 332
Rieske proteins 325
rigid wire-like metal diynediyl complexes 152

Robin-Day classes 17, 93, 98, 240–247
Robin-Day's classification 5, 16, 30, 69, 94, 232, 464
  of MV compounds 16
$Ru(dppe)_2Cl]^+$ (dppe = 1,2-bis(diphenylphosphino)ethane) moieties 371
$Ru^{III}Ru^{IV}$ mixed valent systems 135–139
$Ru^{II}Ru^{I}$ and $Ru^{I}Ru^{0}$ mixed valent systems 139–141
$Ru^{II}Ru^{III}$ mixed valent systems
  deprotonated 2-thiouracil 133
  diruthenium(II) complexes 130
  oxido bridged diruthenium complex 134
  pyrazine derived bridges 124–130
$Ru(dppe)_2$X-based termini 163–165
Rydberg orbitals 47

s

scanning tunneling microscopy break junction (STMBJ) technique 355
single-chain magnets 473, 476
single-molecule magnets 473, 474
spin-density distributions 107, 127, 138, 159
spiro cation 109
star-shaped tri-Ru complex 453
state-interaction pair-density functional theory (SI-PDFT) 109
stereoisomerism
  ammonia 75–79
  aromaticity in benzene 80–81
  BNB molecule 75
  proton-transfer reactions 79
  pyridine 74–75
strong-coupling regime 51
superexchange model 2
superexchange pathway 122
switchable cyanide bridged MMCT systems 465–472
symmetric diruthenium complex 383

t

TDDFT 97, 102, 105, 107, 108
terephthalaldehyde (PDA) 407
tetracyanoquinodimethane (TCNQ) 297, 298
N,N, N',N'-tetrakis(4-aminophenyl)-1,4-benzenediamine (TPDA) 407
tetrakis(4-methoxyphenyl)benzidine (MeO-TPD) 379
tetrakis(dianisylamine)-substituted tetraazacyclophane 314
tetranuclear and higher nuclear cyanide-bridged mixed-valence complex 284–290
tetraoxolene-based compounds 400–405
tetraruthenium metallacycle 352

tetrathiafulvalene (TTF)   297
    based compounds   399–400
thermal expansion (TE)   478–480
thiazolo[5,4-*d*]thiazole (TzTz)-based
    compounds   397–399
thieno[3,2-*b*]thiophene-2,5-
    dicarbaldehyde (TTDA)   407
thienylene bridging ligand   244
thioacetyl-terminated diruthenium alkynyl
    complex   372
three-redox-centered MV system   314
through-bond IVCT   394
tppz bridged diruthenium complexes   128
transition-metal complexes   100
transition metal-ethynyl complexes
    dinuclear Group-8 (Os) and Group-9 (Co)
        metal-polyyndiyl complexes   170–171
    dinuclear group-6 (Cr, Mo) metal-ethynyl
        complexes   168–169
    dinuclear group-7 (Mn, Re)
        metal-polyynediyl complexes
        169–170
transition-metal MV systems   97
transition state theoretic model (TST)   30
triarylamine-based (TAA) MV systems   99
triarylamine derivatives   353, 368, 385
triarylamine-dithienylethene-acetylide ligands
    373
triarylamine-terminated dithienylethene
    derivatives   369
2,5,8-tri(4-pyridyl)-1,3-diazaphenalene
    (TPDAP)   406
trinuclear cyanide-bridged MV complexes
    276–284
triphenylamine derivatives   437–439
triphenylamine dimer   372
tris(4-aminophenyl)amine (TAPA)   407

TTF-based ligand tetrathiafulvalene
    tetrabenzoate (TTFTB)   399
two coupled diabatic potential-energy surfaces
    46
2D honeycomb chloranilate frameworks   403
2D tetraoxolene honeycomb frameworks   401

u

unimolecular mixed-valence D-B-A systems   8
urea-bridged cyclometalated diruthenium
    complex   378
ureido pyrimidinedione (UPy) derivatives   355

v

valence tautomerism (VT)   463, 464
valency oscillation   4
vanadium (V) nitrogenase   328–329
variable optical attenuator (VOA)   453
vertical reorganization energy   395
vertical transition energy (IVCT)   53, 240
vinyl-functionalized ruthenium-amine
    complex   450
viologen derivatives   435–437
Vivianite   4

w

wild-type (WT)   71, 72
Wurster-type redox system   298

x

XE cyanide-bridged MMCT complexes
    photochromis   480
XE photochromis   480

z

Zimmerman's system   355